Environmental Analysis of Contaminated Sites

Ecological and Environmental Toxicology Series

Series Editors

Jason M. Weeks
National Centre for Environmental Toxicology, WRc-NSF, UK

Sheila O'Hare
Scientific Editor
Hertfordshire, UK

Barnett A. Rattner
Patuxent Wildlife Research Center, USGS
Laurel, MD, USA

The fields of environmental toxicology, ecological and ecotoxicology are rapidly expanding areas of research. This explosion of interest within the international scientific community demands comprehensive and up-to-date information that is easily accessible to both professionals and an increasing number of students with an interest in these subject areas.

Books in the series will cover a diverse range of relevant topics ranging from taxonomically based handbooks of ecotoxicology to current aspects of international regulatory affairs. Publications will serve the needs of undergraduate and postgraduate students, academics and professionals with an interest in these developing subject areas.

The Series Editors will be pleased to consider suggestions and proposals from prospective authors or editors in respect of books for future inclusion in the series.

Published titles in the series

Environmental Risk Harmonization: Federal and State Approaches to Environmental Hazards in the USA
Edited by Michael A. Kamrin (ISBN 0 471 972657)

Handbook of Soil Invertebrate Toxicity Tests
Edited by Hans Løkke and Cornelius A. M. van Gestel (ISBN 0 471 971030)

Pollution Risk Assessment and Management: A Structured Approach
Edited by Peter E. T. Douben (ISBN 0471 972975)

Statistics in Ecotoxicology
Edited by Tim Sparks (ISBN 0 471 96851 X CL,
ISBN 0 471 972991 PR)

Demography in Ecotoxicology
Edited by Jan E. Kammenga and Ryszard Laskowski (ISBN 0 471 490024)

Forecasting the Environmental Fate and Effects of Chemicals
Edited by Philip S. Rainbow, Steve P. Hopkin and Mark Crane (ISBN 0 471 491799)

Ecotoxicology of Wild Mammals
Edited by Richard F. Shore and Barnett A. Rattner (ISBN 0 471 974293)

Forthcoming titles in the series

Behaviour in Ecotoxicology
Edited by Giacomo Dell'Omo (ISBN 0 471 968528)

PAHs: An Ecological Perspective
Edited by Peter E. T. Douben (ISBN 0 471 560243)

Environmental Analysis of Contaminated Sites

Edited by

Geoffrey I. Sunahara
Biotechnology Research Institute,
National Research Council of Canada,
Montreal, Quebec, Canada

Agnès Y. Renoux
Biotechnology Research Institute,
National Research Council of Canada,
Montreal, Quebec, Canada

Claude Thellen
Centre d'Expertise en Analyse Environnementale du Quebec,
Ministere de l'Environnement du Quebec,
Sainte-Foy, Quebec, Canada

Connie L. Gaudet
Environmental Quality Branch,
Environment Canada,
Ottawa, Ontario, Canada

Adrien Pilon
Biotechnology Research Institute,
National Research Council of Canada,
Montreal, Quebec, Canada

JOHN WILEY & SONS, LTD

Baffins Lane, Chichester,
West Sussex PO19 1UD, UK

National 01243 779777
International (+44) 1243 779777
e-mail (for orders and customer service enquiries): cs-books@wiley.co.uk
Visit our Home Page on http://www.wiley.co.uk
or http://www.wiley.com

Other Wiley Editorial Offices

John Wiley & Sons, Inc., 605 Third Avenue,
New York, NY 10158-0012, USA

Wiley-VCH GmbH, Pappelallee 3,
D-69469 Weinheim, Germany

John Wiley & Sons Australia, Ltd., 33 Park Road, Milton,
Queensland 4064, Australia

John Wiley & Sons (Asia) Pte, Ltd., 2 Clementi Loop #02-01,
Jin Xing Distripark, Singapore 129809

John Wiley & Sons (Canada) Ltd., 22 Worcester Road,
Rexdale, Ontario M9W 1L1, Canada

Cover photograph provided by Parks Canada-J. Mercier (1998).

Library of Congress Cataloging-in-Publication Data

Environmental analysis of contaminated sites
edited by Geoffrey I. Sunahara . . . [et al.].
p.cm. — (Ecological & environmental toxicology series)
Includes bibliographical references and index.
ISBN 0-471-98669-0 (alk. paper)
1. Environmental toxicology. 2. Soil pollution I. Sunahara, Geoffrey I. (Geoffrey Isao), 1953- II. Series.

RA1226 .E56 2001
651.9′02 — dc21 2001045405

British Library Cataloguing in Publication Data

A catalogue record for this book is available from the British Library

ISBN 0 471 98669 0

Typeset in 10/12pt Garamond from the author's disks by Laser Words Private Limited, Chennai, India.
Printed and bound in Great Britain by T. J. International Ltd, Padstow, Cornwall
This book is printed on acid-free paper responsibly manufactured from sustainable forestry, in which at least two trees are planted for each one used for paper production.

Contents

List of contributors

Serge Barbeau
Ville de Montréal — Service des travaux publics et de l'environnement — Laboratoire, 999 rue de Louvain Est, Montréal, Québec, Canada H2M 1B3

Christian Bastien
Centre d'Expertise en Analyse Environnementale du Québec, Ministère de l'Environnement et de la Faune, 2700 rue Einstein, Sainte-Foy, Québec, Canada G1P 3W8

Sylvie Bisson[1*]
Centre d'Expertise en Analyse Environnementale du Québec, Ministère de l'Environnement du Québec, 1665 boul. Wilfrid-Hamel, Édifice 2, bureau 1.03, Québec, Québec, Canada G1N 3Y7[1*]

Alan Blankenship[2*]
224 National Food Safety and[2*] Toxicology Center, Michigan State University, East Lansing, MI 48824, USA

Jacques Bureau
NSERC Industrial Chair in Site Remediation and Management, École Polytechnique de Montréal, 2900 Édouard-Montpetit, C.P. 6079, succ. Centre ville, Montréal, Québec, Canada H3 CC3A7

Claude Chamberland
Products Safety and Environment, Shell Canada Products, 400 4th Ave. S.W., Calgary, Alberta, Canada T2P 2H5

Raynald Chassé
Centre d'Expertise en Analyse Environnementale du Québec, Ministère de l'Environnement du Québec, 1665 boul. Wilfrid-Hamel, Édifice 2, bureau 1.03, Québec, Québec, Canada G1N 3Y7

[1*] Present address: Ministère de l'Industrie et du Commerce, Édifice Capitanal, 100 rue Laviolette, bureau 321, Trois-Rivières, Québec, Canada G9A 5S9.
[2*] Present address: ENTRIX Inc., 4295 Okemos Road, Suite 101, Okemos, MI 48864, USA

Ronald T. Checkai
US Army Edgewood RD&E Center, SCBRD-RTL (E3150), 5232 Fleming Road, Aberdeen Proving Ground, MD 21010-5423, USA

Larry D. Claxton
Environmental Carcinogenesis Division, US Environmental Protection Agency, MD-68, Research Triangle Park, NC 27711, USA

Yvon Courchesne
QSAR Inc., 1650 Champlain St., Trois-Rivières, Quebec, Canada G9A 4S9

Trudie Crommentuijn
Ministry of Housing, Spatial Planning and the Environment, Directorate-General for Environmental Protection, Department of Soil Protection (IPC 625), P.O. Box 30945, 2500 GX The Hague, The Netherlands

Elisabeth M. Dirven-Van Breemen
Laboratory for Ecotoxicology, National Institute of Public Health and the Environment, Antonie van Leeuwenhoeklaan 9, P.O. Box 1, 3720 BA Bilthoven, The Netherlands

Kirby C. Donnelly
Veterinary Anatomy and Public Health, Texas A&M University, MS4458, College Station, TX 77843-4458, USA

Yves Dudal
NSERC Industrial Chair in Site Remediation and Management, École Polytechnique de Montréal, 2900 Édouard-Montpetit, C.P. 6079, succ. Centre ville, Montréal, Québec, Canada H3 CC3A7

Luc Dussault
Biogénie Inc., 350 rue Franquet, Sainte-Foy, Québec, Canada G1P 4P3

Clive A. Edwards
Department of Entomology, Ohio State University, 1735 Neil Avenue, Columbus, OH 43210, USA

Ernestine Elkenbracht
Quintens advies en management, Regulierenring 13b, 3981 LA Bunnik, The Netherlands

Stephen Esposito
Jacques Whitford Environment Ltd., 3 Spectacle Lake Drive, Dartmouth, Nova Scotia, Canada B3B 1W8

Shannon S. Garcia
Veterinary Anatomy and Public Health, Texas A&M University, MS4458, College Station, TX 77843-4458, USA

Connie L. Gaudet
Environment Canada, Guidelines and Standards Division, 351 St. Joseph Blvd., Hull, Quebec, Canada K1A 0H3

Renée Gauthier
Ministry of Environment — Québec, Marie-Guyart Building — 9th Floor, 675 boulevard René-Levesque Est, Québec, Québec, Canada G1R 5V7

S. Elizabeth George
Environmental Carcinogenesis Division, US Environmental Protection Agency, MD-68, Research Triangle Park, NC 27711, USA

John P. Giesy
224 National Food Safety and Toxicology Center, Michigan State University, East Lansing, MI 48824, USA

Bruce M. Greenberg
Department of Biology, University of Waterloo, Waterloo, Ontario, Canada N2L 3G1

Chantal Guay
Ville de Montréal — Service des travaux publics et de l'environnement — Laboratoire, 999 rue de Louvain Est, Montréal, Québec, Canada H2M 1B3

Jalal Hawari
Biotechnology Research Institute, National Research Council — Canada, 6100 Royalmount Ave., Montreal, Quebec, Canada H4P 2R2,

Laurens Henzen
TNO Nutrition and Food Research Institute, Department of Environmental Toxicology, P.O. Box 6011, 2600 JA Delft, The Netherlands

Jan-Willem Kamerman
Province of Gelderland, Department of Environment and Water, Sub Department of Soil Remediation, P.O. Box 9090, 6800 GX Arnhem, The Netherlands

Thomas Knacker
ECT Oekotoxikologie GmbH, Böttgerstr. 2-14, D-65439 Flörsheim am Main, Germany

Petra Kreule
Soil Quality Management, Tauw Milieu Consultancy, Handelskade 11, P.O. Box 133, NL-7400 AC Deventer, The Netherlands

Roman Kuperman
Geo-Centres, Inc., Gunpowder Branch, P.O. Box 68, Aberdeen Proving Ground, MD 21010-0068, USA

Anne-Marie Lafortune
Centre d'Expertise en Analyse Environnementale du Québec, Ministère de l'Environnement du Québec, 1665 boul Wilfrid-Hamel, Édifice 2, bureau 1.03, Québec, Québec, Canada G1N 3Y7

Greg Linder
Heron Works Farm, 5400 Tacoma Street, NE, Salem, OR 97305, USA

Sylvain Loranger
QSAR Inc., Room 800, 360 St-Jacques St. W., Montreal, Quebec, Canada H2Y 1P5

Louis Martel
Centre d'Expertise en Analyse Environnementale du Québec, Ministère de l'Environnement du Québec, 1665 boul. Wilfrid-Hamel, Édifice 2, bureau 1.03, Québec, Québec, Canada G1N 3Y7

Charles A. Menzie
Menzie-Cura & Associates, 1 Courthouse Lane, Chelmsford, MA 01824, USA

Jos Notenboom
National Institute of Public Health and Environmental Protection (RIVM), Antonie van Leeuwenhoeklaan 9, P.O. Box 1, 3720 BA Bilthoven, The Netherlands

Sylvain Ouellet
Federal Provincial Relations Branch, Strategic Policy and Partnerships, Policy and Communications, Environment Canada, 10 Wellington Street, 22nd floor, Hull, Quebec, Canada K1A 0H3

Simon Paradis
Network Development, Shell Canada Products Ltd., 400 4th Ave. S.W., Calgary, Alberta, Canada T2P 2H5

Graeme I. Paton
Department of Plant & Soil Science, University of Aberdeen, Aberdeen, AB2 3 Old Aberdeen, Scotland

Stan J. Pauwels
Abt Associates Inc., 55 Wheeler St., Cambridge, MA 02138-1168, USA

Adrien Pilon
Biotechnology Research Institute, National Research Council — Canada, 6100 Royalmount Ave., Montreal, Quebec, Canada H4P 2R2

Yvan Pouliot
Biogénie Inc., 350 rue Franquet, Sainte-Foy, Québec, Canada G1P 4P3

Judith Raes
Soil Quality Management, Tauw Milieu Consultancy, Handelskade 11, P.O. Box 133, NL-7400 AC Deventer, The Netherlands

Agnès Y. Renoux*
Biotechnology Research Institute, National Research Council — Canada, 6100 Royalmount Ave., Montreal, Quebec, Canada H4P 2R2*

Pierre Yves Robidoux
Biotechnology Research Institute, National Research Council — Canada, 6100 Royalmount Ave., Montreal, Quebec, Canada H4P 2R2

Jörg Römbke
ECT Oekotoxikologie GmbH, Böttgerstr. 2-14, D-65439 Flörsheim am Main, Germany

Lorraine Rouisse
SANEXEN Services Environnementaux Inc., 579 LeBreton, Longueuil, Québec, Canada J4 GG1R9

Sébastien Sauvé
QSAR Inc., Room 800, 360 St-Jacques St. W., Montreal, Quebec, Canada H2Y 1P5

Gladys L. Stephenson
ESG International, 361 Southgate Dr., Guelph, Ontario, Canada N1 GG3M5

Geoffrey I. Sunahara
Biotechnology Research Institute, National Research Council — Canada, 6100 Royalmount Ave., Montreal, Quebec, Canada H4P 2R2

* Present address: Sanexen Environmental Services Inc., Longueuil, Quebec, Canada.

Claus Svendsen
Centre for Ecology & Hydrology, Monks Wood, Abbots Ripton, Huntingdon, Cambridgeshire PE28 2LS, UK

Claude Thellen
Centre d'Expertise en Analyse Environnementale du Québec, Ministère de l'Environnement du Québec, 2700 rue Einstein, Sainte-Foy, Québec, Canada G1P 3W8

Sonia Thiboutot
Defense Research Establishment Valcartier, Canadian Ministry of National Defense, 2459 Pie IX Boulevard North, Val Bélair, Quebec, Canada G3J 1X5

Jean-Pierre Trépanier
SANEXEN Services Environnementaux Inc., 579 LeBreton, Longueuil, Québec, Canada J4G G1R9

Ferry van den Oever
Soil Quality Management, Tauw Milieu Consultancy, Handelskade 11, P.O. Box 133, NL-7400 AC Deventer, The Netherlands

Cornelis A.M. van Gestel
Institute of Ecological Science, Vrije Universiteit, De Boelelaan 1087, 1081 HV Amsterdam, The Netherlands

Suzanne Visser
Department of Biological Sciences, University of Calgary, Calgary, Alberta, Canada T2N 1N4

Jason M. Weeks
National Centre for Environmental Toxicology WRc-NSF Ltd, Henley Road, Medmenham, Marlow, Bucks SL7 2HD, UK

Gérald J. Zagury
NSERC Industrial Chair in Site Remediation and Management, Chemical Engineering Department, École Polytechnique de Montréal, 2900 Édouard-Montpetit, C.P. 6079, succ. Centre ville, Montréal, Québec, Canada H3 CC3A7

Series Foreword

This is a book that everyone involved in the development and management of contaminated sites should read. The editors of this informative and comprehensive book have managed to bring within one cover, details of the scientific tools and approaches for contaminated site assessment, including a host of effects based methodologies. This book has been written as a reference source and as such encompasses risk assessment procedures and addresses socio-economic issues and those of relevance to both ecological (and human) health assessments. This book provides an overview of the diversity of interrelated issues that one needs to understand in order to address the concerns and issues pertaining to contaminated sites. Furthermore, the text is exemplified using case study examples to unravel the complexity and interwoven nature of the subject area. It is hoped that this book will go someway towards the construction of better and more informed pathways of communication between all the various stakeholders involved in this complex issue.

We are delighted to have this latest contribution as part of the *Ecological and Environmental Toxicology Series* and wish to thank the editors for their considerable efforts in producing a text that grew and matured from the original seed.

This book crosses many different boundaries and disciplines and will be of interest to chemists, biologists, developers, toxicologists, managers, lawyers, risk assessors, regulators and those being regulated.

Jason M. Weeks
Sheila O'Hare
Barnett A. Rattner

Editors' Preface

This book evolves from a workshop entitled '*Toxicity Testing Applied to Soil Ecotoxicology*', held in Montreal in 1995. This forum of invited ecotoxicologists, contaminated site risk assessors and managers from government, universities and the private sector was organized by the National Research Council of Canada, Environment Canada and the Quebec Ministry of Environment. The overall objective of this workshop was to give an overview of ecotoxicity testing applied to contaminated site management, and to update the needs of management relative to contaminated sites. Through discussion groups, a number of recommendations were made: (1) to improve the documentation on the conduct of terrestrial toxicity bioassays in order to take account of soil sampling, handling and interpretation requirements; (2) to increase the fundamental knowledge of soil in order to interpret with more accuracy the potential effects on the target organisms; and (3) to develop a framework allowing an optimal use of terrestrial bioassays during contaminated site assessment.

It became evident that the contaminated site sector is becoming more and more complex, at the technical, risk assessment (ecological and human health concerns) and risk management levels. In addition, this evolving trend is being experienced by a growing number of countries in North America (Canada and the USA) and in Europe (the European Union). This book was written to help clarify this dynamic situation for the non-specialist reader (graduate students, government regulators and environmental firm consultants) and those wishing to keep abreast of these developments. We describe the technical as well as the theoretical aspects of environmental risk assessment and management using real-life examples (case studies).

The science-based evaluation of contaminated sites represents one of the major international challenges facing those in the field of environmental risk assessment and management. Appreciating the different aspects (strengths, weaknesses and uncertainties) of this issue is one of the first steps towards meeting this challenge. It is hoped that this book can serve as a tool for the reader to accept and use the scientific and theoretical approaches towards the successful restoration of a contaminated site.

Topics addressed herein include: the integration of terrestrial ecotoxicity testing with respect to a chemical's behavior in soil, the recent developments in contaminated soil risk assessment, and the use of advanced scientific data for ecological and human health risk assessment and contaminated site management purposes. We also describe the interrelationship between the laboratory and field ecotoxicologist, the ecotoxicologist and human health risk assessor and different international environmental regulatory agencies, and illustrate how their work is integrated by a common objective — the successful evaluation and management of rehabilitated contaminated sites.

The Editors wish to thank all of the contributors to this book, as well as the many scientific experts who assisted in the peer-review process of the following chapters. We are most grateful to Claire Couzinie-Holmiére, Helen E. Blaho and Kelly Wilson for their tireless assistance with preparation of the manuscript.

Geoffrey I. Sunahara
Agnés Y. Renoux
Claude Thellen
Connie L. Gaudet
Adrien Pilon

Abbreviations

$[^1O_2]$	singlet-state oxygen
^1H NMR	proton nuclear magnetic resonance
2,4-DANT	2,4-diaminonitrotoluene
2,6-DANT	2,6-diaminonitrotoluene
2-ADNT	2-aminodinitrotoluene
2-OH-ADNT	2-hydroxy-aminodinitrotoluene
4-ADNT	4-aminodinitrotoluene
4-OH-ADNT	4-hydroxy-aminodinitrotoluene
AChE	acetylcholinesterase
AhR	aryl hydrocarbon receptor
ASTM	American Society of Testing and Materials
ATP	adenosine triphosphate
ATSDR	Agency for Toxic Substances and Disease Registry (an agency of the US Department of Health and Human Services)
BAF	bioaccumulation factor
B(a)P	benzo(*a*)pyrene
BBA	Biologische Bundesanstalt für Land- und Forstwirtschaft, Deutschland (Federal Agency for Agriculture and Forestry, Germany)
BBodSchG	Bundesbodenschutzgesetz (German Soil Protection Act)
BBSK	Bodenbiologische Standortklassifikation (Soil Biological Site Classification)
BCF	bioconcentration factor
BEAST	benthic assessment of sediment
BenzROD	benzyloxyreorufin-*o*-dealkylase
BMF	biomagnification factor
BOD	biological oxygen demand
BP	biopile
BS	bioslurry
BSAF	biota to sediment accumulation factor
BTEX	benzene, toluene, ethylbenzene and *m*,*p*,*o*-xylenes (monocyclic aromatic hydrocarbons)
C_w^{sat}	aqueous saturation concentration of the contaminant
C/N	carbon/nitrogen
CCME	Canadian Council of Ministers of the Environment

CDNB	1-chloro-2,4-dinitrobenzene
CEAEQ	Centre d'expertise en analyse environmentale du Québec (Quebec Centre for Environmental Analyses, Canada)
CEC	cation exchange capacity
ChE	cholinesterase
Chl *a*	chlorophyll *a* fluorescence
CHO	Chinese hamster ovary
CMHC	Canadian Mortgage and Housing Corporation
cPAH	carcinogenic PAH
CPPI	Canadian Petroleum Products Institute
CV	coefficient of variation
CYPIA1	cytochrome P450 IA1 monooxygenase enzyme
cyt	cytoP450
DDT	dichlorodiphenyltrichloroethane
DNA	deoxyribonucleic acid
DNB	dinitrobenzene
DNT	dinitrotoluene
EC	equivalent carbon number index
EC	European Community (old name of the EU)
EC_{50}	effective concentration 50 (median)
EC_x	concentration that results in an effect in a specified percentage x of the population
EcD	ecological dose
ECOD	ethoxycoumarin-*o*-dealkylase
EEC	European Economic Community
EPH	extractable petroleum hydrocarbon
ERA	environmental risk assessment or ecological risk assessment or ecotoxicological risk assessment, depending on jurisdiction
EROD	ethoxyresorufin-*o*-dealkylase
ERT	enchytraeid reproductionl Test
ETHA	ethacrynic acid
EU	European Union
EV	exposure value
f_{OM}	soil fraction of organic matter
FBRC	Federal Biological Research Centre for Agriculture and Forestry
GC/MS	gas chromatography/mass spectroscopy
GST	glutathione S-transferase
GTE	Technical Evaluation Committee (MENV)
HC	hazardous concentration
HCH	hexachlorocyclohexane
HDPE	high-density polyethylene

HGPRT	hypoxanthine - guanine phosphoribosyl transferase
HI	hazard index
HMX	octahydro-1,3,5,7-tetranitro-1,3,5,7-tetrazocine
HQ	hazard quotient
HR	hazard ratio
IARC	International Agency for Research on Cancer
IC_p	inhibiting concentration for a specified percentage effect *p*
ICM	isolation, containment and monitoring
ICP - AES	inductively coupled plasma - atomic emission spectrometry
IELCR	incremental excess lifetime cancer risk
IRIS	Integrated Risk Information System (USEPA)
ISM	integrated soil microcosm
ISO	International Standards Organization
K_d	soil-water distribution coefficient
K_h	Henry's law constant
K_oc	octanol - carbon coefficient
K_ow	octanol - water partition coefficient
K_pd	plant - soil partition coefficient
K_pw	skin permeability
K_sp	solubility product
LC_{50}	concentration causing 50% lethality
LC_x	concentration that results in lethality in a specified percentage *x* of the population
LD_{50}	dose that is lethal to 50% of a test population
LOAEL	lowest observed adverse effect level
LOEC	lowest observed effect concentration
LUFA	Landwirtschaftliche Untersuchungs- und Forschungsanstalt, Deutschland (Agricultural Testing and Research Institution, Germany)
MAFF	Ministry of Agriculture, Forestry and Fisheries (Great Britain)
MAH	monocyclic aromatic hydrocarbon
MEFQ	Ministère de l'Environnement et de la Faune du Quebec (Ministry of Environment of Québec, Canada)
MENV	Ministère de l'Environnement du Québec (Quebec Ministry of Environment, Canada)
MFO	mixed function oxidase
MGPR	manufactured gas plant residue
MSSS	Ministère de la Santé et des services sociaux (Health and Social Services Ministry of Quebec, Canada)
MT	metallothionein
MW	monitoring well

NAP	Normaal Amsterdams peil (Amsterdam ordnance datum, The Netherlands)
NCI	National Cancer Institute (US)
NCP	non-conventional pollutant
NOAEC	no observed adverse effect concentration
NOAEL	no observed adverse effect level
NOEC	no observed effect concentration
NPL	National Priorities List (list of Superfund sites recognized by the USEPA)
NRRT	neutral red retention time
NRTEE	National Round Table on Environment and Economy
NTP	National Toxicology Program
O_3	ozone
OC	organic carbon
OECD	Organization for European Co-operation and Development
OM	organic matter
OP	organophosphorus pesticide
ORP	oxido-reduction potential
P450	cytochrome P450 monooxygenases
PAH	polycyclic aromatic hydrocarbon
PCB	polychlorinated biphenyl
PCB52	2,2′,5,5′-tetrachlorobiphenyl
PCDD	polychlorinated dibenzodioxin
PCDF	polychlorinated dibenzofuran
PCDH	polychlorinated diaromatic hydrocarbon
PCN	polychlorinated naphthalene
PCP	pentachlorophenol
PCR	polymerase chain reaction
PDF	probability density function
PEIDTE	Prince Edward Island Department of Technology and Environment (Canada)
PentROD	pentoxyreorufin-*o*-dealkylase
pH-KCl	Soil pH measured in 1 M KCl
PIRI	Partnership in RBCA Implementation
*p*NP	*p*-nitrophenol
POE	point of exposure
ppb	parts per billion
ppm	parts per million
QA/QC	quality assurance/quality control
QSAR	quantitative structure activity relationship
RA/RM	risk assessment/risk management
RBCA	risk-based corrective action
RDX	1,3,5-trinitro-1,3,5-triazacyclohexane

RH	relative (air) humidity
RI	risk index
RME	reasonable maximum exposure
RMK	risk reduction, environmental merit and cost
RV	reference value
SCE	sister chromatid exchange
SECOFASE	Development, Improvement and Standardization of Test Systems for Assessing Sublethal Effects of Chemicals on Fauna in the Soil Ecosystem
SIR	substrate-induced respiration method
SOP	standard operating procedure
SSE	selective sequential extraction
SSTL	site-specific target level
STT	short-term test
TAT	2,4,6-triaminotoluene
TCA	trichloroacetic acid
TCA	tolerable concentration in air
TCDD	2,3,7,8-tetrachlorodibenzo-*p*-dioxin
TCDD-EQ	TCDD-equivalent
TEF	toxic equivalency factor
TEQ	TCDD toxic equivalent
TFA	trifluoroacetic acid
TIA	total immune activity
TK	thymidine kinase
TME	terrestrial model ecosystem
TNB	trinitrobenzene
TNT	2,4,6-trinitrotoluene
TPH	total petroleum hydrocarbon
TPHCWG	Total Petroleum Hydrocarbon Criteria Working Group
TRV	toxicity reference value
UBA	Umweltbundesamt, Deutschland (Federal Environmental Agency, Germany)
UCL	upper confidence limit
US	United States
USEPA	United States Environmental Protection Agency
WHO	World Health Organization

PART ONE

Ecotoxicity Tools and Novel Approaches

1

Introduction

AGNÈS Y. RENOUX AND GEOFFREY I. SUNAHARA

Biotechnology Research Institute, National Research Council of Canada, Montreal, Quebec, Canada

1.1 BACKGROUND

The accumulation of pollutants in soil is well known, and its re-discovery was relatively recent (i.e., within the last three decades). Ratcliffe (1967) was the first to establish the link among the land application of the pesticide DDT, the reduced population of sparrow hawk (*Accipiter nisus*) and peregrine falcon (*Falco peregrinus*), and eggshell thinning. As evidenced by these studies, it is possible to resolve that the presence of a chemical in the soil may be associated with a hazardous effect. However, this correlation is still an important challenge to ecotoxicologists, especially if one wishes to predict the risk to potential biological receptors, including humans, at a specific contaminated site. To understand why the use of toxicological tools has become a key element in the analysis of contaminated sites, it is necessary to examine the accumulation of pollutants in soil, their bioavailability and pathways for organism exposure.

1.1.1 ACCUMULATION OF POLLUTANTS AT CONTAMINATED SITES

Because of its high retention capacity, soil is very vulnerable to contaminant accumulation. Organic or inorganic contamination may stem from an accidental (overflow) or deliberate spill, or from a medium to long-term accumulation, such as that following agricultural land application, atmospheric deposit and groundwater flow (Sheppard *et al.* 1992). As a result, there is a high spatial heterogeneity in the contaminant distribution. Furthermore, the variability in soil composition associated with numerous physico-chemical properties can affect the chemical form in which contaminants appear. Chemically, contaminants could be found in soils as distinct particles, solutes in interstitial water, volatilized in air, or adsorbed to mineral and organic particles (Eijsackers 1994).

Environmental Analysis of Contaminated Sites. Edited by G. I. Sunahara, A. Y. Renoux, C. Thellen, C. L. Gaudet and A. Pilon

1.1.2 BIOAVAILABILITY

Soil contaminant concentrations (external dose) provide little information on their bioavailability; that is, the contaminant fraction passively or actively transferred from soil (external dose) to a biological receptor (internal dose) where it can be transformed, accumulated and/or induce a response or undesired health effect (biologically effective dose). Water-soluble forms of contaminants are known to be the most bioavailable. The main route of uptake of hydrophobic contaminants is via soil pore-water for terrestrial invertebrates (Belfroid *et al.* 1996) and for plants (van Gestel *et al.* 1996). Studies have shown that the toxicity of organic compounds in soil could be determined by using their concentrations in the interstitial water, as predicted by the use of adsorption coefficients (van Gestel *et al.* 1991, Hulzebos *et al.* 1993). The organic matter, by strongly influencing the sorption of a compound, is the most important factor for controlling the bioavailability (Belfroid *et al.* 1996, van Gestel *et al.* 1996).

The effect of aging on bioavailability is controversial. As they persist or age in soil, organic compounds may become less bioavailable in short-term experiments (Alexander 2000). This sequestration is believed to be reversible, however, at a slow rate of mass transfer from the solid state to the organisms (Baveye and Bladon 1999, Jagger *et al.* 2000). This can be the result of a slow microbial release (Gunkel *et al.* 1993, Scheunert *et al.* 1995). Moreover, seemingly non-mobile contaminants linked with solid particles can also act as toxicants, simply by ingestion of the contaminated particles (Landrum and Robbins 1990, Belfroid *et al.* 1996, Forbes *et al.* 1998). A more detailed description of the bioavailability for organic and inorganic compounds is found in Chapter 2 of this book. The factors influencing bioavailability are discussed, and recommendations are given for the handling of samples and the experimental assessment of toxicity and bioavailability.

1.1.3 EXPOSURE PATHWAYS

Organisms may be exposed to soil pollutants through two different pathways: (a) direct contact with the soil (soil ingestion, contact, or inhalation); (b) after transfer of contaminants from the soil compartment to another environmental compartment such as groundwater or air. Its chemical nature and interaction with the soil constituents (mobility) influence the fate of a contaminant. Exposed organisms can also affect previously non-exposed organisms through ecological linkages such as symbiotic association, food chain transfer, etc. Thus, the impact of a contaminated soil on its environment is multi-factorial.

1.2 LIMITATIONS OF CHEMICAL ASSESSMENT

Chemical analyses are systematically used to characterize a site or to demonstrate treatment efficiency. However, the complexity of the interactions between the pollutants, the soil constituents and the biological receptors dramatically complicates the interpretation of these chemical evaluations. Consequently, the chemical concentrations cannot systematically predict the toxicity of a specific chemical.

As explained earlier in this chapter, the degree of bioavailability of contaminants in soil can extensively modify a toxic response. As a result, the sole use of chemical data to characterize the ecotoxicological hazard and the risk associated with a contaminated site is extremely limiting, and perhaps dangerous in some cases. In addition, chemical interactions may occur as a result of a multiple contamination. Many studies have shown that the co-presence of contaminants can lead, in a non-predictable manner, to a synergistic, antagonistic or additive interaction (Hass *et al.* 1981, Donnelly *et al.* 1988, Davol *et al.* 1989, Donnelly *et al.* 1990). The appearance of toxicologically unknown metabolites is another limitation to the use of chemical analyses. Physico-chemical and microbiological processes can lead to the reduction or elimination of the compound(s) of concern (mineralization) or to the production of by-products or persistent metabolites. The toxicity of these metabolites is often neglected. The degradation or detoxification by-products are sometimes more toxic than the compound of concern (Alexander 1981, Donnelly *et al.* 1987). This effect, called *activation*, can be illustrated with the soil microbial epoxidation of aldrine into dieldrine, a product which is not only more toxic than its parent compound, but also more persistent in the environment (Lichtenstein and Schulz 1960).

1.3 THE USE OF TOXICOLOGICAL TOOLS IN ASSESSING THE IMPACT OF CONTAMINANTS

Ecotoxicological assessment (also called environmental hazard assessment) is carried out through biological, toxicological and/or ecological analyses to quantify any injury, damage or disturbance caused by a source of contamination. These data are necessary for thorough and accurate ecological risk analysis or for environmental site management. The following series of chapters found in Part One describes the available tools which could be used for environmental hazard assessment. These laboratory and field-based tools are biological systems at different organizational levels (cell, organism, population and ecosystem) which can be used to detect hazardous compounds or conditions. This is accomplished by measuring a toxic effect using a bioassay, a biomarker or a bioindicator in a laboratory-based or field-based study.

1.3.1 BIOASSAYS

A *bioassay* or *test of toxicity* is a biological analytical tool, which can quantify the toxicity of a contaminant or an environmental sample using specific organisms. In these laboratory-based studies, field samples can be analyzed in the laboratory. Because of the complexity of interactions occurring in the field, the effects of environmental factors would be confounding to an accurate interpretation of results. Toxicity testing in the laboratory can reduce the number of these factors, and allow the determination of the cause and effects of specific chemicals or site samples. Soil bioassays are described in this book in three chapters. Stephenson *et al.* (Chapter 3) give an overview of the existing single-species test methods available for soil and groundwater assessment. They also discuss practical considerations, such as the importance of constituting a battery of bioassays. The use of multi-species bioassays is described by Kuperman *et al.* (Chapter 4). These methods integrate species from multi-trophic levels of organization (microorganisms, invertebrates and plants) to quantify the consequences of contaminant toxicity through the trophic chain, and disturbance in biological diversity and in nutrient cycling processes in terrestrial ecosystems. The interpretation of the data obtained with multi-species bioassays is also discussed in this chapter. Particular attention was given to the assessment of phytotoxicity by Greenberg (Chapter 8). In this latter chapter, many types of plant bioassays are described, and the impact of different groups of contaminants on plants is discussed.

1.3.2 BIOMARKERS

Biomarkers of exposure reflect the internal dose of a contaminant in a specific organism exposed in the laboratory or in the field, and can estimate its bioavailability. Problems of detection limits and concentration in organisms, which are not always linked with an affected function, are often encountered. Blankenship and Giesy (Chapter 9) describe the use of biomarkers of exposure in birds and mammals at a PCB-contaminated site.

Biomarkers of effect reflect a biological alteration, which leads to an undesired effect, such as a specific disease in an organism exposed in the laboratory or in the field. The use of biomarkers is limited to certain classes of chemicals, and is thus highly specific (e.g., endocrine disrupters). These tools resolve the problem of determining the internal dose by providing a measure of the toxicant's effect. Svendsen *et al.* (Chapter 7) discuss the utility of biomarkers and summarize the available soil invertebrate and microbial measurements that may be utilized to undertake a contaminated site assessment. Plant biomarkers are also summarized and discussed by Greenberg (Chapter 8). Biomarkers should not be mistaken for *bioindicators*, which are effects on local populations ecologically surveyed at a specific site.

1.3.3 FIELD-BASED STUDIES

Field-based studies are designed to address potential hazards as suggested by the results of laboratory studies. Mesocosm or enclosure experiments can link laboratory and field studies results since they can provide greater experimental control on environmental factors. Römbke and Notenboom (Chapter 10) provide ecotoxicological field methods and illustrate their use with different case studies.

1.3.4 DNA DAMAGE

DNA damage may lead to carcinogenicity and congenital abnormalities. Particular attention is given to carcinogenicity, which is an important public health concern. Claxton and George (Chapter 6) discuss the challenges associated with the identification of carcinogens at contaminated sites and the management of this information during risk assessment. Garcia and Donnelly (Chapter 5) present the short-term tests available for the detection of mutagens, and illustrate their use with three case studies.

REFERENCES

Alexander M (1981) Biodegradation of chemicals of environmental concern. *Science*, **211**, 132-138.

Alexander M (2000) Aging, bioavailability, and overestimation of risk from environmental pollutants. *Environmental Science and Technology*, **34**(20), 4259-4265.

Baveye P and Bladon R (1999) Bioavailability of organic xenobiotics in the environment: a critical perspective. In *Bioavailability of Organic Xenobiotics in the Environment*, Baveye P, Block J-C and Goncharuk V (eds), NATO ASI Series, Kluwer Academic Publishers, Dordrecht, pp. 227-248.

Belfroid AC, Sijm DTHM and van Gestel CAM (1996) Bioavailability and toxicokinetics of hydrophobic aromatic compounds in benthic and terrestrial invertebrates. *Environmental Reviews*, **4**, 276-299.

Davol P, Donnelly KC, Brown KW, Thomas JC, Estiri M and Jones DH (1989) Mutagenic potential of runoff water from soils amended with three hazardous industrial wastes. *Environmental Toxicology and Chemistry*, **8**, 189-200.

Donnelly KC, Davol PD, Brown KW, Estiri M and Thomas JC (1987) Mutagenic activity of two soils amended with a wood-preserving waste. *Environmental Science and Technology*, **21**, 57-64.

Donnelly KC, Brown KW, Estiri M, Jones DH and Safe SH (1988) Mutagenic potential of binary mixtures on nitro-polychlorinated dibenzo-*p*-dioxins and related compounds. *Journal of Toxicology and Environmental Health*, **24**, 345-356.

Donnelly KC, Brown KW, Anderson CS, Barbee GC and Safe SH (1990) Metabolism and bacterial mutagenicity of binary mixtures of benzo(*a*)pyrene and polychlorinated aromatic hydrocarbons. *Environmental Molecular Mutagenesis*, **16**, 238-245.

Eijsackers H (1994) Ecotoxicology of soil organisms: seeking the way in a pitch-dark labyrinth. In *Ecotoxicology of Soil Organisms*, Donker MH, Eijsackers H and Heimbach F (eds), Lewis, Boca Raton, FL, pp. 3-32.

Forbes TL, Forbes VE, Giessing A, Hansen R and Jure LK (1998) Relative role of pore water versus ingested sediment in bioavailability of organic contaminants in marine sediments. *Environmental Toxicology and Chemistry*, **17**(12), 2453-2462.

Gunkel J, Ronnpagel K and Ahlf W (1993) Suitability of microbial bioassays for bound contaminants. *Acta Hydrochimica et Hydrobiologica*, **21**(4), 215-220.

Hass BS, Brooks EE, Schumann KE and Dornfeld SS (1981) Synergistic, additive, and antagonistic mutagenic responses to binary mixtures of benzo(*a*)pyrene and benzo(*e*)pyrene as detected by strains TA98 and TA100 in the *Salmonella*/microsome assay. *Environmental Mutagenesis*, **3**, 159-166.

Hulzebos EM, Adema DMM, Dirvenvanbreemen EM, Henzen L, Vandis WA, Herbold HA, Hoekstra JA, Baerselman R and Vangestel CAM (1993) Phytotoxicity studies with *Lactuca sativa* in soil and nutrient solution. *Environmental Toxicology and Chemistry*, **12**(6), 1079-1094.

Jagger T, Sanchez FAA, Muijs B, van der Velde EG and Posthuma L (2000) Toxicokinetics of polycyclic aromatic hydrocarbons in *Eisenia andrei* (oligochaeta) using spiked soil. *Environmental Toxicology and Chemistry*, **19**(4), 953-961.

Landrum PF and Robbins JA (1990) Bioavailability of sediment-associated contaminants to benthic invertebrates. In *Sediments: Chemistry and Toxicity of In-Place Pollutants*, Baudo R, Giesy JP and Muntau H (eds), Lewis, Ann Arbor, MI, pp. 237-263.

Lichtenstein EP and Schulz KR (1960) Epoxidation of aldrin and heptachlor in soils as influenced by autoclaving, moisture, and soil types. *Journal of Economic Entomology*, **53**(2), 192-197.

Ratcliffe DA (1967) Decrease in eggshell weight in certain birds of prey. *Nature (London)*, **215**(July 8), 208-210.

Scheunert I, Attar A and Zelles L (1995) Ecotoxicological effects of soil-bound pentachlorophenol residues on the microflora of soils. *Chemosphere*, **30**(10), 1995-2009.

Sheppard SC, Gaudet C, Sheppard MI, Cureton PM and Wong MP (1992) The development of assessment and remediation guidelines for contaminated soils, a review of the science. *Canadian Journal of Soil Science*, **72**(4), 359-394.

van Gestel CAM, Ma WC and Smit CE (1991) Development of QSARs in terrestrial ecotoxicology: earthworm toxicity and soil sorption of chlorophenols, chlorobenzenes and dichloroanilines. *The Science of the Total Environment*, **109/110**, 589-604.

van Gestel CAM, Adema DMM and Dirven-van Breemen EM (1996) Phytotoxicity of some chloroanilines and chlorophenols, in relation to bioavailability in soil. *Water, Air and Soil Pollution*, **88**, 119-132.

2

Sample Handling and Preparation for Estimation of Mobility, Bioavailability and Toxicity of Contaminants in Soils

GÉRALD J. ZAGURY[1], YVES DUDAL[1], JACQUES BUREAU[1], CHRISTIAN BASTIEN[2] AND RAYNALD CHASSÉ[2]

[1]NSERC Industrial Chair in Site Remediation and Management, Chemical Engineering Department, École Polytechnique de Montréal, Montreal, Quebec, Canada
[2]Centre d'expertise en analyse environnementale du Quebec, Ministère de l'Environnement du Québec, Ste-Foy, Québec, Canada

2.1 CONTAMINANT BEHAVIOUR IN SOILS

2.1.1 SOIL COMPOSITION

Soil is basically a complex mixture of mineral particles associated with organic matter, water, air, gas and possibly contaminants, which all interact in a dynamic equilibrium that may modify each other's intrinsic characteristics. The inorganic phase consists of clay minerals, iron, manganese and aluminium oxides and hydroxides, amorphous aluminosilicates and carbonates, sulphates, sulphides and phosphates. Non-living organic matter consists of soluble substances (e.g., amino acids, proteins, carbohydrates, organic acids and lignin) and insoluble substances (e.g., humic acids).

Soil is not only a support medium for plants and microorganisms, but also the life medium of many living organisms, which take advantage of the presence of nutrients provided by the recycling process of organic matter. Interactions taking place between living and non-living components are able to modify the behaviour of potential contaminants, and should be addressed in the perspective of the toxicity evaluation of contaminated soils.

Environmental Analysis of Contaminated Sites. Edited by G. I. Sunahara, A. Y. Renoux, C. Thellen, C. L. Gaudet and A. Pilon

2.1.2 CONTAMINANT NATURE

The mobility of contaminants and therefore their potential bioavailability and toxicity is governed by their own physicochemical properties, by the physicochemical and biological characteristics of the soil environment as well as by physiological properties of organisms.

2.1.2.1 Contaminant Physicochemical Properties

Properties of Organic Contaminants

The basic physicochemical properties of organic contaminants required to assess their fate within soils are: their aqueous solubility, Henry's law constant and octanol-water partition coefficient. Once these three parameters are known, they will help to define the distribution of the organic compound between the three phases of the soil: the aqueous phase, the solid phase and the gaseous phase. First, the aqueous solubility, usually given in moles per litre at 25 °C, represents the concentration of a saturated aqueous solution of the contaminant (C_w^{sat}). One characteristic of organic compounds is that their polarity, and thus their aqueous solubility, is very low, explaining why it is expressed as a logarithmic value ($-\log C_w^{sat}$). Aqueous solubility decreases with increasing values of $-\log C_w^{sat}$. The Henry's law constant (K_h, L atm mol^{-1} at 25 °C) links the partial pressure of the contaminant on top of an aqueous solution with the actual concentration of the solution. This parameter reflects the ability of a water-solubilized contaminant to volatilize. Positive values of $\log K_h$ indicate a tendency to volatilize, whereas negative values indicate a relative stability in water. Finally, the octanol-water partition coefficient (K_{ow}, mol L^{-1} octanol per mol L^{-1} water at 25 °C) expresses the hydrophobicity of the contaminant through its affinity for octanol (organic phase) rather than water. It is given as $\log K_{ow}$ because most organic contaminants are hydrophobic, leading to high K_{ow}. Table 2.1 gives values of these three parameters for a series of major organic contaminants.

Table 2.1 Physicochemical properties of major organic contaminants

Family	Name	Formula	$-\log C_w^{sat}$	$\log K_h$	$\log K_{ow}$
Hydrocarbons	*n*-octane	C_8H_{18}	5.20	3.47	5.18
	n-hexadecane	$C_{16}H_{34}$	7.80	2.07	—
BTEX	benzene	C_6H_6	1.64	0.74	2.13
	toluene	C_7H_8	2.25	0.83	2.69
PAH	naphthalene	$C_{10}H_8$	3.61	−0.37	3.36
	benzo(*a*)pyrene	$C_{20}H_{12}$	8.22	−2.92	6.50
PCB	hexachloro-BP	$C_{12}H_6Cl_4$	7.70	−0.51	6.31
	decachloro-BP	$C_{12}Cl_{10}$	10.55	−1.73	8.23
Pesticides	atrazine	$C_8H_{14}ClN_5$	3.81	—	2.56
	2,3,7,8-TCDD (dioxin)	$C_{12}H_4O_2Cl_4$	10.3	−1.3	6.64

Schwarzenbach *et al.* (1993). Reprinted by permission of John Wiley & Sons, Inc.

Properties of Inorganic Contaminants

Solubility product equilibrium concentration can be used to qualitatively predict the behaviour of metal ion species in soil solutions. The reaction of a divalent metal ion M^{2+}, with a divalent ligand L^{2-}, to give a solid phase can be expressed as:

$$M^{2+}\ (aq) + L^{2-}\ (aq) \leftrightarrow ML\ (s)$$

If the solid is a pure phase, the solubility product, K_{sp}, can be written:

$$[M^{2+}][L^{2-}] = K_{sp}$$

Precipitates form if the solubility product, K_{sp}, for the reaction is exceeded. This means that precipitates with low solubility products will dissolve only with difficulty in soil solution. Some ligands important to the precipitation of metal ions include hydroxide, carbonate, silicate, phosphate and sulphide (Allen 1993).

Vapour pressure is also an important property for heavy metals such as Hg, which has a relatively high vapour pressure (1.3×10^{-3} mm Hg at 25 °C) or metalloids such as As and Se. Contaminants with relatively low vapour pressure and high solubility in water are less likely to vaporize and become gaseous.

2.1.2.2 Soil Physicochemical Characteristics

The main soil variables involved in the mobility of contaminants are: pH, oxido-reduction potential (ORP), organic matter (OM), clay mineral, hydrous oxide, carbonate and salt contents. For example, carbonates, exchangeable cations and OM will contribute to the soil buffer capacity whereas clay minerals, OM and salt content will influence the cation exchange capacity (CEC). Buffer capacity gives an indication of the soil resistance to a pH variation. According to Federer and Hornbeck (1985), it can be expressed as the number of moles of H^+ ions needed to lower the initial pH of 1 kg of soil by one pH unit. CEC (usually expressed as milliequivalents per 100 g of soil) is defined as the sum of the exchangeable cations of a soil. Changes in soil environmental conditions may enhance or diminish contaminant mobility by changing its chemical form. The oxidation state of several metals of environmental concern will determine their relative mobility, availability and toxicity. Soil biological activity may be responsible for important changes in environmental conditions.

2.1.2.3 Soil Biological Activity

In recent years, the use of new molecular biology techniques on soil samples has shown the extreme diversity and isolation of the different microbial population in soils (Matheson *et al.* 1997). On one soil particle, a few million bacteria can be found which have no interaction with others found on the next particle. Their main substrate is the soil organic matter, mostly composed of humic and fulvic acids. In case of an organic contaminant, the ubiquity of the soil microflora is responsible for the homogenous response consisting of the growth of a contaminant-degrading consortium throughout the site. Furthermore, soil bacteria have recently been held responsible for the formation of covalent bonds between an organic contaminant and humic matter, leading to bound residues (Richnow *et al.* 1997). When looking at soil biological activity, it is important to also take into account the presence of invertebrates, capable of modifying the contaminant distribution within the soil (i.e., soil ingestion by earthworms). By changing the environmental conditions, the soil organisms can also enhance or diminish contaminant mobility, bioavailability and toxicity.

2.1.3 SOIL–CONTAMINANT INTERACTION

2.1.3.1 Soil Interaction with Organic Contaminants

Once the nature of the soil and the physicochemical properties of the contaminant are known, it is possible to assess the distribution of the contaminant between the different phases of the soil. The non-polarity of some organic contaminants, like the petroleum hydrocarbons molecule, permits only a weak interaction with the clay particle surfaces (van der Waals). However, in the case of positively charged organic molecules (by protonation), it is expected that they will be adsorbed on the soil (clay) solid surfaces, depending on the CEC of the soil. The pH of the medium will interfere with ionizable organic compounds, like phenols, by greatly influencing their aqueous solubility. Nevertheless, the main recipient of the organic contamination, is the organic matter. It has been shown that organic matter is composed of dense and enlarged compartments (Young and Weber 1995). The enlarged organic matter absorbs organic contaminant, but its desorption is fast and easy. The dense fraction absorbs contaminants much more strongly and their desorption is diffusion-limited. The process of aging results in an increase in the fraction of contaminant sorbed in this dense compartment (Luthy *et al.* 1997). Nevertheless, the fraction of total organic matter in the soil (f_{OM}) helps to evaluate the soil–water distribution coefficient (K_d) of hydrophobic organic contaminants, according to:

$$\log K_d = f_{OM} \times \log K_{OM}, \quad \text{with} \quad \log K_{OM} = a \log K_{ow} + b$$

where K_{OM} is the organic matter–water distribution coefficient. The parameters a and b are contaminant-specific and can be found in many studies (Schwarzenbach *et al.* 1993). This distribution is rather stable and is considered to be thermodynamically controlled. Other soil–contaminant interactions have a minor influence on this pseudo-equilibrium. The movement of water, despite its slow rate, carries the water-soluble fraction of the contaminant further down the soil profile, changing the distribution slowly. Additionally, the intrinsic biodegradation that occurs acts as a sink and thus debalances the pseudo-equilibrium. Nevertheless, the overall distribution of the organic contaminant is still mainly driven by the soil organic matter content.

2.1.3.2 Soil Interaction with Heavy Metals

In soils, metals can be found in several pools:

1. Dissolved in the soil solution (as either free ion or a soluble complex);
2. Occupying exchange sites on inorganic soil constituents;
3. Specifically adsorbed (short-range chemical forces such as ionic or covalent bonding) on inorganic soil constituents;
4. Associated (complexation or adsorption) with insoluble soil organic matter or organic colloids;
5. Precipitated as pure or mixed solids;
6. Present in the structure of primary and/or secondary minerals.

Soil particulate phases frequently have high concentrations of metals relative to concentrations of metals in the dissolved phase. The concentration in the soil solution is governed by a number of interrelated processes, including inorganic and organic complexation, oxidation–reduction, precipitation/dissolution reactions and adsorption/desorption reactions (McLean and Bledsoe 1992). Metal speciation or partitioning rather than total concentration is the key to understanding its mobility and also its potential bioavailability and toxicity.

2.2 LABORATORY ESTIMATION OF MOBILITY, BIOAVAILABILITY AND TOXICITY OF CONTAMINANTS IN SOILS

2.2.1 MOBILITY, BIOAVAILABILITY AND TOXICITY

The definition of these three terms helps us to know exactly what a specific test or series of tests will actually measure. Dickson *et al.* (1994) have proposed a conceptual approach to mobility, bioavailability and toxicity. First, mobility is the result of physicochemical interactions between the contaminant and the environment. Hydrogeological models that describe these interactions are of great help in assessing what fraction of the contaminant is mobile. Second, they introduce the environmental bioavailability that can be defined as the extent of biota exposure to the contaminant in the environment (Hrudey *et al.* 1996). Finally, the pharmacological bioavailability is the fraction of contaminant that will be absorbed within the organism and may lead to a toxic response. A similar definition of the bioavailable fraction has been given as the fraction of total chemical in the organism's exposure environment (water, food, sediment) that is available for absorption (McKim 1994). The biological response to the contaminant distribution or the pharmacological bioavailability is thus organism-dependent (Guerin and Boyd 1992, Davies *et al.* 1999). Indigenous invertebrates or small mammals (e.g., moles) may absorb contaminant linked to soil particles or organic matter that is not immediately bioavailable. But the contaminant exposed to specific physicochemical conditions in the gastric or intestinal environment may become bioavailable. In this case, oral bioavailability can be defined as the fraction of a contaminant that reaches the central blood compartment from the gastrointestinal tract (Ruby *et al.* 1999). Biological organisms (either microbial or higher organisms) exposed to the contaminant can also be looked at as sinks, constantly changing the contaminant distribution described above (Baveye and Bladon 1999). Although biomarkers could be incorporated into bioavailability tests, no integrated bioavailability test has yet been developed.

The biological tool to assess the environmental bioavailability could either be a contaminant-degrading microorganism or any organism susceptible to express a toxic response when exposed to the contaminant, or to accumulate the contaminant to the extent there could be a toxic response (Kelsey and Alexander 1997). The responses in these two cases are very different (biodegradation of the contaminant, toxic response or bioaccumulation), but they both measure a bioavailable fraction (see Figure 2.1). The first one may be called mass bioavailability, for its role in decreasing the mass of contaminant through biodegradation. The second one may be called toxic bioavailability, because it may potentially lead to the expression of a toxic response (Dudal *et al.* 1999). The toxicity is the actual response (i.e., lethality, decrease in growth, lack of reproduction, etc.) of indigenous biological organism exposed to the environmentally bioavailable fraction of contaminant.

From these approaches to mobility and bioavailability, the toxic response can be due to two different fractions of contaminant: an easily available or mobilizable fraction and an intrinsic fraction which represents the total

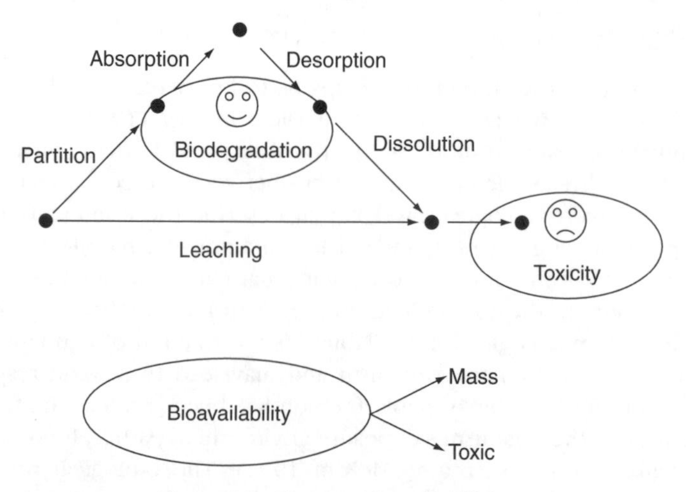

Figure 2.1 The two bioavailabilities.

contamination extractable with solvents, acids or bases (without destruction of the matrix). These fractions are not readily mobile but can represent a potentially ecotoxicological risk in the long term. The ratio of these two fractions evolves with time and is referred to as the aging phenomenon (Kelsey *et al.* 1997). Thus, different sample preparation procedures will lead to an assessment of these different fractions.

This theoretical approach was developed in a context useful to bioremediation action. However, in an ecosystem the response of communities is more complex. For example, the indigenous organisms exposed to the bioavailable fraction of a contaminant can decrease the mass of the contaminant by biodegradation, but they can also convert it into toxic metabolites. These new substances can modify community structure or function, even if they don't affect organisms. The contaminant concentration in the soil is also critical in that it may lead to a biodegradation response only if the concentration remains in the tolerance range (or adaptative range) of microorganisms so the microflora are able to metabolize it. As the concentration increases, a toxic stress concomitant with biodegradation will appear, thus reducing biodegradation efficiency, and ultimately no biodegradation will occur. In other words, biodegradation response and toxic response are not a matter of black or white, but a matter of environmental concentrations, absorption dose and ability of microorganisms to deal with contaminants.

The amount of mobilizable fraction of the contaminant is principally dependent on physicochemical interactions. Since no standard test has yet been developed to assess mass or toxic bioavailability, some research is being carried out to link mobility test responses with bioavailability (Kelsey and Alexander 1997, Loibner *et al.* 1997). Some mobility studies have been correctly correlated with biodegradation or toxicity measurements, opening a way to the assessment of bioavailability (Sauvé *et al.* 1998, Reid *et al.* 1999). However, if bioavailability is defined as the fraction of the total chemical content in the organism's exposure environment (water, food, sediment) that is available for absorption, a real biological response is not necessarily needed (i.e., toxicity) but the evaluation of a solvent-mobilizable fraction (chemically available) does not seem to be enough to assess the environmentally bioavailable fraction. In many cases, the metal dissolved in water but bound to dissolved organic matter, or to an inorganic ligand, or to mobile colloidal material, will not be immediately bioavailable.

2.2.2 FROM THE FIELD TO THE LABORATORY

From what has been established above, it appears that during sampling and handling, great care has to be given to the preservation of the pseudo-equilibrium describing the distribution of the contaminant in the soil. No amount of care in the preparation and assessment of mobility and toxicity can overcome the problems of inappropriate sampling in the field. The bioavailable fraction, for example, is usually composed of at least the water-mobilizable fraction, which can easily be lost during sampling and transport, through volatilization, water run-off or sorption on polymer container walls (Cseh *et al.* 1989). For organic contaminants, sampling tools should be made of stainless steel and should be cleaned adequately. Samples should preferably be stored in Teflon or glass containers with Teflon liner caps, leaving the smallest headspace possible. For inorganic contaminants, sampling tools should be in high-density polyethylene (HDPE). In this case, soil samples should be stored in nitric acid cleaned polypropylene or HDPE containers with polyethylene caps. When samples contain mixed contamination, borosilicate glass containers may be used or subsampling avoiding container walls may be performed in the laboratory. Soil samples should be stored immediately at 4 °C. If necessary, appropriate conservation agents should be added to samples. It is generally recommended to minimize storage and to perform toxicological assessment immediately after sampling, if possible, or within 36 h (Klemm *et al.* 1994). However, some published protocols consider toxicity testing within 7 days (CEAEQ 1999). Freezing a sludge elutriate sample has been shown to decrease its toxicity (Renoux *et al.* 1999) and should not be recommended. In a study on the effects of preservation techniques on metal partitioning in sediments determined by a sequential extraction procedure (Rapin *et al.* 1986), freezing and short-term wet storage at 4 °C of the sample were acceptable preservation techniques, whereas drying (freeze-drying and oven-drying at 105 °C) significantly affected the metal partitioning. Besides these general principles, soil sampling and handling precautions should be adapted depending on the analysis and testing to be performed.

2.2.3 SAMPLE PREPARATION FOR ASSESSMENT OF MOBILITY AND BIOAVAILABILITY

Because the free metal ion (which is soluble) is in general the most bioavailable and toxic form of the metal, numerous studies use various chemical extractants in an attempt to identify bioavailable metal (McLean and Bledsoe 1992, Sauvé *et al.* 1998). As for organics, and with the strong limitations mentioned earlier, the soluble fraction (chemically available) is used to assess the bioavailable fraction. Preparation of a contaminated soil sample for a contaminant mobility test usually includes homogenization, drying and sieving steps that can modify the contaminant distribution in the soil. Drying the soil at 105 °C for a 24-h period volatilizes a fraction of the organic contaminant, and keeping only the under 10, 4 or 2-mm fraction of the soil decreases slightly the contamination. Sieving increases the surface/volume ratio and maximizes the soil/solvent contact. In the case of an inorganic contamination, nylon or stainless steel sieves are recommended. Even though metals are less volatile, oven-drying at 105 °C can affect their partitioning (Rapin *et al.* 1986, Tessier and Campbell 1988). Therefore, air-drying at 40 °C long enough to allow subsequent homogenization and sieving is recommended.

The solvents used for assessing the mobility of organic substances can be water or other extracting solvents such as methanol, dichloromethane, supercritical-CO_2, cyclodextrins, etc. (Kelsey *et al.* 1997, Cuypers *et al.* 1999). The ability to mobilize the organic contaminant is expected to be very different with the use of these hydrophobically different solvents, considering the distribution of the contaminant within the soil organic matter. Even when using water, some variations will occur when considering its pH, for the mobility of ionizable organic compounds. Once a solvent is chosen, the procedure used to isolate the extract still has a major influence on the extent of contaminant mobilized. Closed, semi-open or open systems have shown very different results. Closed systems consist of reaching equilibrium between the sample and the solvent for contaminant distribution. The solid-to-liquid ratio (usually from 1/10 up to 1/100) will influence this result, which is dictated by the solubility of the contaminant, as seen above. Semi-open systems are based on the same concept of reaching an equilibrium, but this time sequential contacts are planned, allowing the total adsorbed fraction to desorb, but usually not the absorbed one (Harms and Zehnder 1995). Open systems are flow-through systems, saturated or not, and give a close-to-the-field approach as well as an idea of the evolution of contaminant desorption with time. Thus, different amounts of recovered contaminant as well as different rates of recovery can be obtained with different sample preparation and handling, leading to different results in terms of mobility and bioavailability.

In order to evaluate the retention form of metals in soils, and therefore their potential mobility and thus potential bioavailability, selective sequential extraction (SSE) procedures are often used. The procedure of Tessier *et al.* (1979) defines the following five fractions: soluble and exchangeable metal (extracted with a magnesium chloride solution), bound to carbonates (leached by an acetic acid/acetate buffer), bound to iron – manganese oxides (extracted with hydroxylamine hydrochloride), bound to organic matter or sulphides (released by nitric acid, hydrogen peroxide and ammonium acetate), and residual metal fraction (dissolved by acid attack). Despite strong criticisms dealing with the interpretation of these sequential extraction procedures (Nirel and Morel 1990), their use has continued to be recognized as a valuable tool, provided that discrimination and care are taken (Tessier and Campbell 1987, 1988, 1991, Zagury *et al.* 1997, 1999, Lo and Yang 1998, Wasay *et al.* 1998, Ho and Evans 2000). These techniques can provide a good indication of metal partitioning in sludge, soils and sediments and also provide a pragmatic estimation of their potential mobility. These sequential fractionations are however insufficient in describing metal bioavailability. Separation of the extracted fraction after single, multiple or sequential extraction is often carried out using filtration or centrifugation. Once again, these technical steps can lead to a substantial loss of the contaminant, according to the kind of material and vessel used. Centrifugation speed will affect the metal concentration in the supernatant (Ravishankar *et al.* 1993) because a significant portion of the metals can be associated with colloidal matter. Importantly, very hydrophobic contaminants will always prefer a surface to water. Preparation and handling of contaminated soil samples require great care in order to minimize redistribution or loss of contaminant and therefore inaccurate assessment of mobility and bioavailability.

2.2.4 SAMPLE PREPARATION FOR SOLID-PHASE TOXICITY TESTING

Numerous direct toxicity tests require specific environmental conditions, which generally reflect the optimal conditions of tested organisms throughout the exposure period. Parameters such as temperature, moisture, pH and dissolved oxygen should be adjusted to an optimal range prior to and during the exposure period (Greene *et al.* 1989, ASTM 1994, 1995, Klemm *et al.* 1994). However, contaminated soils may be characterized by significant deviation from these optimal conditions, and adjustments should be performed to avoid a false positive toxicological response. These adjustments may interfere drastically with the dynamic chemical equilibrium between contaminants and substrate matrix, and may mask or enhance the toxic response of the tested organisms. The dilemma one faces is to take into account the respective effects of artificial changes to the soil samples. This is generally achieved by using multiple control and reference soils, including the initial non-adjusted soil sample. When reference *in situ* soils are not available, artificial media may be useful as long as phenomena such as aging are not of concern in the study and as far as the conceptual design of the test used does not cause direct interference with the matrix (Brouwer *et al.* 1990, True and Heyward 1990, Svenson *et al.* 1996).

In certain circumstances, physical properties of tested soils may directly interfere with the execution of toxicity tests. Excess of fine particles such as clay or silt which cause high compaction, too high or too low humidity of samples, extremely low pH or extreme heterogeneity of grain size require pretreatment of samples, which again may contribute to severe modifications of sample integrity. For example, adjustment of grain size by mixing samples with sand contributes to increased porosity of the samples, which is favourable to the burrowing of worms, but may also modify volatilization of some contaminants, causing a modification of the contaminant bioavailability. Moreover, increasing the soil porosity may increase water circulation, acting as an extracting solvent and increasing the contaminant bioavailability. Sieving may modify the contaminant distribution in the soil, therefore modifying the toxic response demonstrated by living organisms.

Compromise between sample integrity, optimal experimental conditions and environmental representativeness is a matter of analyst expertise as well as direct study objectives. Unlike the risk assessment in which a realistic approach is favoured, the hazard assessment which mainly relies on comparison toxicities between samples of similar properties allows a toxicological evaluation of soil contamination with more flexibility in sample preparation and modification.

2.2.5 SAMPLE PREPARATION FOR INDIRECT LIQUID-PHASE TOXICITY TESTING

In order to complement toxicity information from direct solid-phase toxicity testing, and due to the increasing number of liquid-phase toxicity tests available, additional liquid-phase testing of soil extracts with water, acidified water and solvents like methanol may provide useful information (Maxam *et al.* 2000). As discussed earlier, the mixing of contaminated soil with solvents drastically modifies the equilibrium between the toxic substances and the soil. Although the environmental realism of extracts derived from these preparations is considerably lowered, information derived from such preparations contributes to a better evaluation of the potential toxicity resident in the soil samples, provided the results are interpreted with caution.

Recent studies in closed systems (Bispo 1998, CEAEQ 2000) have shown that soil leaching with water performed during 2 h followed by settling and/or filtration may demonstrate higher toxicity than the identical leaching performed during 24 h. The reduction of toxicity in the latter situation is attributed to the limiting effects of biological and chemical degradation, chemical reduction or contaminant trapping, but also to the final filtration of the leachate where trapping of soluble contaminants by organic matter may occur. In other words, the usual protocol (like AFNOR X31-210; AFNOR 1992) suggesting a longer exposure of sample to water may underestimate the sample toxicity.

Liquid extraction of contaminants from soil may generate disadvantages in terms of contaminant concentrations. To avoid dilution of the contaminants in the liquid phase, some attempts were recently made to increase the soil/liquid ratio from 1/10 to 1/5 and 1/1 in order to reduce the toxic threshold of the tested solution (CEAEQ 2000). Results showed that the sensitivity was not significantly different between ratios 1/10 and 1/5, but was significantly higher for the soil/liquid ratio of 1/1 in one out of four extraction periods tested in this study. Even if a soil/liquid ratio of 1/1 seems more representative of natural soil solutions, toxicity results were not conclusive. Furthermore, the use of such a ratio posed severe constraints upon conducting the toxicity tests because of the small volume of water extract available.

Although the significance of leaching with organic solvents for toxicity assessment is questionable as the environmental realism of such extracts is considerably reduced, non-aqueous leaching performed with solvents such as methanol is of particular interest in case of contamination with hydrophobic organic substances. According to CEAEQ (2000), if detected in solvent extracts, toxicity increased 10 to 100-fold when compared to aqueous leaching. In the same study, parallel successive leaching has shown that toxicity of water extracts remains high and constant even after four contacts with the contaminated soil, suggesting a potential for long-term release of the contaminant. Furthermore, because interfering processes such as bacterial degradation, trapping by dissolved organic matter and pH re-equilibrium may occur in aqueous extracts, sample conservation at 4 °C is highly recommended and toxicity tests should be performed within 7 days. Due to the inherent toxicity of organic solvents, conservation of extracts is generally of less concern than aqueous extracts.

Recent developments in semi-open leaching procedures provide evidence of temporal contaminant gradient simulating the kinetics of contaminant release from the soil. The picture given by continuous aqueous leaching provides a better appreciation of the process occurring naturally *in situ*. However, because of the relative complexity of the required settings, this procedure is actually dedicated to very specific studies, and is not performed on a regular basis. A protocol with an integrated approach with aqueous, solvent and acid extractions is proposed by CEAEQ (1999) for the estimation of the potential mobility of contaminants in soil.

2.3 CONCLUSIONS

Chemical extractions have shown interesting results for the assessment of potential mobility and bioavailability, regarding organic and inorganic contaminants. However, even though chemical methods still have to be developed in order to standardize the quantification of the bioavailable fraction, biological organisms are necessary to accurately assess contaminant bioavailability. It has to be noted that the fraction of a contaminant that is absorbed within an organism is organism-dependent.

The carrying out of toxicity tests with solid phase may cause interference problems due to the suboptimal characteristics of the matrix. The inevitable adjustments of the soil sample performed in order to avoid a false positive toxicological response may interfere with the dynamic equilibrium between contaminants and soil matrix. Experimental conditions applied for soil extraction may greatly influence mobility, bioavailability and toxicity results. Different results can be obtained with different soil sample preparation and handling. Consequently, results should be interpreted with caution.

REFERENCES

AFNOR (1992) *Qualité des sols. Déchets: essai de lixiviation.* X31-210, Association française de normalisation, France, pp. 1–13.

Allen HE (1993) The significance of trace metal speciation for water, sediment and soil quality criteria and standards. *The Science of the Total Environment*, **Supplement** 1993, 23–45.

ASTM (1994) *Standard Practice for Conducting Early Seedling Growth Tests.* ASTM Standard E 1598-94 American Society for Testing and Materials, Philadelphia, PA.

ASTM (1995) *Standard Guide for Conducting a Laboratory Soil Toxicity Test with Lumbricid Earthworm Eisenia foetida.* ASTM Standard E 1676-95 American Society for Testing and Materials, Philadelphia, PA.

Baveye P and Bladon R (1999) Bioavailability of organic xenobiotics in the environment: a critical perspective. In *Bioavailability of Organic Xenobiotics in the Environment — Practical*

Consequences for the Environment, Baveye P, Block JC and Goncharuk VV (eds), Kluwer Academic Publishers, Dordrecht, pp. 227-248.

Bispo A (1998) *Contribution à l'élaboration d'une méthodologie pour évaluer les dangers et les risques liés aux matériaux solides contaminés*, Institut National Polytechnique de Lorraine, France, 138 pp.

Brouwer H, Murphy T and McArdle L (1990) A sediment-contact bioassay with *Photobacterium phosphoreum*. *Environmental Toxicology and Chemistry*, **9**, 1153-1158

CEAEQ (1999) *Protocole de lixiviation applicable aux tests de toxicité pour la caractérisation du potentiel de mobilité des contaminants dans les sols et les déchets solides*, Centre d'expertise en analyse environnementale du Québec, Ministère de l'Environnement du Québec, Canada MA. 500-Lix. 1.0.

CEAEQ (2000) *Développement d'une approche méthodologique pour la caractérisation du potentiel de mobilité des contaminants dans les sols*, Centre d'expertise en analyse environnementale du Québec, Ministère de l'Environnement du Québec, Canada.

Cseh T, Sanschagrin S, Hawari J and Samson R (1989) Adsorption-desorption characteristics of polychlorinated biphenyls on various polymers commonly found in laboratories. *Applied and Environmental Microbiology*, **55**, 3150-3154.

Cuypers MP, Grotenhuis JTC and Rulkens WH (1999) Prediction of PAH bioavailability in soils and sediments by persulfate oxidation. In *Bioremediation Technology for PAH Compounds*, Leeson A and Alleman BC (eds), Batelle Press, Columbus, OH, pp. 241-246.

Davies NA, Edwards PA, Lawrence MAM, Taylor MG and Simkiss K (1999) Influence of particle surfaces on the bioavailability to different species of 2,4-dichlorophenol and pentachlorophenol. *Environmental Science and Technology*, **33**, 2465-2468.

Dickson KL, Giesy JP, Parrish R and Wolfe L (1994) Summary and conclusions. In *Bioavailability: Physical, Chemical, and Biological Interactions*. A SETAC special publication, Hamelink JL, Landrum PF, Bergman HL and Benson WH (eds), Lewis, Boca Raton, FL, pp. 221-230.

Dudal Y, Deschênes L and Samson R (1999) Quantifying the intrinsic bioremediation potential and the hazard index of organic xenobiotics in aquifers based on their bioavailability rates. In *Bioavailability of Organic Xenobiotics in the Environment — Practical Consequences for the Environment*, Baveye P, Block JC and Goncharuk VV (eds), Kluwer Academic Publishers, Dordrecht, pp. 147-151.

Federer CA and Hornbeck JW (1985) The buffer capacity of forest soils in New England. *Water, Air and Soil Pollution*, **26**, 163-173.

Greene JC, Bartels CL, Warren-Hicks WJ, Parkhurst BR, Linder GL, Peterson SA and Miller WE (1989) *Protocol for Short Term Toxicity Screening of Hazardous Waste Sites*. EPA/600/3-88/029, US Environmental Protection Agency, Environmental Research Laboratory, Corvallis, OR.

Guerin WF and Boyd SA (1992) Differential bioavailability of soil-sorbed naphthalene to two bacterial species. *Applied and Environmental Microbiology*, **58**, 1142-1152.

Harms H and Zehnder AJB (1995) Bioavailability of sorbed 3-chlorodibenzofuran. *Applied and Environmental Microbiology*, **61**, 27-33.

Ho MD and Evans GJ (2000) Sequential extraction of metal contaminated soils with radiochemical assessment of readsorption effects. *Environmental Science and Technology*, **34**, 1030-1035.

Hrudey SE, Chen W and Rousseaux CG (1996) *Bioavailability in Environmental Risk Assessment*, CRC Press, Boca Raton, FL, pp. 7-8.

Kelsey JW and Alexander M (1997) Declining bioavailability and inappropriate estimation of risk of persistent compounds. *Environmental Toxicology and Chemistry*, **16**, 582-585.

Kelsey JW, Kottler BD and Alexander M (1997) Selective chemicals extractants to predict bioavailability of soil-aged organic chemicals. *Environmental Science and Technology*, **31**, 214-217.

Klemm DJ, Morrison GE, Norberg-King TJ, Peltier WH and Heber MA (1994) *Short-term methods for estimating the chronic toxicity of effluents and receiving waters to marine and estuarine organisms*. EPA/600/4-91/003, US Environmental Protection Agency, Washington, DC.

Lo IMC and Yang XY (1998) Removal and redistribution of metals from contaminated soils by a sequential extraction method. *Waste Management*, **18**, 1-7.

Loibner AP, Gartner M, Schlegl M, Hautzenberger I and Braun R (1997) PAHs: rapid estimation of their bioavailability in soil. In *In situ and On site Bioremediation*, Vol. 5, Leeson A and Alleman BC (eds), Batelle Press, Columbus, OH, pp. 617-622.

Luthy RG, Aiken GR, Brusseau ML, Cunningham SD, Gschwend PM, Pignatello JJ, Reinhard M, Traina SJ, Weber Jr. WJ and Westall JC (1997) Sequestration of hydrophobic organic contaminants by geosorbents. *Environmental Science and Technology*, **31**, 3341-3347.

Matheson VG, Mumakata-Marr J, Hopkins GD, McCarty PL, Tiedje JM and Forney LJ (1997) A novel means to develop strain-specific DNA probes for detecting bacteria in the environment. *Applied and Environmental Microbiology*, **63**, 2863-2869.

Maxam G, Rila JP, Dott W and Eisentraeger A (2000) Use of bioassays for assessment of water-extractable ecotoxic potential of soils. *Ecotoxicology and Environmental Safety*, **45**, 240-246.

McKim JM (1994) Physiological and biochemical mechanisms that regulate the accumulation and toxicity of environmental chemicals in fish. In *Bioavailability: Physical, Chemical, and Biological Interactions*. A SETAC special publication, Hamelink JL, Landrum PF, Bergman HL and Benson WH (eds), Lewis, Boca Raton, FL, pp. 179-201.

McLean JE and Bledsoe BE (1992) Behavior of metals in soils. *EPA Groundwater Issue*. EPA/540/S-92/018, US Environmental Protection Agency, Washington, DC.

Nirel PMV and Morel FMM (1990) Pitfalls of sequential extractions. *Water Research*, **24**, 1055-1056.

Rapin F, Tessier A, Campbell PGC and Carignan R (1986) Potential artifacts in the determination of metal partitioning in sediments by a sequential extraction procedure. *Environmental Science and Technology*, **20**, 836-840.

Ravishankar BR, Tyagi RD and Narasiah KS (1993) Influence of colloids on metal concentrations in sludge leachates. In *Proceedings of the 9th Eastern Region Conference of the Canadian Association on Water Quality*, Sherbrooke, Quebec, pp. 37-38.

Reid BJ, Jones KC and Semple KT (1999) Can bioavailability of PAHs be assessed by a chemical means? In *Bioremediation Technology for PAH Compounds*, Leeson A and Alleman BC (eds), Batelle Press, Columbus, OH, pp. 253-258.

Renoux AY, Tyagi RD and Samson R (1999) Irradiation and freezing conservation effects on toxicity measurements of sewage sludge elutriates. *Water Quality Research Journal of Canada*, **34**, 589-597.

Richnow HH, Seifert R, Hefter J, Link M, Francke W, Schaefer G and Michaelis W (1997) Organic pollutants associated with macromolecular soil organic matter: mode of binding. *Organic Geochemistry*, **26**, 745-758.

Ruby MV, Schoof R, Brattin W, Goldade M, Post G, Harnois M, Mosby DE, Casteel SW, Berti W, Carpenter M, Edwards D, Cragin D and Chappell W (1999) Advances in evaluating the oral bioavailability of inorganics in soil for use in human health risk assessment. *Environmental Science and Technology*, **33**, 3697-3705.

Sauvé S, Dumestre A, McBride M and Hendershot W (1998) Derivation of soil quality criteria using predicted chemical speciation of Pb^{2+} and Cu^{2+}. *Environmental Toxicology and Chemistry*, **17**, 1481-1489.

Schwarzenbach RP, Gschwend PM and Imboden DM (1993) *Environmental Organic Chemistry*, John Wiley & Sons, New York, pp. 56-156.

Svenson A, Edsholte E, Ricking M, Remeberer M and Rottrop J (1996) Sediment contaminants and Microtox® toxicity tested in a direct contact exposure test. *Environmental Toxicology and Water Quality*, **11**, 293-300.

Tessier A and Campbell PGC (1987) Partitioning of trace metals in sediments: relationship with bioavailability. *Hydrobiologia*, **149**, 43-52.

Tessier A and Campbell PGC (1988) Comments on the testing of the accuracy of an extraction procedure for determining the partitioning of trace metal in sediments. *Analytical Chemistry*, **60**, 1475-1476.

Tessier A and Campbell PGC (1991) Comment on 'Pitfalls of sequential extractions'. *Water Research*, **25**, 115-117.

Tessier A, Campbell PGC and Bisson M (1979) Sequential extraction procedure for the speciation of particulate trace metals. *Analytical Chemistry*, **51**, 844-851.

True CJ and Heyward AA (1990) Relationships between Microtox® test results, extraction methods and physical and chemical composition of marine sediments samples. *Toxicity Assessment*, **5**, 171-173.

Wasay SA, Barrington S and Tokunaga S (1998) Retention form of heavy metals in three polluted soils. *Journal of Soil Contamination*, **7**, 103-119.

Young TM and Weber Jr. WJ (1995) A distributed reactivity model for sorption by soils and sediments. 3. Effects of diagenetic processes on sorption energetics. *Environmental Science and Technology*, **29**, 92-97.

Zagury GJ, Colombano SM, Narasiah KS and Ballivy G (1997) Neutralisation of acid mine tailings by addition of alkaline sludge from pulp and paper industry. *Environmental Technology*, **18**, 959-973.

Zagury GJ, Dartiguenave Y and Setier JC (1999) Ex-situ electroreclamation of heavy metals contaminated sludge: pilot scale study. *Journal of Environmental Engineering ASCE*, **125**, 972-978.

3

Toxicity Tests for Assessing Contaminated Soils and Ground Water

GLADYS L. STEPHENSON[1], ROMAN G. KUPERMAN[2], GREG L. LINDER[3] AND SUZANNE VISSER[4]

[1]*ESG International Inc., Guelph, Ontario, Canada*
[2]*Geo-Centres Inc., Gunpowder Branch, Aberdeen Proving Ground, MD, USA*
[3]*Heron Works Farm, Salem, OR, USA*
[4]*Department of Biology, University of Calgary, Calgary, Alberta, Canada*

3.1 INTRODUCTION

Inorganic and organic chemicals or compounds (e.g., metals and associated complexes, or humic materials and their associated products, respectively) occur naturally in soils at levels that can vary by orders of magnitude. Many of these substances are natural constituents of rock and of soils. However, there are sites associated with different anthropogenic land uses (e.g., mining, agriculture, forestry and disposal practices) where there are elevated levels of these substances, as well as substances that do not occur naturally in soil environments. The sources for pollutants are numerous, widespread and variable; ultimately, soil and ground water serve as sinks for these substances.

Assessing the toxicity of contaminated soils is a relatively 'new' regulatory concern. In Canada, results of toxicity tests are used to derive national soil quality criteria, establish site-specific, risk-based, clean-up objectives (e.g., remediation targets) and assess the efficacy of remediation technologies. An important regulatory driver at many contaminated sites in the US is ecological risk assessment (ERA). The ERA process might involve site-specific toxicity assessment of soils and surface or ground water to determine the magnitude and extent of impact. The toxicity tests used to assess site soils were first summarized by the US Environmental Protection Agency (1989, 1992), with the European perspective summarized by Van Straalen and Løkke (1997). In contrast to aquatic toxicity testing, where test protocols are standardized and widely used, testing protocols for soils are less well developed and not applied to the same extent. Typically, however, there is a reliance on single-species

Environmental Analysis of Contaminated Sites. Edited by G. I. Sunahara, A. Y. Renoux, C. Thellen, C. L. Gaudet and A. Pilon

toxicity tests, conducted primarily in the laboratory, to determine the toxicity of surface water, ground water and soil.

The development of soil toxicity test methods lags behind that for water. In the early 1980s, acute toxicity tests for plants (seedling germination and root elongation) and earthworms (survival) were developed jointly by the European Economic Community (EEC) and the Organization for European Co-operation and Development (OECD), mainly to satisfy the regulatory requirement for registration of new chemicals. At the end of the 1980s, Greene *et al.* (1989) recommended a battery of tests to assess soils at contaminated sites. The battery included a 14-day acute earthworm toxicity test with whole soils together with phytotoxicity tests with both whole soils (e.g., lettuce seed germination) and soil leachate (e.g., lettuce root elongation). Additionally, aquatic toxicity tests, including water flea survival, algal growth and fathead minnow survival tests, were recommended for assessing the toxicity of soil eluates (Greene *et al.* 1989). A review of the toxicity tests available for toxicity assessment of contaminated sites was compiled by Keddy *et al.* (1992). Unfortunately, most of these tests cannot determine the effects of soil contamination on the structure and function of soil communities and the organisms comprising these communities.

Since the early 1990s, a limited number of terrestrial toxicity tests have been developed, or improved, with standardization by different agencies and organizations occurring in recent years. In the forefront of test method standardization are: the International Standards Organization (ISO), the American Society for Testing and Materials (ASTM), Environment Canada, the OECD and the US Environmental Protection Agency (USEPA). Also, a multinational initiative called SECOFASE (Development, Improvement and Standardization of Test Systems for Assessing **S**ublethal **E**ffects of **C**hemicals **o**n **F**auna in the **S**oil **E**cosystem) was sponsored by the European Union (EU) to develop test systems for the early detection and evaluation of sublethal effects of chemicals on organisms in soil ecosystems. The SECOFASE test methods have been summarized by Løkke and Van Gestel (1998).

Current terrestrial toxicity test methods were designed to assess the toxicity of specific chemicals (i.e., mostly pesticides) in soils using spiking methodologies whereby a soil is amended with the chemical. In addition, artificial soils have been used as a 'soil surrogate' to develop most of these test protocols. Only a few test methods have been modified for use with contaminated site soils. Greene *et al.* (1989) were the first to recommend a test protocol based on a dilution series whereby the contaminated soil was diluted with a clean control soil.

The objectives of this chapter are to present a brief overview of the tests available to evaluate the toxicity of whole soils and ground water, and to review the procedures for: preparing test soils, selecting measurement endpoints and deriving statistical endpoints from the entire concentration–response curve to assess an effect. The procedures for collection, handling and storage of soils and ground water have been discussed by Zagury *et al.* (see Chapter 2). Toxicity tests with soil eluates and soil pore water have not been included, and those tests relevant to soil processes and ecosystem function have been discussed elsewhere in this book (see Kuperman *et al.*, Chapter 4 and Svendsen *et al.*, Chapter 7).

3.2 SINGLE-SPECIES TERRESTRIAL TOXICITY TESTS

Contaminants may influence populations of soil-inhabiting plants and invertebrates in different ways. These include: (1) direct acute toxicity; (2) chronic toxicity such as effects on growth and/or reproduction; (3) indirect toxicity by altering soil structure or fertility; (4) indirect toxicity by adversely affecting nutrient cycling or food supplies; or (5) by killing predators and parasites. It is difficult to determine the relative contribution to toxicity of each of these factors, but one approach is to use a standardized battery of tests. The test battery approach for soil assessments is preferred because it increases the accuracy of the toxicity evaluation by employing test species that are representative of the diversity of soil fauna and flora in terms of sensitivities to contaminants, exposure pathways, taxonomic and functional groups within and among different trophic levels, etc. (Van Gestel and Van Straalen 1994).

The question that usually confronts an ecotoxicologist in any assessment is 'Which of the available terrestrial toxicity tests constitutes an appropriate test battery?'. Important considerations, beyond the objectives and goals of the assessment, are that the test battery includes the following:

1. Species from different taxonomic groups to represent a range of sensitivities;
2. Species that represent the spectrum of ecological functions found in healthy soil communities — these should include primary producers, herbivores, predators, saprovores, bacterivores and fungivores;
3. Species, or representative surrogate species, that are ecologically relevant to the geographical area of concern;
4. Species that actively move through the soil and have contact with contaminants;
5. Species sensitive to a wide range of contaminants;
6. Species that reflect different routes of exposure (e.g., ingestion, inhalation, dermal absorption, uptake from soil solution);
7. Species that occupy different soil microhabitats (e.g., nematodes inhabit the water film around soil particles, whereas microarthropods inhabit air-filled soil pores);
8. Species amenable to life-cycle tests to help identify the most vulnerable developmental stage of the test organisms (e.g., egg, juvenile or adult survival and reproductive success).

In addition to the above ecotoxicological selection criteria, the terrestrial toxicity test methods should accommodate a number of practical considerations.

Some of these were originally proposed by Léon and Van Gestel (1994) and have been discussed in detail by Van Gestel (1998). They include:

Culture of test species

- Easy to maintain in laboratory culture
- High reproduction rate
- Short generation time
- Relatively insensitive to environmental changes
- Reference test protocol for production of control charts to monitor culture health and changing sensitivities

Practical considerations

- Low cost of conducting the test (staff, time and equipment)
- Length of test duration (acute versus chronic)
- Easy to standardize
- Non-controversial test species (not rare or endangered species)

Test evaluation criteria

- Reproducible results
- Clearly defined validity criteria
- Measurement endpoints that can be statistically analysed to produce statistically valid results
- Sufficient literature information on the performance of the test

Terrestrial toxicity tests, with representative test species, that have been standardized and that generate reproducible, statistically valid results, and are performed following GLP (Good Laboratory Practices) or ISO/IEC Guide 25, impart a greater confidence in the data. As a result, there may be less uncertainty or controversy associated with the decisions and recommendations that are based on these data.

3.2.1 OVERVIEW OF SOIL TOXICITY TEST METHODS

The suite of available single-species toxicity methods with whole soils covers a broad range of effect measurements for different components of soil ecosystems. The suite of equally important 'functional assays' (i.e., tests that evaluate a process essential to terrestrial ecosystem function) are described elsewhere in this book (see Kuperman *et al.*, Chapter 4 and Svendsen *et al.*, Chapter 7). The challenge to investigators is to choose the most appropriate test battery from among the test methods, test organisms and test conditions available. Important considerations, in addition to those mentioned previously, are:

- The nature, magnitude and extent of the contamination;
- Physicochemical characteristics of the site soils and water;

- The goals and objectives of the assessment;
- The assessment endpoints;
- The acceptability of test species to the regulatory authorities.

Whole-soil toxicity tests with terrestrial invertebrates (e.g., earthworms, springtails, mites, nematodes and isopods) and plants are discussed in the following subsections. Whole-soil toxicity tests with micro-organisms (e.g., bacteria, protozoan or bioluminescence assays) are discussed by Svendsen *et al.* (see Chapter 7). Also, toxicity tests with soil extracts (e.g., eluates, leachate and pore water) and those tests involving terrestrial vertebrates (e.g., shrews, meadow voles and amphibians) have not been included in this review, although the latter could be particularly important when the interrelationships between surface water, soils and ground water are a management issue for a particular habitat or site. Soil extract tests are generally restricted to mechanistic evaluations for specific exposure pathways and are occasionally, but not routinely, conducted to assess 'soil' toxicity.

3.2.1.1 Earthworm Survival Toxicity Test (Acute)

The 14-day lethality test (OECD 1984a, Greene *et al.* 1989, US Environmental Protection Agency 1989, ISO 1995, ASTM 1997b) uses either *E. fetida* (Savigny 1826) or *Eisenia andrei* (Bouché 1972). These species were selected because they are relatively easy to culture in the laboratory. Questions remain, however, regarding the use of these species in soil toxicity testing because they are restricted in distribution to compost and manure piles and are not representative of soil earthworm communities. In North America, *Lumbricus terrestris* (Linnaeus 1758) (Canadian Night Crawler or Dew Worm) is considered a more ecologically relevant earthworm species than *E. fetida* (Environment Canada 1994). A background document on the development of earthworm toxicity tests for assessment of contaminated soils was produced by Environment Canada (1998a) with a guidance document imminent. There is currently no standard guide or annex for *L. terrestris* in the 1999 Annual Book of ASTM Standards; however, there is guidance for a bioaccumulation test with *E. fetida* (ASTM 1999b) and draft guidance for *L. terrestris*.

The OECD (1984a) and ISO (1995) recommend that the toxicity of chemicals and substances be evaluated by spiking a soil with a chemical to produce an exposure series of different concentrations. Römbke *et al.* (1998) also provide guidance for conducting an acute toxicity test with *Enchytraeus albidus* for chemicals in artificial soil. However, the guidance provided by Greene *et al.* (1989), US Environmental Protection Agency (1989), ASTM (1997b) and Environment Canada (1998a) addresses both the toxicity of chemicals (e.g., spiked-soil tests) and the toxicity of contaminated site soils. The latter can be diluted with a clean control or reference soil to generate an exposure series of different levels of contamination (see Section 3.2.3) or it (the undiluted contaminated site soil) can be compared directly to a control or reference soil.

3.2.1.2 Lumbricid Earthworm Reproduction Toxicity Test (Chronic)

A 21-day reproduction test was proposed by Van Gestel *et al.* (1989) to evaluate chronic toxicity. This protocol is similar to the standard 14-day acute survival assay; the main difference is that the earthworms are fed during the test. After 21 days, surviving earthworms are recorded and cocoons are recovered and counted. In contrast to the earthworm acute toxicity test, the chronic test has greater ecological relevance and justifies the additional effort. Since this test was first developed, other reproduction test methods have been published or investigated. There is a 28- to 56-day sublethal test where growth (wet mass) and survival of adults and cocoon production are measured after 28 days. The cocoons are then segregated from the soil and placed onto moistened filter paper for an additional 28-day incubation period. The measurement endpoints of the second phase of the test are the number of hatched juveniles, the number of hatched cocoons and the number of infertile, or unhatched, cocoons (ISO 1998a). Alternatively, the adults can be removed from the test units (e.g., vesicles) and measured, leaving the cocoons in the test soil for an additional 28-day period, at which time the number of unhatched cocoons and juveniles are recovered and counted (BBA 1994, Environment Canada 1998a). A more ecologically relevant test species, *Aporrectodea caliginosa,* has also been proposed for use in this sublethal test, with some modifications to the recommended test procedures (Kula and Larink 1997, 1998). The ISO (1998a) has recently published the reproduction test with *E. fetida*.

The earthworm reproduction tests were developed mainly to test the toxicity of chemicals or substances spiked into artificial or reference control soils. The application of these test methods to site soil assessment has received less attention. Nevertheless, the test procedures have been modified and adapted to accommodate assessment of site soils (Environment Canada 1998a, ISO 1998a).

3.2.1.3 Enchytraeid Reproduction Test

A potential alternative to the lumbricid earthworm reproduction test with *E. fetida* is the reproduction test with enchytraeid pot worms. The enchytraeid reproduction test (ERT) was developed to determine the effects of test chemicals on the reproductive potential of *E. albidus*, as representative of the soil fauna. This species has several advantages over *E. fetida*: (1) it is more ecologically relevant because it lives in a variety of soils and is associated with soil pore water, whereas *E. fetida* inhabits compost or manure piles and is not a 'true' soil species; (2) it is relatively easy to culture; (3) it has a shorter generation time; and (4) it requires much less soil (1/10th) per test unit, thereby reducing the quantity of contaminated material required and the expenses incurred for disposal.

Enchytraeid worms, including *E. albidus*, have been used to assess soil quality for over three decades (Weuffen 1968, Kaufman 1975, Huhta 1984, Römbke 1989, Römbke *et al.* 1994, Römbke and Federschmidt 1995, Römbke *et al.* 1998). Recently, efforts to standardize the test methods culminated in an International Ringtest jointly sponsored by the EU and the German Environmental Protection Agency (UBA). More than 100 tests by participants from 14 countries were performed. In the spring of 1998, the results of these tests were reviewed and a draft guideline was prepared for distribution within the OECD, ISO and ASTM. The current guidance only addresses the toxicity of chemically amended control soils. A modified ERT method that covers the evaluation of both chemicals and contaminated soils is available in draft form (ISO 1999).

A sublethal toxicity test with a second enchytraeid species, *Cognettia sphagnetorum* (Vejdovsky 1877), has also been developed in Europe to evaluate chemicals in a control soil (Rundgren and Augustsson 1998). This species inhabits a variety of soil environments (deciduous and coniferous forest soils, grassland, meadow and peat soils) and is unique because it reproduces asexually by fragmentation. The fragments can then differentiate to develop into individual worms. The test organisms are exposed to different concentrations of a chemical in a control soil (modified LUFA 2.2); the measurement endpoints include percent mortality, growth as reflected by the number of fragments produced, or rate of growth (e.g., fragmentation rates). Measurements are recorded weekly over a 10-week period. Reproductive success is determined by counting the number of surviving fragments (Rundgren and Augustsson 1998). Again, this test has been developed to assess the sublethal toxicity of chemicals in control soils and these methods must be modified before they can be used to evaluate the toxicity of contaminated soils.

3.2.1.4 Earthworm Avoidance-response Test

Recently, several studies have investigated using earthworm behavioural tests, specifically the avoidance response of earthworms to contaminated soils (Gibbs *et al.* 1996, Environment Canada 1998a, Hund 1998, Stephenson *et al.* 1998). This novel approach to toxicity assessment is useful because the results are generated within 72 h and have been 'predictive' of the results from much longer reproduction tests (Environment Canada 1998a, Stephenson *et al.* 1998). The test can be used with both *E. fetida* and *L. terrestris*. Although the development of the test is in its infancy, it shows great promise. Achazi *et al.* (1996) have also used the avoidance behaviour of pot worms (Enchytraeidae: *Enchytraeus crypticus*) to assess soil contaminants.

3.2.1.5 Nematode Life-cycle Test

The soil-dwelling, bacterivorous, rhabditid nematode, *Caenorhabditis elegans*, is a useful surrogate species for the bacterivore components of soil food web (Donkin and Dusenbery 1993). This organism can provide a homogeneous test population that minimizes variation due to individual differences. It is easy to culture, has a short life-cycle and an extensively mapped genome. In addition, mass cultures of genetically identical individuals can be produced because it is capable of self-fertilization (Wood 1988).

Plectus acuminatus is another potentially useful species of a free-living bacterivorous nematode. This species can be cultured in standard artificial soil, which enables a comparison with other standardized test systems using earthworms and springtails (Kammenga *et al.* 1996, Kammenga and Riksen 1998). Both species can be used in acute and chronic toxicity tests.

3.2.1.6 Collembolan Reproduction Test

The Collembola (springtails) are essential to nutrient cycling and the breakdown of organic matter in soil and, as such, are probably the most studied group of soil microarthropods. Several species have been used for soil toxicity testing (Crommentuijn 1994, Smit 1997, Environment Canada 1998b) and methods are well developed for *Isotoma tigrina* (Kiss and Bakonyi 1992, Wiles and Krogh 1998) and *Folsomia fimetaria* (Løkke 1995, Petersen and Gjelstrup 1995, Wiles and Krogh 1998). These two species are more ecologically relevant than *F. candida* because they are not restricted to sites with humus-rich soils. *F. candida* is the species used most often in ecotoxicity evaluations. The test with *F. candida* measures the effects of soil contamination on adult and juvenile survival and reproduction (Thompson and Gore 1972). This parthenogenetic species is sensitive to a wide range of disturbances and is a sentinel species for change to soil communities. This is partly because *F. candida* belongs to the fungal energy channel (all food chains originating from fungi) which is more sensitive to disturbances than the bacterial or root energy channels (Moore and DeRuiter 1993). The folsomid test can determine the effects of soil contamination on the structure of soil microarthropod community and it was standardized by ISO in 1998 (ISO 1998b).

Onychiurus folsomi (Onychiuridae) is a more ecologically relevant species than *F. candida* to Canadian soil environments and a standard test method to assess the toxicity of chemicals and contaminated site soils is being developed (Environment Canada 1998b).

3.2.1.7 Oribatid Mite Test

Oribatids are one of the most abundant groups of soil microarthropods and comprise hundreds of species. Denneman and Van Straalen (1991) described a reproduction toxicity test using the parthenogenetic species *Platynothrus peltifer*. This test assesses the effects of chemicals in food on oribatid egg production and is not a 'true' soil toxicity test because the test organisms are exposed to a contaminated paste of algae on a filter paper. The test with *P. peltifer* and artificial soil is described in detail by Van Gestel and Doorenekamp (1998) (as cited in Løkke and Van Gestel 1998). The low rate of egg production in this species (only two eggs per week) and its long life-cycle (one year) make it impractical for use in risk assessment, although it was the most sensitive soil invertebrate tested for cadmium, and more sensitive than collembolans to copper and lead (Van Straalen *et al.* 1989, Denneman and Van Straalen 1991). Therefore, even though the addition of oribatid mites to a terrestrial test battery is likely, more research is needed to standardize procedures for culturing or collecting the test organisms and to identify optimal test conditions and methods for site soil assessment.

3.2.1.8 Isopod Test

Terrestrial isopods or woodlice are an important group of crustacean arthropods (suborder Oniscidea; order Isopoda) in metal research because of their unique ability to bioaccumulate high levels of metals (Hopkin 1989). Isopods are easy to culture and do not require special conditions. The disadvantage of this group for soil toxicity testing is their relatively long (one year) life-cycle. The most suitable test species identified to date is the parthenogenic *Trichoniscus pusillus* (Van Straalen and Van Gestel 1993). The use of isopods as a test group for soil toxicity testing is restricted to a few studies (Hopkin 1989, Donker and Bogert 1991, Drobne 1997). To date, there is no standardized isopod toxicity test method to assess contaminated soils. The current test methods were developed to assess effects of individual chemicals on survival, growth and reproduction of *Porcellio scaber* (Isopoda: Porcellionidae) (Hornung *et al.* 1998) and need to be modified in order to be useful for site-specific soil assessments.

3.2.1.9 Seedling Emergence Test

Several plant toxicity tests have been standardized recently (ASTM 1999a,c OECD 1999). The earliest tests (Greene *et al.* 1989, US Environmental Protection Agency 1989, Keddy *et al.* 1992) were modelled after the OECD terrestrial plant growth test (OECD 1984b) which is currently being revised. A seedling emergence test for the evaluation of contaminated soils has been developed by ASTM (1999a). Basically, the test involves exposing seeds of recommended test species to potentially contaminated site soils, or to a dilution series (i.e., site soils amended with control soils) and measuring the number of seedlings that emerge from the soil to a minimum height of 3 mm. The test differs from, and should not be mistaken for, a seedling germination test. Seedling germination is a different measurement endpoint than seedling emergence. The test is amenable to a variety of plant species. The duration of the test can vary from 5 to 10 days, depending on the species; however, 5–7 days is typical for most species. Generally, seedling emergence is not as sensitive an endpoint as growth metrics, which are obtained from early seedling growth tests or definitive toxicity tests (Environment Canada 1998c). Environment Canada is in the process of producing a test method that uses emergence as only one of seven measurement endpoints.

3.2.1.10 Early Seedling Growth Test

The early seedling growth test was one of the first ASTM standard guides for testing with plants. It differs from the seedling emergence test in that the test duration is usually longer (i.e., greater than 14 days) and the measurement endpoints include shoot and root length, shoot and root wet and dry mass, seedling emergence and seedling survival at the end of the test. Modifications of test procedures to accommodate contaminated site soils are found in Environment Canada (1998c). Modification of test methods and conditions to accommodate different test species is required (Environment Canada 1998c, ASTM 1999a).

3.2.1.11 Brassica Life-cycle Test

A life-cycle test with *Brassica rapa* (a genetically modified species) is a test that goes from seed to seed in about six weeks. It was developed at the University of Wisconsin and appears as a standard protocol in an annex of the ASTM guide (ASTM 1999a). Recently, a draft guideline for testing contaminated site soils has been developed in Germany. The recommended test species is either *B. rapa* or *Avena sativa* (Kalsch and Römbke 1998). The major advantage of these chronic tests is that they assess toxicity of sublethal levels of contaminants over the entire life span of the plant species.

3.2.2 GROUND-WATER TOXICITY TESTS

Notenboom and Boessenkoll (1993) developed a toxicity test specific for ground water; however, the method was never standardized. There are no toxicity tests developed and standardized to assess contaminated ground water. However, existing aquatic test methods designed primarily to assess toxicity of surface waters have been used to evaluate both soil eluates and ground water. The battery of standard tests most commonly used in aquatic assessments comprises tests with fish (usually fathead minnow or rainbow trout), freshwater invertebrates (waterfleas, rotifers), micro-organisms (BOD, SOS Chromotest, Microtox™), aquatic vascular plants and algae (Environment Canada 1990a,b, 1992a–c, 1996, 1997, ASTM 1995a,b, 1997a,c, 1999c). An overview of some of these test methods can be found in Keddy *et al.* (1992) and Van Straalen and Løkke (1997).

3.2.2.1 Fathead Minnow (*Pimephales promelas*) Survival and Growth Test

The Canadian fathead minnow test protocol (Environment Canada 1992c) involves exposing larvae (<24-h old) to ground water (10 organisms in 500 ml of water) in 1-L glass beakers for 7 days. In the 7-day static test with daily renewal of test solutions, the fish larvae are fed brine shrimp and daily observations of survivorship are recorded. The measurement endpoints are the number of larvae surviving at the end of the test, body length, and wet and dry weight.

3.2.2.2 Rainbow Trout Tests (*Oncorhynchus mykiss*)

The 96-h acute lethality test for rainbow trout is a static test that involves placing fish, at a loading rate of <0.5 g fish per litre, in ground water and measuring the number of organisms surviving at the end of the test. Usually, this involves exposing 10 fish in a 20-L test unit (e.g., bucket) to test solution for 96 h (Environment Canada 1990a). Other toxicity tests involving the use of early life stages of rainbow trout are also available, including an embryo test suitable for frequent or routine monitoring, an embryo/alevin test for measuring the effects of toxicants on multiple phases of development and an embryo/alevin/fry test for definitive investigations (Environment Canada 1996).

3.2.2.3 Waterflea (*Daphnia magna* or *Ceriodaphnia dubia*) Survival, Growth and Reproduction Tests

Ten neonates of *C. dubia* are individually, or in groups of 10 individuals, exposed to either a series of ground-water dilutions in glass vials, or the undiluted ground water, for 7 ± 1 days (60% of control organisms must have produced a minimum of three broods within 8 days). The test unit must accommodate a loading rate of 15 ml of test solution per organism. The test solutions are renewed daily and the organisms are fed algae and a prepared mixture of yeast, cerophyll and trout chow. The test organisms are transferred to new test solutions each day; prior to solution renewal, pH, dissolved oxygen, conductivity and temperature are measured. Adult survivorship and production of young are also recorded daily. The measurement endpoints are adult survivorship, total or mean number of live neonates produced per treatment, and fecundity (mean number of live neonates produced per adult) (Environment Canada 1992a).

The acute toxicity test with *D. magna* is a 48-h test where organisms are exposed individually, or in groups of 10 to undiluted ground water, or a dilution series of the ground water, in either glass vials or glass beakers (Environment Canada 1990b). The test solutions are renewed daily and the test organisms are not fed during the test. The measurement endpoint is adult survivorship at the end of the test.

3.2.2.4 Algal Growth Inhibition Tests

Tests with *Selenastrum capricornutum* have been used to assess the toxicity to plants of both ground water and surface waters. The algal species is exposed to either undiluted ground water, or a series of ground-water dilutions. Two approaches can be used: one uses polystyrene microplates (Environment Canada 1997); the other uses glass flasks or bottles (US Environmental Protection Agency

1989). At the end of the test (72 h), the density of cells in either the microplate wells, or subsamples from a flask, is determined electronically using a particle counter (e.g., Coulter counter) or manually using a haemocytometer. Alga growth in the ground water or test solutions with ground water is compared to that in the uncontaminated control treatment to determine if significant growth inhibition has occurred. This test is frequently used to assess the toxicity of soil aqueous extracts; the limitations of this application are discussed by Hund (1997).

3.2.2.5 Bacterial Test

One of the most frequently used toxicity tests applied to ground-water samples uses luminescent bacteria (*Photobacterium phosporeum*) and is marketed under the trade name Microtox™. There are explicit instructions for conducting the tests and specific procedural modifications have been recommended by different agencies that require the test in their regulatory framework (e.g., Environment Canada 1992b). Comparative tests are now available (e.g., Lumistox™) which also use light production by bacteria to assess potential toxicity of ground water.

Because these tests use a marine bacterium, their ecological relevance is often questioned. Numerous studies have compared the results of the Microtox™ assay with those of more traditional tests (alga, terrestrial plant, daphnia, earthworm, fish) and, as one might expect, the results were contradictory. The main advantage of the test is that it is rapid, relatively inexpensive, uses small sample volumes and is not labour intensive; therefore, many ground-water samples can be processed in a relatively short period of time. Recent developments in bioluminescence technologies suggest that this assay will likely become an important component of any terrestrial test battery, as described by Svendsen *et al.* (see Chapter 7).

3.2.3 SITE SOIL DILUTION OR AMENDMENT

There are three approaches to toxicity assessment of contaminants in soils: (1) to evaluate the toxicity of specific levels of toxicants that are added to a control soil (i.e., spiked-soil tests); (2) to collect a contaminated site soil and dilute or amend it with different amounts of a clean control soil (e.g., usually field-collected with physicochemical characteristics similar to the site soil) (Environment Canada 1998a–c), on the basis of dry weight, to establish a dilution series of different levels of contamination; and (3) to collect soil samples along a concentration gradient directly from the site and use the soil in the samples with low levels of contamination (e.g., non-toxic) to dilute the soils in samples that have high levels of contamination and are toxic (e.g., hotspot). The first two approaches require an uncontaminated control soil to serve as the test soil for the experimental control and/or as the diluent soil. It can be an artificial soil formulated from constituents of sand, clay and peat, or a field-collected reference soil that matches the physicochemical characteristics of the site soil (Environment Canada 1998a–c). Alternatively, the control soil might be a blend of natural soils with different physicochemical characteristics (Sheppard and Evenden 1998). The third approach overcomes some of the problems associated with selection of an appropriate reference control soil. Examples of the latter approach are described in Posthuma (1997) and McMillen *et al.* (2001).

3.2.4 MEASUREMENT ENDPOINTS, STATISTICAL ENDPOINTS AND THE ADVANTAGES OF USING THE ENTIRE DOSE OR CONCENTRATION–RESPONSE RELATIONSHIP TO ASSESS TOXICITY

When conducting toxicity tests, physical, chemical and/or biological measurements (e.g., shoot length) are made to assess toxicity or to estimate a statistical endpoint (e.g., IC_p for shoot length) that is indicative of a toxic effect.

Two types of data are generated from toxicity tests: quantal (e.g., number of adults alive at the end of an exposure period) and continuous (e.g., shoot length). The objectives of the toxicity test and the nature of the data dictate the type of analyses that must be used to derive statistical endpoints. For example, when a dose or concentration–response is observed, there are specific regression models (e.g., probit) that can be applied to quantal data to estimate an EC_x (median effective concentration) that are not applicable to continuous data which require different models and procedures to generate the IC_p (inhibiting concentration for a specified percent effect). The limitation of most of the approaches used to generate these statistical estimates is that only the linear portion of the dose or concentration–response curve is used in the effect estimation. Therefore, the data at the beginning and end of the concentration–response curve for a continuous variable are excluded or simply ignored. However, there are procedures (e.g., non-linear regression procedures) that can be applied to non-linear relationships, that use the entire exposure concentration–response relationship to assess toxicity (Draper and Smith 1981, Stephan and Rogers 1985, Bruce and Versteeg 1992, Moore and Caux 1997, Stephenson *et al.* 2000). The non-linear regression approach has several advantages over traditional analytical procedures. These include: (1) preclusion of the *a priori* selection of significance levels for a test; (2) preclusion of the need to identify a no observed effect concentration; (3) parameter estimates that are, to some extent, independent of the number of replicates and test concentrations used; (4) using the entire concentration–response relationship to consider the biological relevance of the size of a perturbation or effect; and (5) increased sensitivity to detecting toxic effects of low concentrations of a substance.

Non-linear regression techniques have a number of intrinsic problems which are not found in linear regression analysis and include: (1) parameters of the models cannot be estimated using matrix algebra, they must be estimated using iterative procedures; (2) in order for the iterative procedures to converge, parameter estimates must be accurate; (3) collinearity among parameters which usually limits the estimation of the model parameters and standard errors, and prevents extrapolation of results to the other data; (4) weighting the data to address heteroscedasticity in non-linear models. Draper and Smith (1981) discuss measures that can be taken to minimize or overcome these limitations. The application of four non-linear regression models, parameterized to include any IC_p or EC_x and their associated 95% confidence limits, to evaluate over 232 concentration-response relationships resulting from toxicity tests with plants exposed to contaminants in soils is discussed in detail by Environment Canada (1998c) and Stephenson *et al.* (2000).

Often the results of a toxicity assessment of polluted soil are not described by a classic exposure concentration-response relationship, and the observed responses must be directly compared to either a non-polluted control or a clean reference sample using parametric or non-parametric procedures.

3.3 SUMMARY

Whole-soil and ground-water toxicity test methods for site soil assessment have been developed primarily by modifying the procedures in test methods that are currently used to assess the toxicity of a chemically-spiked soil or ground-water sample. An overview of these toxicity test methods included a discussion of the ways in which contaminants in soil can affect the inhabiting plant and invertebrate populations, important considerations for selecting an appropriate test battery and what constitutes an appropriate test battery for site soil assessment. Three approaches to site soil evaluation with multi-concentration exposure scenarios were identified, and the value of using the entire concentration-response curve for derivation of statistical endpoints was highlighted.

REFERENCES

Achazi RK, Chroszcz G, Pilz C, Rothe B, Steudel I and Throl C (1996) Der Einfluss des pH-Wertes und von PCB 52 auf Reproduktion und Besiedlungsaktivität von terrestrischen Enchytraeen in PAK-, PCB-, und schwermetallbelasteten Rieselfeldböden. *Verhandlungen der Gesellschaft für Ökologie*, **26**, 37-42.

ASTM (1995a) *Standard Guide for Conducting Static 96-h Toxicity Tests with Microalgae*. ASTM Standard E 1218-90, American Society for Testing and Materials, West Conshohocken, PA, pp. 573-584.

ASTM (1995b) *Standard Guide for Conducting Early Life-stage Toxicity Tests with Fishes*. ASTM Standard E 1241-92, American Society for Testing and Materials, West Conshohocken, PA, pp. 585-612.

ASTM (1997a) *Standard Guide for Conducting Daphnia magna Life-cycle Toxicity Tests*. ASTM Standard E 1193-96, American Society for Testing and Materials, West Conshohocken, PA, pp. 474-491.

ASTM (1997b) *Standard Guide for Conducting a Laboratory Soil Toxicity Test with Lumbricid Earthworm Eisenia fetida*. Draft Revision of the ASTM Standard E 1676-95, American Society for Testing and Materials, West Conshohocken, PA, pp. 1055-1071.

ASTM (1997c) *Standard Guide for Conducting Acute Toxicity Tests on Aqueous Ambient Samples and Effluents with Fishes, Macroinvertebrates, and Amphibians*. ASTM Standard E 1192-97, American Society for Testing and Materials, West Conshohocken, PA, pp. 461 - 473.

ASTM (1999a) *Standard Guide for Conducting Terrestrial Plant Toxicity Tests*. ASTM Guide E 1963-98, American Society for Testing and Materials, West Conshohocken, PA, pp. 1481 - 1500.

ASTM (1999b) *Standard Guide for Conducting Laboratory Soil Toxicity or Bioaccumulation Tests with the Lumbricid Earthworm Eisenia fetida*. ASTM Guide E 1676-97, American Society for Testing and Materials, West Conshohocken, PA, pp. 1062 - 1079.

ASTM (1999c) *Standard Practice for Conducting Early Seedling Growth Tests*. Draft Revision of the ASTM Practice E 1598-94, American Society for Testing and Materials, West Conshohocken, PA, pp. 1000 - 1006.

BBA (1994) *Richtlinien für die Zulassung von Pflanzenschutzmitteln im Prüfungsverfahren, Teil VI, Auswirkungen von Pflanzenschutzmitteln auf die Reproduktion und das Wachstum von Eisenia fetida/Eisenia andrei (Guidelines for the Testing of Plant Protection Products within Registration, Part VI, Effects of Plant Protection Products on Reproduction and Body Weight of Eisenia fetida/Eisenia andrei)*, Saphir-Verlag, Ribbesbüttel, 12 pp.

Bruce RD and Versteeg DJ (1992) A statistical procedure for modelling continuous toxicity data. *Environmental Toxicology and Chemistry*, **11**, 1485 - 1494.

Crommentuijn T (1994) *Sensitivity of Soil Arthropods to Toxicants*. PhD Thesis, Netherlands Integrated Soil Research Programme/Ministry of Housing, Physical Planning and Environment, Amsterdam, 141 pp.

Denneman CAJ and Van Straalen NM (1991) The toxicity of lead and copper in reproduction tests using the orbatid mite *Platynothrus peltifer*. *Pedobiologia*, **35**, 305 - 311.

Donker MH and Bogert C (1991) Adaptation to cadmium in three populations of the isopod *Porcellio scaber*. *Comparative Biochemistry and Physiology*, **100c**, 143 - 146.

Donkin SG and Dusenbery DB (1993) A soil toxicity test using the nematode *Caenorhabditis elegans* and an effective method of recovery. *Archives of Environmental Contamination and Toxicology*, **25**, 145 - 151.

Draper NR and Smith H (1981) *Applied Regression Analysis* 2nd edition, John Wiley & Sons, Toronto, 709 pp.

Drobne D (1997). Terrestrial isopods — a good choice for toxicity testing of pollutants in the terrestrial environment. *Environmental Toxicology and Chemistry*, **16**, 1159 - 1164.

Environment Canada (1990a) *Reference Method for Determining Acute Lethality of Effluents to Rainbow Trout*. EPS Report 1/RM/13, Environmental Protection Series, Environment Canada, Ottawa, 18 pp.

Environment Canada (1990b) *Biological Test Method: Acute Lethality Test Using Daphnia sp*. EPS Report 1/RM/12, Environmental Protection Series, Environment Canada, Ottawa, 57 pp.

Environment Canada (1992a) *Biological Test Method: Toxicity Test of Reproduction and Survival Using the Cladoceran Ceriodaphia dubia*. EPS Report 1/RM/21, Environmental Protection Series, Environment Canada, Ottawa, 57 pp.

Environment Canada (1992b) *Biological Test Method: Toxicity Test Using Luminescent Bacteria (Photobacterium phosphoreum)*. EPS Report 1/RM/24, Environmental Protection Series, Environment Canada, Ottawa, 61 pp.

Environment Canada (1992c) *Biological Test Method: Test of Larval Growth and Survival Using Fathead Minnows*. EPS Report 1/RM/22, Environmental Protection Series, Environment Canada, Ottawa, 70 pp.

Environment Canada (1994) *Assessment of Soil Toxicity Test Species for Canadian Representativeness*. Report TS-28, Method Development and Applications Section, Technology Development Directorate, Environment Canada, Ottawa, 70 pp. (Appendices).

Environment Canada (1996) *Biological Test Method: Toxicity, Tests Using Early Life Stages of Salmonid fish (Rainbow Trout, Coho Salmon, or Atlantic Salmon)*. EPS Report 1/RM/Draft, Environmental Protection Series, Environment Canada, Ottawa, 101 pp.

Environment Canada (1997) *Biological Test Method: Growth Inhibition Test Using the Freshwater Alga Selenastrum capricornutum*. EPS Report 1/RM/25, Environmental Protection Series, Environment Canada, Ottawa, 41 pp.

Environment Canada (1998a) *Development of Earthworm Toxicity Tests for Assessment of Contaminated Soils*. Report Prepared by Aquaterra Environmental for the Method Development and Applications Section, Environmental Technology Centre, Environment Canada, Ottawa, 52 pp. (Appendices).

Environment Canada (1998b) *Development of a Reproduction Toxicity Test with Onychiurus folsomi for Assessment of Contaminated Soils*. Report Prepared by Aquaterra Environmental for the Method Development and Applications Section, Environmental Technology Centre, Environment Canada, Ottawa, 253 pp.

Environment Canada (1998c) *Development of Plant Toxicity Tests for Assessment of Contaminated Soils*. Report Prepared by Aquaterra Environmental for the Method Development and Applications Section, Environmental Technology Centre, Environment Canada, Ottawa, 75 pp. (Appendices).

Gibbs MH, Wicker LF and Stewart AJ (1996) A method for assessing sublethal effects of contaminants in soils to the earthworm, *Eisenia foetida*. *Environmental Toxicology and Chemistry*, **15**, 360-368.

Greene JC, Bartels CL, Warren-Hicks WJ, Parkhurst BR, Linder GL, Peterson SA and Miller WE (1989) *Protocols for Short Term Toxicity Screening of Hazardous Waste Sites*, US Environmental Protection Agency, Corvallis, OR, 600/3-88-029/102 pp.

Hopkin S (1989) *Ecophysiology of Metals in Terrestrial Invertebrates*, Elsevier Applied Science, London, 366 pp.

Hornung E, Farkas S and Fischer E (1998) Tests on the isopod *Porcellio scaber*. In *Handbook of Soil Invertebrate Toxicity Tests*, Løkke H and Van Gestel CAM (eds), John Wiley & Sons, Chichester, pp. 207-226.

Huhta V (1984) Response of *Cognettia sphagnetorum* (Enchytraeidae) to manipulation of pH and nutrient status in coniferous forest soil. *Pedobiologia*, **27**, 245-260.

Hund K (1997) Algal growth inhibition test — feasibility and limitations for soil assessment. *Chemosphere*, **35**, 1069-1082.

Hund K (1998) Earthworm avoidance test for soil assessment: alternative for acute and reproduction test. In *Contaminated Soil '98. Sixth International FZK/TNO Conference on Contaminated Soil, Edinburgh*, 17-21 May, pp. 1039-1040.

ISO (1995) *Soil Quality — Effects of Pollutants on Earthworms (Eisenia fetida). Part 1: Determination of Acute Toxicity Using Artificial Soil Substrate*. International Standard ISO 11268-1: 1993(E), International Standards Organization, Geneva, 6 pp.

ISO (1998a) *Soil Quality — Effects of Pollutants on Earthworms (Eisenia fetida). Part 2: Determination of Effects on Reproduction*. Draft International Standard ISO 11268-2: 1998(E), International Standards Organization, Geneva, 16 pp.

ISO (1998b) *Soil Quality — Effects of Soil Pollutants on Collembola (Folsomia candida) Method for the Determination of Effects on Reproduction*. International Standard ISO 11267, International Standards Organization, Geneva, 16 pp.

ISO (1999) *Enchytraeid Reproduction Test*. Working Draft 15677, International Standards Organization, Geneva, 19 pp.

Kalsch W and Römbke J (1998) Zur chronischen Wirkung von TNT auf die Stoppelrübe *Brassica rapa* im Labortest. In *Ökotoxikologie — Ökosystemare Ansätze und Methoden*, Oehlmann J, Markert B (eds), Verlag Urban & Fischer, Stuttgart, Jena.

Kammenga JE and Riksen JAG (1998) Test on the competition between the nematodes *Plectus acuminatus and Heterocephalobus pauciannulatus*. In *Handbook of Soil Invertebrate Toxicity Tests*, Løkke H and Van Gestel CAM (eds), John Wiley & Sons, Chichester, pp. 227-238.

Kammenga JE, Van Koert PHG, Riksen JAG, Korthals GW and Bakker J (1996) A toxicity test in artificial soil based on the life-history strategy of the nematode *Plectus acuminatus*. *Environmental Toxicology and Chemistry*, **15**, 722-727.

Kaufman ES (1975) Certain problems of phenol intoxication of *Enchytraeus albidus* from the view-point of stress. *Journal of Hydrobiology*, **11**, 44-46 (in Russian with English summary).

Keddy C, Greene JC and Bonnell MA (1992) *A Review of Whole Organism Bioassays for Assessing the Quality of Soil, Freshwater Sediment, and Freshwater in Canada*. CCME Subcommittee on Environmental Quality Criteria for Contaminated Sites, National Contaminated Sites Remediation Program, Canadian Council of Ministers for the Environment, Ottawa, 293 pp.

Kiss I and Bakonyi G (1992) Guideline for testing the effects of pesticides on *Folsomia candida* Willem (Collembola): Laboratory test. *Bulletin of the International Organization for Biological and Integrated Control of Noxious Animals and Plants, West Palaearctic Regional Section (IOBC/WPRS)*, **15**, 131-137.

Kula H and Larink O (1997) Development and standardization of test methods for the prediction of sublethal effects of chemicals on earthworms. *Soil Biology and Biochemistry*, **29**, 635-639.

Kula H and Larink O (1998) Tests on the earthworms *Eisenia fetida* and *Apporrectodea caliginosa*. In *Handbook of Soil Invertebrate Toxicity Tests*, Løkke H and Van Gestel CAM (eds), John Wiley & Sons, Chichester, pp. 95-112.

Léon CD and Van Gestel CAM (1994) *Selection of a Set of Laboratory Ecotoxicity Tests for the Effects Assessment of Chemicals in Terrestrial Ecosystems*. Discussion Paper D94004, Department of Ecology and Ecotoxicology, Vrije Universiteit, Amsterdam.

Løkke H (ed) (1995) *Effects of Pesticides on Meso- and Microfauna in Soil*, Ministry of the Environment and Energy, Bekaempelsesmiddelforskning fra Miljøstyrelsen, **8**, 185 pp.

Løkke H and Van Gestel CAM (1998) Soil toxicity tests in risk assessment of new and existing chemicals. In *Handbook of Soil Invertebrate Toxicity Tests*, Løkke H and Van Gestel CAM (eds), John Wiley & Sons, Chichester, pp. 3-19.

McMillen SJ, Van Gestel CAM, Lanno RP and Linder GL (2001) Biological measures of bioavailability. In *Contaminated Soils: From Soil-Chemical Interactions to Ecosystem Management*, Lanno RP (eds), Society of Environmental Toxicology and Chemistry (SETAC), Pensacola, FL, pp. 269-316.

Moore DRJ and Caux PY (1997) Estimating low toxic effects. *Environmental Toxicology and Chemistry*, **16**, 794-801.

Moore JC and DeRuiter PC (1993) Assessment of disturbance on soil ecosystems. *Veterinary Parasitology*, **48**, 75-85.

Notenboom J and Boessenkoll J-J (1993). Acute toxicity testing with the groundwater copepod *Parastenocaris germanica* (Crustacea). In *Proceedings 1st International Conference on Groundwater Ecology*, US Environmental Protection Agency & American Water Research Agency (1992), Washington, DC, pp. 301-309.

OECD (1984a) *OECD Guidelines for Testing of Chemicals: Earthworm Acute Toxicity Test*. Guideline No. 207, Organization for Economic Co-operation and Development, Paris, France, 9 pp.

OECD (1984b) *OECD Guidelines for Testing of Chemicals: Terrestrial Plant Growth Test*. Guideline No. 208, Organisation for Economic Co-operation and Development, Paris, France, 11 pp.

OECD (1999) Draft of the Proposed Revision to *Guidelines for Testing of Chemicals: Plant Toxicity Tests*. Guideline No. 204, 1984, Organization for Economic Co-operation and Development, Paris, France.

Petersen H and Gjelstrup P (1995) Development of a semi-field method for evaluation of laboratory tests as compared with field tests. In *Effects of Pesticides on Meso- and Microfauna in Soil*, Ministry of the Environment and Energy, Bekaempelsesmiddelforskning fra Miljøstyrelsen, **8**, 67-142.

Posthuma L (1997) Effects of toxicants on population and community parameters in field conditions, and their potential use in the validation of risk assessment methods. In *Ecological Risk Assessment of Contaminants in Soil*, Van Straalen N and Løkke H (eds), Chapman & Hall, Padstow, pp. 85-126.

Römbke J (1989) *Enchytraeus albidus* (Enchytraeidae, Oligochaeta) as a test organism in terrestrial laboratory systems. *Archives of Toxicology*, **13**, 402-405.

Römbke J and Federschmidt A (1995) Effects of the fungicide carbendazim on Enchytraeidae in laboratory and field tests. *Newsletter on Enchytraeidae*, **23**, 301-309.

Römbke J, Knacker T, Förster B and Marcinkowski A (1994) Comparison of effects of two pesticides on soil organisms in laboratory test, microcosms and in the field. In *Ecotoxicology of Soil Pollution*, Donker MH, Eijsackers H and Heimbach F (eds), Lewis, Chelsea, MI, pp. 229-240.

Römbke J, Moser T and Knacker T (1998) Enchytraeid reproduction test. In *Advances in Earthworm Ecotoxicology*, Sheppard S, Bembridge J, Holmstrup M and Posthuma L (eds), SETAC Press, Pensacola, FL, pp. 83-97.

Rundgren S and Augustsson AK (1998) Test on the enchytraeid *Cognettia sphagnetorum* (Vejdovsky) 1877. In *Handbook of Soil Invertebrate Toxicity Tests*, Løkke H and Van Gestel CAM (eds), John Wiley & Sons, Chichester, pp. 73-94.

Sheppard SC and Evenden WG (1998) An approach to defining a control or diluent soil for ecotoxicity assays. In *Environmental Toxicology and Risk Assessment*, Vol.7, Little EE, DeLonay AJ and Greenberg BM (eds). ASTM STP 1333, American Society for Testing and Materials, West Conshohocken, PA, pp. 215-226.

Smit E (1997) *Field Relevance of the Folsomia candida Soil Toxicity Test*. PhD Thesis, Vrije Universiteit, Utrecht, 157 pp.

Stephan CE and Rogers JW (1985). Advantages of using regression analysis to calculate results of chronic toxicity tests. In *Aquatic Toxicology and Hazard Assessment*, Bahner RC. and Hansen DJ (eds). ASTM STP 891, American Society for Testing and Materials, Philadelphia, PA, pp. 328-338.

Stephenson GL, Kaushik A, Kaushik NK, Solomon KR, Steele T and Scroggins RP (1998) Use of an avoidance-response test to assess the toxicity of contaminated soils to earthworms. In *Advances in Earthworm Ecotoxicology*, Sheppard S, Bembridge J, Homstrup M and Posthuma L (eds), SETAC Press, Pensacola, FL, pp. 67-81.

Stephenson GL, Koper N, Atkinson GF, Solomon KR and Scroggins R (2000) Use of nonlinear regression techniques for describing concentration-response relationships for plant species exposed to contaminated site soils. *Environmental Toxicology and Chemistry*, **19**, 229-242.

Thompson AR and Gore FL (1972) Toxicity of twenty-nine insecticides to *Folsomia candida*: laboratory studies. *Journal of Economic Entomology*, **65**, 1255-1259.

US Environmental Protection Agency (1989) *Protocols for Short Term Toxicity Screening of Hazardous Waste Sites*, United States Environmental Protection Agency, Environmental Research Laboratory, Corvallis, OR, 600/3-88-029/102 pp.

US Environmental Protection Agency (1992) *Evaluation of Terrestrial Indicators for Use in Ecological Assessments at Hazardous Waste Sites*, United States Environmental Protection Agency, Office of Research and Development, Washington, DC, EPA/600/R-92/183 pp.

Van Gestel CAM (1998) Evaluation of the development status of ecotoxicity tests on soil fauna. In *Handbook of Soil Invertebrate Toxicity Tests*, Løkke H and Van Gestel CAM (eds), John Wiley & Sons, Chichester, pp. 57-68.

Van Gestel CAM and Doorenekamp A (1998) Tests on the orbatid mite *Platynorthrus peltifer*. In *Handbook of Soil Invertebrate Toxicity Tests*, Løkke H and Van Gestel CAM (eds), John Wiley & Sons, Chichester, pp. 113-130.

Van Gestel CAM and Van Straalen NM (1994) Ecotoxicological test systems for terrestrial invertebrates. In *Ecotoxicology of Soil Organisms*, Donker MH, Eijsackers H and Heimbach F (eds), Lewis, Boca Raton, FL, pp 205-228.

Van Gestel CAM, Van Dis WA, van Breeman EM and Sparenburg PM (1989) Development of a standardized reproduction toxicity test with the earthworm species *Eisenia fetida andrei* using copper, pentachlorophenol, and 2,4-dichloroaniline. *Environmental Toxicology and Safety*, **18**, 305-312.

Van Straalen NM and Løkke H (eds) (1997) *Ecological Risk Assessment of Contaminants in Soil*, Chapman & Hall, London, 354 pp.

Van Straalen NM and Van Gestel CAM (1993) *Ecotoxicological Test Methods Using Terrestrial Arthropods*. D93002, Department of Ecology and Ecotoxicology, Vrije Universiteit, Amsterdam.

Van Straalen NM, Schobben JHM and de Goede RGM (1989) Population consequences of cadmium toxicity in soil microarthropods. *Ecotoxicology and Environmental Safety*, **17**, 194 - 204.

Weuffen W (1968) Zusammenhänge zwischen chemischer Konstitution und keimwidriger Wirkung. *Archives of Experimental Veterinary Medicine*, **22**, 127 - 132.

Wiles JA and Krogh PH (1998) Tests with the Collembolans *Isotoma viridis*, *Folsomia candida*, and *Folsomia fimetaria*. In *Handbook of Soil Invertebrate Toxicity Tests*, Løkke H and Van Gestel CAM (eds), John Wiley & Sons, Chichester, pp. 131 - 156.

Wood WB (1988) *The Nematode Caebirgabditis elegans*, Cold Spring Harbor Laboratory Press, Cold Spring Harbor, NY, pp. 1 - 16.

4

Multispecies and Multiprocess Assays to Assess the Effects of Chemicals on Contaminated Sites

ROMAN KUPERMAN[1], THOMAS KNACKER[2], RONALD CHECKAI[3] AND CLIVE EDWARDS[4]

[1] *Geo-Centers Inc., Aberdeen Proving Ground, MD, USA*
[2] *ECT Oekotoxikologie GmbH, Germany*
[3] *US Army Edgewood Chemical Biological Center, Aberdeen Proving Ground, MD, USA*
[4] *Soil Ecology Laboratory, Ohio State University, Columbus, OH, USA*

4.1 INTRODUCTION

Maintaining soil quality, fertility and structure is essential to protect and sustain biodiversity and ecological integrity in terrestrial ecosystems. Central to achieving this goal is the need for a greatly improved understanding of the potential effects of chemical contaminants on the structure and function of ecosystems. Chemical contaminants of soils can exert their effects directly, through toxicity to soil organisms, or indirectly, by altering specific interactions and by disrupting soil food webs. In practice, soil organisms and carbon and nutrient cycling processes are so intimately linked that effects of chemicals on any one of these aspects are likely to impact many of the others. Furthermore, the intensity and duration of the environmental effects of chemicals may strongly depend upon those processes that influence the activity, fate, persistence and movement of chemical contaminants through the soil ecosystem and into soil organisms and plants. Ultimately, these effects can interfere with key soil processes that may be important to the regulation, flow and internal cycling of carbon and nutrients in ecosystems (Parmelee *et al.* 1993, Edwards and Bohlen 1995, Kuperman and Carreiro 1997, Kuperman *et al.* 1998).

The effects of chemicals on particular species or groups of soil organisms, dynamic soil processes and less frequently on whole soil systems have been used to evaluate the potential environmental impact of chemicals reaching soil, using both laboratory and field experiments. Methods for the laboratory assessment of the toxicity of contaminated soils to selected species of soil biota

Environmental Analysis of Contaminated Sites. Edited by G. I. Sunahara, A. Y. Renoux, C. Thellen, C. L. Gaudet and A. Pilon

are described in Chapter 3. Field evaluation methods for site contamination are discussed in detail by Römbke and Notenboom (see Chapter 10). The purpose of this chapter is to discuss methods which can help bridge the gap between laboratory assessment and field evaluation methods.

The objective of assessing risks of a chemical to the environment is to determine whether the integrity of an ecosystem, rather than an individual organism, is perturbed when such a chemical stressor is introduced into the system, except when threatened and endangered species are considered. Historically, ecological effects have usually been assessed using the results of single species laboratory tests. These tests do not take into account interactions between populations of individual organisms and their environment.

Laboratory and semi-field microcosms and theoretical computer simulation models can bridge the gap between single species laboratory tests and field studies. Giesy and Odum (1980) defined microcosms as artificially bounded subsets of naturally occurring environments, which are replicable. These authors considered that, to qualify as a microcosm, an experimental unit should include several trophic levels. Thus, a simple pot system for studying uptake of a chemical by plants from a synthetic medium or sterile soil would not be a microcosm, whereas a system with a homogeneous soil or intact soil core could be classified as a microcosm if natural communities of microorganisms were allowed to develop and influence nutrient cycling and chemical transformation processes. Multispecies test systems use a range of endpoints, any of which can be selected, depending on the priorities given to specific ecosystem structures and functions of particular sites.

In a special issue of *Ecology*, several articles (Drake *et al.* 1996, Ives *et al.* 1996, Jaffee 1996, Lawton 1996, Moore *et al.* 1996, Verhoef 1996) reviewed the pros and cons of microcosm experiments in ecology. These authors concluded that a combination of microcosm studies, fieldwork and theory is necessary to explore ecological principles. This approach also applies to ecological problems such as the fate and effects of chemicals in the environment.

4.2 THE ROLE OF MULTISPECIES AND MULTIPROCESS TOXICITY TESTING IN RISK ASSESSMENT OF CONTAMINATED SOILS

Microcosm studies can be used to improve risk assessment of chemicals to soil environment by incorporating ecological principles into risk assessment methodologies. These systems provide: (1) data that enhance the interpretation of single species test results; (2) data on ecosystem functions (which have been neglected thus far in ecological risk assessments of contaminated sites); (3) indirect as well as synergistic or compensatory effects of chemicals at the ecosystem level; and (4) significantly improved assessment of the fate of contaminants in terrestrial ecosystems. Förster *et al.* (1996) demonstrated that multispecies test systems are effective tools for the analysis of indirect effects of chemicals on integrative ecosystem processes such as organic matter decomposition. Toxicological data from a microcosm study can be used to delineate areas where significant ecological effect could occur by incorporating the microcosm no observed effect concentration (NOEC) data into site spatial analysis. Microcosms are designed to allow both a diagnostic approach for contaminated sites and a prognostic approach for risk assessment of chemicals.

Microcosm systems have their greatest utility for assessing environmental risks in a tiered approach to risk assessment. For example, the established procedure to register new chemicals and to assess the environmental risk of new and existing chemicals within the European Union is based on no terrestrial ecotoxicological studies in Tier 0 and only two laboratory single species studies in Tier 1. The extrapolation of toxicity results from such a limited data set to entire ecosystems contains a high degree of uncertainty. To improve this highly restricted risk assessment approach for soil systems, Pedersen and Samsoe-Petersen (1994) as well as Römbke *et al.* (1996) have developed ecotoxicological test strategies that propose a tiered approach requiring either microcosm or field studies in higher tiers. Microcosms are less time- and labor-intensive than field studies, and can be used to validate the results of single species toxicity tests and facilitate interpretation and extrapolation of this information to the ecosystem level. Many factors determine whether or not multispecies tests are used in a risk assessment at a contaminated site. However, the risk manager of the site should at least consider such tests.

4.3 DESIGN OF MULTISPECIES TEST SYSTEMS

The designs of multispecies test systems can be as diverse as the hypotheses they are employed to test. Reviews by Morgan and Knacker (1994), Fraser and Keddy (1997) and Sheppard (1997) summarize the use of microcosms in ecotoxicology and community ecology research. These authors showed that for one set of questions, site-specific microcosms could be used to assess localized effects in a very specific area. For other questions, more generalized systems with a number of components and pathways common to most ecosystems could be used.

Microcosms can be structurally simple or complex, large or small, synthetic (defined) or natural. They are limited in size, time and mass, and in both biotic and abiotic components. Microcosms have boundaries, which restrict interaction with the rest of the ecosystem. Microcosms do not mimic natural systems exactly at all levels of organization and are incapable of self-perpetuation over long periods. To be useful as an ecotoxicological tool, microcosms should minimize variability without reducing realism significantly. However, achieving replicability in the behavior of microcosms in space and time often results in decreased similarity to natural systems. To validate such microcosms, all simplifications made should be evaluated by field-microcosm comparisons.

Chemical toxicity can be tested using two basic types of microcosm design: (1) a small integrated soil microcosm (ISM) with sieved soil, selected introduced and indigenous invertebrates, single plant species, highly replicated – controlled laboratory studies, and (2) a larger terrestrial model ecosystem (TME) with intact soil cores, indigenous invertebrates and mixed plant flora with greater biodiversity, minimally replicated within studies designed to simulate field conditions. Such multispecies test systems can offer high resolution of the ecotoxicological testing of chemicals in soil ecosystems at two levels of complexity.

4.3.1 INTEGRATED SOIL MICROCOSM

The ISM consists of well-defined, thoroughly-mixed, sieved, field-collected soil containing endogenous microorganisms, nematodes, microarthropods with soil packed gently into a plastic cylinder and seeds planted into the soil (Edwards *et al.* 1997). This type of microcosm is best suited for determining a concentration-response relationship and/or NOEC/lowest observed effect concentration (LOEC) levels for sites with a single chemical contamination problem. In such cases, a dilution series ranging from the highest concentration predicted or found in the field, to the 'clean' soil level can be used. Reference soils with similar physical and chemical characteristics but with no contamination, or background concentration level of contaminants, should be collected in a nearby area shortly before the start of the test.

Each microcosm unit is 7.5 cm (inside diameter) × 15 cm high, constructed from a commercially available high-density polyethylene (HDPE) pipe. Approximately 1 kg (oven-dry mass basis) of prepared soil is used in each microcosm and all units are kept at 16-18 °C in a continuous light chamber or greenhouse. Ten plant seeds of one species are sown in the top 0.5 cm of the microcosm soil and after one to two weeks are thinned to four per microcosm. In addition to the indigenous soil microorganisms, microarthropods and nematodes present in the soil, three small epigeic earthworms such as *Aporrectodea tuberculata* (Eisen.) or similar species (with total mass of 1-2 g), common in soils of the area, are added to each microcosm. There are six replicate microcosms for each treatment and sampling date. Microcosms are sampled destructively after 5, 10, 20 and 40 days.

A layer of mixed-bed, ion-exchange resins is placed at the bottom of each microcosm cylinder separated from the bottom of the soil core by a thin layer of glass wool (Figure 4.1). This allows free passage of the soil leachate from the microcosm, and acts as a partial barrier to prevent root growth out of the soil core bottom, while collecting nutrient ions leaching from the microcosm. For leaching studies of contaminants the resins are replaced by funnels closed by taps.

Water is added to each microcosm two or three times a week to maintain a soil moisture content of 40-60% of field capacity. Excess water is added weekly to leach through the soil core. The leachate from each microcosm is collected into a dish for nutrient and chemical analysis. Soil samples for microbial biomass, litter decomposition, enzyme activity, bait-lamina tests, nutrient leaching and test compound(s) concentration measurements are taken weekly. At the end of the

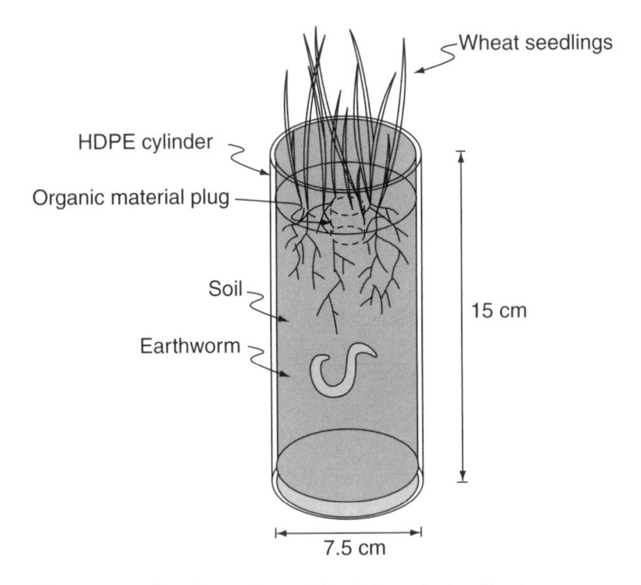

Figure 4.1 Configuration of the integrated soil microcosm.

experiment, microbial biomass, numbers of microarthropods, nematodes and earthworms may be assessed. The type and number of endpoints depends on the assessment endpoints selected and the resources available to investigators.

In the cases of mixed chemical contamination, a dilution series can be prepared using the site soil as a test mixture 'spike'. The reference soil is collected as described before, but the dilution series is made with defaunated site soil instead of a chemical compound. Methods for defaunating of soil include deep freezing, prolonged heating (several days after the constant mass was achieved) at 60–70 °C or radiation treatment. In any case, the treatment chosen should not interfere with or modify contaminant properties. For example, heating may not be appropriate if contaminants are volatile or semi-volatile organic chemicals. It is important to ensure that the amount of fresh soil used, containing endogenous soil biota, remains constant throughout the range of treatments. The target concentrations are attained by mixing similarly treated site soil and reference soil, in appropriate proportions, followed by combining these mixtures with untreated fresh soil to achieve a final concentration range corresponding to that used for single contaminants.

4.3.2 TERRESTRIAL MODEL ECOSYSTEM

The TME designed for determining the potential fate and ecological effects of new and existing chemicals can also be used in risk assessment of contaminated sites. The main modification to the original procedure (Knacker and Römbke 1997) involves selection of the appropriate range of test concentrations based on contamination information ascertained during earlier phases of the site characterization. In contrast with ISM, determination of the concentration – response relationship, which requires multiple concentration levels, may not be practical using the TME. The TME approach is best suited for determining NOEC and LOEC levels.

The principles of the test are as follows. Soil cores from the site containing biota of interest are treated with the test contaminant using a surface spray under controlled conditions. Any effects on ecosystem level parameters are determined by comparing the results from the treated TME with those obtained from non-treated soil cores. Because TME uses an intact soil core, its utility for assessing the ecological effects of soil contamination is limited to cases where the treatments can be simulated or contaminants can be reliably extracted from the site soil and introduced into TME as elutriates in the appropriate dilution series.

The basic TME design consists of intact (non-homogenized) soil cores extracted hydraulically from the test site and encased in the HDPE tube (Figure 4.2). The materials required for a single model ecosystem are: a 40-cm long HDPE tube (17.5 cm diameter), an 18.6-cm diameter porcelain Buchner funnel, a thin layer of inert gauze to fit between the Buchner funnel and the bottom of the soil core, and silicone tubing to connect the Buchner funnel to an Erlenmeyer flask, which acts as a collection vessel. An important modification to this design was proposed by Checkai *et al.* (1993a) which included fitting each

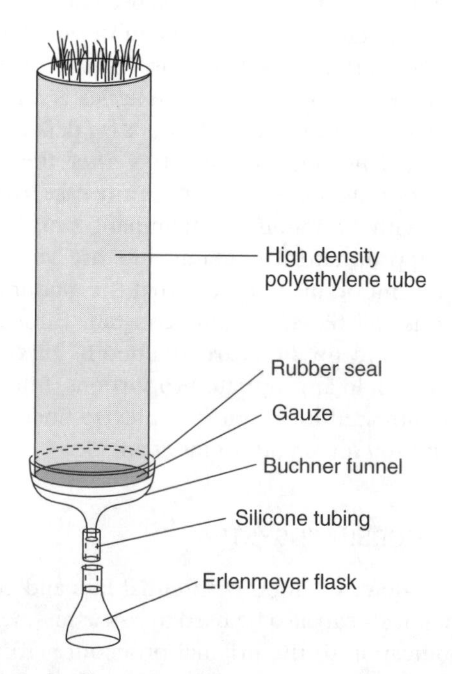

Figure 4.2 Configuration of the terrestrial model ecosystem.

soil column with a porous ceramic plate and a polyethylene endcap containing fittings for Teflon tubing, so that tension could be applied at the bottom of each soil column (30–35 kPa) to mimic field conditions. This modification prevents the build-up of water within columns, which can change the chemical, physical and biological properties of soil (Figure 4.3).

Extraction of the soil core is done gently with hydraulic equipment to minimize the disturbance to the natural vegetation, soil microflora and fauna, and compression of soil inside the cores (Figure 4.4). All soil cores are placed in special containers (Figure 4.5) and kept under temperature-, moisture- and

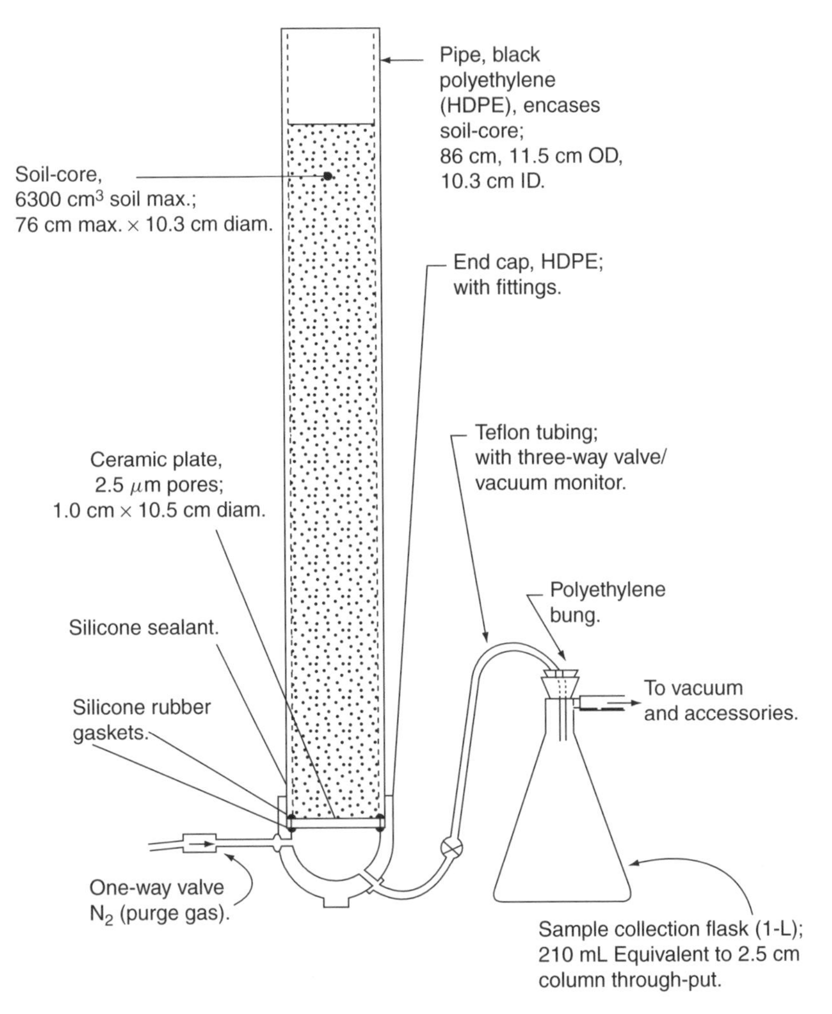

Figure 4.3 Soil core column fitted with a polyethylene endcap to apply tension at the bottom of each soil column (30–35 kPa) mimicking field conditions (from Checkai *et al.* 1993b).

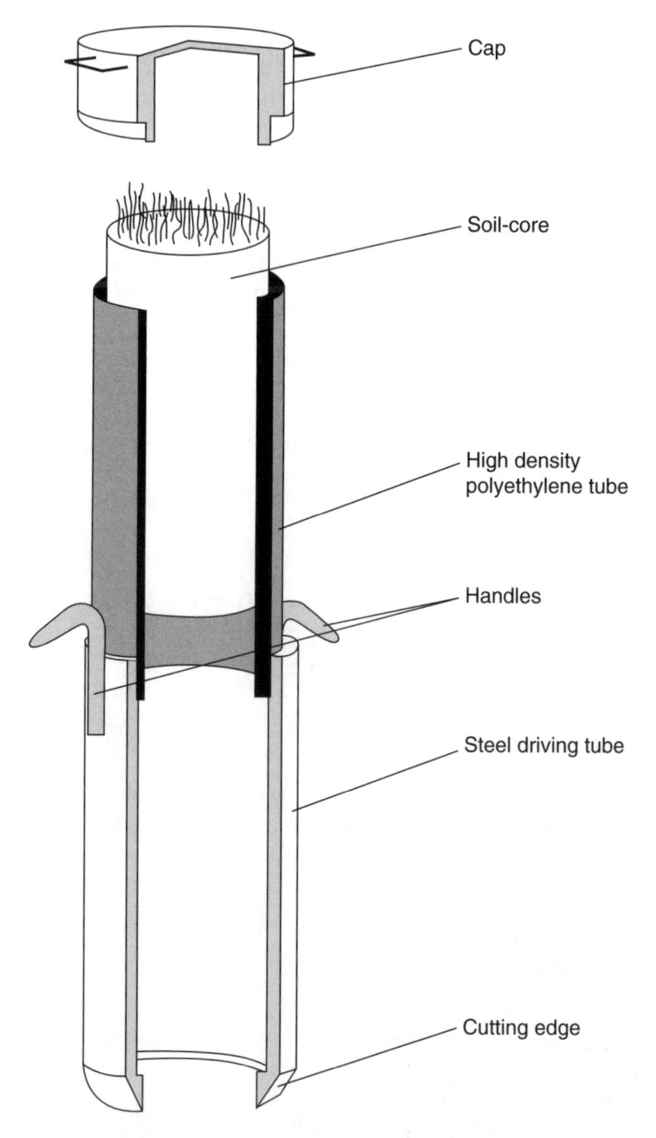

Figure 4.4 Soil core extraction apparatus.

light-controlled conditions (i.e., in a greenhouse or controlled environment growth chamber). Containers used in the study should be of equal size and volume with the same configuration. Each soil core is watered via special rain-heads (Figure 4.6) and leachates can be sampled from funnels at the bottom of each core. The study usually lasts for 16 weeks from first application of the test compounds. Soil cores (minimum of five replicates per treatment) are treated identically to estimate the variability within the TMEs of the same treatment

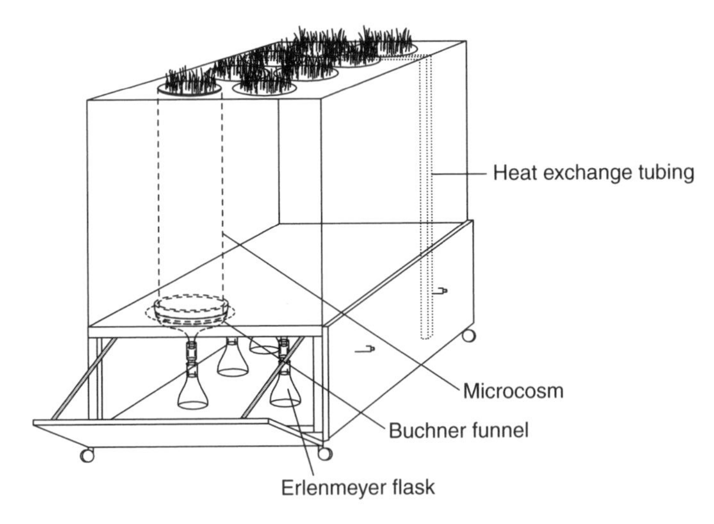

Figure 4.5 Microcosm cart designed to hold eight model ecosystems and adequate insulating material. Each cart has tubes running through it, which are connected via a heat exchanger to a cooling unit.

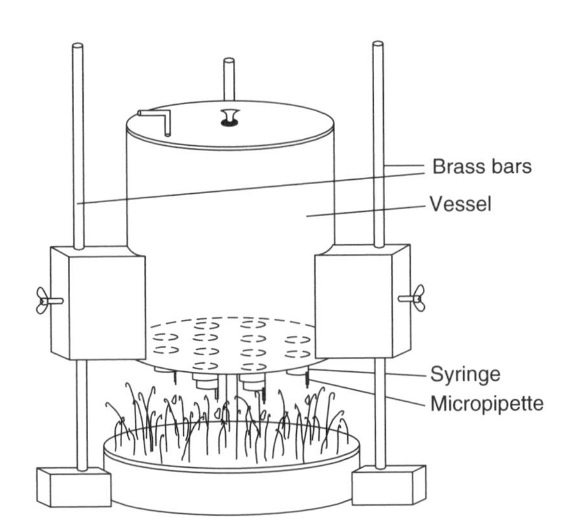

Figure 4.6 Rain-head for watering the model ecosystems. A rain-head is made of Plexiglas, 16.5 cm high × 14 cm diameter with 10 evenly spaced holes for inserted micropipettes.

level. A minimum of four treatment levels should be used, one close to the measured or expected environmental concentration, one negative control and two with concentrations to span the range between the field level and the negative control. During and at the end of the investigation period, samples (leachates, soil, plants, invertebrates, litter) are collected and the fate and

Table 4.1 Measurements in integrated soil microcosms and terrestrial model ecosystems.

Measurement	ISM	TME
Ecosystem structure		
Microbial activity & diversity	microbial biomass	microbial biomass & diversity
Nematode communities	numbers	numbers in trophic groups
Earthworm populations	numbers & biomass	numbers, biomass & diversity
Microarthropod populations & diversity	numbers	numbers & diversity
Enchytraeid populations	numbers & biomass	numbers, biomass & diversity
Plant populations	dry weight biomass	dry weight biomass & diversity
Ecosystem processes		
Mineralization	microbial respiration	microbial respiration & mineralization
Soil chemistry	C, N, P, pH	C, N, P, pH
Plant nutrient uptake	C, N	C, N, P
Organic matter decomposition	loss from litterbags	loss from litterbags
Biological activity (bait-lamina test)	rate of loss of organic matter	rate of loss of organic matter
Fate of chemicals		
Degradation pathways	fate in soil	fate in soil
Leaching	amount leached	amount leached
Uptake into earthworms	amount in earthworms	amount in earthworms
Uptake into plants	amount in test plants	amount in indigenous plants
Volatilization	optional test	optional test

effect endpoints are assessed (Table 4.1). A detailed description of site selection methods, TME equipment, soil core extraction technique, TME maintenance under laboratory or greenhouse conditions, test substance application, leachate collection, strategies on endpoint selection, sampling strategies and statistical evaluation of results is given in UBA (1994) and Knacker and Römbke (1997).

4.4 ASSESSMENT AND MEASUREMENT ENDPOINTS

The purpose of the ecological assessment of soil contamination with chemicals is to determine the fate of chemicals and the extent of damage to soil biota and biologically mediated processes they regulate. The adverse effects on these biota and processes can ultimately threaten sustainable functioning of a terrestrial ecosystem. A summary of recommended measurements is shown in Table 4.1. The choice of scientifically defensible assessment and measurement endpoints is based on protection of these critical measurable structural and functional biological parameters in the soil ecosystem.

4.4.1 FATE ENDPOINTS

The overall fate of the test chemical is assessed in the leachate throughout the study period, and in the soil/litter layers and in any plant and/or invertebrate material at the end of the study. From these data, fate endpoints such as mass balance, bioconcentration factors, persistence and mobility of chemicals are estimated. At the end of the TME study, the soil core is divided into at least three depths for sampling. The concentration of chemicals and their degradation products are determined for all depths.

Leaching of the model ecosystems is performed at certain points of the study. The leachate collected is used to examine losses of the contaminant and essential elements (nutrients). To determine the amount of contaminant taken up by plants in model ecosystems, plants are harvested at the end of the study period and, possibly, once or twice during the study, depending on the types of plants grown. Provided that the chosen field sites contain sufficient biota (i.e., suitable size and number), the accumulated amount of the contaminant can be measured in the soil biomass.

4.4.2 EFFECT ENDPOINTS: BIOLOGICAL DIVERSITY MEASUREMENTS

The effects of chemical contamination on species abundance are regarded as a structural attribute of the ecosystem. The species of soil biota should be selected to allow the assessment of the effects of chemicals on different components of the soil ecosystem. These must include species from different taxonomic groups to represent a range of sensitivities, which often correlate with physiologically determined mechanisms varying among taxa. Species from primary producers, herbivore, predator, saprovore, bacterivore and fungivore functional groups should be included to represent the spectrum of ecological functions in the soil community. To ensure contact with chemicals, species actively moving through soil are preferred. Different exposure routes to chemicals (ingestion, inhalation, dermal absorption, uptake from soil solution) are assessed by selecting species which provide information on toxicology in different soil microsites (e.g., nematodes inhabit the water film around soil particles, whereas microarthropods inhabit air-filled soil pores).

In general, these requirements can be achieved by selecting species from the following six groups: plants, earthworms, enchytraeids, Collembola, mites and nematodes. Plants growing in the model ecosystems are carefully monitored during the study for changes in physical appearance. Their productivity can be measured as total above-ground biomass. Abundance of soil invertebrates is assessed in subsamples of soil from microcosms using methods described in Edwards (1991). A microbial community in soil is usually assessed as microbial biomass (carbon or nitrogen) because identification and enumeration of soil microorganisms using direct count methods can require highly skilled experts, which might not be feasible for some situations. Assessment of a microbial community can be done indirectly by quantifying its activity in the assays described below. Having selected appropriate species, the investigators should sample the model ecosystems accordingly and quantify the species abundances. Changes in dominance should help identify any shifts in community structure.

4.4.3 EFFECT ENDPOINTS: MEASUREMENT OF SOIL PROCESSES

Indicators of soil processes include litter decomposition, carbon mineralization and net nitrogen mineralization assays, and assays to quantify the potential activity of soil enzymes involved in nutrient cycling processes in terrestrial ecosystems. These indicator processes are chosen because they are easily measured and integrate the combined activities of organisms within the soil ecosystem.

Organic matter decomposition is one of the most integrating processes within the soil ecosystem because it involves complex interactions of soil microbial and faunal activity with the soil chemical environment. Species from different taxonomic groups contribute to the breakdown of plant and animal debris. Any disturbance which alters organic matter decomposition can result in nutrient losses and a decline in soil fertility. Therefore, an assessment of how chemical contamination may alter rates of organic matter decomposition and rates of nutrient retention and release is critical to understanding its impact on overall ecosystem structure and function. A small amount of litter for decomposition studies can be used in the microcosms. Bogomolov *et al.* (1996) used 0.2–0.3 g of wheat straw confined within a litterbag of fiberglass screen (1.6 mm × 1.8 mm) and could detect significant treatment effect on litter mass loss after 40 days.

Carbon mineralization potential can be determined using the basal and/or substrate-induced respiration method (SIR). The addition of substrate (usually glucose) to a soil sample induces a maximal respiratory response from the soil microbial biomass, measured as CO_2 evolution. Microbial biomass is the major acting agent for most soil biogeochemical processes in terrestrial ecosystems, and interacts with the primary productivity of ecosystems by regulating nutrient availability and degradation pathways of soil contaminants.

Substrate-induced respiration can be measured using a soil respiration measuring system with continuous gas flow (Cheng and Coleman 1989). An additional advantage of this method is that selective biotic inhibitors, such as streptomycin and cyclohexamide, can be added to the glucose solution to determine the contribution of fungal and bacterial groups to overall soil respiration.

Unlike carbon mineralization, nitrification is one of the most sensitive processes related to soil function because conversion of ammonia to nitrate is carried out only by a limited number of bacterial genera (mostly *Nitrosomonas* and *Nitrobacter* spp.), whereas carbon mineralization is carried out by all heterotrophic organisms in soil. Nitrogen utilization requires distinct enzymatic steps for extracellular hydrolysis, uptake, deamination and intracellular catabolism, each of which may be regulated differently (Smith *et al.* 1989). The effects of soil contamination on overall nitrogen mineralization provide important information regarding changes in the availability of this critical nutrient in the ecosystem.

The dynamics of soil nitrogen transformations are determined by measuring changes in the sizes of soil nitrogen pools. Rates of net nitrogen mineralization are calculated as the difference in soil inorganic N at the beginning and end of a particular time period, plus the amount of N taken up into plants and lost through leaching during that period. Leaching losses of inorganic nitrogen are measured in soil leachate samples using a nitrogen analyzer.

The potential activities of the following soil extracellular enzymes can be quantified: beta-1,4-glucosidase (C-acquiring enzyme), *n*-acetylglucosaminidase (N-acquiring enzyme) and acid phosphatase (P-acquiring enzyme). Assays are conducted using soil slurries suspended in an acetate buffer. The activities of all enzymes can be measured using the spectrophotometric method described in Sinsabaugh and Linkins (1990), which uses substrates bound to the chromogen, *p*-nitrophenol (pNP). These substrates are pNP-beta-D-glucopyranoside, pNP-*n*-acetylglucosaminide and pNP-phosphate. Urease is important in biochemical transformations of nitrogen in the soil. Nitrogen limitation to microbial and plant growth may be important when soil water and organic carbon sources are abundant. The intracellular dehydrogenase enzyme plays an important role in cellular respiration, and therefore can provide a good index of the respiratory status and biomass of soil microorganisms.

4.5 INTERPRETATION AND USE OF DATA FROM MULTISPECIES ASSAYS

The interpretation of data from microcosm assays is determined by the original hypotheses formulated by the risk assessor prior to commencement of the study. In contrast to single species tests with clearly defined effect measurements (survival, growth, reproduction, etc.), change in the soil biota community structure can be more complex and difficult to interpret. Soil contaminants may have disproportionate effects on different groups of soil biota leading to changes in the trophic structure of the soil community. These changes may not only result in a decrease in abundance of specific groups of soil biota, but also in an increase, analogous to the hormetic effect commonly observed in single species tests (Sheppard 1997). Parmelee *et al.* (1993) reported an increase in nematode abundance at intermediate copper concentrations (100 and 200 $\mu g\ g^{-1}$) due to a negative effect on nematode predators, which were more sensitive to this metal.

The interpretation of data is influenced by the intended purposes of the microcosms. This would differ depending on whether microcosms are used as screening tools or if they are used as tools for improved understanding of the ecotoxicological effects of soil pollution. When microcosms are used as a screening tool to examine multiple effects endpoints during the early stages of risk assessment, the results can identify the specific single species and/or process toxicity assays that can be targeted for further assessment. For example, if the results of the microcosm study show significant reduction in earthworm and springtail numbers, the inclusion of corresponding single species toxicity tests (the standard earthworm test or springtail reproduction test described by Stephenson *et al.*, see Chapter 3) into a test battery would be beneficial.

More often, data from microcosm studies are used to gain insight into the mechanisms of contamination effects on the soil community. Most measurement endpoints included in the study (Table 4.1) provide valuable information on the potential effects of chemicals on soil biota populations and processes in the field and thus can be used independently. The 'safe' concentrations for site assessment or remediation are determined by comparing measurement endpoints in treated and control microcosms using appropriate statistical tools. The highest concentration treatment with no effect (no statistically significant difference from control) can be regarded as the microcosm NOEC value for that measurement endpoint. The measurement endpoints with the lowest NOEC value can become the 'safe' concentration for that site. The use of summary indices such as the hazardous concentration HC_5 (Kooijman 1987, Van Straalen and Denneman 1989), defined as the concentration at which 95% of the effect parameters do not exceed the microcosm NOEC value, should be avoided. Their use in this extrapolation method would violate the underlying assumption of independence of the population sample from other measurement endpoints in the model system. The challenge for future research is to develop the best statistical tools that lead from direct and indirect measurement endpoints in model ecosystems to assessing overall responses of soil ecosystems to site chemical contamination.

4.6 SUMMARY

The use of multispecies assays such as integrated soil microcosm and terrestrial model ecosystem in assessing contaminated sites offers holistic tools for risk assessment by combining measurements of soil organism populations, soil processes, plant growth, bioaccumulation and persistence of contaminants into a single testing system. Both ISM and TME approaches offer a suite of organismal and process level measurement endpoints at different levels of biological and ecological organization. They can provide a much broader understanding of mechanisms by which soil contamination can affect the structure and function of soil ecosystem. Structural and functional measurement endpoints of microcosms should be qualitatively validated and quantitatively calibrated relative to analogous field conditions. This would allow the determination of the extent to which the data from microcosms are predictive of effects observed in field studies. The added effort required in conducting microcosm studies compared to single species tests will be beneficial if any community and ecosystem level detrimental effects are identified.

REFERENCES

Bogomolov DM, Chen S-K, Parmelee RW, Subler S and Edwards CA (1996) An ecosystem approach to soil toxicity testing: a study of copper contamination in laboratory microcosms. *Applied Soil Ecology*, **4**, 95-105.

Checkai R, Wentsel R and Phillips T (1993a) Controlled environment soil-core microcosm unit for investigating fate, migration, and transformation of chemicals in soils. *Journal of Soil Contamination*, **2**(3), 229-243.

Checkai R, Major M, Nwanguma R, Amos J, Philips C, Wenstel R and Sadusky M (1993b) *Transport and fate of nitroaromatic and nitramine explosives in soils from open burning/open detonation operations: Milan army munition plant (MAAP)*. U.S. Army Edgewood Research, Development and Engineering Center. ERDEC-TR-136. Aberdeen Proving Ground, Maryland, USA, p. 238.

Cheng W and Coleman DC (1989) A simple method for measuring CO_2 in a continuous airflow system: modifications to the substrate-induced respiration technique. *Soil Biology and Biochemistry*, **21**, 385-388.

Drake JA, Huxel GR and Hewitt CL (1996) Microcosms as models for generating and testing community theory. *Ecology*, **77**, 670-677.

Edwards CA (1991) The assessment of populations of soil-inhabiting invertebrates. In *Modern Techniques in Soil Ecology*, Crossley Jr. DA, Coleman DC, Hendrix PF, Cheng W, Wright DH, Beare MH and Edwards CA (eds), Elsevier Science publishers, New York, pp. 145-176.

Edwards CA and Bohlen PJ (1995) The effects of contaminants on the structure and function of soil communities. *Acta Zoologica Fennica*, **196**, 48-49.

Edwards CA, Knacker TT, Pokarzhevskii AA, Subler S and Parmelee R (1997) Use of soil microcosms in assessing the effects of pesticides on soil ecosystems. In *Proceedings of the International Symposium on Environmental Behavior of Crop Protection Chemicals*. IAEA-SM-343/3, International Atomic Energy Agency, Vienna, pp. 435-352.

Förster B, Eder M, Morgan E and Knacker T (1996) A microcosm study of the effects of chemical stress, earthworms and microorganisms and their interactions upon litter decomposition. *European Journal of Soil Biology*, **32**(1), 25-33.

Fraser LH and Keddy P (1997) The role of experimental microcosms in ecological research. *Trends in Ecology and Evolution*, **12**(12), 478-481.

Giesy Jr. JP and Odum EP (1980) Microcosmology: introductory comments. In *Microcosms in Ecological Research*, Giesy Jr. JP (ed), Technical Information Center, US Department of Energy, pp. 1-13.

Ives AR, Foufopoulos J, Klopfer ED, Klug JL and Palmer TM (1996) Bottle or big-scale studies: how do we do ecology? *Ecology*, **77**, 681-685.

Jaffee BA (1996) Soil microcosms and the population biology of nematophagous fungi. *Ecology*, **77**, 690-693.

Knacker T and Römbke J (1997) Terrestrische Mikrokosmen. *Environmental Science and Pollution Research*, **9**(4), 219-222.

Kooijman SALM (1987) A safety factor for LC_{50} values allowing for differences in sensitivity among species. *Water Resources*, **21**, 269-276.

Kuperman R and Carreiro M (1997) Soil heavy metal concentrations, microbial biomass and enzyme activities in a contaminated grassland ecosystem. *Soil Biology and Biochemistry*, **29**(2), 179-190.

Kuperman R, Williams G and Parmelee R (1998) Spatial variability in the soil food webs in a contaminated grassland ecosystem. *Applied Soil Ecology*, **9**, 509-514.

Lawton JH (1996) The ecotron facility at Silwood park: the value of 'big bottle' experiments. *Ecology*, **77**, 665-669.

Moore JC, DeRuiter PC, Hunt HW, Coleman DC and Freckman DW (1996) Microcosms and soil ecology: critical linkages between field studies and modeling food webs. *Ecology*, **77**, 694-705.

Morgan E and Knacker T (1994) The role of laboratory terrestrial model ecosystems in the testing of potentially harmful substances (Mini Review). *Ecotoxicology*, **3**, 213-233.

Parmelee RW, Wentsel RS and Phillips CT (1993) Soil microcosm for testing the effects of chemical pollutants on soil fauna communities and trophic structure. *Environmental Toxicology and Chemistry*, **12**, 1477-1486.

Pedersen F and Samsoe-Petersen L (1994) *Discussion Paper Regarding Guidance for Terrestrial Effect Assessment*. Water Quality Institute, Horsholm, Germany.

Römbke J, Bauer C and Marschner A (1996) Hazard assessment of chemicals in soil. Proposed ecotoxicological test strategy. *Environmental Science and Pollution Research*, **3**, 78-82.

Sheppard S (1997) Toxicity testing using microcosms. In *Soil Ecotoxicology*, Tarradellas J, Bitton G and Rossel D (eds), CRC Press, Boca Raton, FL, pp. 345-373.

Sinsabaugh RL and Linkins AE (1990) Enzymic and chemical analysis of particulate organic matter from a boreal river. *Freshwater Biology*, **23**, 301-309.

Smith MS, Rice CW and Paul EA (1989) Metabolism of labeled organic nitrogen in soil: regulation by inorganic nitrogen. *Soil Science Society of America Journal*, **53**, 768-773.

UBA (1994) *UBA Workshop on Terrestrial Model Ecosystems*. Texte 54/94, Umweltbundesamt, Berlin, p. 89.

Van Straalen NM and Denneman CAJ (1989) Ecotoxicological evaluation of soil quality criteria. *Ecotoxicology and Environmental Safety*, **18**, 241-251.

Verhoef HA (1996) The role of soil microcosms in the study of ecosystem processes. *Ecology*, **77**, 685-690.

5

Genotoxicity Analysis of Contaminated Environmental Media

SHANNON S. GARCIA AND KIRBY C. DONNELLY

Department of Veterinary Anatomy and Public Health, Texas A&M University, College Station, TX, USA

5.1 INTRODUCTION

The release of genotoxic chemicals to the environment represents a potential threat to human and ecological health. Examples of genotoxic chemicals include combustion by-products, such as benzo(*a*)pyrene and other polycyclic aromatic hydrocarbons (PAHs); industrial chemicals, such as vinyl chloride; and naturally occurring compounds, such as aflatoxin (McCann and Ames 1977, Miller and Miller 1981). Most often, these chemicals are contained in complex mixtures, which are released into the environment. Examples of complex mixtures include cigarette smoke, urban air, automobile exhaust, wood smoke, coal tar, pyrolyzed amino acids from the cooking of protein containing foods, hazardous industrial wastes, and agricultural runoff (DeMarini 1991). The genotoxicity analysis of contaminated environmental media requires a careful selection of sampling, extraction and analytical procedures. Although chemical analysis is often used as a first tier of environmental analyses, short-term tests (STTs), including mutation assays, provide a valuable tool for investigating the genotoxicity of complex environmental mixtures.

In general, these STTs are designed as a tool for the detection of mutagens. Genotoxins, including mutagens, are compounds capable of altering the genetic code (DNA). Mutagenesis includes the induction of DNA damage and alteration of the genetic code ranging from base-pair changes to changes in chromosome structure and number (Hoffman 1996). Increases in the mutation rate in human germ cells, such as eggs or sperm, can lead to an increased incidence of genetic disease, while mutations in somatic cells is one of the causes of cancer. Many carcinogens have been found to be mutagenic. While the exact mechanisms of carcinogenesis have not been fully determined, it is generally seen to be a multistep process. Mutagenesis often plays a role in this process. The objective

Environmental Analysis of Contaminated Sites. Edited by G. I. Sunahara, A. Y. Renoux, C. Thellen, C. L. Gaudet and A. Pilon

of this chapter is to briefly detail a few of the more commonly used STTs and to describe how these tests may be useful in determining the genotoxicity of contaminated environmental media.

5.2 USE OF STTs AS A SCREEN FOR RODENT CARCINOGENS

In vivo rodent carcinogenicity tests are time-consuming and expensive. Lave and Omenn (1986) estimated the cost of testing one chemical in a 2–4-year long-term animal bioassay to be greater than US $1 million. The National Cancer Institute (NCI) protocol for rodent carcinogenicity studies requires long-term (2 year) exposure of both sexes of two species of rodents (usually rats and mice) at high doses with 50 animals per treatment group. While these tests are necessary, it is simply not possible to test every chemical in long-term carcinogenicity tests. However, fast and relatively inexpensive tests for mutagenic activity have been developed. It may be possible, therefore, to screen chemicals in STTs and then further test those compounds exhibiting mutagenic activity in the rodent carcinogenicity tests. Tests capable of quickly and reliably detecting mutagens would be useful to those responsible for reducing exposures to genotoxic compounds and, consequently, in reducing the risk of cancer or genetic disorders. Over the past 20 years, more than 200 assays to detect mutagens have been developed (Hoffman 1996). Both *in vitro* and *in vivo* tests have been designed, with systems ranging in complexity from microbial assays to mammalian cell lines to live, whole-animal assays. By using a battery of tests, a broad range of genetic damage including genetic mutation, chromosome damage and other endpoints indicative of mutagen exposure, such as unscheduled DNA synthesis, DNA strand breaks and sister chromatid exchange, can be detected.

Several of these tests have been extensively reviewed by the US Environmental Protection Agency (USEPA) Gene-Tox program (USEPA 1980, 1996,1997). The National Toxicology Program (NTP) has conducted extensive reviews of STTs to validate the selection of these assays (Tennant *et al.* 1987, Tennant and Zeiger 1993). Validation of genotoxicity assays involves determining both an assay's sensitivity and specificity. For the purpose of assay validation, tests are conducted using a certain number of known rodent carcinogens and non-carcinogens. Sensitivity is defined as the percentage of chemicals correctly identified as carcinogens, whereas specificity is the percentage of chemicals correctly identified as non-carcinogenic. The most accurate tests would be both sensitive and specific. Tennant *et al.* (1987) evaluated four STTs commonly used to predict carcinogenicity: the Ames *Salmonella*/mammalian microsome assay; assays using cultured Chinese hamster ovary (CHO) cells to detect chromosome aberrations and sister chromatid induction; and the mouse lymphoma cell mutagenesis assay. In these studies, the *Salmonella* assay was found to have the highest specificity (91%) and positive predictivity (proportion of positives that are carcinogens, 89%) but the lowest sensitivity (48%). The cell culture assays for sister chromatid induction and the mouse lymphoma assay had the highest sensitivities (69% and 72%, respectively) and the lowest positive predictivities (Tennant and Zeiger 1993). For the most part, the assays detected the same carcinogens, with the cell culture assays giving more false positives. This indicates that a battery of tests may be no more effective than the *Salmonella* assay alone at detecting rodent carcinogens. Zeiger (1998) determined that combinations of *in vitro* genetic toxicity tests (*Salmonella* assay, chromosome aberration assay, mouse lymphoma assay) did not improve on the predictivity of the individual tests with regards to carcinogenicity. Moreover, Zeiger recommended that if a chemical was mutagenic in *Salmonella*, a positive result in any other *in vitro* or *in vivo* test did not increase the probability that the chemical was carcinogenic; likewise, a negative in any other test did not diminish the implication of the positive response in *Salmonella*. However, it is known that different assays may be more accurate with regard to certain classes of chemicals (Claxton *et al.* 1988). For instance, *Salmonella* appears able to detect carcinogens accurately from the following classes: polycyclic aromatic hydrocarbons, aromatic amino/nitro-type compounds, natural electrophiles (including reactive halogens) and minor groups of structurally alerting chemicals (Claxton *et al.* 1988, Ashby 1992). DeMarini *et al.* (1990) and DeMarini and Brooks (1992) have used the *E. coli* prophage induction assay to detect genotoxic chlorinated compounds. Thus, much like the detector on a gas chromatograph is selected to detect a certain class of compounds, STTs may be selected for analysis of complex mixtures based on knowledge of the characteristics of the contaminants of concern.

5.3 STTs FOR GENOTOXICITY

The *Salmonella* assay (Ames *et al.* 1975, Maron and Ames 1983) is one of the most widely used short-term tests for mutagenicity. The *Salmonella* assay was one of the first tests to show a correlation between mutagenicity and carcinogenicity (Li and Loretz 1991). A *Salmonella typhimurium* strain auxotrophic for histidine is the test organism, and a compound's mutagenicity is determined by its ability to cause reverse mutations to wild-type. Different *Salmonella* strains have been developed which detect base-pair substitution or frame-shift mutations. Other microbial mutagenicity tests used extensively include the *E. coli* WP2 tryptophan reversion assay and tests using the yeast, *S. cerevisiae*. Endpoints detected by the different *S. cerevisiae* strains used in mutagenicity testing include forward and reverse mutations as well as mitotic recombination and aneuploidy (Li and Loretz 1991).

Due to the differences in complexity between microbial and human genes, it is unlikely that microbial tests alone can accurately predict human carcinogenesis. Essentially, every mechanism capable of causing genotoxic mutations can be investigated using mammalian cell lines. Endpoints for these types of assays include gene mutation, sister chromatid exchange (SCE), chromosome aberration and unscheduled DNA synthesis. Mutations of the hypoxanthine-guanine phosphoribosyl transferase (HGPRT) and thymidine kinase (TK) genes can be measured in mutation assays (Li and Loretz 1991, Hoffman 1996). The HGPRT assay is often performed using CHO cells, whereas the TK assay uses the L5178Y mouse lymphoma cell line (Li and Loretz 1991). Both assays monitor forward mutations where mutation at the HGPRT or TK gene loci results in the inability of the cell to synthesize the respective protein. In short, wild-type cells are exposed to a potential mutagen; the cells are cultured and any residual HGPRT or TK protein is depleted and a selective agent is applied. Wild-type cells with HGPRT or TK will form toxic metabolites and die; however, mutants deficient in HGPRT or TK will survive and grow to form colonies, which can be quantified.

The CHO cell line can also be used to study chromosome aberrations (Li and Loretz 1991). Cells are treated with a mutagenic compound and arrested during metaphase. The metaphase cells are collected and examined; data collected may include cytotoxicity (reduction in proportion of cells in division), number of chromosomes, chromatid and chromosome exchanges, and endoreduplication (doubling without separation) (Li and Loretz 1991). This assay can also be performed *in vivo* by treating an animal (commonly rodents) and then collecting cells for analysis. Because bone marrow possesses a large number of rapidly dividing cells, it is frequently used. Advantages of whole-animal experiments over the *in vitro* assays include mammalian metabolism, DNA repair and pharmacodynamics (Hoffman 1996).

In vivo cytogenetic assays, such as the mouse micronucleus assay, have been proposed as complements to the *Salmonella* assay (Gatehouse and Tweats 1988, Shelby 1988). The mouse micronucleus assay is a widely used assay that measures chromosome fragments as indicators of DNA damage. Mature erythrocytes do not normally contain DNA since the nucleus is expelled during development. Immature erythrocytes with chromosome aberrations can lead to the development of mature erythrocytes containing micronuclei. Micronuclei contain chromatin and represent chromosomes or chromosome fragments not incorporated in the nucleus during mitosis. During anaphase, these lagging chromosomal fragments may not be incorporated in the daughter nuclei. In this assay, erythrocytes from treated mouse bone marrow (or peripheral blood) are isolated and micronuclei are counted. The micronucleus assay has the advantage of being easier to perform and less time-consuming than metaphase analysis.

5.4 COMPLEX MIXTURES AND GENOTOXICITY TESTING

Humans are usually exposed to mixtures of chemicals rather than single chemicals. Complex mixtures of chemicals are present in food, occupational settings and the environment. Complex mixtures may be composed of tens, hundreds, even thousands of chemicals. Usually, only a fraction of the components of a complex mixture are identified and even fewer are quantifiable. For many of these complex mixtures, the data on exposure and toxicity is extremely limited. Although complex mixtures constitute the majority of human exposure to hazardous chemicals, the study of single chemicals still consumes the majority of toxicology resources (Cassee *et al.* 1998). Many researchers have been reluctant to tackle the problem of complex mixture toxicity due to the complicated experimental designs necessary to determine effects and the high cost of the analytical chemistry. Mechanistic studies are complicated due to the possibility of interactions between components. Yang and Rauckman (1987) provide an example that illustrates the difficulty encountered when trying to assess complex mixtures. Based on their estimates, a 25-chemical mixture has '$2^n - 1$' or 33 554 431 combinations, and the cost to perform systematic toxicity testing according to NTP subchronic protocols using only one species could be as much as US $3 trillion. Due to their simplicity, short duration, low cost and small sample size requirements, short-term tests for genotoxicity are being investigated as tools for assessing environmental and workplace exposures and monitoring treatment processes for hazardous waste.

Some of the conclusions reached concerning single chemicals and short-term genotoxicity tests may be applicable to complex mixtures. An analysis of data found in the IARC Monograph Series proposed recommendations for complex mixtures that are similar to those for single agents. Mixtures with unknown carcinogenic potential showing sufficient evidence of activity in short-term genotoxicity assays should be considered a hazard to human health (Bartsch and Malaveille 1990). Bartsch and Malaveille (1990) also suggest that a mixture be classified as genotoxic if it produces a positive response in a test for DNA damage, mutations and chromosomal effects. One of the tests must involve mammalian cells (*in vivo* or *in vitro*) and include two of the three DNA damage endpoints previously listed. Mixtures lacking sufficient evidence would still need testing in a rodent bioassay for carcinogenicity, as there are no validated STTs for non-genotoxic carcinogens.

One approach to determine mixture toxicity is to treat a given mixture as a single chemical. One advantage of using this approach is that the mixture being tested is comparable to the mixture of chemicals associated with human exposures and not a partial mixture of selected components. However, this method does not address interactions between components and raises the question of whether extrapolations can be made from high (experimentally effective) doses to low (environmental and realistic) doses. At high concentrations, the contributions of the different compounds towards the toxicological effect may be different proportionally from a low dose of the same mixture (Cassee *et al.* 1998). There is a great deal of uncertainty regarding the interactions of non-mutagenic components of a mixture with the genotoxic components. These non-mutagenic components could be toxic to bacteria and liver homogenates, and indirect mutagens may compete for important enzyme sites (Anderson 1990). Another concern is the production of artifacts during collection or extraction, or alterations in the relative concentration of components during the solvent reduction process. These concerns are best addressed by following validated and standardized protocols. Another important issue in genotoxicity analyses is that of STT sensitivity. Complex mixtures often contain several mutagens, and the most active mutagen in an STT may not be the most important chemical from a human health viewpoint. In addition, health effects may be due to interactions between several compounds present in the mixture (Anderson 1990).

Bioassay-directed fractionation of complex mixtures is another approach used to assess the genotoxicity of complex mixtures. In this process, the biologically active components in a mixture are identified by combining fractionation, biological testing and chemical analysis (Claxton *et al.* 1996). Mixtures are separated into component fractions based on chemical properties. These fractions can then be tested for activity in biological assays. The fractionation and chemical identification processes are continued until the toxic components are identified. One advantage of this procedure is that it may identify toxic components whose activity has been masked by relatively inert compounds present in the mixture. Brooks *et al.* (1998) recently used bioassay-directed fractionation to evaluate soils contaminated with PAHs derived from creosote before and after various bioremediation treatments. Azaarenes, which are mutagens, were found in the mutagenic fractions. Donnelly *et al.* (1987c) separated extracts from soils amended with wood-preserving waste into acid, base and neutral fractions. In this study, the base fractions were found to have the maximum levels of mutagenicity.

In a component-based approach, the toxicity of each component is used to determine the toxicity of the whole mixture. This approach is often used in assessing mixtures at hazardous waste sites. The toxicity of a mixture is determined by estimating the toxicity of each component. A hazard quotient (HQ; where HQ = monitored level:allowable level) is calculated, then a hazard index (HI) is determined by summing all the HQs for the mixture. This approach does not take into account component interactions on mixture toxicity (Mumtaz *et al.* 1993). This approach assumes additivity between compounds and therefore is only valid in cases where the compounds in a mixture induce similar toxic effects by the same mechanism. Often, it is difficult to obtain mechanistic information for each of the mixture components.

5.5 COLLECTION AND EXTRACTION OF COMPLEX MIXTURES

A critical first step in the analysis of contaminated environmental media is the collection of a representative sample. Contaminated media, almost by definition, exhibit extreme heterogeneity. A sampling plan should be developed to determine the media to be sampled, procedures to be employed, and the volume, number and frequency of samples that are needed. Statistical considerations should also be incorporated into a sampling plan to provide the most accurate assessment of a specific environment. For specific media at a specific site, a variety of factors will influence the number and type of samples to be collected. The design of the overall sampling plan will be influenced by site characteristics including types of contaminants present, age of the site and affected media. Factors such as total budget for sampling and analysis, and time restraints may also influence the design of a sampling plan.

The objectives of sampling will have a significant effect on the frequency of sampling, and also on the number of individual samples to be collected. For example, the number and frequency of samples to be collected for a study designed to characterize the vertical and lateral extent of a release will be very different from sampling designed only to determine if genotoxic chemicals have been released into the environment. A common use of bioassays is to monitor the efficacy of remedial procedures. In these cases, samples are often collected over different time intervals, or before and after a specific treatment method. In monitoring remediation, the chemical composition and concentration of a complex mixture are likely to change over time due to volatilization, biodegradation and other physical or chemical processes. Thus, at some sites, both spatial and temporal considerations will influence the sampling protocol.

The types of environmental media which may be subjected to genotoxicity analysis include air, soils, surface and groundwater, effluents and leachates. The equipment and procedures to be used at any given site will be dependent on the media being sampled and the analytical method (chemical or biological) to be used. Methods for the collection and extraction of various environmental media are described by the USEPA (1980). Methods for extraction and chemical analysis as well as sample preservatives and holding times can also be found in SW846 (USEPA 1997). Detailed methods for extraction and genotoxicity analysis of soils and solid waste have been described previously (Donnelly *et al.* 1987a,b, Collie and Donnelly 1997). Useful information regarding the development of sampling plans may be found in guides by Mason (1992) and USEPA (1996). Safety of personnel should also be an important concern whenever environmental media is sampled for genotoxicity analysis. Following collection, samples should be preserved using appropriate techniques, transported to the laboratory and extracted as soon as possible. In some cases, it may be necessary to homogenize the sample at the laboratory to obtain a more representative sample for analysis.

5.6 CASE STUDIES

The following text describes results obtained from the genotoxicity analysis of three very different sites. The objectives of sampling and, thus, the sampling plan were different for each of these sites. Information will be provided to briefly describe the advantages and limitations of genotoxicity analysis at these sites, as well as a discussion of the interpretation of the results from these analyses. These research studies were conducted by the Texas A&M University

Superfund Basic Research Program funded through the National Institute of Environmental Health Sciences.

5.6.1 GENERAL ENVIRONMENTAL ASSESSMENT

The purpose of sampling at the first site was to determine if genotoxic chemicals were released into the environment. Due to the size of the area being investigated, sampling was designed only for detection and not for characterization. Sampling was conducted using a biased sampling design. Samples were collected from areas meeting one of three criteria: (1) they were downgradient of potential sources of a release, (2) they appeared to have been affected by a contaminant release (as indicated by stained soil or stressed vegetation), or (3) they were in close proximity to a potentially exposed population. The major limitation of this type of sampling design is that it may miss the most severely contaminated areas, or could miss contaminated areas altogether. The major advantage biased sampling offers is to provide a rapid preliminary evaluation of a large area.

Based on results from an earlier reconnaissance survey of sites in Azerbaijan, nine locations were selected for sample collection (data are shown in Figure 5.1a–c for 10 samples from these nine areas plus one background sample from a national park in the United States). Elevated levels of genotoxicity were detected in samples collected from the first area (stream bed adjacent to

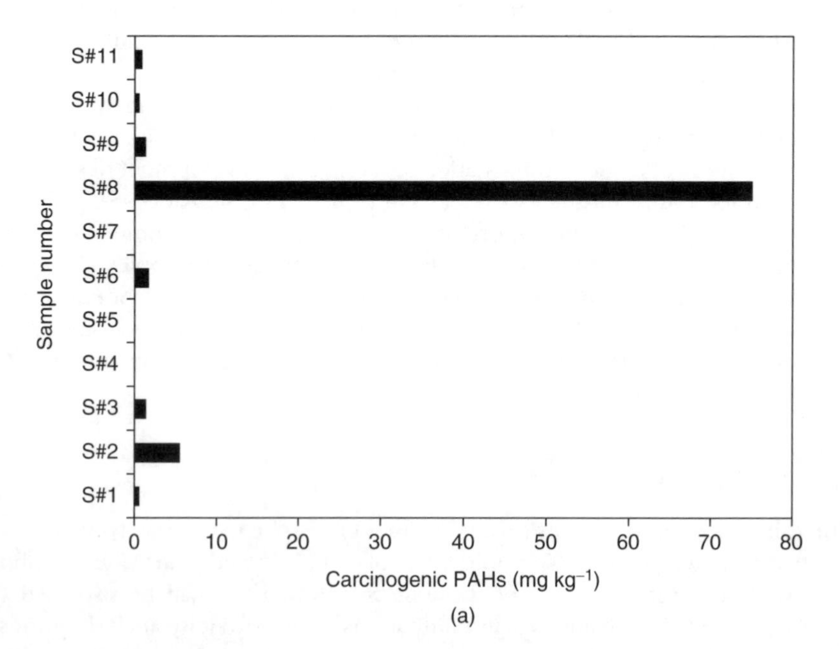

Figure 5.1 (a) Carcinogenic PAHs detected in environmental samples from Azerbaijan and control site (S#11).

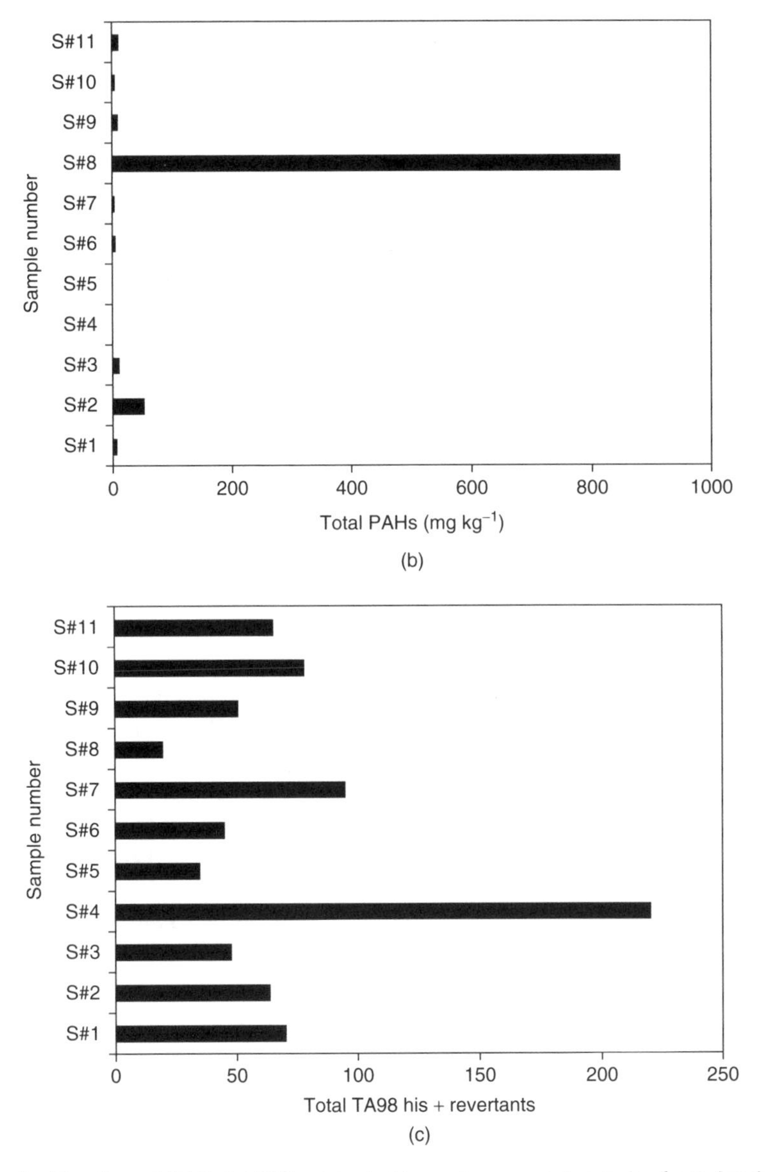

Figure 1 (*Continued*) (b) Total PAHs detected in environmental samples from Azerbaijan and control site (S#11) (c) Genotoxic potential of extracts of environmental samples from Azerbaijan and control site (S#11).

a petrochemical facility) during the reconnaissance survey and consequently multiple samples (samples S#1–3) were collected during the second sampling event. Other samples collected included a surface soil from an area adjacent to a petrochemical plant (sample S#4), a surface soil from an oil production area near a large apartment building (sample S#8), and two soils collected near apartments in a highly populated area of the city (samples S#9 and S#10). Sample S#11 was the background soil sample from the national park. Samples were extracted with hexane:acetone according to USEPA (1997) Method 3541.

Only two samples (S#2 and S#8) appeared to have elevated levels of PAHs (Figure 5.1b) or carcinogenic PAHs (Figure 5.1a). Both soil samples S#4 and S#8 were visibly stained soils. Sample S#8 was collected adjacent to a producing well and was visibly contaminated with oil. Samples S#4, 5, 6, 7 and 10 had lower levels of PAHs than the extract of the sample collected from a pristine area. However, results from genotoxicity testing in the *Salmonella* mutagenicity assay (procedure of Maron and Ames 1983) provide a very different ranking of the sites when compared to chemical analysis. Only two of the samples (S#1 and S#4) induced a positive genotoxic response (a doubling of revertants at two consecutive dose levels). Total TA98 revertants for each of the samples is shown in Figure 5.1 c. In addition, the extract of sample S#4 was clearly the most genotoxic of all samples collected. Sample S#4 was collected from a small pile of darkly stained material outside an abandoned petrochemical facility. A total of 0.1 mg kg^{-1} carcinogenic PAHs was detected in this sample, and benzo(*a*)pyrene concentrations were below the method detection limit. It is notable that the sample collected in this area induced a genotoxic response with and without metabolic activation. This is not generally observed with PAHs and suggests that the genotoxic chemicals may be products of PAH degradation or other partially oxidized compounds. Additional fractionation and chemical characterization would be useful to determine the identity of the genotoxic chemicals in these samples. The results from this general environmental assessment demonstrate the utility of a combined biological and chemical testing approach. Although chemical analysis identified the samples with the highest level of carcinogenic PAHs (cPAHs), biological tests detected elevated levels of genotoxic chemicals in a sample with minimal levels of cPAHs.

5.6.2 AREA-SPECIFIC ASSESSMENT

A series of samples was collected from sediments of a freshwater lake located adjacent to an abandoned manufactured gas plant (Randerath *et al.* 1999). During operation, manufactured gas plant residue (MGPR) was buried at several locations in and around the plant property. Routine monitoring data revealed that contaminants from the buried MGPR had migrated under the site to the lake sediments. The objective of sampling at this location was to collect sediment samples from several locations adjacent to the site in order to estimate the lateral extent of the release. A total of six samples (202, 204, 206, 208–210) were collected at varying distances from the shoreline. An increasing sample number indicates an increasing distance from the shoreline. Samples were extracted using a Tecator Soxtec with methylene chloride and methanol. The sediment extracts were divided into aliquots for chemical analysis by gas chromatography/mass spectrometry (GC/MS) and bioassay using the *Salmonella* microsome assay as described in the previous section and ^{32}P post-labeling assay for DNA adducts both *in vitro* and *in vivo*. *In vitro* DNA treatments used DNA isolated from rat lung and followed the procedure of Randerath *et al.* (1992) with modifications. *In vivo* treatments using female ICR mice were performed as described by Reddy and Randerath (1986). All of the samples were analyzed in the microbial genotoxicity assay, whereas only samples 202, 206 and 209 were evaluated using GC/MS and DNA post-labeling.

Sample 202 was collected closest to the source of contamination and had the highest concentration of benzo(*a*)pyrene (B(a)P) and B(a)P-equivalents (based on the toxic equivalency factors (TEFs) as described by Nisbet and LaGoy 1992). Both B(a)P and B(a)P-equivalent concentrations were lower in sample 206 and appreciably lower in sample 209 when compared to sample 202 (Figure 5.2). Chemical analysis revealed that considerable levels of toxic and carcinogenic contaminants persisted in the adjacent lake sediments. Sample 202 gave a weak genotoxic response in the microbial assay, which may indicate a cytotoxic or inhibitory effect. Samples 204 and 206 appeared to induce the maximum genotoxic response, and induced a doubling of revertant colonies at a dose of 0.05 mg plate^{-1} (Table 5.1). Samples 208, 209 and 210 also induced a genotoxic response, although a doubling of *Salmonella* revertants was not observed until the dose reached 0.1 mg plate^{-1}. All of the samples required

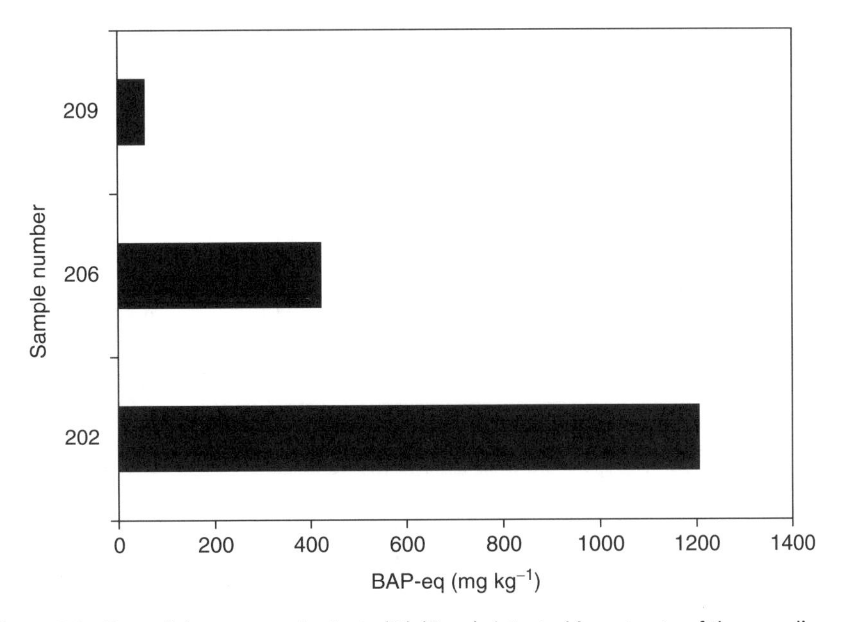

Figure 5.2 Benzo(*a*)pyrene equivalents (B(a)P-eq) detected in extracts of three sediment samples.

Table 5.1 Bacterial mutagenicity, as measured with *S. typhimurium* TA98, of methylene chloride extracts of selected sediment samples from MGPR site[a]. Data expressed as mean±standard deviation

Sample no.	DOSE (mg extract per plate)	Mean TA98 his + revertants	
		−S9	+S9
202	0	24±5	35±11
	0.05	19±5	63±38
	0.1	28±4	96±69*
	0.25	22±4	94±71*
	0.5	28±3	83±56*
	1.0	39±6	84±57*
204	0	24±5	35±11
	0.05	23±5	121±6*
	0.1	27±8	168±11*
	0.25	26±8	181±13*
	0.5	29±4	168±21*
	1.0	32±14	198±38*
206	0	28±4	31±5
	0.05	25±2	124±23*
	0.1	26±7	197±34*
	0.25	27±4	224±8*
	0.5	29±5	243±10*
	1.0	31±13	221±26*
208	0	25±2	31±5
	0.05	21±4	52±8
	0.1	26±1	84±9*
	0.25	24±5	127±16*
	0.5	26±2	144±8*
	1.0	32±5	177±19*
209	0	28±4	31±5
	0.05	19±4	37±1
	0.1	25±1	74±9*
	0.25	20±6	100±9*
	0.5	23±5	123±20*
	1.0	28±3	165±42*
210	0	28±4	31±5
	0.05	18±3	40±5
	0.1	26±7	71±20*
	0.25	21±4	103±13*
	0.5	27±1	127±4*
	1.0	24±5	170±24*

[a]Each sample was tested on duplicate plates in two independent experiments.
*Indicates a positive result (or a doubling of revertants above the solvent (DMS) control). Samples were tested without (−S9) and with (+S9). S9 is a metabolic activation cocktail made from rat liver homogenate used to mimic mammalian metabolism.

metabolic activation (+S9) to produce a genotoxic response. The requirement of activation is a standard characteristic of PAHs.

DNA post-labeling showed decreasing levels of adducts with increasing distance from the source of pollution (Randerath *et al.* 1999). While the post-labeling data do not directly correlate with microbial genotoxicity data, the results suggest the response of sample 202 may have been masked by cytotoxic chemicals in the extract. The results indicate that genotoxic chemicals were present in all sediment extracts. The results from both the post-labeling and the microbial genotoxicity assay suggest that neither the benzo(*a*)pyrene concentration nor the B(a)P-equivalents provide an accurate indication of genotoxic potential.

5.6.3 ASSESSMENT OF REMEDIAL TECHNOLOGIES

Sampling and analysis conducted at the third area were designed to monitor the utility of bioremediation for the detoxification of a wood-preserving waste-contaminated soil. Previously, this Superfund site had been a railroad tie treating plant. Creosote and zinc compounds were used in the preservation process. Waste management practices resulted in soil and groundwater contamination. Efforts to remediate the contaminated soil involved bioremediation using a land treatment unit. During the degradation process, the components of a complex mixture undergo a broad range of oxidation and/or reduction processes that alter their structure. These alterations may change the solubility of chemicals and may result in an increase or decrease in genotoxic potential (Donnelly *et al.* 1987c, Davol *et al.* 1989, Aprill *et al.* 1990). A major advantage of STTs is their ability to detect changes in genotoxicity that may occur as a result of the degradation process. Test systems used to monitor remediation of complex mixtures should be capable of detecting a range of genotoxic chemicals and a range of toxic or genotoxic endpoints. For example, the *Salmonella* /microsome assay detects point mutations and is sensitive to PAHs (McCann and Ames 1977), whereas the *E. coli* prophage induction assay, which detects DNA strand breaks, is sensitive to chlorinated phenols (DeMarini *et al.* 1990). Additional limitations of most STTs are their inability to detect metals and insensitivity to genotoxic compounds when present in mixtures with high concentrations of cytotoxic compounds (this may explain the negative response observed for sample 202 above).

A series of 12 soil samples was collected from 12 cells in a land treatment unit and extracted as described in Section 5.6.2. For comparison purposes, two samples of untreated waste were also obtained. The results presented in Figure 5.3 compare the weighted activity determined from results in the *Salmonella* mutagenicity assay for each of these soil extracts. The weighted activity accounts for both the specific mutagenic activity and the residue weight of the soil, and provides a more quantitative measure of potential soil genotoxicity. For the soils collected in the fall of 1995, the maximum weighted activity was 600 revertants per gram of soil; the extracts of soils from three cells failed to induce a genotoxic response. As described for the previous site (Section 5.6.2), this may reflect a masking of genotoxicity due to the presence of elevated concentrations of cytotoxic low-molecular-weight PAHs. Overall, the weighted activity of the extracts from soils collected in the

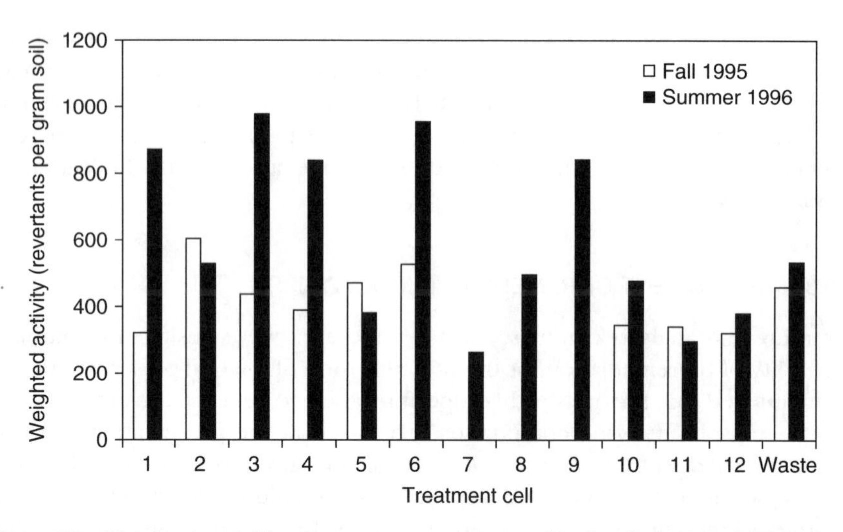

Figure 5.3 Weighted activities (revertants per gram soil) of soil extracts from a land treatment cell and waste pile.

spring of 1996 appeared to be greater than was observed in soils collected in 1995. The minimum response for samples collected in the spring of 1996 was approximately 300 revertants per gram, whereas five of 12 samples induced a response that was greater than 600 revertants per gram. These data suggest that although degradation has reduced the concentration of carcinogenic PAHs, the relative genotoxicity is unchanged or possibly increased.

A more detailed comparison of two samples from this facility is provided in Table 5.2. The data compares two samples, a soil extract from Cell 7 collected in the spring of 1996 after approximately 9 months of treatment (it should be noted that these include winter months during which degradation was limited), and an extract of an untreated waste sample. The results from chemical

Table 5.2 Selected properties of soil from Cell 7 and untreated waste

Concentration (mg kg^{-1})	Cell 7	Waste
Fluorene	0.32	5.06
Anthracene	6.99	9.91
Pyrene	65.73	60.34
Benzo(*a*)pyrene	10.62	10.61
Benzo(*g,h,l*)perylene	4.3	3.57
Specific activity (net revertants)	110	82
Weighted activity (revertants per gram dry soil)	978	541

Specific activity = total TA98 revertants − solvent control.
Weighted activity = TA98 revertants per gram dry soil.

analysis indicate that the concentrations of low-molecular-weight PAHs were reduced appreciably, whereas the high-molecular-weight and genotoxic PAH concentrations were relatively unchanged during the initial treatment process. These data are comparable to the results of the genotoxicity analyses. The specific activity of soil from Cell 7 was 48 net revertants per milligram, whereas the extract of the waste induced 82 net revertants per milligram. These results suggest that biodegradation initially attacks low-molecular-weight (and non-genotoxic) PAHs. Consequently, the genotoxic potential of the residual chemicals in soils or groundwater may be equal to, or greater than, the genotoxic potential of the mixture before biodegradation.

5.7 CONCLUSIONS

The most common method of estimating the human health risk associated with a contaminated environment uses toxicity values for the contaminants of concern combined with an estimate of contaminant intake. Unfortunately, this approach may overlook unidentified components, components which do not have toxicity values, or potential component interactions. Using STTs to develop an estimate of risk overcomes each of these obstacles, but also has inherent limitations. Biological testing will always be associated with a high degree of variability. Thus, although two samples may induce a response that is significantly different numerically, it may be difficult to determine the biological significance of these values. In addition, many adverse effects are highly dose-dependent. While biological testing may provide the most accurate measurement of these effects, some interactions or potential adverse effects may be masked or not observed in certain dose ranges. For example, high concentrations of cytotoxic chemicals may mask the presence of genotoxic chemicals. Another limitation of STTs is the concern over false positives (genotoxic non-carcinogens) and false negatives (non-genotoxic carcinogens). This is one reason why chemical analysis must still be a component of any risk assessment. It is possible that as new tests are developed, a battery of biological tests could be used to detect both genotoxic and epigenetic endpoints. However, a combined approach using both chemical analysis and STTs provides the most accurate information from which to assess human health risk.

5.8 FUTURE DIRECTIONS OF GENOTOXICITY TESTING

The ability of STTs to predict carcinogenic potential is a critical issue to the application of these procedures for the analysis of complex mixtures. It is obvious that bioassays, which detect genetic toxicity, will be incapable of detecting chemicals or mixtures that affect the carcinogenic process through epigenetic mechanism(s). However, as our knowledge of the mechanisms of cancer induction improve, additional test systems are being developed to detect different epigenetic endpoints. A testing protocol for any complex environmental mixture should be designed to detect a range of toxic endpoints, as well as a variety of different classes of chemicals and interactions. In order to design an effective test battery, it is critical to have some knowledge of the nature of the chemicals present in a complex mixture. Alternatively, evaluations of unknown mixtures for toxicity or potential carcinogenicity should employ a series of bioassays, which detect a range of genetic and non-genetic endpoints (Lewtas 1990). Perhaps in the future, as more STTs capable of detecting non-genotoxic carcinogens are developed, a better understanding of the risk posed by exposure to complex mixtures will be gained.

5.9 ACKNOWLEDGEMENTS

The authors express their sincere thanks to the following: Ling-yu He for help with soil extractions, Tom McDonald who provided GCMS analysis, and Shanna Collie, Stephanie East, Karl Markiewicz, Bill Reeves and Carol Swartz for assisting in the collection and analysis of samples discussed in Section 5.6. The research presented here was funded by EPA grant R825408 and NIEHS grants P42 ES04917 and P30 ES09106.

REFERENCES

Ames BN, McCann J and Yamasaki E (1975) Methods for detecting carcinogens and mutagens with the *Salmonella*/microsome mutagenicity tests. *Mutation Research*, **31**, 347-364.

Anderson D (1990) The use of short-term tests in detecting carcinogenicity of complex mixtures. In *Complex Mixtures and Cancer Risk*, Vainio H, Sorsa M and McMichael AJ (eds), International Agency for Research on Cancer, Lyon, France, pp. 89-100.

Aprill W, Sims RC, Sims JL and Matthews JE (1990) Assessing detoxification and degradation of wood preserving and petroleum wastes in contaminated soil. *Waste Management Research*, **8**, 45-65.

Ashby J (1992). Use of short-term tests in determining the genotoxicity or nongenotoxicity of chemicals. In *Mechanisms of Carcinogenesis in Risk Identification*, Vainio H, Magee PN, McGregor DB and McMichael AJ (eds), International Agency for Research on Cancer, Lyon, France, pp. 135-164.

Bartsch H and Malaveille C (1990) Screening assays for carcinogenic agents and mixtures: an appraisal based on data in the IARC Monograph Series. In *Complex Mixtures and Cancer Risk*, Vainio H, Sorsa M and McMichael AJ (eds), International Agency for Research on Cancer, Lyon, France, pp. 65-74.

Brooks LR, Hughes TJ, Claxton LD, Austern B, Brenner R and Kremer F (1998) Bioassay-directed fractionation of mutagens in bioremediated soils. *Environmental Health Perspectives*, **106**, 1435-1440.

Cassee FR, Groten JP, Van Bladeren PJ and Feron VJ (1998) Toxicological evaluation and risk assessment of chemical mixtures. *Critical Reviews in Toxicology*, **28**, 73-101.

Claxton LD, Stead AG and Walsh D (1988) An analysis by chemical class of *Salmonella* mutagenicity tests as predictors of animal carcinogenicity. *Mutation Research*, **205**, 197-225.

Claxton LD, Houk VS and George SE (1996) Integration of complex mixture toxicity and microbiological analyses for environmental remediation research. In *Ecotoxicity and Human Health*, deSerres FJ and Bloom AD (eds), CRC Press, Boca Raton, FL, pp. 87-122.

Collie SL and Donnelly KC (1997) Measurement of mutagenic activity in contaminated soils. In *Bioremediation Protocols*, Sheehan D (ed), Humana Press Inc., Totowa, NJ, pp. 127-151.

Davol P, Donnelly KC, Brown KW, Thomas JC, Estiri M and Jones DH (1989) Mutagenic potential of runoff water from soils amended with three hazardous industrial wastes. *Environmental Toxicology and Chemistry*, **8**, 189-200.

DeMarini DM (1991) Environmental mutagens/complex mixtures. In *Genetic Toxicology*, Li AP and Heflich RH (eds), CRC Press, Boca Raton, FL, pp. 285-302.

DeMarini DM and Brooks HG (1992) Induction of prophage lambda by chlorinated organics: detection of some single-species/single-site carcinogens. *Environmental and Molecular Mutagenesis*, **19**, 98-111.

DeMarini DM, Brooks HG and Parkes DG (1990) Induction of prophage lambda by chlorophenols. *Environmental and Molecular Mutagenesis*, **15**, 1-9.

Donnelly KC, Brown KW and Kampbell D (1987a) Chemical and biological characterization of hazardous industrial waste: prokaryotic bioassays and chemical analysis of a wood-preserving bottom-sediment waste. *Mutation Research*, **180**, 31-42.

Donnelly KC, Brown KW and Scott BR (1987b) Chemical and biological characterization of hazardous industrial waste: eukaryotic bioassay of a wood-preserving bottom sediment. *Mutation Research*, **180**, 43-53.

Donnelly KC, Davol P, Brown KW, Estiri M and Thomas JC (1987c) Mutagenic activity of two soils amended with a wood-preserving waste. *Environmental Science and Technology*, **21**, 57-64.

Gatehouse DG and Tweats DJ (1988) Further debate of testing strategies. *Mutagenesis*, **3**, 95-102.

Hoffman GR (1996) Genetic toxicology. In *Casarett and Doull's Toxicology: The Basic Science of Poisons*, Klaassen CD (ed), McGraw-Hill, New York, pp. 269-300.

Lave LB and Omenn GS (1986) Cost-effectiveness of short-term tests for carcinogenicity. *Nature*, **324**, 29-34.

Lewtas J (1990) Future directions in research on the genetic toxicology of complex mixtures. In *Genetic Toxicology of Complex Mixtures*, Waters MD, Daniel FB, Lewtas J, Moore MM and Nesnow S (eds), Plenum Press, New York, pp. 353-361.

Li AP and Loretz LJ (1991) Assays for genetic toxicity. In *Genetic Toxicology*, Li AP and Heflich RH (eds), CRC Press, Boca Raton, FL, pp. 119-141.

Maron DM and Ames BN (1983) Revised methods for the *Salmonella* mutagenicity test. *Mutation Research*, **113**, 173-215.

Mason BJ (1992) *Preparation of soil sampling protocols. Sampling techniques and strategies.* EPA/600/R-92/128, Exposure Assessment Research Division, Environmental Monitoring Systems Lab, Las Vegas, NV.

McCann J and Ames BN (1977) The *Salmonella*/microsome mutagenicity test: predictive value for animal carcinogenicity. In *Origins of Human Cancer*, Hiatt HH, Watson JD, and Winsten JA (eds), Cold Spring Harbor Laboratory Press, Cold Spring Harbor, NY, pp. 1431-1449.

Miller EC and Miller JA (1981) Mechanisms of chemical carcinogenesis. *Cancer*, **47**, 1055-1064.

Mumtaz MM, Sipes IG, Clewell HJ and Yang RSH (1993) Risk assessment of chemical mixtures: biologic and toxicologic issues. *Fundamental and Applied Toxicology*, **21**, 258-269.

Nisbet ICT and LaGoy PK (1992) Toxic equivalency factors (TEFs) for polycyclic aromatic hydrocarbons (PAHs) *Regulatory Toxicology and Pharmacology*, **16**, 290-300.

Randerath E, Danna TF and Randerath K (1992) DNA damage induced by cigarette smoke condensate *in vitro* as assayed by ^{32}P-postlabeling. Comparison with cigarette smoke-associated DNA addcut profiles *in vivo*. *Mutation Research*, **268**, 139-153.

Randerath K, Randerath E, Zhou GD, Supunpong N, He LY, McDonald TJ and Donnelly KC (1999) Genotoxicity of complex PAH mixtures recovered from contaminated lake sediments as assessed by three different methods. *Environmental and Molecular Mutagenicity*, **34**, 303-312.

Reddy MV and Randerath K (1986) Nuclease P1-mediated enhancement of sensitivity of ^{32}P-postlabeling test for structurally diverse DNA adducts. *Carcinogenesis*, **7**, 1543-1551.

Shelby MD (1988) The genetic toxicity of human carcinogens and its implications. *Mutation Research*, **204**, 3-15.

Tennant RW and Zeiger E (1993) Genetic toxicology: current status of methods of carcinogen identification. *Environmental Health Perspectives*, **100**, 307-315.

Tennant RW, Margolin BH, Shelby MD, Zeiger E, Haseman JK, Spalding J, Caspary W, Resnick M, Stasiewicz S, Anderson B and Minor R (1987) Prediction of chemical carcinogenicity in rodents from *in vitro* genetic toxicity assays. *Science*, **236**, 933-941.

USEPA (1980) *Samplers and Sampling Procedures for Hazardous Waste Streams*. EPA600/2-80-018, US Environmental Protection Agency, Washington, DC.

USEPA (1996) *Sampling for Hazardous Materials (165.9) Student Manual*. EPA/540/R-96/035, Office of Emergency and Remedial Response, US Environmental Protection Agency, Washington, DC.

USEPA (1997) *Test Methods for Evaluating Solid Waste: Physical/Chemical Methods*. SW846, Office of Solid Waste, US Environmental Protection Agency, Washington, DC.

Yang RSH and Rauckman EJ (1987) Toxicological studies of chemical mixtures of environmental concern at the National Toxicological Program: health effects of groundwater contaminants. *Toxicology*, **47**, 15-34.

Zieger E (1998) Identification of rodent carcinogens and noncarcinogens using genetic toxicity tests: premises, promises, and performance. *Regulatory Toxicology and Pharmacology*, **28**, 85-95.

6

Challenges and Approaches for Identifying Carcinogens in Contaminated Media

LARRY D. CLAXTON AND S. ELIZABETH GEORGE

Environmental Carcinogenesis Division, US Environmental Protection Agency, Research Triangle Park, NC, USA

6.1 WHY THE CHALLENGES EXIST AND THEIR IMPORTANCE

Although bringing many benefits to society, industrial development and modern agriculture have led to undeniable adverse effects on public health, environmental quality and ecosystem integrity. Most of these adverse effects result from the exposure to the generated waste products. A variety of relatively effective methods exist to treat waste water, soils and volatile emissions; however, the treatment processes may not eliminate all toxicants and may actually generate other toxic components. The public has called upon government and industry to properly evaluate, regulate and administer the clean-up of toxic wastes released into the environment (Johnson 1995, Johnson and DeRosa 1997). Due to the high incidence of cancer, and because cancer can be associated with long-term exposure to minute levels of carcinogens, one primary concern of both the public and health scientists is the presence or generation of carcinogens. Therefore, there is a need to have rapid and reliable methods to evaluate and compare the carcinogenic potential of contaminated media.

6.1.1 PUBLIC HEALTH CONCERNS

The most recent approximation for worldwide cancer mortality estimated that more than 5.1 million individuals died of cancer in 1990 (Parkin *et al.* 1999, Pisani *et al.* 1999). Malignant neoplasms (cancer) were the second leading cause of death in the United States in 1997. Nearly 540 000 US citizens died of cancer in that year (Kramarow *et al.* 1999). Although there is disagreement as to what percentage of these cancers is due to environmental carcinogens, even a small percentage could account for thousands of deaths per year. With this said, environmental carcinogens are a public health concern only to the degree to

Environmental Analysis of Contaminated Sites. Edited by G. I. Sunahara, A. Y. Renoux, C. Thellen, C. L. Gaudet and A. Pilon

which humans are exposed. This process begins with identifying carcinogens in environmental media. However, the extent of carcinogen identification depends upon the precise purpose of the study. For example, the industrial toxicologist may want to identify a specific carcinogen that is occurring in a waste stream so that the manufacturing or control processes can be changed to eliminate the carcinogen. Parties responsible for monitoring a waste site may need to know only the total carcinogenic load within the contaminated soil, or they may need to understand more precisely what carcinogens are present so that an appropriate clean-up alternative can be chosen. The analyst, therefore, should understand the options for identifying potential carcinogens.

6.1.2 ENVIRONMENTAL CARCINOGENS AND THEIR SOURCES

Humans are potentially exposed to environmental carcinogens in air, soil, sediment, water and contaminated biota (Johnson and DeRosa 1997). Because carcinogens have been identified in all types of media, identifying the sources and types of carcinogens associated with these sources is important. Some contaminated sites contain carcinogens (e.g., arsenic) which occur naturally in the environment. Most contaminated sites of public health concern, however, contain anthropogenic carcinogens. Because types, numbers and sources of environmental carcinogens are so extensive, the following discussion will focus on the challenges posed by environmental carcinogens.

Metals, including the carcinogenic metals such as arsenic and cadmium (Hayes 1997), are components of the Earth's crust and they cycle through the environment naturally. With precipitation, metals leach out of the rocks and can enter the groundwater, surface waters and marine environments (Wade *et al.* 1993). Arsenic is found naturally in arsenopyrite formations (Lahermo *et al.* 1998). In areas where there is a high concentration of arsenic in the rock, arsenic concentrations in groundwater can exceed 3 ppm (Wade *et al.* 1993). Normal chromium levels are 2 ppb (Cary 1982) and cadmium exists in natural deposits at concentrations of 1 ppb (Friberg *et al.* 1986). In addition, anthropogenic sources of metals may upset the delicate balance resulting in environmental accumulation and toxicity. For example, in Sweden, industrial copper mining and chemical industries have contributed to increased levels of lead and arsenic in contaminated soils (Lin *et al.* 1998).

Metals can also be translocated into plants. Ingestion of some of these metals in small amounts can satisfy human micronutrient requirements, but in large amounts may cause toxicity. In the marine environment, metals (including arsenic) are taken up by phytoplankton and can be bioaccumulated or excreted (Phillips 1990). For example, arsenosugars are formed from arsenic derivatives in seaweed, bioaccumulate in the marine organisms and can ultimately enter the human food chain. The volatilization of metals and metal complexes is also possible. Arsine (AsH_3) can be released into the atmosphere as a gas or absorbed to particles. Arsenite and arsenate have been detected on particles isolated from

Los Angeles, CA air (Rabano *et al.* 1989). Once airborne, contaminated particles can be distributed and ultimately inhaled.

Polycyclic aromatic hydrocarbons (PAHs) contaminate aquatic sediments and soils as byproducts of wood preservation, coking operations, petroleum refining, combustion sources and manufactured gas plant residues (Hites and Gschwend 1982, Mueller *et al.* 1989, Pyy *et al.* 1997, Rodriguez *et al.* 1997). Chemicals found in the emissions from these industries have been linked to human and animal cancers (Bertrand *et al.* 1987, Weyand *et al.* 1995, Rodriguez *et al.* 1997). PAHs and their nitrated derivatives, primarily generated from fossil fuel combustion emissions, are prevalent in urban air (Tokiwa and Ohnishi 1986). Tokiwa *et al.* (1998) described the identification of several PAHs (including 1-nitropyrene, 1,3-dinitropyrene, 2-nitrofluoranthene, benzo[*a*]pyrene, benzo[*k*]fluoranthene, and benzo[*ghi*]perylene) in human lung cancer biopsies. Such studies, therefore, confirm that humans are exposed to environmental carcinogens.

The waste streams of industrial facilities present another source of environmental carcinogens. In 1996, 250.7 million pounds of Occupational Safety and Health Administration (OSHA)-designated carcinogens were released into the environment, representing 10.3% of the total releases (USEPA 1998). The majority of the carcinogens were emitted into the air (184.0 million pounds), including approximately 1 million pounds of vinyl chloride. Using the *Salmonella* assay, McGeorge *et al.* (1985) detected mutagens in the effluents of petroleum refineries, organic compound industries, resin manufacturers and dye manufacturers. Organic chemical manufacturers, metal refining operations, petroleum refineries, and pulp and paper mills released effluents that were genotoxic in the SOS Chromotest (White *et al.* 1996) and the *Salmonella* mutagenicity assay (Claxton *et al.* 1998).

6.1.3 FINDING CAUSE AND EFFECT

The most convincing evidence for carcinogenicity comes from human data using epidemiological methods. However, environmental epidemiology often cannot overcome the challenges associated with identifying, understanding and avoiding confounding factors. The greatest drawback, however, to cancer epidemiology is the fact that populations of people must have been exposed to the carcinogen for a long enough time period (>5 years usually) before the overt disease can be detected. Other drawbacks include the necessity of epidemiology studies to have large population surveys and the need for adequate quantitative exposure information, which is hard to obtain. Hazard identification and risk assessment, therefore, often employ toxicological information based on studies using non-human systems. Presently, the most highly regarded tests employ other mammalian species; however, the usefulness of these toxicological tests is limited by extrapolation issues (e.g., rodent to human and high dose to low dose extrapolations), synergism and antagonism issues, costs and time needed to perform and analyze the tests (IARC 1979, 1980, Huff 1999). Except in rare cases, the use of these tests to evaluate a particular site or process is prohibited by cost and time constraints. Because many compounds (but obviously not all compounds) function through mutagenic mechanisms in the cancer process, the analyst can take advantage of simpler and less costly genotoxicity tests to identify these chemicals and to compare mutagenic potency and modes of action.

6.2 THE CHALLENGES

6.2.1 THE KNOWN CARCINOGEN CHALLENGE

Because the number of compounds known to cause cancer in humans is a limited subset of carcinogens identified by toxicological tests, and because identifying environmental carcinogens is very difficult, most environmental regulations prescribe the use of chemical methods for the identification and quantification of some subset of known carcinogens. Recognition that toxic substances can be found in many unregulated waste sites in the United States prompted political actions that resulted in the Comprehensive Environmental Response, Compensation and Liability Act (CERCLA). Commonly known as Superfund, the act provided broad federal authority to respond directly to releases or threatened releases of hazardous substances that may endanger public health or the environment. On 17 November 1999, there were 1218 sites on the National Priorities List (NPL) awaiting a clean-up (www.epa.gov/superfund/sites/query/queryhtm/nplfin.htm). The human carcinogens arsenic, benzene, chromium and vinyl chloride and presumptive human carcinogens trichloroethylene, tetrachloroethylene, cadmium, polychlorinated biphenyls, chloroform, 1,2-dichloroethane, methylene chloride, carbon tetrachloride, polycyclic aromatic hydrocarbons, di-(2-ethylhexyl) phthalate, benzo[*a*]pyrene and beryllium are the most prominent listed carcinogens identified at NPL sites (Johnson and DeRosa 1997). Trichloroethylene was the most prevalent chemical occurring at 213 (of 530) NPL sites examined by ATSDR (Johnson and DeRosa 1997).

The key challenge associated with the analysis of discrete carcinogens concerns the development and application of analytical techniques. A central objective in these techniques is having the ability to sample the appropriate media and recover the carcinogen(s) of interest. These methods should provide a representative sample, be highly reproducible, give high recovery rates and produce a minimum of artifacts. When monitoring, tradeoffs in sample collection may be necessary to provide adequate coverage in a cost-effective manner (Moeller 1992). For example, rather than sample and process a larger number of poorly selected samples, it is better to analyze a small number of samples taken at key representative sites. After recovery, the qualitative and quantitative chemistry must be accurate enough to identify the target compound(s) amidst many very similar compounds. To ensure accuracy, analytical methods should also employ appropriate standards and use sample preparation and analysis methods that enhance sensitivity. Although a thorough discussion of these topics is beyond the scope of this text, Zagury *et al.* (see Chapter 2) can provide some insight.

6.2.2 THE MIXTURE CHALLENGE

Environmental carcinogens typically occur within complex mixtures. Fay and Mumtaz (1996) have surveyed HazDat, a database containing NPL site chemical information, and report that the most prevalent combinations of chemicals typically include the presumptive human carcinogen trichloroethylene and/or carcinogenic metals. A survey by Johnson (1995) reported similar findings for binary mixtures that are prevalent water contaminants.

The toxicity of a carcinogen may be impacted by the presence of other compounds. The combinations of compounds may result in interactions exhibiting additivity, synergy, potentiation or inhibition. For example, Arochlor 1254 and pentachlorophenol potentiate the formation of 2,6-dinitrotoluene hepatic DNA adducts and urine mutagens in Fisher 344 rats (Chadwick *et al.* 1991, 1993). When compared to animals receiving a benzo[*a*]pyrene equivalent dose, female A/J mice had an augmented tumor response when exposed to non-benzo[*a*]pyrene PAHs in a manufactured gas plant residue in a chronic ingestion study (Weyand *et al.* 1995). Synergy among unidentified manufactured gas plant PAHs has been suggested as a mechanism to explain the non-additivity response observed in a complementary DNA adduct study (Rodriguez *et al.* 1997).

Wood-preserving waste facilities further illustrate the complexity of environmental carcinogen contamination. Typically, wood was preserved using creosote, pentachlorophenol or a chromium-arsenic-copper treatment. Coal tar creosote is composed of 150 to 200 chemicals and typically includes 85% polycyclic aromatic hydrocarbons (e.g., naphthalene, anthracene, chrysene, benzo[*a*]pyrene), 10% phenolic compounds (e.g., phenol, cresols, xylenols, trimethylphenol) and 5% N-, S- and O-heterocyclics (e.g., carbazole, acridine, methylquinolines, benzo[*b*]thiophene, dibenzofuran) (Mueller *et al.* 1989). A random survey of NPL wood-preserving sites indicates that chemical contamination is site-specific. Exposure assessments show that contamination of ground and surface water adjacent to many of these sites has already occurred or is likely.

Superfund sites also demonstrate the complexity of environmental carcinogen contamination. In 1994, NPL sites included waste storage/treatment facilities or landfills (42%), abandoned manufacturing facilities (31%), waste recycling facilities (8%), mining sites (5%) and government properties (4%) (Johnson 1995). Most of these sites, therefore, could reasonably be expected to contain more than one environmental carcinogen affecting more than one environmental media (Johnson 1995).

6.2.3 THE ENVIRONMENTAL TRANSFORMATION CHALLENGE

Both abiotic and biotic influences can impact the chemical state and toxicity of chemical mixtures. Abiotic reactions (e.g., photooxidation, photolysis, hydrolysis and redox reactions) can result in the activation or degradation of chemicals. Photooxidation of PAHs results in the formation of nitro-PAHs, which tend to be more genotoxic. For example, phenanthrene is transformed by photooxidation to the more mutagenic nitrophenanthrene lactones (nitro-6H-dibenzo[*b*,*d*]pyran-6-ones) (Arey *et al.* 1992). Hydrolysis of the mutagenic soil fumigant nematicide, 1,3-dichloropropene, results in its conversion to the ultimate mutagenic product, 3-chloro-2-hydroxypropanal (Schneider *et al.* 1998). However, photolysis transforms other chemicals to inactive degradation products. Inactivation of the halobenzonitrile pesticides occurs in the presence of sunlight (Millet *et al.* 1998). Oxidation–reduction processes can also influence bioavailability and toxicity. For example, AsIII (reduced) moves through the environment more readily and is more toxic than AsV (oxidized) (Korte and Fernando 1991).

Biotic influences, such as biodegradation and bioconversion, impact the chemical state. Consortia of microorganisms can completely mineralize many compounds. Following microbial reductive dehalogenation of dibenzo-*p*-dioxins and dibenzofurans, fungi and bacteria can oxidize (oxidase and dioxygenase pathways) these compounds for carbon and energy sources (Wittich 1998). Heitkamp and Cerniglia (1988) reported mineralization of PAHs, including 1-nitropyrene, phenanthrene and naphthalene, by a bacterial sediment isolate. Following hydrolysis, microorganisms degrade carbofuran (Mabury *et al.* 1996). However, some compounds are more recalcitrant and either incomplete products of metabolism or the parent compounds may accumulate in the soil. Higher ring number PAHs, such as benzo[*a*]pyrene and pyrene, are less easily degraded (Heitkamp and Cerniglia 1988). The rate of polychlorinated biphenyl congener transformation is dependent on the position and number of chlorines present. Transformation, in turn, influences accumulation of the parent compound or resulting metabolites (Billingsley *et al.* 1997).

The ability of abiotic forces to impact chemical transformations is dependent on the properties of the chemical and environmental conditions (e.g., soil properties) (Farrington 1991). Chemicals can be sequestered into clay micropores or bind to organic constituents in soil and sediment. Phyllosilicate clay binds aflatoxins, reducing their toxicity (Phillips *et al.* 1995). The explosive 2,4,6-trinitrotoluene is readily reduced to the triamino derivative, but this compound tends to bind irreversibly to organic components in the soil (Rieger and Knackmuss 1995). Pentachlorophenol adsorbs to sediment organic matter (Pignatello *et al.* 1983). If a chemical, such as PAH, is poorly water soluble, it is less likely to be leached out of soil into groundwater or partition into the water column from sediment, so it may be less available for biodegradation (Hughes *et al.* 1997).

6.2.4 THE UNKNOWN CARCINOGEN CHALLENGE

As pointed out previously, most assessments for environmental carcinogens have focused upon a relatively small number of known human carcinogens. Many unidentified and unmonitored carcinogens have received little scrutiny. In one USEPA study, Bramlett *et al.* (1987) examined the composition of leachates from 13 hazardous waste sites. Only 4% of the total organic carbon in the leachates was characterized for chemical structure by the gas chromatography/mass spectroscopy methods used. However, more than 200 separate compounds were identified within this 4%. Although an extensive and state-of-the-art effort was given to identifying the components of the leachates, the toxicity of the unidentified 96% is unknown. Therefore, the total toxicity of leachates in these types of situations cannot be reliably estimated from chemical analysis alone.

In a report on hazardous wastes, the National Research Council (US) referred to these unidentified, unregulated pollutants as non-conventional pollutants (NCPs). They stated that NCPs 'are a potentially important source of hazardous exposure' and 'represent a risk of unknown magnitude' (National Research Council 1991). Little has changed in the past decade. Many sites contain hundreds of NCPs. The numbers and amounts of carcinogens within the NCPs are not known.

6.2.5 RISK ASSESSMENT: NEEDS AND LIMITS

In 1994, Lee Thomas, a former EPA administrator, said 'One of the real problems or mistakes EPA has made, and I was part of it, is taking a concept like risk assessment and using it like it was a scalpel when it's about as accurate as a meat ax' (Minard 1996). He was illustrating that risk assessment is an inexact tool that should be used by decision makers primarily to develop an appropriate context for their decision making. In reality, risk assessments can range from being highly qualitative (the ax approach) to fairly quantitative (the scalpel approach). For environmental cancer issues, the scalpel approach equates to having definitive human hazard assessment information and data (i.e., human epidemiology and supporting toxicology) with accurate and extensive exposure information on each of the environmental carcinogens present. When this is unavailable and unreasonable to attain, surrogate information or decision-making processes must be used to place a risk into proper context.

In the risk management process, how and when bioassays should be incorporated can only be determined on a case-by-case basis. For example, in a relatively small crude oil remediation project, a single bioassay on composited samples taken at the start and end of the project may be sufficient to qualitatively demonstrate that the carcinogenic component has been removed. On the other hand, extensive use of bioassays (including bioassay-directed fractionation) may be needed in very large projects. In the largest and/or demonstration projects, bioassays may be used: (1) in planning stages (e.g., pilot-scale projects), (2) before site remediation (to establish treatment and control site baseline values), (3) during remediation (to monitor transient impacts and understand the rate at which potentially carcinogenic components are being degraded), and (4) at the end of the project (to establish acceptable residual levels of contaminants). When there is a need to develop a 'risk value' for an environmental substance, comparative risk methods can integrate bioassay data to estimate the carcinogenic potency of a substance by direct comparison to known standards (Nesnow 1990).

Use of bioassays in both risk assessment and risk management efforts will help risk managers to use risk assessment and risk management more like a scalpel, by making the decision processes more precise and reliable.

6.3 APPROACHES FOR EVALUATING MEDIA FOR CARCINOGENS

To determine the most appropriate strategy for identifying carcinogens, one must ask a number of questions. What types of carcinogens would be expected to be present at this site? Which carcinogens were identified previously? To what level are people likely to be exposed to carcinogens and by what route (dermal exposure, inhalation or ingestion)? Do natural abiotic and/or biotic mechanisms alter the amount or availability of the carcinogens? In what manner will the remediation effort affect the exposure of humans to the carcinogenic components? This final question concerning the effect of remediation efforts is critical. Remediation efforts could either sequester carcinogens or increase the bioavailability of the toxicants. Remediation efforts may degrade some or all carcinogens present, but also have the potential to metabolize these or other components to a carcinogenic moiety.

Those responsible for remediating sites or wastes that may contain carcinogens, therefore, need useful tools for identifying and assessing the level of carcinogens before, during and after remediation efforts. Ideally, these tools should meet the criteria of being cost-effective, reliable, relatively rapid and acceptable to responsible parties (e.g., site administrator, corporate management and the involved public). If there are relatively few environmental carcinogens, these criteria generally could be met using analytical methods. However, assessment of some types of contamination (e.g., mixtures of PAHs) can be made more reliable by incorporating bioassays that relate to carcinogenicity.

The extent to which analytical chemistry and/or bioassay methods are used is a situation-by-situation decision that depends upon knowledge concerning the contaminants present (e.g., types, mode of action, amount of contaminant, degradability), the likelihood of human exposure (e.g., media to which humans are exposed, future use of the site, bioavailability of known contaminants) and the cost of clean-up options versus the cost of analysis. The purpose of the following discussion is to introduce the reader to practical and useful options that can be considered when evaluating the presence of carcinogens in environmental media.

6.3.1 ANALYTICAL CHEMISTRY

Analytical chemistry is an indispensable tool for the evaluation of carcinogens in environmental samples. The most obvious use is to identify specific organic or inorganic chemicals that occur on lists of known carcinogens. Once identified, the quantization of any identified carcinogens indicates the maximum extent to which populations could be exposed. Depending upon the medium (i.e., air, soil sediment or water), the method of collection, sample preparation (e.g., extraction of a soil sample) and the level of qualitative and quantitative analysis needed, the analytical methods will vary. It is beyond the scope of this chapter to provide guidance in this area (see Chapter 2). Many types of procedures (e.g., gas chromatography, high pressure liquid chromatography) and modifications of these procedures exist. Standard procedures for most commonly examined carcinogens can be gained from the USEPA, state agencies, consultants and contracting services, and university academicians (USEPA 1989). For monitoring, a variety of sources exist for helping one to determine which chemicals have known carcinogenic effects. The Integrated Risk Information System (IRIS), prepared and maintained by the USEPA, is an electronic database containing information on human health effects of environmental chemicals. IRIS is available via an internet connection (http://www.epa.gov/ncea/iris.htm).

Detailed knowledge of the chemical composition of environmental media can also be used to estimate the persistence, bioaccumulative potentials, toxicity, and general fate and transport properties of the carcinogenic components. These efforts will help the risk assessor to know which carcinogens are transported via one or more media (including air, soils, sediments, surface water and groundwater) to potential receptors (through inhalation, dermal contact, ingestion and/or via food sources). With this knowledge, appropriate site and background samples can be collected from the affected media.

One primary drawback with chemical methods for identifying carcinogens is the limitation of being able to identify only a small number of known environmental carcinogens. For example, even when there is evidence that a large number of PAHs are present in an environmental sample, the number of carcinogens generally identified and quantified is limited to a relatively small number (Department of Health and Human Services 1999). Therefore, in many situations, not all environmental carcinogens are identified, and routine analytical services will not make an attempt to identify other carcinogens present.

6.3.2 BIOASSAY ANALYSIS

Genotoxicity tests are formally incorporated into the regulation of pesticides, pharmaceuticals and other manufactured chemicals; however, many countries, including the US, have not formally incorporated the use of genetic or carcinogenicity bioassays into regulations controlling air emissions, wastewater effluents or contaminated soil. These assays are most often used to characterize the carcinogenic potential of individual chemicals. However, they may also be used for comparative assessments, identifying multiple carcinogens in complex environmental mixtures and providing information concerning the mechanisms of action (Claxton *et al.* 1996, DeMarini 1998, Schoen 1998). In Chapter 5, Garcia and Donnelly provide a description of genotoxicity assays that can be used on a routine basis to evaluate the carcinogenic potential of contaminated sites.

6.3.3 INTEGRATED BIOASSAY AND CHEMICAL ANALYSIS

Recent studies in our laboratory (Brooks *et al.* 1998, Hughes *et al.* 1998) illustrate how the most commonly used genotoxicity test, the *Salmonella* mutagenicity assay, can be used to examine for the presence and comparative level of potential mutagens before and after remediation processes. Brooks *et al.* (1998) also illustrated how the *Salmonella* assay is used in bioassay-directed chemical analysis to identify specific carcinogens and species of unidentified carcinogens. In these studies, creosote-contaminated soil from a US Superfund site was evaluated for carcinogenic potential (by use of the spiral *Salmonella* assay and chemical analysis) before and after four independent pilot-scale remediation efforts: bioslurry, biopile, compost and land treatments. All four remediation processes reduced the total PAH concentrations (48–74%), suggesting that all four processes were capable of lowering carcinogenic potential. In contrast, the mutagenicity assay did not indicate the same definitive loss in carcinogenic potential. Indeed, for two of the processes (biopile [BP] treatment and bioslurry [BS] treatment), the mutagenicity assay revealed a potential increase in carcinogenic activity. Diagnostic use of the assay provided two key observations and two potential explanations. Although the organics from the untreated soil were mutagenic only when an exogenous mammalian metabolism system was used, the organics associated with the BP and BS treatments were mutagenic without this activation system. Secondly, after the bioremediation process, specialized tester strains indicated that a biologically detectable level of nitroaromatic compounds was present. These two observations indicate that the remediation process either: (1) produced nitroaromatic mutagens while reducing the level of PAH genotoxicants and/or (2) mutagenic amendments (e.g., activated sludge containing nitroaromatic compounds) were unknowingly added to the remediation process. To examine these observations more closely, some of the samples underwent bioassay-fraction/chemical-identification procedures. In this process, organic extracts were fractionated using reverse-phase, high-performance liquid chromatography (HPLC). Each of the 40 fractions was examined with one of the more sensitive *Salmonella* tester strains (YG1041), providing a mutagenicity profile of the HPLC fractions (referred to as a mutagram). The mutagrams of the BS and BP extracts were both qualitatively and quantitatively different from those of the untreated soil organics. The mutagenic fractions were analyzed by gas chromatography/mass spectrometry (GC/MS).

Of the 11 BS and BP mutagenic fractions analyzed by GC/MS, only three fractions contained any of the priority pollutant PAHs. Many of the compounds identified in other fractions were azaarenes, several of which are known mutagens. This analysis confirmed that the bioremediation process introduced another class of mutagens and potential carcinogens.

6.4 DEVELOPING APPROACHES FOR MEETING THE CHALLENGES

The management of carcinogenic risks implies the capacity to control exposure to cancer-causing agents. Effective control depends on the detection and identification of the responsible agents, the capability to estimate or monitor human exposure, and the ability to determine the risk emanating from the exposure. An understanding of the underlying mechanisms by which substances cause cancer provides the rationale for: (1) use of genotoxicity testing as a surrogate for cancer bioassay, (2) use of bioassays to assess relative or comparative risks, and (3) improvements in risk assessments based on mechanistic knowledge. Mutations are clearly involved in the processes by which most known, but not all, cancer-causing agents act. During the last decade, specific genes have been detected that, when mutated, result in advancing the cancer process (i.e., oncogenes, tumor suppressor genes, etc.). Therefore, in the environmental situation when complex mixtures or compounds elicit a mutagenic response, it is prudent to assume that the mixture or compound has the potential to initiate or promote the carcinogenic process. Coupled together, chemical and bioassay analysis provides the most effective tool for monitoring the environment before, during and after remediation efforts for carcinogens.

For the purposes of remediation and land reclamation actions, therefore, how can bioassays and analytical methods be incorporated in a useful risk assessment effort? The methods chosen should: (1) identify a broad range and extensive number of potential environmental carcinogens, (2) be cost-effective for decision-making purposes, (3) be managerially effective (i.e., provide a manageable level of understandable data in a reasonable time frame), and (4) allay the fears and suspicions of the public (i.e., be an inherently clear and logical approach to understanding and monitoring the environmental situation while being comprehensive in its scope). Integrated short-term assays and analytical chemistry methods can typically accomplish each of these goals. Epidemiology and long-term chronic cancer bioassays, generally, are too costly (however, there may be exceptions), require too much time, have little utility for monitoring purposes, have too many confounders and can be managerially ineffective. A testing protocol using short-term bioassays coupled with analytical chemistry can be: (1) a relatively cost-effective means to identify a broad range of potential carcinogens, (2) used for monitoring purposes, (3) summarized in manageable and understandable forms, (4) used for comparative assessments, and (5) done within an acceptable time frame. Once relevance to the mechanisms of cancer is established, short-term assays are usually acceptable to the general public. Analytical chemistry can be used for the detection and identification of well-established carcinogens, whereas bioassays are employed to identify a breadth of carcinogens for which analytical methods are lacking. Biological analysis may also provide insight into the potential interactions of the components of the complex mixture.

6.5 RECOMMENDATIONS AND FUTURE DIRECTIONS

Before bioassays are incorporated into risk analysis and risk management, responsible parties should have a clear understanding of: (1) what is **and is not** indicated by both 'positive' and 'negative' responses (i.e., mutagenic and non-mutagenic, or carcinogenic and non-carcinogenic) in a bioassay, (2) how the bioassay responds to the different classes of environmental substances likely to be present within the study, (3) what assay modifications (choice of method, strains and substrains of the organism, etc.) are most informative, (4) whether or not the bioassay is likely to only be redundant to chemical analysis, (5) what types of controls and other quality control efforts are needed, (6) how the quantitative data from the bioassay will be analyzed and whether or not the likely variance will allow reliable comparisons, and (7) if the bioassay can be done within the existing cost and time constraints.

Future research needs to emphasize how diagnostic and mechanistic toxicological methods can be used with complex environmental samples and should broaden the capabilities of presently-available assays. For example, specific mammalian metabolizing systems can be incorporated into the bacterial strains used for the *Salmonella* mutagenicity assay. This would indicate that these strains could be used not only to detect mutagenic potential, but also to define the type of mammalian metabolism required. This information could help make the assay much more informative. The future will also see a continual merging of molecular assay methodologies (e.g., gene array analysis) with both environmental and human monitoring techniques that are reliable and affordable.

6.6 ACKNOWLEDGEMENTS

This document has been reviewed in accordance with US Environmental Protection Agency policy and approved for publication. Mention of trade names or commercial products does not constitute endorsement or recommendation for use.

REFERENCES

Arey J, Harger WP, Helmig D and Atkinson R (1992) Bioassay-directed fractionation of mutagenic PAH atmospheric photooxidation products and ambient particulate extracts. *Mutation Research*, **281**, 67–76.

Bertrand JP, Chau N, Patris A, Mur JM, Pham QT, Moulin JJ, Morviller P, Auburtin G, Figueredo A and Martin J (1987) Mortality due to respiratory cancers in the coke oven plants of the Lorraine coal mining industry (Houilleres du Bassin de Lorraine). *British Journal of Industrial Medicine*, **44**, 559–565.

Billingsley KA, Backus SM, Juneson C and Ward OP (1997) Comparison of the degradation patterns of polychlorinated biphenyl congeners in Aroclors by *Pseudomonas* strain LB400 after growth on various carbon sources. *Canadian Journal of Microbiology*, **43**, 1172–1179.

Bramlett J, Furman C, Johnson A, Ellis WD and Nelson N (1987) *Composition of leachates from Actual Hazardous Waste Sites*. Project Report EPA/600/2-87/043, US Environmental Protection Agency, Washington, DC.

Brooks L, Hughes TJ, Claxton LD, Austern B, Brenner R and Kremer F (1998) Bioassay-directed fractionation and chemical identification of mutagens in bioremediated soils. *Environmental Health Perspectives*, **106**(Suppl. 6), 1435–1440.

Cary EE (1982) Chromium in air, soil, and natural waters. In *Biological and Environmental Aspects of Chromium*, Langard S (ed), Elsevier Science Publishers, Amsterdam, pp. 49–64.

Chadwick RW, George SE, Chang J, Kohan MJ, Dekker JP, Long JE, Duffy MC and Williams RW (1991) Potentiation of 2,6-dinitrotoluene genotoxicity in Fischer 344 rats by pretreatment with pentachlorophenol. *Pesticide and Biochemical Physiology*, **39**, 168–181.

Chadwick RW, George SE, Kohan MJ, Williams RW, Allison JC, Hayes YO and Chang J (1993) Potentiation of 2,6-dinitrotoluene genotoxicity in Fischer 344 rats by pretreatment with Aroclor 1254. *Toxicology*, **80**, 153–171.

Claxton LD, Houk VS and George SE (1996) Integration of complex mixture toxicity and microbiological analyses for environmental remediation research. In *Ecotoxicity and Human Health: A Biological Approach to Environmental Remediation*, deSerres F and Bloom A (eds), CRC Press/Lewis, Boca Raton, FL, pp. 87–122.

Claxton LD, Houk VS and Hughes TJ (1998) Genotoxicity of industrial wastes and effluents. *Mutation Research*, **410**, 237–243.

DeMarini DM (1998) Use of mutagenicity for predicting carcinogenicity. In *Carcinogenicity: Testing, Predicting, and Interpreting Chemical Effects*, Kitchin K (ed), Marcel Dekker, New York, pp. 209–226.

Department of Health and Human Services (1999) Agency for Toxic Substances and Disease Registry [ATSDR-155] Notice of the revised priority list of hazardous substances that will be the subject of toxicological profiles. *Federal Register*, **64**(203), 56 792–56 794.

Farrington JW (1991) Biogeochemical processes governing exposure and uptake of organic pollutant compounds in aquatic organisms. *Environmental Health Perspectives*, **90**, 75–84.

Fay RM and Mumtaz MM (1996) Development of a priority list of chemical mixtures occurring at 1188 hazardous waste sites, using the HazDat database. *Food Chemistry and Toxicology*, **34**, 1163–1165.

Friberg L, Kjellstrom T and Nordberg G (1986) Cadmium. In *Handbook of the Toxicology of Metals*, Vol. II, Friberg L, Nordberg G and Vouk V (eds), Elsevier Science Publishers, Amsterdam, pp. 130–184.

Hayes RB (1997) The carcinogenicity of metals in humans. *Cancer Causes Control*, **8**, 371–385.

Heitkamp MA and Cerniglia CE (1988) Mineralization of polycyclic aromatic hydrocarbons by a bacterium isolated from sediment below an oil field. *Applied and Environmental Microbiology*, **54**, 1612–1614.

Hites RA and Gschwend PM (1982) The ultimate fates of polycyclic aromatic hydrocarbons in marine and lacustrine sediments. In *Polynuclear Aromatic Hydrocarbons: Physical and Biological Chemistry*, Cooke M, Dennis A and Fisher G (eds), Battelle Press, Columbus, OH, pp. 357–365.

Huff J (1999) Value, validity, and historical development of carcinogenesis studies for predicting and confirming carcinogenic risks to humans. In *Carcinogenicity: Testing, Predicting, and Interpreting Chemical Effects*, Kitchin K (ed), Marcel Dekker, New York, pp. 21–124.

Hughes JB, Beckles DM, Chandra SD and Ward CH (1997) Utilization of bioremediation processes for the treatment of PAH-contaminated sediments. *Journal of Industrial Microbiology and Biotechnology*, **18**, 152–160.

Hughes TJ, Claxton LD, Brooks L, Warren S, Brenner R and Kremer F (1998) Genotoxicity of bioremediated soils from the Reilly Tar Site, St. Louis Park, Minnesota. *Environmental Health Perspectives*, **106**(Suppl. 6), 1427–1433.

IARC (1979) *Monographs on Cancer Risks, Supplement No. 1: Chemicals and Industrial Processes Associated with Cancer in Humans*, International Agency for Research on Cancer, Lyons, 71 pp.

IARC (1980) *Monographs on Cancer Risks, Supplement No. 2: Long-Term and Short-Term Screening Assays for Carcinogens: A Critical Appraisal*, International Agency for Research on Cancer, Lyons, 426 pp.

Johnson BL (1995) Nature, extent, and impact of superfund hazardous waste sites. *Chemosphere*, **31**, 2415–2418.

Johnson BL and DeRosa C (1997) The toxicologic hazard of Superfund hazardous-waste sites. *Reviews on Environmental Health*, **12**, 235–251.

Korte NF and Fernando Q (1991) A review of arsenic III in groundwater. *Critical Review of Environmental Contamination*, **21**, 1–39.

Kramarow E, Lentzner H, Rooks R, Weeks J and Saydah S (1999) *Health, United States, 1999, Health and Aging Chartbook: Health, United States*, National Center for Health Statistics, Hyattsville, MD.

Lahermo P, Alfthan G and Wang D (1998) Selenium and arsenic in the environment in Finland. *Journal of Environmental Pathology, Toxicology, and Oncology*, 1998, 205–216.

Lin Z, Harsbo K, Ahlgren M and Qvarfort U (1998) The source and fate of Pb in contaminated soils at the urban area of Falun in central Sweden. *The Science of the Total Environment*, **209**, 47–58.

Mabury SA, Cox JS and Crosby DG (1996) Environmental fate of rice pesticides in California. *Reviews of Environmental Contamination and Toxicology*, **147**, 71–117.

McGeorge LJ, Louis JB, Atherholt TB and McGarrity GJ (1985) Mutagenicity analyses of industrial effluents: results and considerations for integration into water pollution control programs. In *Short-Term Bioassays in the Analysis of Complex Environmental Mixtures*, Vol. IV, Waters M, Sandhu S, Lewtas J, Claxton L, Strauss G and Nesnow S (eds), Plenum Press, New York, pp. 247–257.

Millet M, Palm WU and Zetzsch C (1998) Abiotic degradation of halobenzonitriles: investigation of the photolysis in solution. *Ecotoxicology and Environmental Safety*, **41**, 44–50.

Minard RA (1996) CRA and the States: history, politics, and results. In *Comparing Environmental Risk: Tools for Setting Government Priorities*, Davies JC (ed), Resources for the Future, Washington, DC, pp. 23–61. (Note: This is referenced as 'Thomas, Lee, 1994. Personal interview with the author, 27 January, Washington, DC'.)

Moeller DW (1992) *Environmental Health*, Harvard University Press, Cambridge, MA, pp. 229–252.

Mueller JG, Chapman PJ and Pritchard PH (1989) Creosote-contaminated sites. *Environmental Science and Technology*, **23**, 1197–1201.

National Research Council (1991) Committee on Environmental Epidemiology. In *Environmental Epidemiology*, National Academy Press, Washington, DC, pp. 8–10.

Nesnow S (1990) Mouse skin tumors and human lung cancer: relationships with complex environmental emissions. *Complex Mixtures and Cancer Risk, IARC Scientific Publication*, **104**, 44–54.

Parkin D, Pisani P and Ferlay J (1999) Estimates of the worldwide incidence of twenty-five major cancers in 1990. *International Journal of Cancer*, **80**, 827–841.

Phillips D (1990) Arsenic in aquatic organisms: a review emphasizing chemical speciation. *Aquatic Toxicology*, **16**, 151–186.

Phillips TD, Sarr AB and Grant PG (1995) Selective chemisorption and detoxification of aflatoxins by phyllosilicate clay. *Natural Toxins*, **3**, 204–213.

Pignatello JJ, Martinson MM, Steiert JG, Carlson RE and Crawford RL (1983) Biodegradation and photolysis of pentachlorophenol in artificial freshwater streams. *Applied and Environmental Microbiology*, **46**, 1024–1031.

Pisani P, Parkin DM, Bray F and Ferlay J (1999) Estimates of the worldwide mortality from twenty-five major cancers in 1990. *International Journal of Cancer*, **83**(1), 18–29.

Pyy L, Makela M, Hakala E, Kakko K, Lapinlampi T, Lisko A, Yrjanheikki E and Vahakangas K (1997) Ambient and biological monitoring of exposure to polycyclic aromatic hydrocarbons at a coking plant. *The Science of the Total Environment*, **199**, 151-158.

Rabano ES, Castillo NT, Torre KJ and Solomon PA (1989) Speciation of arsenic in ambient aerosols collected in Los Angeles. *Journal of the American Pollution Control Association*, **39**, 76-80.

Rieger P-G and Knackmuss H-J (1995) Basic knowledge and perspectives on biodegradation of 2,4,6-trinitrotoluene and related nitroaromatic compounds in contaminated soil. In *Biodegradation of Nitroaromatic Compounds*, Spain J (ed), Plenum Press, New York, pp. 1-18.

Rodriguez LV, Dunsford HA, Steinberg M, Chaloupka KK, Zhu L, Safe S, Womack JE and Goldstein LS (1997) Carcinogenicity of benzo[*a*]pyrene and manufactured gas plant residues in infant mice. *Carcinogenesis*, **18**, 127-135.

Schneider M, Quistad GB and Casida JE (1998) 1,3-Dichloropropene epoxides: intermediates in bioactivation of the promutagen 1,3-dichloropropene. *Chemical Research in Toxicology*, **11**, 1137-1144.

Schoen D (1998) A renaissance for genotoxicity testing? Test results are driving a rebirth of interest in the use of the technique to evaluate pollution. *Environmental Science and Technology*, **32**, 498A-501A.

Tokiwa H and Ohnishi Y (1986) Mutagenicity and carcinogenicity of nitroarenes and their sources in the environment. *CRC Critical Reviews in Toxicology*, **17**, 23-60.

Tokiwa H, Nakanishi Y, Sera N, Hara N and Inuzuka S (1998) Analysis of environmental carcinogens associated with the incidence of lung cancer. *Toxicology Letters*, **99**, 33-41.

USEPA (1989) *Risk Assessment Guidelines for Superfund, Vol. 1, Human Health Evaluation Manual (Part A)*. EPA/540/1-89/002/, Office of Emergency and Remedial Response, US Environmental Protection Agency, Washington, DC, 298 pp.

USEPA (1998) *Toxic Release Inventory Public Data Release Report*. EPA/745-R-98-005, Office of Prevention, Pesticides, and Toxic Substances, US Environmental Protection Agency, Washington, DC, 472 pp.

Wade MJ, Davis BK, Carlisle JS, Klein AK and Valoppi LM (1993) Environmental transformation of toxic metals. *Occupational Medicine*, **8**, 574-601.

Weyand EH, Chen YC, Wu Y, Koganti A, Dunsford HA and Rodriguez LV (1995) Differences in the tumorigenic activity of a pure hydrocarbon and a complex mixture following ingestion: benzo[alpha]pyrene vs manufactured gas plant residue. *Chemical Research in Toxicology*, **8**, 949-954.

White PA, Rasmussen JB and Blaise C (1996) Comparing the presence, potency, and potential hazard of genotoxins extracted from a broad range of industrial effluents. *Environmental and Molecular Mutagenesis*, **27**, 116-139.

Wittich R-M (1998) Degradation of dioxin-like compounds by microorganisms. *Applied Microbiology and Biotechnology*, **49**, 489-499.

7

Soil Biomarkers (Invertebrates and Microbes) for Assessing Site Toxicity

CLAUS SVENDSEN[1], GRAEME PATON[2] AND JASON M. WEEKS[1,*]

[1]Centre for Ecology & Hydrology, Monks Wood, Abbots Ripton, Huntingdon, Cambridgeshire, UK
[2]Department of Plant & Soil Science, University of Aberdeen, Aberdeen, Scotland, UK

7.1 INTRODUCTION

This chapter summarizes the available soil invertebrate and soil microbial measurements that may be utilized to undertake a contaminated site assessment. Only well-established techniques for which a documented history exists in the scientific literature are included (and this list is not exhaustive). For reasons of convenience, this chapter is divided artificially into two main parts. The first examines invertebrate biomarkers and the second looks at soil microbial processes at various biological levels of organization.

7.1.1 WHY DO WE NEED SOIL BIOMARKERS AND BIOINDICATORS?

Environmental protection and environmental contamination have been issues of growing concern in recent years. Specifically, potential impacts of environmental contaminants on sustainable land use and agricultural production have received increased attention. Soil acts as a short-term sink for many pollutants whilst having a limited function as a geographical transport medium. Soil may, however, eventually serve as an important source of contamination for groundwaters and the atmosphere. The most significant difference when comparing aquatic and atmospheric pollution with soil pollution is the much increased technical difficulty, both in assessing and dealing with soil pollution. Soil represents a more complex and diverse matrix than either water or air, due to the high degree of heterogeneity in soils. Soil pollution does, however, have the

* Present address: WRc-NSF Ltd., NCET, Henley Road, Medmenham, Marlow, Buckinghamshire, UK.

Environmental Analysis of Contaminated Sites. Edited by G. I. Sunahara, A. Y. Renoux, C. Thellen, C. L. Gaudet and A. Pilon

advantage of being logistically simpler to tackle due to the limited mobility of pollutants within the system. There remains a need for better detection and assessment of soil contamination.

Major legislative efforts have been made establishing standard test protocols and assessment systems for both new and existing chemicals (for a review see Løkke and Van Gestel 1998) and in the development of soil quality standards to protect the soil environment. No matter how well-developed and sophisticated such regulatory processes become, they will always be largely based on prediction and extrapolation, and will therefore never negate the need for follow-up environmental monitoring. Therefore, there is a clear need for new approaches and regulatory guidelines on how to correctly assess and classify existing polluted sites and how to set soil quality standards that are ecologically relevant. This is particularly the case for soils due to the paucity of work done in this media compared to either water or air, and the greater inherent complexity in assessing both actual and bioavailable levels of pollutants. Pre-release regulatory procedures are in place to protect soils, yet no system of direct effects monitoring is in place. It is within such monitoring systems that the development of biomarker and bioindicator approaches could prove to have great potential.

7.1.2 PREDICTIVE VALUE FOR ECOSYSTEM HEALTH CHANGES AND BIOLOGICAL MONITORING

Biomarkers and some bioindicators represent one end of a range of biological responses that may be considered for use in environmental monitoring. This response continuum begins with effects at the molecular level, through to the intact organism, population or community structures, and eventually the structure and function of ecosystems (see Figure 7.1). As one moves up through these levels of organization, several changes are observed. First, the time scale of response increases, moving from seconds or minutes to years or even decades, as we change our scope from molecular to ecosystem effects. Second, the degree of importance increases (we are probably more concerned with the continued function of ecosystems than with subtle molecular changes). Finally, the ability to make causal links between events and effects decreases as one moves through this continuum. For these reasons, environmental monitoring (e.g., biomarkers) must be applied at the lower end of the range if the response is to be used for anything other than stating the obvious. It is here that chemicals are known to exert their effects (Moore 1985) and consequently responses are sensitive and rapid, whilst being reasonably easy to interpret (Figure 7.1).

7.1.3 UTILITY OF BIOMARKERS AND CAVEATS FOR THEIR USE

The incentive to apply biomarkers and bioindicators in environmental monitoring is high. They constitute the first set of tools with which one can measure

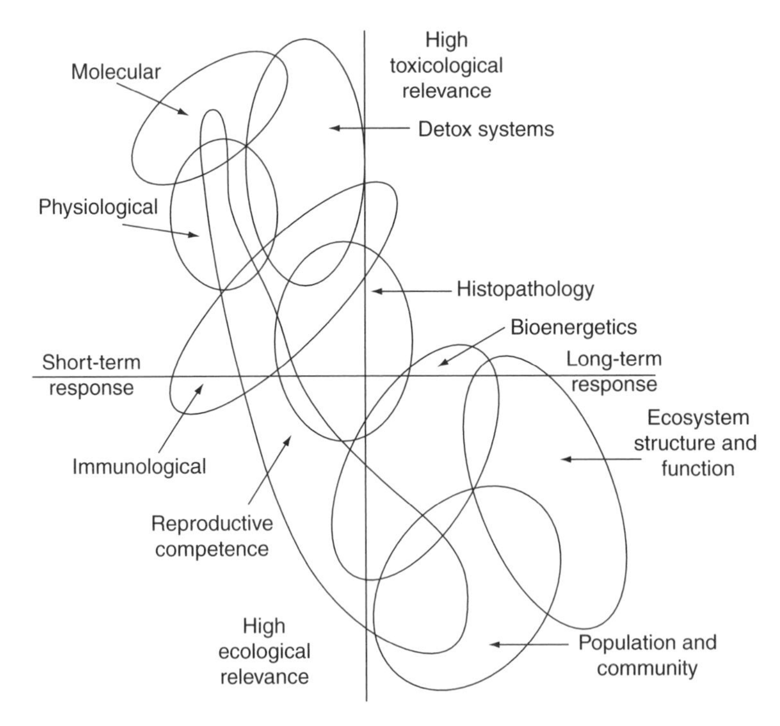

Figure 7.1 The three major factors that vary as one moves through the spectrum of the possible endpoints to monitor in the biological response range. 1. Time after exposure to manifested response occurs (increases from left to right). 2. Importance of response – molecular versus population effects (increases from top to bottom). 3. The ability to assign causal links – DNA adduct versus tree death (decreases from top left to bottom right). Modified after Adams *et al.* (1989). Reprinted with permission from Elsevier Science.

the actual effects that chemicals are having on biota in the field. They provide sensitive and biologically relevant measures, without the need to make assumptions or extrapolate data, and often offer good or at least some indication of the causative agent. Used properly, biomarkers have many potential benefits, but their improper application or interpretation carries the risk of both false negatives and false positives. As with nearly all biological measurements, there is prerequisite information that we must consider, check or control before we can make reliable use of biomarkers and bioindicators. Information required to develop criteria to maximize the outcome of any biomarker monitoring includes the following:

- The degree to which the magnitude of the response depends upon the dose of the exposure (i.e., the dose-response relationship) should be known.
- The time it takes from the onset of exposure until a response is manifested, how long the response is maintained during exposure and the persistence

after the end of exposure (responsiveness, transience and reversibility) should be known.

- The sensitivity of the biomarker response should be higher than those of the organism life-history parameters, and could originate from either very short response times following initial exposure or the ability to detect changes at lower concentrations.
- Some biomarkers and indicators are more applicable to certain groups of organisms for various reasons (e.g., size, physiology, metabolic characteristics of the host organism) and thus have inherent biological specificity.
- Furthermore, it is required to understand the stressor specificity of the biomarker. Non-specific responses give a measure of general stress resulting from a broad range of pollutants (e.g., several classes of pollutants, for example heavy metals, polycyclic aromatic hydrocarbons (PAHs), etc.) or severe natural stress, whereas specific responses (e.g., DNA adducts) can be linked directly to a specific pollutant.
- As for all biological measurements, some inherent variability will always accompany biomarker measurements resulting from season, temperature, pH, sex, weight and handling. Such inherent variability in any biomarker measurement should preferably be within such limits that it does not interfere with the interpretation of any toxic effects (e.g., signal-to-noise ratio). To enable us to assess which responses are due only to contaminants, we need to build up comprehensive baseline data of the expected range of biological responses under normal ranges for all conditions likely to be encountered.
- To be useful as an assessment or even predictive tool there should be some link between the biological significance of the measured responses and effects at higher levels of organization, or the likelihood of such effects occurring.
- Methodological concerns such as precision, reproducibility, ease of use and cost must be considered.

7.2 CHOICE OF SPECIES

Ideal soil invertebrates or microbial measures for use in any monitoring or assessment should meet the following criteria:

- Widely available and easy to sample. This includes obvious issues such as a good geographical range and temporal (i.e., not seasonal) presence.
- Robust, easy to transport and maintain in the laboratory.
- Sensitive to a wide range of chemicals.
- An ecologically important species that has been well characterized with respect to its ecology, toxicology and basic biology.
- Reasonably sedentary or have a small home range, if monitoring is to be site-specific.

7.3 SOIL INVERTEBRATE BIOMARKERS

There are few terrestrial organisms that meet all of the above criteria in relation to determining soil pollution effects, and most fail completely on one or several counts. Several soil invertebrate species, however, may serve as useful targets in which to assess soil health. Invertebrates represent more than 95% of all animal species and are mostly ubiquitous and abundant. They carry out essential functions in most soil ecosystems, both in terms of breaking down organic matter and as important links in food chains. The ecological and scientific relevance of soil invertebrates is reflected in their choice as preferred test species for standard toxicity testing. Soil-dwelling species (e.g., earthworms) cover a relatively small area (home range) and, through their behaviour and feeding habits, tend to integrate any spatial heterogeneity within a defined area. Furthermore, soil invertebrates are easy to handle and maintain, and there are few ethical and legal considerations discouraging their use compared with, for example, small mammals.

The following sections outline some specific biomarkers in selected soil invertebrates in detail, with particular attention to biomarkers demonstrated in earthworms.

7.3.1 DNA DAMAGE

DNA damage has been well researched in humans, other mammals and fish, but relatively little has been done in terrestrial invertebrates. DNA damage takes two major forms, DNA adducts and strand-breakages. In earthworms, only three species: *Lumbricus terrestris* (both adducts and strand-breakages) (Van Schooten *et al.* 1995, Verschaeve and Gilles 1995, Walsh *et al.* 1995), *Lumbricus castaneus* (strand-breakage) (Verschaeve *et al.* 1993) and *Eisenia fetida* (strand-breakage, but with few experimental results reported) (Verschaeve and Gilles 1995, Salagovic *et al.* 1996) have been examined in any detail. Exposure to mixtures of dioxins, X-rays, mitocin C and soil contaminated with a range of organic compounds such as coke, benzene and aniline has been tested with respect to strand-breakage using a single cell electrophoretic assay (Verschaeve *et al.* 1993, Verschaeve and Gilles 1995, Salagovic *et al.* 1996). For these exposures an increased number of strand-breakages were found, concomitant with an increasing degree of contamination.

Earthworms exposed to industrially contaminated soils (including PAHs and other organic compounds) were tested for DNA adducts using the ^{32}P-postlabelling technique (Van Schooten *et al.* 1995, Walsh *et al.* 1995). These studies showed an increased number of adducts measured in worms exposed to contaminated soil. A time-dependent increase in DNA adduct formation under continued exposure was reported by Van Schooten *et al.* (1995). In invertebrates, strand-breakage may be repaired rather rapidly, but in organisms with active metabolism for a particular carcinogen, DNA adducts generally have longer biological half-lives than the substrate carcinogens (Shugart *et al.*

1992). Thus, measurement of the levels of specific DNA adducts may well provide a good and relevant biological indicator of prior exposure to certain environmental carcinogens (Van Schooten *et al.* 1995). The other advantage of adducts over strand-breakages is that adducts usually give a good indication of the causative agent, whereas strand-breakages provide none (Shugart *et al.* 1992). However, studies linking DNA alterations and life-history endpoints, or effects on population size or community structure, or the influences of natural variation on the occurrence of strand-breaks and formation of DNA adducts in earthworms or other soil invertebrate groups have not been undertaken.

7.3.2 METALLOTHIONEINS

7.3.2.1 Gastropods

Apart from fruit flies and nematodes, which due to their small size and ecology are less convenient for site assessment, the most thoroughly investigated terrestrial invertebrate metallothioneins (proteins involved with metal binding and storage) (MTs) are those of snails and slugs (Dallinger 1996). In contrast to vertebrates, snails possess organ- and metal-specific MT isoforms, which react differently to various extrinsic environmental stimuli. The midgut MT isoforms of terrestrial snails are induced rapidly in response to cadmium (Berger *et al.* 1995), with the isoforms becoming nearly exclusively loaded with cadmium. However, the copper-binding MT isoform in the snail mantle is non-inducible by either cadmium or copper (Dallinger *et al.* 1997). This strong differential responsiveness makes the snail MT a multiple biomarker system, with potential utility in risk assessment and terrestrial ecotoxicology (Dallinger 1996). Because of possible species-specific peculiarities in MT structure and function, it is suggested only to use MTs from *Helix pomatia* and *Arianta arbustorum*, which have been thoroughly investigated for biomarker purposes (Dallinger and Berger 1993). A cadmium saturation assay (Bartsch *et al.* 1990) has been adopted to specifically detect and quantify the cadmium-binding MT pool in the midgut gland of the snails (Berger *et al.* 1995), with both accuracy and reproducibility comparable to spectrophotometric quantification of MT (Dallinger 1998). This allows the MT concentration in the midgut gland to serve as a specific measure of cadmium exposure under both laboratory and field conditions. Additionally, as (Cd)-MT can accumulate over extended periods of time, it can serve as a biomarker for both recent and previous Cd exposure. Whilst being non-inducible by cadmium and copper, the copper-specific mantle isoform seems to be susceptible to ionizing radiation, some organic chemicals and other stressors (e.g., exposure to cold) (Dallinger and Berger, unpublished data), all of which may cause a substantial reduction of mantle MT concentration. MTs may also confer protection against oxidizing agents and free radicals, being 'sacrificed' in a process of thiolate oxidation and metal release (Thornalley and Vašák 1985, Thomas *et al.* 1986).

Thus, such decreases in MTs may have potential as a non-specific marker of stress due to adverse climatic conditions, ionizing radiation or any kind of xenobiotic effects causing oxidative stress. It remains to be assessed how variable such a parameter would be under normal conditions, and to what extent it may be meaningful for biomarker purposes under varying conditions of stress and toxicant exposure in the field.

7.3.2.2 Lumbricid Earthworms

Various metal-binding proteins have been identified in earthworms. Some resemble metallothionein found in other invertebrates, although generally containing less cysteine(s), while others are quite distinct from MTs (e.g., containing large quantities of aromatic amino acids) (Yamamura *et al.* 1981, Bauer-Hilty *et al.* 1989, Morgan *et al.* 1989, Nejmeddine *et al.* 1992, Dallinger 1996). MTs, metallothionein-like proteins and metal-binding proteins have been identified in a number of earthworm species such as *E. fetida, Dendrodrillus rubidus, Dendrobaena octaedra, Aporectodea (Allobophora) caliginosa* and various *Lumbricus* species (i.e., *terrestris, rubellus, variegatus* and *castaneus*) (Suzuki *et al.* 1980, Furst and Nguyen 1989, Morgan *et al.* 1989, Ramseier *et al.* 1990, Bengtsson *et al.* 1992, Nejmeddine *et al.* 1992, Dallinger 1998). One of the first to report a metal-binding protein in earthworms was Suzuki *et al.* (1980), who exposed *E. fetida* to 1 to 500 mg Cd kg^{-1} and observed a dose-response induction of three cadmium-binding proteins with a concurrent increase in the cadmium concentration in the worms. One of the proteins was characterized as a metallothionein-like protein (Yamamura *et al.* 1981). Other studies also reported induction of metal-binding proteins in earthworms after exposure to cadmium, but never by other metals (Suzuki *et al.* 1980, Yamamura *et al.* 1981, Bauer-Hilty *et al.* 1989, Furst and Nguyen 1989, Morgan *et al.* 1989, Ramseier *et al.* 1990, Nejmeddine *et al.* 1992, Dallinger 1998). Morgan *et al.* (1989) reported zinc but not copper to be present in the cadmium-binding proteins. Bengtsson, *et al.* (1992), when sampling *D. octaedra* in the vicinity of brass mills of different ages, found that Cu was bound to two molecular size fractions and Zn to a number of large and medium-sized fractions, which partly overlapped the three fractions binding Cd. Two of the Cd-binding fractions bound 73-89% of the Cd content, and one was more abundant in worms from the longer-term exposure site. Later studies have shown that there are two different proteins, one binding copper and cadmium, and one with a distinctly lower weight that bound lead and zinc (Sturzenbaum 1997). By fingerprinting amplified messenger RNA (mRNA) species by the technique of directed differential display, it has been possible to pinpoint genes that were up-regulated in a population of earthworms adapted to exceptionally high levels of heavy metals (Sturzenbaum *et al.* 1998a). Combining standard gel chromatographic techniques and novel molecular methodologies (directed differential display and quantitative PCR), it has been possible to isolate and sequence two isoforms of the first true earthworm metallothionein. Both proteins are characteristically high in cysteine residues and possess no significant aromatic residues. Metal responsiveness was confirmed by determining metallothionein-specific expression profiles in earthworms exposed to soils of differing heavy metal concentration. Analysis of the derived amino acid sequence of isoform 2 identified two putative N-glycosylation signal sequences, suggesting that the two isoforms may have different subcellular distributions and functions (Sturzenbaum *et al.* 1998b). Also, the primary structure of three MT isoforms from *E. fetida* has recently been elucidated (Gruber *et al.* 1998). The MTs of *E. fetida* are present as mainly Cu-binding proteins at a low level in uncontaminated animals, but are strongly induced by Cd exposure, with up to 65% of the Cd becoming bound to the protein isoforms. In a study with *L. rubellus* with different metal pre-exposure histories it was found that copper and cadmium, but not zinc, were bound to a metallothionein homologue. Also, whereas cadmium was a strong inducer of MTs, copper did not induce MT synthesis efficiently in earthworm tissues, but copper may bind avidly to cadmium-induced MTs by substituting for cadmium in the molecule as the Cu–thionein is a more stable complex (Marino *et al.* 1998).

7.3.2.3 Enchytraeid Earthworms

Willuhn *et al.* (1994a) showed that when the oligochaete *Enchytraeus buchholzi* was exposed to cadmium in solution it was rapidly accumulated by the worm, with no detectable acute toxic effects at exposure concentrations below 4 mg Cd l^{-1}. Such subtoxic cadmium concentrations, however, caused the induction of an mRNA species coding for a non-metallothionein 33-kDa protein. Further studies of the effects of sublethal cadmium concentrations on gene expression in *E. buchholzi* revealed that accumulation of cadmium in large amounts coincided with the induction of an mRNA encoding a novel Cys-rich (27%) non-metallothionein protein (25 kDa). The gene was not constitutively expressed in untreated worms, but was rapidly induced by cadmium (Willuhn *et al.* 1994b). Investigation of the Cd-specificity and Cd-sensitivity of the gene induction showed that the amount of mRNA correlated with the environmental as well as the intra-worm Cd concentration, and that even the subtoxic concentration of 100 μg Cd l^{-1} induced cys-rich protein gene expression after 2 h exposure. Other heavy metals or stress conditions induced only little or no mRNA (Willuhn *et al.* 1996a). The same occurred for another Cd-responsive gene from *E. buchholzi* that encoded a putative aldehyde dehydrogenase, the expression of which was strongly enhanced by Cd, but remained unaffected by other stressors such as Zn, Hg and H_2O_2. The Cd-specific induction of this aldehyde dehydrogenase suggested that Cd-detoxification in *E. buchholzi* required low intracellular concentrations of aldehydes, that were otherwise known to target sulphydryl groups, thus inactivating the Cd-binding capacity of the metal-binding protein (Willuhn *et al.* 1996b).

The above studies suggest that such genes may be promising candidates for monitoring bioavailable Cd at subtoxic levels in terrestrial environments. The Cd-induced synthesis of these transcriptionally regulated proteins might be a pre-indicator for Cd intoxication in enchytraeids. Transcription is switched on long before Cd becomes toxic to the worms, and can therefore be taken as a warning signal of prospective irreversible Cd-damage at an early and still reversible stage. However, it remains for levels of transcription to be linked to observed adverse effects on survival or reproduction to put these biomarker responses into context.

7.3.2.4 Other Species and New Developments

Other species have been shown to contain non-MT metal-binding proteins and glycoproteins including the nematode *Caenorhabditis elegans*, the grasshopper *Aiolopus thalassinus* and the stonefly *Pteronarcys californica* (Clubb *et al.* 1975, Lastowski-Perry *et al.* 1985, Slice *et al.* 1990, Schmidt and Ibrahim 1994). Attempts to prove the existence of MTs in terrestrial isopods for years, however, have met limited success (Donker *et al.* 1990). Instead, a cadmium-binding protein has been isolated from metal-exposed isopods, which is probably a glycoprotein involved in metal storage and/or detoxification (Dallinger 1993).

One of the latest developments in metal-binding moities was the discovery of elevated levels of free amino acids after metal exposure. When Gibb *et al.* (1997) employed high-resolution ^{1}H-NMR spectroscopy to investigate the biochemical effects of Cu(II) in earthworms, the most marked metabolic response was the elevation of endogenous whole-body free histidine in the animals, which positively correlated with increasing copper exposure and total body copper burdens. Histidine forms thermodynamically stable copper complexes over a wide range of physicochemical conditions, and it was proposed that the elevation of free histidine provided an energetically 'low cost' detoxification mechanism. Elevation of free histidine may provide a novel molecular biomarker of Cu(II) exposure in such environmental situations (Gibb *et al.* 1997).

It is now possible to assess MT synthesis at genetic, molecular or biochemical levels. The specific induction of particular MT isoforms synthesized in response to specific metals or chemicals may provide information on particular chemical exposures in the field. Care must, however, be taken to consider any inherent variability from both extrinsic and intrinsic parameters that could interfere with such measurements.

7.3.3 ENZYMES

Some of the most popular biochemical parameters used as biomarkers result from the interaction of xenobiotics with enzyme systems. Such effects can result in either the induction of detoxification enzymes, switching between isozymes or the inhibition of enzymes. Although many enzymes have been studied in the context of toxicology and ecotoxicology, only a few have been thoroughly investigated in soil invertebrates.

7.3.3.1 Detoxication Enzymes

Mixed Function Oxidase

One of the most widespread and important detoxification enzyme systems is the cytochrome P450 or mixed function oxidase (MFO) system. The most promising set of investigations of these enzymes in soil invertebrates for the purposes of environmental monitoring has been in earthworms. Stenersen *et al.* (1973) showed that carbofuran was metabolized to 3-hydroxycarbofuran by *L. terrestris*, and Nelson *et al.* (1976) demonstrated that aldrin was converted to dieldrin. Both of these observations suggest the involvement of a P450 pathway (Berghout *et al.* 1991, Stenersen *et al.* 1992). The actual presence of cytochrome P450 monooxygenases in earthworms was more clearly documented by the isolation and characterization of the P450-dependent monooxygenase system from the midgut of *L. terrestris* by Berghout *et al.* (1989, 1991) and from whole-body microsomes of *E. fetida* by Achazi *et al.* (1998). Although induction of ethoxyresorufin-O-dealkylase (EROD) is one of the most commonly used means to measure P450 activity, no studies have observed EROD activity in earthworms. Ethoxycoumarin-O-dealkylase (ECOD) activity has been observed by Liimatainen and Hanninen (1982) and Stenersen (1984) in *L. terrestris* and by Eason *et al.* (1998) in *L. rubellus, A. caliginosa* and *Eisenia andrei*. These results, however, are in conflict with the findings of Berghout *et al.* (1990, 1991) which showed no ECOD activity, but found benzyloxyreorufin-O-dealkylase (BenzROD) activity in *L. terrestris.* Achazi *et al.* (1998) also found no EROD activity, but demonstrated both BenzROD and pentoxyreorufin-O-dealkylase (PentROD) in *E. fetida*. These discrepancies may be due to the use of different fractions of the worms and/or homogenates (Eason *et al.* 1998). The additional identification of non-inducible forms of P450 in *Dendrobaena veneta* (*Eisenia veneta*) (Milligan *et al.* 1986) and *E. fetida* (Achazi *et al.* 1998), along with some species differences in actual activity reported by Eason *et al.* (1998), suggest caution in using P450 data for earthworms. *A. caliginosa* have lower inherent P450 activity than *L. rubellus* (Eason *et al.* 1998), and were more sensitive to aldicarb, carbaryl and carbofuran (compounds normally metabolized by the P450 system) than *L. rubellus* (Stenersen 1979). This suggests that *L. rubellus* may be protected by its higher P450 levels.

From these studies, it is clear that more research is required before earthworm P450 activities can be used as a tool for environmental monitoring. P450 levels found in earthworms are generally very low compared with those of other organisms (Eason *et al.* 1998), but if methodological standardization is achieved and inducibility can be demonstrated, P450 activity should be a potentially valuable biomarker. The P450 system addresses effects of important organic contaminants that are otherwise hard to assess and have potential to monitor recovery of treated or remediated sites, as the biochemical effects are reversible. Again, this marker remains to be linked to negative effects on reproduction or earthworm fitness.

Glutathione S-transferase

The other detoxification enzyme system that has been well researched in earthworms is glutathione S-transferase (GST). GST was initially found in eight species of earthworms (Stenersen *et al.* 1979, Stenersen and Øien 1981). A multitude of compounds were able to induce GST, but Stenersen and Øien (1981) also found that a range of compounds inhibited the conjugation between glutathione and 1-chloro-2,4-dinitrobenzene (CDNB) in *E. fetida*, thereby inhibiting the metabolism of CDNB. There have been differing reports on the inducibility of GST in earthworms. Stokke and Stenersen (1993) found no induction of GST in *E. andrei* following exposure to three classic inducers (*trans*-stilbene oxide, 3-methylcholanthrene and 1,4-bis[2-(3,5-dichloropyridoxyl)] benzene) for a period of 3 days. Later exposure of both *E. andrei* and *E. veneta* by Borgeraas *et al.* (1996) to *trans*-stilbene oxide, 3-methylcholanthrene and phenobarbital for three weeks similarly did not show elevated activity of GST, when measured with CDNB and ethacrynic acid (ETHA). Hans *et al.* (1993), however, found induction of GST in *Pheretima posthuma* by aldrin, endosulphan and lindane. The induction measured by Hans *et al.* (1993) was shown to be transient, with GST activity being elevated only during the first one to three weeks of exposure. Thereafter the GST activity returned to near control levels under prolonged exposure. Hans *et al.* (1993) suggested this transient effect may be related to the biotransformation and elimination of the pesticides. One other explanation may result from there being between five and six isoenzymes of GST in the *Eisenia* species, all showing different substrate specificities toward the substrates CDNB, 1,2-dichloro-4-nitrobenzene, ETHA and cumene hydroperoxide (Borgeraas *et al.* 1996). It may be possible that the exposure compounds both affect and induce isoenzymes for which the assay substrates have low affinities, or as shown by Stenersen and Øien (1981) that the exposure compound inhibits the metabolism of the assay substrate, thus resulting in no induction.

These problems are manageable with more research into method development in line with that of Borgeraas *et al.* (1996) for GST and SaintDenis *et al.* (1998) for a range of antioxidant defence enzymes. GST activity has been shown to be unaffected by exposure to zinc and cadmium in *E. fetida* (Grelle and Descamps 1998). In wolf spiders, however, GST activity decreased with increased feeding rate, but was unaffected by sex and body weight (Nielsen and Toft 1998). GST induction has potential to serve as a valid biomarker of exposure for soil invertebrates once the mechanism, the mode of action and any inherent variation is better understood, and the methodology more developed.

7.3.3.2 Enzyme Inhibition: Acetylcholinesterase and Cholinesterase

Direct inhibition of enzyme activity can in theory occur for any enzyme. A well-known and well-developed biomarker that is based on enzyme inhibition is acetylcholinesterase (AChE) or cholinesterase (ChE) inhibition. Acetylcholine is used by invertebrates for the transmission of nerve signals just as it is in vertebrates. Two classes of chemical compounds, organophosphorus (OP) and carbamate pesticides, were specifically developed with the aim of inhibiting AChE by blocking the enzymes binding site for acetylcholine and serving as a chemical warfare agent and later as an insecticide (Walker *et al.* 1996). It is therefore no surprise that both classes of compound cause depression of total ChE activity in invertebrates and that their impact at the biochemical level can be expressed as a change in ChE activities (Edwards and Fischer 1991). Inhibition of ChE has been studied in various terrestrial invertebrates such as lacewing larvae (Rumpf *et al.* 1997), carabid beetle (Jensen *et al.* 1997) and grasshoppers (Schmidt and Ibrahim 1994), but has been studied in most detail in earthworms.

The initial study of ChE activity in earthworms by Stenersen *et al.* (1973) investigated the effects of seven OP and three carbamate compounds on *L. terrestris*. The results showed that the OP compounds caused a more severe inhibition (>99%) of the ChE activity than the carbamates (30–80%). Additionally, the ChE activities of worms exposed to OPs demonstrated slow (>50 day) or no recovery (malathione 60 day), whereas ChE activities recovered to normal levels within 30 to 40 days after exposure to carbamates. In contrast to the degree of ChE inhibition and ability to recover, the mortality of worms exposed to carbamates was higher than that of OP exposed worms. In the case of carbofuran (a carbamate) mortality was as much as eight times higher than that observed for OP exposures. In a study of five benzimidazole compounds in *L. terrestris*, Stringer and Wright (1976) found no effects on ChE activity of *in vivo* or *in vitro* exposure to the parent compounds, but did observe inhibition of ChE after *in vitro* exposure to metabolic compounds of the benzimidazoles. Stringer and Wright (1976) also observed that mortality was not linked to the ChE inhibition; benomyl, for example, caused no ChE inhibition itself, but mortality was observed after 6 days. The chemical specificity of ChE has been demonstrated in many organisms and regarded historically as a specific marker for OP and/or carbamate exposure. Recent research has challenged this point after observing effects of cadmium, mercury and lead on ChE activity in a grasshopper (Schmidt and Ibrahim 1994) and lead and uranium inhibiting ChE activities in *E. andrei* (Labrot *et al.* 1996, Guilhermino *et al.* 1998). The use of ChE as a specific biomarker for OP and carbamate pesticides should now be questioned. However, in cases potentially involving either OPs or carbamates, depression in ChE activity remains the best biomarker of exposure but not necessarily of effect.

Generally, the use of different enzyme systems presents one of the richest and most versatile fields for using biomarkers to monitor a given problem, particularly for pesticides. The need for a detailed understanding of the mechanistic modes of action is essential, such that any factors influencing either the amount of enzyme present or its activity are known.

7.3.4 STRESS PROTEINS

The use of 'stress proteins' (previously known as 'heat shock proteins') as biomarkers for environmental hazards in invertebrates has primarily focused on the aquatic environment (reviewed by Sanders 1993). Over the past few years, however, interest in the induction of stress protein expression by toxicants in terrestrial invertebrates has grown, and current approaches to establish biotests show promising results.

7.3.4.1 Stress-70 Group

The first indication of the potential use of the stress-70 protein group (named on the basis of the typical molecular weight of 70 kDa for proteins in the group) induction in terrestrial invertebrates was demonstrated by Köhler *et al.* (1992) for the isopod *Oniscus asellus*. Laboratory studies of the effect of various heavy metal mixtures on *O. asellus* demonstrated that the stress-70 response increased with dose until a threshold beyond which further increase in exposure concentrations resulted in declined stress-70 levels (Eckwert *et al.* 1997). Corresponding exposures at various field sites confirmed these findings, thus demonstrating the potential for application of stress-70 in the field (Köhler and Eckwert 1997). Further experiments measured induction of stress-70 proteins in *O. asellus* and another isopod, *Porcellio scaber*, in response to a variety of environmentally relevant compounds (Cd, Pb, Zn, lindane, pentachlorophenol, 2,2′,5,5′-tetrachlorobiphenyl (PCB52) and benzo(*a*)pyrene) (Eckwert *et al.* 1994, 1997, Köhler *et al.* 1999a). This latter study, however, showed that the stress-70 response was transient and disappeared after only days of continued exposure for PCB52 and benzo(*a*)pyrene, thus urging caution in the interpretation of field-collected data (Köhler *et al.* 1999a).

Slugs, *Deroceras reticulatum* and *Arion ater*, following carbamate molluscicide exposure had induced levels of stress-70 proteins (Köhler *et al.* 1992), and exposure to sublethal concentrations of metals (Zn, Cd, Pb) and pentachlorophenol resulted in a dose–response pattern of stress-70 induction (Köhler *et al.* 1994, 1996a, 1999a). The stress-70 response in *D. reticulatum* was proved as an early warning marker for the ecological effects of metal, in an experiment detailing changes in life-history parameters, fecundity, offspring number, longevity and mortality of the slugs linked to the degree of stress-70 induction (Köhler *et al.* 1998).

The diplopod *Julus scandinavius*, when exposed to cadmium, zinc, PCP52 and lindane, showed elevated stress-70 levels (Eckwert *et al.* 1994, Zanger *et al.* 1994, Zanger and Köhler 1996).

Investigations of stress-70 levels in the collembolans *Orchesella bifasciata* and *Tomocerus flavescens* revealed that different populations influenced by different soil metal concentrations developed distinct adaptive stress response mechanisms, and further that stress-70 levels recovered to normal levels when exposure was stopped (Köhler *et al.* 1999b).

7.3.4.2 Stress-60

Exposure to metals (Zn, Pb, Cd) caused a slight increase in the level of stress-60 in the supernatant of homogenates of *D. reticulatum, J. scandinavius, O. asellus* and *P. scaber* (Rahman, Zanger and Köhler, unpublished data), but much reduced in comparison to stress-70 response observed in *O. asellus* (Eckwert *et al.* 1997). The induction of stress-60 was also studied in the nematode *Plectus acuminatus* following exposure to heat, copper and cadmium, and was related to increased concentrations of cadmium and copper (Kammenga *et al.* 1998). For copper, the induction of stress-60 in *P. acuminatus* was three orders of magnitude more sensitive than the EC_{20} for reproduction. For cadmium, stress-60 induction was one order of magnitude more sensitive. These results demonstrate that stress-60 induction occurred at concentration levels that were realistic for the field situation. Thus, stress-60 may be suitable as a potential biomarker of toxicant stress in *P. acuminatus* in the field.

7.3.4.3 Field Usage

Up to now, information on the stress protein status of populations of soil invertebrates taken from their natural habitats has been scarce, but some aspects are promising. Köhler *et al.* (1992) found elevated stress-70 levels in a woodlouse (*O. asellus*) population living on a smelter spoil bank when compared to a population from a pristine site. In a comparative study with 14 populations of the diplopod *J. scandinavius*, stress-70 levels were correlated with the lead and zinc concentrations in soil for all sites, except for a military area (potentially influenced by pollutants other than metals) and two long-term polluted sites near a former open-cast mine (where likely adaptation mechanisms had evolved in resident populations; Kammenga and Simonsen 1997). Tolerance was also shown by a population of isopods (*O. asellus*) living in the above-mentioned mining area (Eckwert and Köhler 1997). Observations like these may lead to the assumption that reliance solely on laboratory-based stress protein studies may be limited in their predicative potency in respect to longer-term contaminated areas, but also that there is good potential for assessing stress levels in the field using these techniques. Although good data may be obtained by using stress proteins, care must be taken to check for the possible confounding effects of any natural stresses influencing any conclusions.

7.3.5 HISTOPATHOLOGICAL AND ULTRASTRUCTURAL CHANGES

Histopathological changes have been examined in diverse groups of terrestrial invertebrate species. Some of these are reported here.

7.3.5.1 Gastropods

In the gastropods *D. reticulatum*, *A. ater* and *H. pomatia*, cellular alterations have mainly been described in the hepatopancreas (especially in resorptive and basophilic cells) following exposure to molluscicides (Triebskorn 1989, Triebskorn and Kunast 1990, Triebskorn *et al.* 1998) and metals (Recio *et al.* 1988, Marigomez *et al.* 1996, Triebskorn and Köhler 1996). Although such studies have mainly been restricted to laboratory experiments with distinct contaminants, some results show potential to develop these techniques for use in the field. As an example, Marigomez *et al.* (1996) investigated the effects of Hg on the structure of the digestive gland of the slug *A. ater*. It was found that the excretory activity in digestive cells was initially enhanced. After prolonged exposure the relative numbers of digestive cells declined until, in extreme cases, the digestive epithelium was mostly comprised of calcium and excretory cells. Similar observations were made on slugs collected from an abandoned copper mine (Marigomez *et al.* 1998). Such results suggests that a 'Slug Watch' monitoring programme could be developed similar to the 'Mussel Watch' programme, which is currently applied to assess environmental quality in coastal and estuarine areas (Marigomez *et al.* 1996, 1998).

7.3.5.2 Oligochaeta

In Oligochaeta (*E. fetida, Enchytraeus* spp.), the deleterious influence of the insecticides cypermethrin and parathion on cells of the chloragogenous tissue were described by Hagens and Westheide (1987), Westheide *et al.* (1989) and Fischer and Molnar (1992). Impact of the pesticide monocrotophos on oogenesis in the earthworm *Eudichogaster kinneari* was histologically shown by Lakhani *et al.* (1991). In *E. fetida,* paraquat intoxication has been observed to cause a depletion of the chloragogenous tissue, and heavy metals can markedly influence the elemental composition of chloragosomes (Fischer and Molnar 1992). Similar changes have been observed in the intracellular compartmentation of cadmium, lead, zinc and calcium in *D. rubidus, L. rubellus* and *A. caliginosa* sampled from unpolluted and metal-contaminated sites (Morgan and Morris 1982, Morgan and Morgan 1998). More recent studies have revealed both histological and ultrastructural damage to the post-clitellar segment of the ganglia of juvenile *E. fetida* caused by the herbicide paraquat and the fungicide triphenyltin (Zsombok *et al.* 1997).

7.3.5.3 Isopods

Isopods (*O. asellus)* collected from contaminated field sites had record high soft tissue concentrations of zinc, cadmium, lead and copper in their hepatopancreas (Hopkin and Martin 1982a). This finding was linked to the demonstration of ultrastructural changes in the form of the presence of metal-containing granules within the hepatopancreas (Hopkin and Martin 1982b). Thus, the hepatopancreas of isopods appeared to be the most important organ to monitor the effects of heavy metal pollution in the isopods *Armadillidium vulgare, O. asellus* and *P. scaber* (Hopkin and Martin 1982b, Prosi *et al.* 1983, Prosi and Dallinger 1988). Here the degree of the ultrastructural alterations was found to be dose-dependent, with low metal concentrations causing reactions of distinct organelles, whilst higher metal concentrations resulted in pathological changes to the epithelium of the hepatopancreas (Köhler *et al.* 1996b).

7.3.5.4 Diplopods and Collembola

In diplopods both midgut and hepatic cells have been shown to be suitable for heavy metal effect monitoring in the laboratory as well as in the field (Köhler and Alberti 1992, Berkus *et al.* 1994). Similar responses have been observed to occur in a dose-dependent manner in laboratory toxicity tests with the collembolan *Tetrodontophora bielanensis* (Pawert *et al.* 1994, 1996).

7.3.5.5 Field Usage

In many of the studies cited here, one of the most common ultrastructural changes observed in response to metal exposure was the formation of metal-containing spheres or granules. Such granules have been found in the digestive and excretory organs of a great number of terrestrial invertebrates (for a detailed review see Hopkin 1989). Many of the studies cited above showed that the formation of these metal-containing granules occurred in a dose-dependent manner (Triebskorn and Kunast 1990, Marigomez *et al.* 1996, 1998, Pawert *et al.* 1996, Triebskorn and Köhler 1996). Köhler and Triebskorn (1998) developed a protocol according to the physiological basis of the ultrastructural responses in order to combine the qualitative and quantitative aspects of the observed ultrastructural alterations into an easy to interpret 'impact index'. Using their protocol, the data clearly demonstrated differential susceptibility: (1) of organelles to the respective metals, (2) of the investigated species to particular metals, and (3) of the monitored tissue to increasing metal concentration. The development of such a systematic protocol, and the breadth of species and contaminants covered in the studies cited above, demonstrate the potential to develop these measurements for use in the field site assessment (Marigomez *et al.* 1996, 1998).

7.3.6 SUBCELLULAR AND CELLULAR FUNCTION

Studying histopathological change is a more integrated approach that reveals the cumulative effect of all the molecular and biochemical perturbations occurring in a tissue or organism. A similar approach can be taken with subcellular and cellular functions, which can be assessed via a multitude of parameters, an approach that has only recently been studied in terrestrial invertebrates.

7.3.6.1 Lysosomal Integrity

At the subcellular level, the lysosomal system has been identified as a particular target for the toxic effects of contaminants (Moore 1990). There are two techniques traditionally used for assessing the stability of the lysosomal membrane, one is cytochemical and the other is biochemical. In stable lysosomes, hydrolases are prevented from reacting with substrates by an intact membrane. Both techniques are based on measuring the activity or amount of lysosomal enzyme that leaks through the lysosomal membrane after either pH or hypo-osmotic shock, respectively. In the cytochemical procedure, frozen unfixed tissue sections are exposed to low pH in a staining medium. The stain is specific to a lysosomal enzyme, hence the time it takes for staining to occur (the latency period) reflects the leakage of lysosomal enzyme to the cytosol and in effect the lysosomal membrane's ability to deal with the low pH insult (Moore 1976). The biochemical technique is based on generating an enriched lysosomal fraction from the sample tissue by homogenization and differential centrifugation, which is then osmotically shocked *in vitro*, followed by determination of the amount of leaked enzymes (Baccino and Zurretti 1975). Neither of these techniques operates under normal physiological conditions when observing the leakage from the lysosomes.

Modifications of the neutral-red technique developed by Lowe *et al.* (1992) were implemented by Svendsen and Weeks (1995) and Weeks and Svendsen (1996) to enable its use with freshwater and terrestrial invertebrates. This new technique operates with minimum treatment stress to the cells as both sample preparation and observation is completed under physiologically relevant *in vitro* conditions. Also, the modified technique has a greatly increased time resolution and therefore higher sensitivity (Svendsen and Weeks 1995, Weeks and Svendsen 1996). The neutral-red retention assay is technically much simpler than both the cytochemical and biochemical techniques described above, and measures cell damage using the fact that only lysosomes in healthy cells permanently retain the cationic dye after initial uptake. Decreased retention of the neutral-red dye within the lysosomal compartment over time is used as a measure of increasing damage to the lysosomal membrane.

In recent studies of the effect of exposure temperature on neutral-red retention response to an increasing range of soil copper concentrations, the response was shown to be unaffected by the differences in exposure temperatures (Svendsen 2000). However, it is important that the effect of possible confounding factors is assessed for each new organism investigated. Lysosomal membrane destabilization has been observed in earthworms after exposure to various heavy metals (Cu, Cd, Zn, Pb, Ni) and some organic compounds (benzo(*a*)pyrene, chlorpyriphos, carbendazim and iprodione) (Scott-Fordsmand 1998, Svendsen 2000).

An additional argument for the use of lysosomal membrane stability as an environmental monitoring tool is that most organisms offer ways in which adequate samples can be obtained via non-destructive protocols and methods (e.g., extraction of cells from body fluids). Also, recent studies on earthworms exposed to copper proved that the sensitivity of the lysosomal membrane stability was relatively higher when compared to the sensitivity of effects at the whole individual level (e.g., growth, survival and reproduction) (Svendsen and Weeks 1997a). These effects could be validated under semi-field conditions (Svendsen and Weeks 1997b). Measurement of lysosomal membrane stability has been shown to be applicable under field conditions in terrestrial systems, when employed on earthworms from an industrial accident site (Svendsen *et al.* 1996).

Thus, lysosomal membrane stability appears to have a very considerable potential for *in situ* assessment of soil quality. With this method being relatively uncomplicated to undertake, requiring little sample preparation and apparatus, there is potential for use in low-cost routine and *ad hoc* screening programmes for unknown stressors.

7.3.6.2 Immune System Function

Immune systems are, in general, highly evolved and responsible for providing organisms with the ability to resist infections from various sources. Although the complexity of these systems has increased with evolution, there are several aspects of immunity that are phylogenetically conserved and thus may serve as pan-specific biomarkers of immune competence (Weeks *et al.* 1992). In terms of monitoring effects from pollution on immunological function, it is known that interaction of environmental chemicals with components of the immune system can both suppress and enhance immune activity (Luster *et al.* 1988). Hence, pollutants have the potential to greatly affect the susceptibility of organisms towards infections and diseases. Immunotoxicological endpoints may therefore provide information that is directly ecologically relevant.

The immune system of earthworms is based in the coelom and involves both coelomocyte cells and the coelomic fluid. Earthworm immunobiology is generally well researched and understood. The earthworm immune system, however, comprises both cellular and humoral components, which in turn have both specific and non-specific elements. The functionality of this system is so similar to that of mammals, that when an invertebrate surrogate system for mammalian immunotoxicity was to be developed, the earthworm was chosen (Goven *et al.* 1988, 1994a). Apart from providing the relevance of the earthworm system as a surrogate, the high degree of similarity meant that the established techniques from mammalian immunotoxicology have been easily transferred.

In the context of the effects of pollutants on the earthworm immune system, the parameters and techniques employed have covered both the cellular and humoral components. Activities or competencies of cellular aspects have been measured through the following parameters: the ability to reject allo- and xenografts and perform wound healing (Cooper and Roch 1992); phagocytosis of non-self material and rosette formation of foreign cells adhering to the cell surface of the coelomocytes (Goven *et al.* 1988, 1993, 1994a, Fitzpatrick *et al.* 1990, 1992); production of reactive oxygen species (H_2O_2 and O_2^-) (Chen *et al.* 1991, Valembois and Lassegues 1995); spontaneous cytotoxicity (Suzuki *et al.* 1995) and elimination of non-pathogenic bacteria (Valembois *et al.* 1985, Roch and Cooper 1991). The humoral aspect of earthworm immunity is based on agglutinins and lytic factors synthesized and secreted by coelomocytes and chloragocytes. Effectiveness of the humoral components has been assessed through the activity of various lytic factors and processes. These include lysozymes, proteases, hemolysis activity and antibacterial properties towards pathogenic bacteria (Goven *et al.* 1994b, Ville *et al.* 1995). An additional and frequently used measure is the ability to form multilayers so-called 'secretory rosettes' of foreign cells around coelomocytes through the action of agglutinins (Rodriguez-Grau *et al.* 1989).

The most used compounds in tests with earthworms have been polychlorinated biphenyles (PCBs) (Goven *et al.* 1988, 1993, 1994a, Rodriguez-Grau *et al.* 1989, Fitzpatrick *et al.* 1990, 1992, Eyambe *et al.* 1991, Suzuki *et al.* 1995, Ville *et al.* 1995, Cikutovic *et al.* 1999). Apart from that the range of compounds tested include the metals Cu, Cd, Hg, Pb and Zn (Goven *et al.* 1994b, Fugere *et al.* 1996), the organic compounds carbaryl, 2,4-dichlorophenoxy acetic acid (2,4D) (Ville *et al.* 1997) and pentachlorophenol (PCP) (Giggleman *et al.* 1998), plus a range of pesticides (Bunn *et al.* 1996).

Importantly, cellular immunity was, in general, always depressed following exposure to any compound, whereas the various humoral parameters, in general, were enhanced with only a few becoming depressed and some being unaffected (Roch and Cooper 1991, Cooper and Roch 1992, Suzuki *et al.* 1995, Ville *et al.* 1995, 1997). The most useful parameter for general screening of compounds or soils for immunotoxic effects on earthworms seems to be the level of phagocytosis and possibly the formation of monolayer rosettes by cell-surface adhesion (Ville *et al.* 1995). These are also technically among the easiest and simplest to measure. From the studies that measured more than one parameter, it is possible to say that humoral parameters were more sensitive than cellular ones, and that the sensitivity of the latter was generally a little higher or comparable with that of mortality (Fitzpatrick *et al.* 1992, Goven *et al.* 1994a,b, Ville *et al.* 1997, Giggleman *et al.* 1998). The apparent low sensitivity of the cellular immune functions is possibly due to built-in redundancy, backup and plasticity of the immune system (i.e., effects are not observed until the last reserves are drained).

Laboratory studies where exposures were derived from a source of complex contamination, namely refuse-derived fuel ash (mainly metals) (Fitzpatrick *et al.* 1992, Goven *et al.* 1994a), indicated that the immune parameters could be used to assess real on-site contamination. Such assessments with *in situ* exposures of worms have been undertaken at a hazardous waste site (mainly herbicides and pesticides) (Goven *et al.* 1994a) and at the site of an industrial-scale plastics fire (metals and a cocktail of organic combustion products) (Svendsen *et al.* 1998).

Both these studies illustrated that these immune system parameters could be successfully applied to actual field exposure scenarios, for example by using the recovery of the immune system as an indication of remediation success (Goven *et al.* 1993, 1994a).

7.4 SOIL MICROBIAL BIOINDICATORS

The second part of this chapter examines the roles played by soil microbes at different levels of biological organization; focusing initially at the soil process level and ultimately providing examples of how individual responses at the molecular level may be used as markers of contaminated soils.

The application of soil microbial bioindicators can be considered at various different ecological stages, including the soil process level, the biological or ecological function level, the organism level and then the molecular level. This review serves to summarize the main components of this systematic approach and will consider aspects at a process level and at an organism level to reflect the true extent of soil microorganisms as bioindicators.

7.4.1 PROCESS LEVEL: OVERVIEW

Soil microbial processes, which include nitrification and sulphur oxidation, are mediated by specific components of the soil microbial biomass. Changes in the biomass composition, mineralization rates and microbial activity may be indicative of the effect of a pollutant. Effects of pollutants may also be identifiable from changes in diversity, extractable biomass and health of soil flora and fauna.

Babich and Stotzky (1983) proposed the idea of 'ecological dose' (EcD), similar to the lethal dose concept, as a method of defining the inhibitions at the soil processes level. The EcD_{50} represents the point at which the concentration of a pollutant caused a 50% inhibition of the soil process being monitored. This concept has been applied to soil respiration and nitrification studies in metal-contaminated sites. Soil process inhibition has been considered as reversible or persistent in terms of toxicity and, by relating delay factors to the monitoring period, it is possible to bring time in as a factor. The choice of process selected for study in terms of soil ecotoxicological assessment is generally a compromise that brings together a range of tests as a function of cost and convenience. Mineralization of organic matter and other functions of carbon and nitrogen cycles are included in most testing programmes because they are sensitive to pollutants and represent processes which determine the availability of nutrients through the food chain.

7.4.1.1 Carbon Mineralization

Carbon mineralization can be studied by measuring the evolution of CO_2 or the consumption of O_2. Respiration can be measured by using an alkaline trap and back titration, infrared gas analysis, gas chromatography or radiorespirometry using a labelled pollutant. As bulk soil respiration measurements have been criticized for being insensitive to pollutants by some authors (e.g., Domsch 1984) a more strategic approach is to evaluate substrate-induced respiration, thus considering a much wider range of metabolic pathways.

7.4.1.2 Nitrogen Cycles

Measurement of fluxes involved in nitrogen cycling requires great care in the sampling and handling of the soil. Quantification of components of the N-cycle requires the selective extraction of N from the soil matrix and the subsequent analysis thereafter. An appropriate extracting solution must allow quantitative extraction in a form that can be analysed, inhibit biochemical reactions that may change the form of N, utilize chemicals that do not interfere with the N assay method and remain stable for an adequate period of time.

7.4.1.3 Nitrogen Fixation

The largest source of fixed nitrogen is from free-living and root-associated terrestrial microorganisms (Paul and Clark 1989). Nitrogen fixation is greatly affected by environmental parameters such as the concentration of inorganic nitrogen and trace elements, pH and water potential, in addition to the growth phase of the organism. Measurements are most commonly carried out using the acetylene reduction assay, but the use of ^{15}N can resolve some of the problems associated with the acetylene assay and enable a more comprehensive understanding of the fluxes involved. Heavy metals in agricultural soil have been found, in general, to have little effect on fixation (e.g., Rother *et al.* 1983) although Tyler (1975) had found that N_2 fixation correlated with pollution levels in a forest soil at comparatively low soil metal concentrations (100 μg Cu g^{-1} soil).

7.4.1.4 Nitrification

The chemoautotrophic bacteria responsible for nitrification rely on sufficient O_2 and a suitable moisture and temperature regime in addition to NH_4^+ and NO_2^-. Several heterotrophic organisms have also been implicated in nitrification. Nitrification rates can be assessed by the measurement of ambient NO_3^- concentrations. More recently UV spectroscopy and nitrate specific ion have offered rapid screening methods, although the most commonly used quantitative technique involves nitrate determination using phenoldisulphonic acid and subsequent colorimetric analysis.

Nitrification has been found to be inhibited by a soil metal concentration of 100 mg kg^{-1} for Cu, Ni and Zn; 100–500 mg kg^{-1} for Pb and Cr; 10–100 mg kg^{-1} for Cd; and 1–10 mg kg^{-1} for Hg (Doelman 1986). Chang and

Broadbent (1982) noted that 400 mg kg^{-1} of Cr(III), Cd, Cu, Zn, Mn and Pb inhibited nitrification. Giashuddin and Cornfield (1978) found Ni more toxic to nitrification than to net N mineralization (100 mg kg^{-1} Ni reduced the accumulation of nitrate by 36%). The degree of toxicity of metals for nitrification was found to be: Hg>Cr(III)>Cd>Ni>Fe(III)>Co, Cu>Sn>Fe(III)>Zn>V>Mn>Pb (Liang and Tatabai 1978, Chang and Broadbent 1982).

The adaptation of mineralizing and nitrifying organisms to high concentrations of pollutants has been demonstrated by Rother *et al.* (1982). These workers found that nitrification of peptone was inhibited by amendments of 1000 mg kg^{-1} Cu or Zn and 10 000 mg kg^{-1} Cd.

7.4.1.5 Denitrification

Denitrification completes the nitrogen cycle by returning nitrogen atoms to the atmosphere and it also mobilizes NO_3^- that has percolated below the root zone of plants. It is perceived as a deterrent to soil fertility, rendering forms of nitrogen unavailable to plants. Bollag and Barabasz (1979) found that denitrification was sensitive to metals in the order Cd>Zn>Cu>Pb, although there is little evidence to suggest that insecticides or fungicides (Bollag and Liu 1990) have an effect on denitrification. Of all the components of the nitrogen cycle, denitrification may be the one most dominated by physicochemical factors. Measurements of denitrification have involved a range of methods, the disappearance of NO_3^- using an ion selective probe and the use of the C_2H_2 inhibition method.

7.4.1.6 Ammonification

A diverse and heterogeneous group of microbes is capable of proteolytic activity. This group includes both aerobic and anaerobic prokaryotes and eukaryotes. Measurement of ammonification is carried out by quantifying changes in NH_4^+ and NO_3^- concentration. A wealth of literature and commercially driven promotions has demonstrated that herbicides, fungicides and insecticides applied at normal field applications do not alter ammonification rates. Indeed, Wainwright and Pugh (1973) showed that the NH_4^+ pool increased after the addition of a range of fungicides. Bewley and Stotzky (1983a) found that 1000 mg kg^{-1} Cd did not affect ammonification in an acidic environment, whereas nitrification was reduced by a concentration of 500 mg kg^{-1} Cd.

7.4.1.7 The Use of Enzymes in Ecotoxicity Assessment

Enzymes catalyse numerous metabolic reactions in microbial cells and, accordingly, their inhibition could be the underlying cause of toxicity. Activities of dehydrogenases, amylases, phosphatases, arylsulphatases and cellulases have all been measured. The measurement of enzymes has enabled relatively inexpensive and rapid assessment of environmental samples. Enzymes in bulk soil have been found to exist in at least 10 different forms depending upon inhibition or stimulation (Burns 1982).

Most studies have considered the dehydrogenase enzymes and the phosphatase enzyme. Specific dyes can be used to trace electron transport activity. These dyes act as artificial hydrogen acceptors and, upon reduction, change colour allowing easy quantification with a spectrophotometer. Bitton and Koopman (1986) found that the dehydrogenase assay was sensitive in the detection of inhibition caused by heavy metals. As well as offering sensitivity, the assay can be carried out with a microbial community or single species, making it flexible in application. These beneficial aspects prompted the USEPA to adopt the use of dehydrogenase and phosphatase assays in 1978 to assess the toxicity of pesticides on non-target organisms (Bitton and Dutka 1986).

7.4.1.8 ATP-based Systems

Adenosine triphosphate (ATP) is a product of catabolic reactions and is rapidly destroyed after cell death, thus making it an ideal assay for differentiating living cells from dead cells. Soil ATP measurements provide an independent measure of soil microbial biomass (Jenkinson and Oades 1979, Burns 1982). Brookes and McGrath (1984) measured ATP in five metal-contaminated plots and five non-contaminated plots and found that ATP concentration declined as pollutant concentration increased. Ziblinski and Wagner (1982) found that ATP concentration was affected by amendments of Cd, Cu and Cr, while Eiland (1985) reported that results were inconclusive when comparing the standard fumigation incubation technique and measurement of extracted ATP. A determination of total microbial biomass based on measuring ATP can be achieved by following the reaction of firefly luciferin (Eiland 1985). This reaction is catalysed by magnesium and luciferase. The potential of using ATP content of actively growing cells to assess toxicity has seldom been systematically evaluated. The range of ATP kits that is widely available has made this assay convenient and relatively simple to carry out.

7.4.1.9 Soil Microbial Biomass Composition and Diversity

Two general microbial aspects of contaminated soil environments can often be observed. Firstly, a reduction in numbers and species diversity, and secondly, the development of pollutant-resistant microbial populations. The fumigation method provides a reliable estimate of soil microbial biomass concentration in contaminated and uncontaminated soils alike, and this was confirmed by comparison with ATP concentrations (Jenkinson and Oades 1979). Brookes *et al.* (1986) employed direct microscopy techniques to count microorganisms and demonstrate the same trends as biomass measurement by Brookes and McGrath (1984).

Bacterial and fungal abundance has conventionally been measured by estimations of colony forming units using plate count techniques. This method has been questioned as far as fungi are concerned because the plate count technique primarily determines spore number and actively growing hyphae have little chance to form a colony (Paul and Clark 1989). It is also considered a doubtful method in the case of bacteria if rich media are used because only a fraction of the total bacteria will be able to grow on such media (Olsen and Bakken 1987). Furthermore, pollutants may change the component of the biomass that is viable but non-culturable (VBNC). Bond *et al.* (1976) found no effect on colony forming units for bacteria and fungi after the addition of 10 mg Cd kg^{-1} of forest litter, although a decrease in soil respiration rate was evident. Soil obtained at 15 km from a lead smelter and containing 28 000 mg kg^{-1} Pb, 972 mg kg^{-1} Cu and 151 mg kg^{-1} Cd had a lower population density of bacteria including actinomycetes and fungi than control soil samples 1000 km from the smelter (Bisessar 1982). Similar decreases in populations were identified 2 km from a zinc smelter with soil containing 80 000 mg kg^{-1} Zn, 1500 mg kg^{-1} Cd and 1100 mg kg^{-1} Pb (Jordan and Lechevalier 1975).

A concentration of 7500 mg kg^{-1} Cu decreased while 7500 mg kg^{-1} Zn increased the total number of fungi in soil (Badura *et al.* 1979a) and the number of *Streptomyces* species was reduced only by Cu (Badura *et al.* 1979b). The addition of a combination of 1000 mg kg^{-1} Cd and 1000 mg kg^{-1} Zn to a glucose-amended soil had the effect of reducing the population density of bacteria including actinomycetes to a greater extent than the population density of fungi (Bewley and Stotzky 1983b,c). Cr(III) added to the soil (10 and 100 mg kg^{-1}) decreased the population density of aerobic and anaerobic bacteria and actinomycetes. Fungal population densities were reduced by concentrations over 100 mg kg^{-1} Cr(III) (Drucker *et al.* 1979).

Soil fungal biomass and fungal species composition of a coniferous forest soil were determined along a steep Cu and Zn gradient (up to 20 000 mg kg^{-1} Cu and Zn) (Nordgren *et al.* 1983). Below 1000 mg kg^{-1} Cu, there was no clear reduction in fungal biomass noticed, although there was a tendency for a reduction in total mycelial length. Fungal species composition was also affected. The frequency of species in the genera *Penicillium* and *Ochiodendron* decreased from about 30% and 20%, respectively, at control sites to only a few percent close to the area of maximum Cu and Zn contamination. Some fungal taxa increased in abundance as metal concentration reached a maximum. For example, *Geomyces* increased from 1% to 10%, *Paecilomyces* from 0% to 10% and sterile forms from 10% to 20% (Nordgren *et al.* 1983).

The use of diversity indices as a means of studying the impact of pollutants does not give the same amount of information as using species composition (Bååth 1989). Nordgren *et al.* (1983) found that there were only minor effects on fungal species diversity except at high contamination levels (such as 10 000 mg Cu kg^{-1} soil in the Gusum area). The monitoring of bacterial diversity is more difficult than for fungi because they are rarely identified to the species level (Bååth 1989). Many authors, however, report that metal pollutants induce a shift towards Gram-negative bacteria.

There has been a demand for some time to evaluate the effect of pollutants on microbial populations. Garland and Mills (1991) made a significant step by employing the Biolog bacterial identification kit as a community-based technique. This simultaneously tests the utilization of 95 different carbon substrates as sole carbon sources, enabling a metabolic fingerprint of microbial communities to be made. This technique, while offering a wide spectrum approach for the assessment of the effects of metals in soil microbial populations, has recently been shown to be more sensitive to metal pollutants than total soil biomass.

7.4.1.10 Acclimation and Resistance to Toxic Shock

The development of resistance to pollutants has been termed acclimation. This characteristic has been observed in the autotrophic component of the nitrifier community, but has been absent among the heterotrophic nitrifying population. This acclimation can be seen as protecting the biomass in the event of a shock load, although the actual degree of protection is uncertain. In studies where acclimation and bacterial type were monitored, it has often been found that acclimation was associated with a reduction in species diversity. Low microbial diversity of industrially-contaminated soils may make them more susceptible to shock loading events.

The study of metal-resistant microorganisms has typically involved the isolation of strains from contaminated environments. The existence of tolerant species does not provide information concerning the selection pressure of a pollutant (Diaz-Ravina *et al.* 1994). The increase of tolerant organisms in polluted environments may be due to genetic changes, physiological adaptations involving no genotype alterations or replacement of sensitive species with more tolerant ones (Bååth 1989). Selection pressure makes isolation of pollutant resistant/xenobiotic degrader organisms an excellent choice of bioindicator of pollutant pressure.

Jordan and Lechevalier (1975) found that near a zinc smelter the soil that was most polluted had an increased proportion of tolerant bacteria, although these tolerant bacteria were also identified in the non-contaminated sites. This characteristic has been found in a range of studied terrestrial environments, including aerobic sediments and soil. Bååth (1989) suggested that this demonstrated species composition is more important than species adaptation. The development of suitably resistant strains may take several years to develop after exposure to the pollutant, and it is also likely that resistant strains may have different growth rates from non-resistant strains.

7.4.2 USE OF SPECIFIC ORGANISM-BASED ASSAYS: OVERVIEW

There are many proposed mechanisms by which pollutants inhibit and eventually kill microbes. Substances such as halogens may lead to the denaturation of proteins, while phenol disrupts bacterial cell membranes causing a leakage of DNA, RNA and other organic materials, with acids and alkalis displacing cations from adsorption sites on the cell walls. Some toxic materials block bacterial chemoreceptors, often leading to an inhibition of the organic decomposition and self-purification processes associated with microbes employed in soil processes. The impact of pollutants must, however, be seen as linked to physicochemical conditions, including the presence of other cations, pH, redox potential, organic matter and clay micelles.

7.4.2.1 The Use of Fungal and Yeast Assays in Ecotoxicity Assessment

Some species of fungi are pathogenic to plants and animals and may colonize and deteriorate surfaces. Accordingly, bioassays have been based upon a range of methods such as measurement of radial growth rates on solid media, growth inhibition in broth, spore germination tests, agar diffusion, respirometry or the measurement of potassium release following the exposure to a pollutant (Gadd 1990, 1992). Dutka and Bitton (1986) noted that, in terms of soil quality testing, these methods were not well developed, but they suggested that such assays could be of particular use in the examination of biodegradation and phytopathology.

7.4.2.2 The Use of Protozoan Assays in Ecotoxicity Assessment

The effect of heavy metals on aquatic protozoa has been extensively studied in simple liquid systems (e.g., Houba and Remacle 1982), and also in the soil environment (Foissner 1987, Forge *et al.* 1993). Foissner (1987) suggested that protozoa might be sensitive indicators of soil pollution. Many protozoa have short generation times and the effect of pollutants on populations of individual species or communities should manifest rapidly. On account of their large size, relative to other microorganisms, they are easier to observe in soil and in pure culture (Forge *et al.* 1993). Schreiber and Brink (1989) studied the effect of pesticides on the growth of *Oikomonas termo* isolated from fresh water and from sewage sludge. Steinberg *et al.* (1990) reported that a species of *Acanthomoeba* was more sensitive than its bacterial prey to PCBs in soil.

7.4.2.3 The Use of Bacterial Assays in Ecotoxicity Assessment

The importance of bacteria as key players in geochemical carbon and nutrient cycling has made them suitable for a range of diverse ecotoxicity assays. Bacteria, primarily involved in the recycling of mineral nutrients and in the mineralization of organic substrates, have short life-cycles and, as a consequence, respond relatively quickly to environmental change. Bacteria are stable, easily maintained at low cost and a large number of cells can be exposed to a pollutant in one study, making them ideal organisms for rapid screening of solutions. The use of bacteria as surrogate assay organisms, it has been argued, is based on the assumption that some biochemical and physiological systems are evolutionary conservative, and pollutants elicit observed effects due to the interactions with biomolecules which are similar in many different organisms.

Bacteria belonging to the genera *Pseudomonas*, *Klebsiella*, *Aeromonas* and *Citrobacter* have been used to assess the effect of pollutants on growth inhibition, respiration or cell viability. These tests utilize the techniques previously mentioned which were used to assess process level impacts (including respiration, enzyme activity, nitrogen cycling) or culturable or direct microscopy techniques.

7.4.2.4 The Bioluminescence Assay

A number of marine organisms are naturally bioluminescent and can be both free-living and associated with higher marine organisms. Biochemically, bioluminescence is considered as a branch of the electron transport system. The principle that a pollutant may inhibit metabolic activity and, as a consequence, decrease bioluminescence output has been used for the detection of pollutants. Hastings *et al.* (1985) reported the use of an *in vivo* bioluminescence assay using the bacterium *Vibrio harveyi* which was mutated using nitrosoguanidine and a *lux D* mutant selected and designated M17. This allowed the detection of tetradecanoic acid down to concentrations of 10 pM.

Commercially available bioluminescent-based bacteria have been available for over a decade. The test kit marketed by Azur and named Microtox™ has been internationally acknowledged as a sensitive, simple, inexpensive, reproducible and rapid ecotoxicity assay. Reviews by Steinberg *et al.* (1995) have considered the range of potential applications of the assay.

In the case of soil and landfill extracts, the assay proved to be effective in most cases for assessing general toxicity. Many of these sites, however, reflect areas that have been chemically characterized and only recently has there been a trend towards studying ecotoxicity. It is generally accepted that Microtox™ proved to be a valid assay, but it should be seen as a component of a battery test system.

One of the limitations of the Microtox™ assay is that it utilizes a marine organism, thus not reflecting the ecological niches of the soil environment. To overcome this problem, the genes responsible for bioluminescence can be fused within soil organisms. This has enabled the development of *lux*-based sensors, which have two distinctive approaches. Firstly, bacteria that have the *lux* genes linked to no specific reporter genes are able to give a general indication of soil health. Because the expression of *lux* is intimately associated with the electron transport system of the organism this type of approach has been widely used for pollution testing with genetically marked soil organisms (Paton *et al.* 1995, 1997). The second approach involves the use of constructs which have the *lux* genes coupled to genes responsible for either catabolic or pollutant-specific pathways.

Genes encoding Hg detoxification were fused with *lux* genes in *Vibrio fischeri* and the detection of Hg was achieved by monitoring an increase in bioluminescence (Gieselhart *et al.* 1991). Selifonova *et al.* (1993) reported that this system was able to detect bioavailable Hg from nanomole to micromole concentrations. Van-Dyk *et al.* (1994) demonstrated that the fusion of *lux* genes to heat shock genes in *E. coli* could enable an assessment of the toxicity of a range of organic and inorganic pollutants.

Heitzer *et al.* (1992, 1994) developed and demonstrated a biosensor to study naphthalene and salicylate bioavailability. The insertion of *lux* genes and the genes responsible for naphthalene and salicylate catabolism into *Pseudomonas fluorescens* enabled a rapid method to assess the quantity and bioavailability of target pollutants. The rapid development of *lux*-sensor technology to detect a broad suite of hydrocarbons and even biphenyls continues to progress. These new developments will have significant roles in the assessment of sites contaminated with a diverse range of contaminants.

7.5 OVERALL CONCLUSION

It is hoped that this rapid summary provides an introductory guide to the wealth of techniques available to the practitioner to measure effects of soil contamination at the invertebrate and microbial levels. Risk assessment tools based on soil invertebrate biomarkers or changes in microbial processing have not yet been developed and this field is still open to debate. The success of the techniques outlined above as tools for risk assessment purposes depends largely on the choice of the assay (both for effect and exposure measurements), which should both be sensitive and reflect the relevant ecological effects of toxicity at a site.

The field of biomarker research in particular has developed rapidly in the last decade and triggered an ongoing debate on the pros and cons of various biomarkers to be included in risk assessment procedures. Of great consideration is the validity of a predictive role for biomarkers in the assessment of risks to soil invertebrate communities and the potential for the circumvention of lasting damage to their populations (Weeks 1998). Weeks *et al.* (1999) set out a step wise approach for the implementation of a biomarker monitoring survey using soil invertebrate species to generate data for risk assessment purposes. There are, however, no rigid guidelines currently used to implement such a scheme. The use of microbial measurements for stringent risk assessment protocol suffers from the same lack of established defined guidelines to interpret the significance of any measured change.

REFERENCES

Achazi RK, Flenner C, Livingstone DR, Peters LD, Schaub K and Scheiwe E (1998) Cytochrome P450 and dependent activities in unexposed and PAH-exposed terrestrial annelids. *Comparative Biochemistry and Physiology C*, **121C**, 339–350.

Adams SM, Shepard KL, Greeley MS, Jimenez BD, Ryon MG, Shugart LR, McCarthy JF and Hinton DE (1989) The use of bioindicators for assessing the effects of pollutant stress on fish. *Marine Environmental Research*, **28**, 459-464.

Bååth E (1989) Effects of heavy metals in soil on microbial processes and populations (a review). *Water, Air and Soil Pollution*, **47**, 335-379.

Babich H and Stotzky G (1983) Developing standards for environmental toxicants: the need to consider abiotic environmental factors and microbe mediated ecological processes. *Environmental Health Perspectives*, **49**, 247-257.

Baccino FM and Zurretti MF (1975) Structural equivalents of latency for lysosome hydrolases. *Biochemical Journal*, **146**, 97-108.

Badura L, Gorka B and Ulfig K (1979a) Oddzialywanie soili cynku I miedzi na drobnoustroje gleby Cz I. reakcje grzybow. *Acta Biology (Katowice)*, **7**, 89-99.

Badura L, Rusecka J and Smylla A (1979b) Oddzialywanie soili cynku I miedzi na drobnoustroje gleby. *Acta Biology (Katowice)*, **9**, 128-142.

Bartsch R, Klein D and Summer KH (1990) The Cd-Chelex assay: a new sensitive method to determine metallothionein containing zinc and cadmium. *Archives of Toxicology*, **64**, 177-180.

Bauer-Hilty A, Dallinger R and Berger B (1989) Isolation and partial characterization of a cadmium-binding protein from *Lumbricus variegatus* (Oligochaeta, Annelida). *Comparative Biochemistry and Physiology*, **94C**, 373-379.

Bengtsson G, Ek H and Rundgren S (1992) Evolutionary response of earthworms to long term metal exposure. *Oikos*, **63**, 289-297.

Berger B, Dallinger R and Thomaser A (1995) Quantification of metallothionein as a biomarker for cadmium exposure in terrestrial gastropods. *Environmental Toxicology and Chemistry*, **14**, 781-791.

Berghout A, Buld J and Wenzel E (1989) Isolation and partial purification of cytochrome P450 from the gut of the earthworm *Lumbricus terrestris*. *Biological Chemistry Hoppe Seyler*, **370**, 614.

Berghout A, Buld J and Wenzel E (1990) The cytochrome P450 dependent monooxygenase system of the midgut of the earthworm *Lumbricus terrestris*. *European Journal of Pharmacology*, **183**, 1885-1886.

Berghout AGRV, Wenzel E, Buld J and Netter KJ (1991) Isolation, partial purification, and characterization of the cytochrome P450 dependent monooxygenase system from the midgut of the earthworm *Lumbricus terrestris*. *Comparative Biochemistry and Physiology*, **100C**, 389-396.

Berkus M, Gräff S, Alberti G and Köhler H-R (1994) The impact of the heavy metals, lead, zinc, and cadmium, on the ultrastructure of midgut cells of *Julus scandinavius* (Diplopoda). *Verhandlungen der Deutschen Zoologischen Gesellschaft*, **87**, 321.

Bewley RJF and Stozky G (1983a) Effects of cadmium and simulated acid rain on ammonification and nitrification in soil. *Archives of Environmental Contamination Toxicology*, **12**, 285-291.

Bewley RJF and Stotzky G (1983b) Effects of cadmium and zinc on microbial activity in soil — influence of clay minerals 1. Metals added individually. *The Science of the Total Environment*, **31**, 41-55.

Bewley RJF and Stozky G (1983c) Effects of cadmium and zinc on microbial activity in soil — influence of clay minerals 2. Metals added simultaneously. *The Science of the Total Environment*, **31**, 57-69.

Bisessar S (1982) Effect of heavy metals on microorganisms in soils near a secondary lead smelter. *Water, Air and Soil Pollution*, **17**, 305-308.

Bitton G and Dutka BJ (1986) Introduction and review of microbial and biochemical toxicity screening procedures. In *Toxicity Testing Using Microorganisms*, Vol. II, Dutka BJ and Bitton G (eds), CRC Press, Boca Raton, FL, pp. 2-7.

Bitton G and Koopman B (1986) Effect of toxicants on dehydrogenases. In *Toxicity Testing Using Microorganisms*, Vol. I, Bitton G and Dutka BJ (eds), CRC Press, Boca Raton, FL, pp. 32-41.

Bollag JM and Barabasz W (1979) Effects of heavy metals on the denitrification process in soil. *Journal of Environmental Quality*, **8**, 196-201.

Bollag JM and Liu SY (1990) Biological transformation processes of pesticides. In *Pesticides in the Soil Environment: Processes Impact and Modelling*, Cheng HH (ed), Soil Science Society of America, Madison, WI, pp. 103-127.

Bond H, Lighthart B, Shimbuku R and Russel L (1976) Some effects of cadmium on coniferous soil and litter microorganisms. *Soil Science*, **24**, 128-287.

Borgeraas J, Nilsen K and Stenersen J (1996) Methods for purification of glutathione transferases in the earthworm genus *Eisenia*, and their characterization. *Comparative Biochemistry and Physiology*, **114C**, 129-140.

Brookes PC and McGrath SP (1984) Effect of metal toxicity on the size of the microbial biomass. *Journal of Soil Science*, **35**, 341-346.

Brookes PC, Heijnen CE, McGrath SP and Vance ED (1986) Soil microbial estimates in soils contaminated with metals. *Soil Biology and Biochemistry*, **18**, 383-388.

Bunn KE, Thompson HM and Tarrant KA (1996) Effects of agrochemicals on the immune systems of earthworms. *Bulletin of Environmental Contamination and Toxicology*, **57**, 632-639.

Burns RG (1982) Enzyme activity in soils: location and possible role in soil microbiology. *Soil Biology and Biochemistry*, **14**, 423-429.

Chang FH and Broadbent FE (1982) Influence of trace metals on some soil nitrogen transformations. *Journal of Environmental Quality*, **11**, 1-4.

Chen SC, Fitzpatrick LC, Goven AJ, Venables BJ and Cooper EL (1991) Nitroblue tetrazolium dye reduction by earthworm (*Lumbricus terrestris*) coelomocytes — an enzyme assay for nonspecific immunotoxicity of xenobiotics. *Environmental Toxicology and Chemistry*, **10**, 1037-1043.

Cikutovic MA, Fitzpatrick LC, Goven AJ, Venables BJ, Giggleman MA and Cooper EL (1999) Wound healing in earthworms *Lumbricus terrestris*: a cellular-based biomarker for assessing sublethal chemical toxicity. *Bulletin of Environmental Contamination and Toxicology*, **62**, 508-514.

Clubb RW, Lords J and Gaufin A (1975) Isolation and characterization of a glycoprotein from stonefly, *Pteronarcys californica*, which binds cadmium. *Journal of Insect Physiology*, **21**, 53-60.

Cooper EL and Roch P (1992) The capacities of earthworms to heal wounds and to destroy allografts are modified by polychlorinated-biphenyls (PCB). *Journal of Invertebrate Pathology*, **60**, 59-63.

Dallinger R (1993) Strategies of metal detoxification in terrestrial invertebrates. In *Ecotoxicology of Metals in Invertebrates*, Dallinger R and Rainbow PS (eds), Lewis, Boca Raton, FL, pp. 245-289.

Dallinger R (1996) Metallothionein research in terrestrial invertebrates: synopsis and perspectives. *Comparative Biochemistry and Physiology*, **113C**, 125-133.

Dallinger R (1998) Application and standardization of biochemical fingerprint techniques in view of risk assessment of toxicants in soil ecosystems. In *Progress Report 1998 of BIOPRINT-II*, Kammenga JE (ed), Ministry of Environment and Energy, National Environmental Research Institute, Silkeborg, Denmark, pp. 21-25.

Dallinger R and Berger B (1993) Function of metallothioneins in terrestrial gastropods. *The Science of the Total Environment*, **Supplement Pt.** 1, 607-615.

Dallinger R, Berger B, Hunziker PE and Kägi JHR (1997) Metallothionein in snail Cd and Cu metabolism. *Nature*, **388**, 237-238.

Diaz-Ravina M, Bååth E and Frostegård A (1994) Multiple heavy metal tolerance of soil bacterial communities and its measurement by a thymidine incorporation technique. *Applied and Environmental Microbiology*, **60**, 2238-2247.

Doelman P (1986) Resistance of soil microbial communities to heavy metals. In *FEMS Symposium No. 33, Microbial Communities in Soil, Copenhagen, 4-8 August 1985*, Elsevier Applied Science, London, pp. 368-398.

Domsch KH (1984) Effects of pesticides and heavy metals on biological processes in soil. *Plant and Soil*, **76**, 367-374.

Donker MH, Koevoets P, Verkleij JAC and Van Straalen NM (1990) Metal binding compounds in the hepatopancreas and haemolymph of *Porcellio scaber* (Isopoda) from contaminated and reference areas. *Comparative Biochemistry and Physiology*, **97C**, 119-126.

Drucker H, Garland TR and Wildung RE (1979) Metabolic response of microbiota to chromium and other metals. In *Trace Metals in Health and Disease*, Kharasch N (ed), Raven Press, New York, pp. 1-25.

Dutka BJ and Bitton G (1986) *Toxicity Testing using Microorganisms*, Vol. II, CRC Press, Boca Raton, FL.

Eason CT, Booth LH, Brennan S and Ataria J (1998) Cytochrome P450 activity in 3 earthworm species. In *Advances in Earthworm Ecotoxicology. Proceedings from the Second International Workshop on Earthworm Ecotoxicology, 2-5 April 1997*, Sheppard S, Bembridge J, Holmstrup M and Posthuma L (eds), SETAC Press, Pensacola, FL, pp. 191-198.

Eckwert H and Köhler H-R (1997) The indicative value of the hsp70 stress response as a marker for metal effects in *Oniscus asellus* (Isopoda) field populations: variability between populations from metal-polluted and uncontaminated sites. *Applied Soil Ecology*, **6**, 275-282.

Eckwert H, Zanger M, Reiss S, Musolff H, Albert G and Köhler H-R (1994) The effect of heavy metals on the expression of hsp70 in soil invertebrates. *Verhandlungen der Deutschen Zoologischen Gesellschaft*, **87**, 325.

Eckwert H, Alberti G and Köhler H-R (1997) The induction of stress proteins (hsp) in *Oniscus asellus* (Isopoda) as a molecular marker of multiple heavy metal exposure. 1. Principles and toxicological assessment. *Ecotoxicology*, **6**, 249-262.

Edwards CA and Fischer SW (1991) The use of cholinesterase measurement in assessing the impact of pesticides on terrestrial and aquatic invertebrates. In *Cholinesterase Inhibiting Insecticides — Their Impact on Wildlife and the Environment*, Mineau P (ed), Elsevier Science Publishers, Amsterdam, pp. 255-275.

Eiland F (1985) *Determination of adenosine triphosphate (ATP) and adenylate-energy charge (AEC) in soil and use of adenine nucleotides as measures of soil microbial biomass and activity*. PhD Thesis, Royal Veterinary and Agricultural University, Copenhagen.

Eyambe GS, Goven AJ, Fitzpatrick LC, Venables BJ and Cooper EL (1991) A noninvasive technique for sequential collection of earthworm (*Lumbricus terrestris*) leukocytes during subchronic immunotoxicity studies. *Laboratory Animals*, **25**, 61-67.

Fischer E and Molnar L (1992) Environmental aspects of the chloragogenous tissue of earthworms. *Soil Biology and Biochemistry*, **24**, 1723-1727.

Fitzpatrick LC, Goven AJ, Venables BJ, Rodriguez J and Cooper EL (1990) Earthworm immunoassays for evaluating biological effects of exposure to hazardous materials. In *In situ Evaluations of Biological Hazards of Environmental Pollutants*, Sandhu SS (ed), Plenum Press, New York, pp. 119-129.

Fitzpatrick LC, Sassani R, Venables BJ and Goven AJ (1992) Comparative toxicity of polychlorinated biphenyls to earthworms *Eisenia foetida* and *Lumbricus terrestris*. *Environmental Pollution*, **77**, 65-69.

Foissner W (1987) Soil protozoa: fundamental problems, ecological significance, adaptations in ciliates and testaceans, bioindicators, and guide to the literature. *Progress in Protistology*, **2**, 69-212.

Forge TA, Darbyshire JF, Berrow ML and Warren A (1993) Protozoan assays of soil amended with sewage sludge and heavy metals using common soil ciliate *Colpoda steinii*. In *Integrated Soil and Sediment Research; a Basis for Proper Protection*, Eijsackers HJP and Hamers J (eds), Kluwer Academic Publishers, Dordrecht, pp. 315-316.

Fugere N, Brousseau P, Krzystyniak K, Coderre D and Fournier M (1996) Heavy metal-specific inhibition of phagocytosis and different *in vitro* sensitivity of heterogeneous coelomocytes from *Lumbricus terrestris* (Oligochaeta). *Toxicology*, **109**, 157-166.

Furst A and Nguyen Q (1989) Cadmium induced metallothionein in earthworms (*Lumbricus terrestris*). *Biological Trace Element Research*, **21**, 81-85.

Gadd GM (1990) Metal tolerance. In *Microbiology of Extreme Environments*, Edwards C (ed), Open University Press, Milton Keynes, pp. 178-210.

Gadd GM (1992) Heavy metal pollutants: environmental and biotechnological aspects. In *Encyclopedia of Microbiology*, Vol. 2, Lederberg J (ed), Academic Press, San Diego, CA, pp. 351-360.

Garland JL and Mills AL (1991) Classification and characterization of heterotrophic microbial communities on the basis of patterns of community-level sole-carbon-source utilization. *Applied and Environmental Microbiology*, **57**, 2351-2359.

Giashuddin M and Cornfield MH (1978) Incubation study on effects of adding various levels of nickel (as sulphate) on nitrogen and carbon mineralisation in the soil. *Environmental Pollution*, **15**, 231-234.

Gibb JOT, Svendsen C, Weeks JM and Nicholson JK (1997) 1 H NMR spectroscopic investigations of tissue metabolite biomarker response to Cu(II) exposure in terrestrial invertebrates: identification of free histidine as a novel biomarker of exposure to copper in earthworms. *Biomarkers*, **2**, 295-302.

Gieselhart L, Osgood M and Holmes DS (1991) Construction and evaluation of a self-luminescent biosensor. *Annals of the New York Academy of Science*, **646**, 53-60.

Giggleman MA, Fitzpatrick LC, Goven AJ and Venables BJ (1998) Effects of pentachlorophenol on survival of earthworms (*Lumbricus terrestris*) and phagocytosis by their immunoactive coelomocytes. *Environmental Toxicology and Chemistry*, **17**, 2391-2394.

Goven AJ, Venables BJ, Fitzpatrick LC and Cooper EL (1988) An invertebrate model for analyzing effects of environmental xenobiotics on immunity. *Clinical Ecology*, **5**, 150-154.

Goven AJ, Eyambe GS, Fitzpatrick LC, Venables BJ and Cooper EL (1993) Cellular biomarkers for measuring toxicity of xenobiotics — effects of polychlorinated-biphenyls on earthworm *Lumbricus terrestris* coelomocytes. *Environmental Toxicology and Chemistry*, **12**, 863-870.

Goven AJ, Fitzpatrick LC and Venables BJ (1994a) Chemical toxicity and host-defence in earthworms — an invertebrate model. *Annals of the New York Academy of Science*, **712**, 280-300.

Goven AJ, Chen SC, Fitzpatrick LC and Venables BJ (1994b) Lysozyme activity in earthworm (*Lumbricus terrestris*) coelomic fluid and coelomocytes — enzyme assay for immunotoxicity of xenobiotics. *Environmental Toxicology and Chemistry*, **13**, 607-613.

Grelle C and Descamps M (1998) Heavy metal accumulation by *Eisenia fetida* and its effects on glutathione-S-transferase activity. *Pedobiologia*, **42**, 289-297.

Gruber C, Berger B, Gehrig P and Dallinger R (1998) *Characterization of metallothioneins in the earthworm Eisenia foetida. In Proceedings of the 8th Annual Meeting of SETAC Europe, Bordeaux, April 1998*, pp. 49-50.

Guilhermino L, Barros P, Silva MC and Soares A (1998) Should the use of inhibition of cholinesterases as a specific biomarker for organophosphate and carbamate pesticides be questioned? *Biomarkers*, **3**, 157-163.

Hagens M and Westheide W (1987) Subletale Schädigungen bei Enchytraeus minutus (Oligochaeta, Annelida) durch das Insektizid Parathion: Veränderungen in der Ultrastruktur von Chloragog- und Darmzellen in Abhängigkeit von der Belastungsdauer. *Verhandlungsberichte der Kolloid Gesellschaft*, **14**, 237-241.

Hans RK, Khan MA, Farooq M and Beg MU (1993) Glutathione-S-transferase activity in an earthworm (*Pheretima Posthuma*) exposed to 3 insecticides. *Soil Biology and Biochemistry*, **25**, 509-511.

Hastings JW, Potrius CJ, Gupta CJ, Kurfurst M and Makemson JC (1985) Biochemistry and physiology of bioluminescent bacteria. *Advances in Microbial Physiology*, **26**, 235-291.

Heitzer A, Webb OF, Thonnard JE and Sayler GS (1992) Specific and quantitative assessment of naphthalene and salicylate bioavailability using a bioluminescent catabolic reporter bacterium. *Applied and Environmental Microbiology*, **58**, 1839-1846.

Heitzer A, Malachowsky K, Thonnard JE, Bienkowski PR, White DC and Sayler GS (1994) Optical biosensor for environmental online monitoring of naphthalene and salicylate bioavailability with an immobilized bioluminescent catabolic reporter bacterium. *Applied and Environmental Microbiology*, **60**, 1487-1494.

Hopkin SP (1989) *Ecophysiology of Metals in Terrestrial Invertebrates*, Elsevier Applied Science, New York and London, pp. 224-280.

Hopkin SP and Martin MH (1982a) The distribution of zinc, cadmium, lead and copper within the woodlouse *Oniscus asellus* (Crustacea, Isopoda). *Oecologia*, **54** 227-232.

Hopkin SP and Martin MH (1982b) The distribution of zinc, cadmium, lead and copper within the hepatopancreas of a woodlouse. *Tissue & Cell*, **14**, 703-715.

Houba C and Remacle J (1982) Factors influencing the toxicity of cadmium to *Tetrahymena pyriformis*: particulate or solid form and degree of complexation. *Environmental Pollution*, **28**, 35-39.

Jenkinson DS and Oades JM (1979) A method for measuring adenosine triphosphate in soil. *Soil Biology and Biochemistry*, **11**, 193-199.

Jensen CS, Garsdal L and Baatrup E (1997) Acetylcholinesterase inhibition and altered locomotor behavior in the carabid beetle *Pterostichus cupreus*. A linkage between biomarkers at two levels of biological complexity. *Environmental Toxicology and Chemistry*, **16**, 1727-1732.

Jordan MJ and Lechevalier MP (1975) Effects of zinc smelter emissions on forest soil microflora. *Canadian Journal of Microbiology*, **21**, 1855-1865.

Kammenga JE and Simonsen V (1997) *Progress Report 1997 of BIOPRINT-II*, Ministry of Environment and Energy, National Environmental Research Institute, Silkeborg, Denmark.

Kammenga JE, Arts MSJ and OudeBreuil WJM (1998) HSP60 as a potential biomarker of toxic stress in the nematode *Plectus acuminatus*. *Archives of Environmental Contamination and Toxicology*, **34**, 253-258.

Köhler H-R and Alberti G (1992) The effect of heavy metal stress on the intestine of diplopods. In *Advances in Myriapodology*, Meyer E, Thaler K and Schedl W (eds), Ber nat-med Verein Innsbruck Supplement, pp. 257-267.

Köhler H-R and Eckwert H (1997) The induction of stress proteins (hsp) in *Oniscus asellus* (Isopoda) as a molecular marker of multiple heavy metal exposure. 2. Joint toxicity and transfer to field situations. *Ecotoxicology*, **6**, 263-274.

Köhler H-R and Triebskorn R (1998) Assessment of the cytotoxic impact of heavy metals on soil invertebrates using a protocol integrating qualitative and quantitative components. *Biomarkers*, **3**, 109-127.

Köhler H-R, Triebskorn R, Stocker W, Kloetzel PM and Alberti G (1992) The 70 Kd heat-shock protein (Hsp-70) in soil invertebrates — a possible tool for monitoring environmental toxicants. *Archives of Environmental Contamination and Toxicology*, **22**, 334-338.

Köhler H-R, Rahman B and Rahmann H (1994) Assessment of stress situations in the grey garden slug, *Deroceras reticulatum*, caused by heavy metal intoxication: semi-quantification of the 70 kD stress protein (hsp70). *Verhandlungen der Deutschen Zoologischen Gesellschaft*, **87**, 328.

Köhler H-R, Rahman B, Graff S, Berkus M and Triebskorn R (1996a) Expression of the stress-70 protein family (HSP70) due to heavy metal contamination in the slug, *Deroceras reticulatum*: an approach to monitor sublethal stress conditions. *Chemosphere*, **33**, 1327-1340.

Köhler H-R, Huttenrauch K, Berkus M, Graff S and Alberti G (1996b) Cellular hepatopancreatic reactions in *Porcellio scaber* (Isopoda) as biomarkers for the evaluation of heavy metal toxicity in soils. *Applied Soil Ecology*, **3**, 1-15.

Köhler H-R, Belitz B, Eckwert H, Adam R, Rahman B and Tronteli P (1998) Validation of hsp70 stress gene expression as a marker of metal effects in *Deroceras reticulatum* (Pulmonata): correlation with demographic parameters. *Environmental Toxicology and Chemistry*, **17**, 2246-2253.

Köhler H-R, Knodler C and Zanger M (1999a) Divergent kinetics of hsp70 induction in *Oniscus asellus* (Isopoda) in response to four environmentally relevant organic chemicals (B[a]P, PCB52, gamma-HCH, PCP): suitability and limits of a biomarker. *Archives of Environmental Contamination and Toxicology*, **36**, 179-185.

Köhler H-R, Eckwert H, Triebskorn R and Bengtsson G (1999b) Interaction between tolerance and 70 kDa stress protein (hsp70) induction in collembolan populations exposed to long-term metal pollution. *Applied Soil Ecology*, **11**, 43-52.

Labrot F, Ribera D, SaintDenis M and Narbonne JF (1996) In vitro and in vivo studies of potential biomarkers of lead and uranium contamination: lipid peroxidation, acetylcholinesterase, catalase and glutathione peroxidase activities in three non mammalian species. *Biomarkers*, **1**, 21-28.

Lakhani L, Bhatnagar BS and Pandey AK (1991) Effect of monocrotophos on the ovary of the earthworm *Eudichogaster kinneari* (Stephenson). A histological and histochemical profile. *Journal of Reproductive Biology and Comparative Endocrinology*, **3**, 39-46.

Lastowski-Perry D, Otto E and Maroni G (1985) Nucleotide sequence and expression of a *Drosophila* metallothionein. *Journal of Biological Chemistry*, **260**, 1527-1530.

Liang CN and Tatabai MA (1978) Effects of trace elements on nitrification in soils. *Journal of Environmental Quality*, **7**, 291-293.

Liimatainen A and Hanninen O (1982) Occurrence of cytochrome P450 in earthworm *Lumbricus terrestris*. In *Cytochrome P450: Biochemistry, Biophysics, and Environmental Implications*, Hietenen E, Larteinen M and Hanninen O (eds), Elsevier Science Publishers, Amsterdam, pp. 255-257.

Løkke H and Van Gestel CAM (1998) Soil toxicity tests in risk assessment of new and existing chemicals. In *Handbook of Soil Invertebrate Toxicity Tests*, Løkke H and Van Gestel CAM (eds), John Wiley & Sons, Chichester, pp. 3-19.

Lowe DM, Moore MN and Evans BM (1992) Contaminant impact on interactions of molecular probes with lysosomes in living hepatocytes from dab *Limanda limanda*. *Marine Ecology Progress Series*, **91**, 135-140.

Luster MI, Munson AE, Thomas PT, Holsapple MP, Fenters JD, White KL, Lauer LD, Germolec DR, Rosenthal GJ and Dean JH (1988) Development of a testing battery to assess chemical-induced immunotoxicity — National Toxicology Program Guidelines for immunotoxicity evaluation in mice. *Fundamental and Applied Toxicology*, **10**, 2-19.

Marigomez I, Soto M and Kortabitarte M (1996) Tissue-level biomarkers and biological effect of mercury on sentinel slugs, *Arion ater*. *Archives of Environmental Contamination and Toxicology*, **31**, 54-62.

Marigomez I, Kortabitarte M and Dussart GBJ (1998) Tissue-level biomarkers in sentinel slugs as cost-effective tools to assess metal pollution in soils. *Archives of Environmental Contamination and Toxicology*, **34**, 167-176.

Marino F, Sturzenbaum SR, Kille P and Morgan AJ (1998) Cu-Cd interactions in earthworms maintained in laboratory microcosms: the examination of a putative copper paradox. *Comparative Biochemistry and Physiology*, **120C**, 217-223.

Milligan DL, Babish JG and Neuhauser EF (1986) Noninducibility of cytochrome P450 in the earthworm *Dendrobaena veneta*. *Comparative Biochemistry and Physiology*, **85C**, 85-87.

Moore MN (1976) Cytochemical demonstration of latency of lysosomal hydrolases in digestive cells of the common mussel, *Mytilus edulis*, and changes induced by thermal stress. *Cell and Tissue Research*, **175**, 279-287.

Moore MN (1985) Cellular response to pollutants. *Marine Pollution Bulletin*, **16**, 134-139.

Moore MN (1990) Lysosomal cytochemistry in marine environmental monitoring. *Histochemical Journal*, **22**, 187-191.

Morgan AJ and Morris B (1982) The accumulation and intracellular compartmentation of cadmium, lead, zinc and calcium in 2 earthworm species (*Dendrobaena rubida* and *Lumbricus rubellus*) living in highly contaminated soil. *Histochemistry*, **75**, 269-285.

Morgan JE and Morgan AJ (1998) The distribution and intracellular compartmentation of metals in the endogeic earthworm *Aporrectodea caliginosa* sampled from an unpolluted and a metal-contaminated site. *Environmental Pollution*, **99**, 167-175.

Morgan JE, Norey CG, Morgan AJ and Kay J (1989) A comparison of the cadmium binding proteins isolated from the posterior alimentary canal of the earthworms *Dendrodrilus rubidus* and *Lumbricus rubellus*. *Comparative Biochemistry and Physiology*, **92C**, 15-21.

Nejmeddine A, Sautiere P, Dhainautcourtois N and Baert JL (1992) Evidence for a charge variant of Cd Bp 14, a cadmium binding protein from *Allolobophora caliginosa* (Oligochaeta, Annelida). *Comparative Biochemistry and Physiology*, **103B**, 929-932.

Nelson PA, Stewart RR, Morelli MA and Nakatsugawa T (1976) Alderin epoxidation in the earthworm, *Lumbricus terrestris* L. *Pesticide Biochemistry and Physiology*, **6**, 243-253.

Nielsen SA and Toft S (1998) Responses of glutathione-S-transferase and glutathione peroxidases to feeding rate of a wolf spider *Pardosa prativaga*. *Atla-Alternatives to Laboratory Animals*, **26**, 399-403.

Nordgren A, Bååth E and Söderström K (1983) Microfungi and microbial activity along a heavy metal gradient. *Applied and Environmental Microbiology*, **45**, 1829-1837.

Olsen RA and Bakken LR (1987) Viability of soil bacteria: optimization of plate counting technique and comparison between total counts and plate counts with different size groups. *Microbial Ecology*, **13**, 103-114.

Paton GI, Campbell CD, Glover LA and Killham K (1995) Assessment of bioavailability of heavy metals using *lux* modified constructs of *Pseudomonas fluorescens*. *Letters in Applied Microbiology*, **20**, 52-56.

Paton GI, Rattray EAS, Campbell CD, Meussen H, Cresser MS, Glover LA and Killham K (1997) Use of genetically modified microbial biosensors for soil ecotoxicity testing. In *Bioindicators of Soil Health*, Pankhurst CS, Doube B and Gupta V (eds), CAB International, pp. 397-418.

Paul EA and Clark FE (1989) *Soil Microbiology and Biochemistry*. Academic Press, San Diego, CA.

Pawert M, Rahmann H and Köhler H-R (1994) The impact of heavy metals on the ultrastructure of the endoplasmic reticulum in midgut epithelial cells of *Tetrodontophora bielanensis* (Collembola). *Verhandlungen der Deutschen Zoologischen Gesellschaft*, **87**, 331.

Pawert M, Triebskorn R, Graff S, Berkus M, Schulz J and Köhler H-R (1996) Cellular alterations in collembolan midgut cells as a marker of heavy metal exposure: ultrastructure and intracellular metal distribution. *The Science of the Total Environment*, **181**, 187-200.

Prosi F and Dallinger R (1988) Heavy metals in the terrestrial isopod *Porcellio scaber* Latreille. I. Histochemical and ultrastructural characterization of metal-containing lysosomes. *Cell Biology and Toxicology*, **4**, 81-96.

Prosi F, Storch V and Janssen HH (1983) Small cells in the midgut glands of terrestrial Isopoda: sites of heavy metal accumulation. *Zoomorphology*, **102**, 53-64.

Ramseier S, Deshusses J and Haerdi W (1990) Cadmium speciation studies in the intestine of *Lumbricus terrestris* by electrophoresis of metal protein complexes. *Molecular and Cellular Biochemistry*, **97**, 137-144.

Recio A, Marigomez JA, Angulo E and Moya J (1988) Zinc treatment of the digestive gland of the slug *Arion ater* l.1. Cellular-distribution of zinc and calcium. *Bulletin of Environmental Contamination and Toxicology*, **41**, 858-864.

Roch P and Cooper EL (1991) Cellular but not humoral antibacterial activity of earthworms is inhibited by Aroclor-1254. *Ecotoxicology and Environmental Safety*, **22**, 283-290.

Rodriguez-Grau J, Venables BJ, Fitzpatrick LC, Goven AJ and Cooper EL (1989) Suppression of secretory rosette formation by PCBs in *Lumbricus terrestris* — an earthworm assay for humoral immunotoxicity of xenobiotics. *Environmental Toxicology and Chemistry*, **8**, 1201-1207.

Rother JA, Millbank JW and Thornton I (1982) Effects of heavy metal additions on ammonification and nitrification in soils contaminated with cadmium, lead and zinc. *Plant and Soil*, **69**, 238-258.

Rother JA, Millbank JW and Thornton I (1983) Seasonal fluctuations in nitrogen fixation (acetylene reduction) by free living bacteria in soils contaminated with cadmium, lead and zinc. *Journal of Soil Science*, **33**, 101-113.

Rumpf S, Hetzel F and Frampton C (1997) Lacewings (Neuroptera: Hemerobiidae and Chrysopidae) and integrated pest management: enzyme activity as biomarker of sublethal insecticide exposure. *Journal of Economic Entomology*, **90**, 102-108.

SaintDenis M, Labrot F, Narbonne JF and Ribera D (1998) Glutathione, glutathione-related enzymes, and catalase activities in the earthworm *Eisenia fetida andrei*. *Archives of Environmental Contamination and Toxicology*, **35**, 602-614.

Salagovic J, Gilles J, Verschaeve L and Kalina I (1996) The comet assay for the detection of genotoxic damage in the earthworms: a promising tool for assessing the biological hazards of polluted sites. *Folia Biologica*, **42**, 17-21.

Sanders BM (1993) Stress proteins in aquatic organisms: an environmental perspective. *Critical Reviews in Toxicology*, **23**, 49-75.

Schmidt GH and Ibrahim NMM (1994) Heavy-metal content (Hg^{2+}, Cd^{2+}, Pb^{2+}) in various body parts — its impact on cholinesterase activity and binding glycoproteins in the grasshopper *Aiolopus thalassinus* adults. *Ecotoxicology and Environmental Safety*, **29**, 148-164.

Schrieber B and Brink N (1989) Pesticide toxicity using protozoan as test organisms. *Biology and Fertility of Soils*, **7**, 289-296.

Scott-Fordsmand JJ (1998) Biomarkers of contaminated soil. PhD Thesis, Division of Zoology, School of Animal and Microbial Sciences, University of Reading, UK.

Selifonova O, Burlage R and Barkay T (1993) Bioluminescent sensors for the detection of bioavailable Hg (II) in the environment. *Applied and Environmental Microbiology*, **59**, 3083-3090.

Shugart L, Bickham J, Jackim G, McMahon G, Ridley W, Stein J and Steinert S (1992) DNA alterations. In *Biomarkers: Biochemical, Physiological, and Histological Markers of Anthropogenic Stress*, Huggett RJ, Kimerle RA, Mehrle Jr. PM and Bergman HL (eds), Lewis, Boca Raton, FL, pp. 125-153.

Slice LW, Freedman JH and Rubin C (1990) Purification, characterization, and cDNA cloning of a novel metallothioneinlike, cadmium binding protein from *Caenorhabditis elegans*. *Journal of Biological Chemistry*, **265**, 256-263.

Steinberg C, Grosjean MC, Bossand B and Faurie G (1990) Influence of PCB's on the predator-prey relationship between bacteria and protozoa in soil. *FEMS Microbial Ecology*, **73**, 607-624.

Steinberg SM, Poziomek EJ, Englemann WH and Rogers KR (1995) A review of environmental applications of bioluminescence measurements. *Chemosphere*, **30**, 2155-2197.

Stenersen J (1979) Action of pesticides on earthworms. Part I: The toxicity of cholinesterase-inhibiting insecticides to earthworms as evaluated by laboratory tests. *Pesticide Science*, **10**, 66-74.

Stenersen J (1984) Detoxication of xenobiotics by earthworms. *Comparative Biochemistry and Physiology*, **78C**, 249-252.

Stenersen J and Øien N (1981) Glutathione S transferases in earthworms (Lumbricidae) substrate specificity, tissue and species distribution and molecular weight. *Comparative Biochemistry and Physiology*, **69C**, 243-252.

Stenersen J, Gilman A and Vardanis A (1973) Carbofuran: its toxicity to and metabolism by earthworm *(Lumbricus terrestris)*. *Journal of Agricultural and Food Chemistry*, **21**, 166-171.

Stenersen J, Gutenberg C and Mannervik B (1979) Glutathione S-transferases in earthworms (Lumbricidae). *Biochemical Journal*, **181**, 47-50.

Stenersen J, Brekke E and Engelstad F (1992) Earthworms for toxicity testing species differences in response towards cholinesterase inhibiting insecticides. *Soil Biology and Biochemistry*, **24**, 1761-1764.

Stokke K and Stenersen J (1993) Non inducibility of the glutathione transferases of the earthworm *Eisenia andrei*. *Comparative Biochemistry and Physiology*, **106C**, 753-756.

Stringer A and Wright MA (1976) The toxicity of benomyl and some related 2-substituted benzimidazoles to the earthworm *Lumbricus terrestris*. *Pesticide Science*, **7**, 459-464.

Sturzenbaum S (1997) *Molecular genetic responses of earthworms to heavy metals*. PhD Thesis, University of Wales, Cardiff.

Sturzenbaum SR, Kille P and Morgan AJ (1998a) Heavy metal-induced molecular responses in the earthworm, *Lumbricus rubellus* genetic fingerprinting by directed differential display. *Applied Soil Ecology*, **9**, 495-500.

Sturzenbaum SR, Kille P and Morgan AJ (1998b) The identification, cloning and characterization of earthworm metallothionein. *FEBS Letters*, **431**, 437-442.

Suzuki KT, Yamamura M and Mori T (1980) Cadmium-binding proteins induced in the earthworm. *Archives of Environmental Contamination and Toxicology*, **9**, 415-424.

Suzuki MM, Cooper EL, Eyambe GS, Goven AJ, Fitzpatrick LC and Venables BJ (1995) Polychlorinated-biphenyls (PCBs) depress allogeneic natural cytotoxicity by earthworm coelomocytes. *Environmental Toxicology and Chemistry*, **14**, 1697-1700.

Svendsen C (2000) Earthworm biomarkers in terrestrial ecosystems. PhD Thesis, Division of Zoology, School of Animal and Microbial Sciences, University of Reading, UK.

Svendsen C and Weeks JM (1995) The use of a lysosome assay for the rapid assessment of cellular stress from copper to the freshwater snail *Viviparus contectus* (Millet). *Marine Pollution Bulletin*, **31**, 139-142.

Svendsen C and Weeks JM (1997a) Relevance and applicability of a simple earthworm biomarker of copper exposure 1. Links to ecological effects in a laboratory study with *Eisenia andrei*. *Ecotoxicology and Environmental Safety*, **36**, 72-79.

Svendsen C and Weeks JM (1997b) Relevance and applicability of a simple earthworm biomarker of copper exposure 2. Validation and applicability under field conditions in a mesocosm experiment with *Lumbricus rubellus*. *Ecotoxicology and Environmental Safety*, **36**, 80-88.

Svendsen C, Meharg AA, Freestone P and Weeks JM (1996) Use of an earthworm lysosomal biomarker for the ecological assessment of pollution from an industrial plastics fire. *Applied Soil Ecology*, **3**, 99-107.

Svendsen C, Spurgeon DJ, Milanov ZB and Weeks JM (1998) Lysosomal membrane permeability and earthworm immune-system activity: field-testing on contaminated land. In *Advances in Earthworm Ecotoxicology. Proceedings from the Second International Workshop on Earthworm Ecotoxicology, 2-5 April 1997*, Sheppard S, Bembridge J, Holmstrup M and Posthuma L (eds), SETAC Press, Pensacola, FL, pp. 225-232.

Thomas JP, Bachowski GJ and Girotti AW (1986) Inhibition of cell membrane lipid peroxidation by cadmium- and zinc-metallothionein. *Biochimica et Biophysica Acta*, **884**, 448-461.

Thornalley PJ and Vašák M (1985) Possible role for metallothionein in protection against radiation-induced oxidative stress. Kinetics and mechanism of its reaction with superoxide and hydroxyl radicals. *Biochimica et Biophysica Acta*, **827**, 36-44.

Triebskorn R (1989) Ultrastructural-changes in the digestive tract of *Deroceras reticulatum* (Muller) induced by a carbamate molluscicide and by metaldehyde. *Malacologia*, **31**, 141-156.

Triebskorn R and Köhler H-R (1996) The impact of heavy metals on the grey garden slug, *Deroceras reticulatum* (Muller): metal storage, cellular effects and semi-quantitative evaluation of metal toxicity. *Environmental Pollution*, **93**, 327-343.

Triebskorn R and Kunast C (1990) Ultrastructural changes in the digestive system of *Deroceras reticulatum* (mollusca, gastropoda) induced by lethal and sublethal concentrations of the carbamate molluscicide cloethocarb. *Malacologia*, **32**, 89-106.

Triebskorn R, Christensen K and Heim I (1998) Effects of orally and dermally applied metaldehyde on mucus cells of slugs (*Deroceras reticulatum*) depending on temperature and duration of exposure. *Journal of Molluscan Studies*, **64**, 467-487.

Tyler G (1975) Heavy metal pollution and mineralisation of nitrogen in forest soils. *Nature*, **255**, 701-702.

Valembois P and Lassegues M (1995) *In vitro* generation of reactive oxygen species by free coelomic cells of the annelid *Eisenia fetida andrei* — an analysis by chemiluminescence and nitro blue tetrazolium reduction. *Developmental and Comparative Immunology*, **19**, 195-204.

Valembois P, Lassegues M, Roch P and Vaillier J (1985) Scanning electron microscopic study of the involvement of coelomic cells in earthworm antibacterial defence. *Cell and Tissue Research*, **240**, 479-484.

Van-dyk TK, Majarain WR, Konstantinov KB, Young RM, Dhurjati PS and Larossa RA (1994) Rapid and sensitive pollutant detection by induction of heat shock gene bioluminescence gene fusions. *Applied and Environmental Microbiology*, **60**, 1414-1420.

Van Schooten FJ, Maas LM, Moonen EJC, Kleinjans JCS and Vanderoost R (1995) DNA dosimetry in biological indicator species living on PAH contaminated soils and sediments. *Ecotoxicology and Environmental Safety*, **30**, 171-179.

Verschaeve L and Gilles J (1995) Single-cell gel-electrophoresis assay in the earthworm for the detection of genotoxic compounds in soils. *Bulletin of Environmental Contamination and Toxicology*, **54**, 112-119.

Verschaeve L, Gilles J, Schoeters J, van Cleuvenbergen R and de Fré R (1993) The single cell gel electrophoresis technique or comet test for monitoring dioxin pollution and effects. In

Organohalogen Compounds II, Fiedler H, Frank H, Hutzinger O, Pazzefal W, Riss A and Safe S (eds), Federal Environmental Agency, Austria, pp. 213-216.

Ville P, Roch P, Cooper EL, Masson P and Narbonne JF (1995) PCBs increase molecular related activities (lysozyme, antibacterial, hemolysis, proteases) but inhibit macrophage-related functions (phagocytosis, wound-healing) in earthworms. *Journal of Invertebrate Pathology*, **65**, 217-224.

Ville P, Roch P, Cooper EL and Narbonne JF (1997) Immuno-modulator effects of carbaryl and 2,4 D in the earthworm *Eisenia fetida andrei*. *Archives of Environmental Contamination and Toxicology*, **32**, 291-297.

Wainwright M and Pugh GLF (1973) The effect of three fungicides on the nitrification and ammonification in soil. *Soil Biology and Biochemistry*, **5**, 577-583.

Walker CH, Hopkin SP, Sibly RM and Peakall DB (1996) *Principles of Ecotoxicology*, Taylor & Francis, London.

Walsh P, Eladlouni C, Mukhopadhyay MJ, Viel G, Nadeau D and Poirier GG (1995) 32 P postlabeling determination of DNA adducts in the earthworm *Lumbricus terrestris* exposed to PAH contaminated soils. *Bulletin of Environmental Contamination and Toxicology*, **54**, 654-661.

Weeks BA, Anderson DP, DuFour AP, Fairbrother A, Goven AJ, Lahvis GP and Peters G (1992) Immunological biomarkers to assess environmental stress. In *Biomarkers Biochemical, Physiological, and Histological Markers of Anthropogenic Stress*, Huggett RJ, Kimerle RA, Mehrle Jr. PM and Bergman HL (eds), Lewis, Boca Raton, FL, pp. 211-234.

Weeks JM (1998) Effects of pollutants on soil invertebrates: links between levels. In *Ecotoxicology*, Schuurmann G and Markert B (eds), John Wiley & Sons, Chichester.

Weeks JM and Svendsen C (1996) Neutral red retention by lysosomes from earthworm (*Lumbricus rubellus*) coelomocytes: a simple biomarker of exposure to soil copper. *Environmental Toxicology and Chemistry*, **15**, 1801-1805.

Weeks JM, Evdokimova G, Ernst WHO, Scott-Fordsman J, Sousa JP, Sergeev VE and Nakonieczny M (1999) 4. Report of the working group on the strategy for the implementation of a biomarker biomonitoring program in the Kola Peninsula, Russia. In *Biomarkers: A Pragmatic Basis for Remediation of Severe Pollution in Eastern Europe*, Peakall DP, Walker CH and Migula P (eds), Kluwer Academic Publishers, Dordrecht, pp. 279-298.

Westheide W, Bethke-Beilfuß D, Hagens M and Brockmeyer V (1989) Enchytraeiden als Testorganismen — Voraussetzungen für ein terrestrisches Testverfahren und Testergebnisse. *Verhandlungsberichte der Kolloid Gesellschaft*, **17**, 793-798.

Willuhn J, Schmittwrede HP, Greven H and Wunderlich F (1994a) Cadmium induced messenger RNA encoding a nonmetallothionein 33 Kda protein in *Enchytraeus buchholzi* (Oligochaeta). *Ecotoxicology and Environmental Safety*, **29**, 93-100.

Willuhn J, Schmittwrede HP, Greven H and Wunderlich F (1994b) cDNA cloning of a cadmium inducible messenger RNA encoding a novel cysteine rich, nonmetallothionein 25 Kda protein in an Enchytraeid earthworm. *Journal of Biological Chemistry*, **269**, 24 688-24 691.

Willuhn J, Otto A, SchmittWrede HP and Wunderlich F (1996a) Earthworm gene as indicator of bioefficacious cadmium. *Biochemical and Biophysical Research Communications*, **220**, 581-585.

Willuhn J, SchmittWrede HP, Otto A and Wunderlich F (1996b) Cadmium-detoxification in the earthworm Enchytraeus: specific expression of a putative aldehyde dehydrogenase. *Biochemical and Biophysical Research Communications*, **226**, 128-134.

Yamamura M, Mori T and Suzuki KT (1981) Metallothionein induced in the earthworm. *Experientia*, **37**, 1187-1189.

Zanger M and Köhler H-R (1996) Colour change: a novel biomarker indicating sublethal stress in the millipede *Julus scandinavius* (Diplopoda). *Biomarkers*, **1**, 99-106.

Zanger M, Harreus D, Alberti G and Köhler H-R (1994) Different methods for qualifying and quantifying hsp 70 in diplopods. *Verhandlungen der Deutschen Zoologischen Gesellschaft*, **87**, 335.

Ziblinski LM and Wagner G (1982) Bacterial growth and fungal genera distribution in soil amended with sewage sludge containing cadmium, chromium and copper. *Soil Science*, **134**, 370–377.
Zsombok A, Molnar L and Fischer E (1997) Neurotoxicity of paraquat and triphenyltin in the earthworm, *Eisenia fetida* sav. A histo- and cytopathological study. *Acta Biologica Hungarica*, **48**, 485–495.

8

Xenobiotic Impacts on Plants and Biomarkers of Contaminant Phytotoxicity

BRUCE M. GREENBERG

Department of Biology, University of Waterloo, Waterloo, Ontario, Canada

8.1 INTRODUCTION

Most ecosystems are dependent on light as the principal source of energy. Radiant energy is harvested by photosynthetic organisms (algae and vascular plants), and stored as reduced carbon that the rest of the biota in the ecosystem can utilize. This ranges from aquatic systems where phytoplankton is at the bottom of the food chain to crops grown for human consumption. Accordingly, the impact of plant toxicity on the environment is observed in a number of ways. Xenobiotic impairment of plant growth and/or development will limit primary productivity, constraining total biological activity in an ecosystem (Fletcher 1991). Accumulation of contaminants by plants represents an entry point of hazardous compounds into the food web, initiating a biomagnification process (Mahanty 1986, Jones *et al.* 1989, Salanitro *et al.* 1997, Thomas *et al.* 1998). A known effect of pollution on crops is lower yields, with associated economic and human health consequences (Fletcher 1991, Ormrod and Petite 1992, Kapustka 1997). Plants can be used as sentinel species for the detection of toxic contamination in the environment (Stephenson *et al.* 1997, ASTM 1998, Chapter 3). Finally, plants can grow in some contaminated soils generating large amounts of biomass for remediation of xenobiotics (Salt *et al.* 1998). These considerations make the study of plant toxicology an important discipline.

Both vascular and non-vascular plants have been used for toxicity testing (Fletcher 1991, ASTM 1994, 1998, Kapustka 1997). In many cases, a plant toxicity assay is required by a regulatory agency for approval of a new chemical or for environmental monitoring. Accordingly, a wide variety of plant assays have been developed, some of which are standardized (ASTM 1991, 1994, 1997, 1998, Kapustka 1997). In addition, a number of plant bioindicators have been developed for rapid assessment of negative impacts at contaminated sites (Byl

Environmental Analysis of Contaminated Sites. Edited by G. I. Sunahara, A. Y. Renoux, C. Thellen, C. L. Gaudet and A. Pilon

and Klaine 1991, Plewa 1991, Dixon and Kuja 1995, Huang *et al.* 1997a, Krugh and Miles 1997, Gensemer *et al.* 1999). As an added benefit, biomarkers reveal a great deal about the underlying mechanisms of toxicity (Huang *et al.* 1997a).

8.2 WHOLE ORGANISM ASSAYS OF PHYTOTOXICITY

Numerous plant toxicity tests have been developed (Kapustka 1997, Stephenson *et al.* 1997, Chapter 3). They include germination, growth, root elongation, reproduction and life-cycle tests. A few plant species are favored (e.g., *Brassica napus*, corn, rice, onions, soy bean, *Lemna*, etc.). Most tests involve early growth or vegetative growth. There are few life-cycle tests and even fewer examples of perennial species tests.

8.2.1 PARAMETERS IMPACTING ON CHEMICAL TOXICITY TO PLANTS

Plants live in diverse environments and they have evolved specific strategies for survival. Because plants are immobile, stress must be dealt with via metabolic action, which has led to evolution of novel biosynthetic pathways. For instance, plants vary widely with respect to tolerance to drought, solar radiation, heat and salt (Powles 1984, Ristic *et al.* 1991, Foyer *et al.* 1994). Even closely related species and different cultivars of the same species have highly varied tolerances to specific stressors (Ristic *et al.* 1991, Gray *et al.* 1996, Renaud *et al.* 1997). Also, sensitivity to stress is dependent on growth and/or developmental stage. Therefore selection of the appropriate phytotoxicity test is crucial (Fletcher 1991, Wang 1991a, Kapustka 1997).

Germination is often used as a simple and rapid assay, however it lacks sensitivity because seed coats are impermeable to many chemicals (Kapustka 1997). Plant growth is usually a better toxicity assay, although it is more time-consuming and can require a great deal of growth chamber or greenhouse space. Reproduction, yield and full life-cycle assays may be the most important endpoints, but they are cumbersome and time-consuming to perform.

The number of plant species that have been used in toxicity testing is extensive (Wang 1991a, Kapustka 1997, Stephenson *et al.* 1997, Chapter 3). An over-riding standard method of toxicity testing with plants was recently developed (ASTM 1998). Also, Stephenson *et al.* (1997) compared numerous plant species for toxicity assessment, and proposed *B. napus* (canola) and *Zea mays* (corn) as good broad range species for toxicity testing. Due to the variety of conditions that plants are grown under (e.g., soil vs. hydroponics, different lighting conditions, etc.), it is often difficult to compare studies. Thus, generalizations about plant sensitivity to specific chemicals should be viewed with caution; possible multi-species and multi-end point tests should be performed.

Apoptosis (or programmed cell death) is a complicating factor in phytotoxicity (Allan and Fluhr 1997, Wu *et al.* 1997). This is an important natural process for plant cells that are destined to become structural elements or senescent. These cells must die before they can carry out their assigned function. Any contaminant that can trigger apoptosis would be highly phytotoxic. At a cellular level, apoptosis is triggered by high cytosolic Ca^{2+} and proceeds mechanistically via an active oxygen mechanism (e.g., singlet O_2, superoxide, hydrogen peroxide or hydroxyl radical) (Foyer *et al.* 1994, Allan and Fluhr 1997). Strikingly, many environmental contaminants cause rapid production of active O_2 (Krylov *et al.* 1997).

8.2.2 SHORT-TERM TOXICITY TESTS

Short-term bioassays based on seed germination have been developed for several species. In cases where a comparison has been made, seed germination was a less sensitive assay than plant growth (Fiskesjoe 1993, Fargasovo 1994). Using soybean and barley, several contaminants were compared in a 5-day test for inhibition of root and shoot growth (Pfleeger *et al.* 1991). The two species had different tolerances to the test compounds. In millet root elongation and fresh weight accumulation assays, root elongation was the more time-consuming measurement (Wang 1985). However, it was more sensitive. Germination and root growth of *Sinapis alba* (mustard) have been studied for toxicity to metals (Fargosovo 1994). Inhibition was observed for Pb, Cd, Hg, As and Cr. *Allium cepa* (onion) is a rapid system to test germination (Fiskesjoe 1993). It is only a 2 to 3-day test, however; it was not as sensitive as the two-week *Allium* shoot growth assay. A rapid rice toxicity test involving germination and early growth in seed trays has been developed (Wang 1994). This system has been tested with metals, organic contaminants and herbicides, with results consistent with other plant assays. The seed tray assay has been extended to *B. napus* with results similar to a soil assay (Ren *et al.* 1996).

8.2.3 LONG-TERM TOXICITY TESTS

Cucumber and wheat toxicity tests have been applied to on-site toxicity evaluation (Mwosu *et al.* 1991). They can be used for preliminary site screening for toxicity or for assessment of the extent of remediation. A similar test has been developed based on growth of prairie grasses (blue grass and wheat grass) (Siciliano *et al.* 1997). The compounds tested were 2-chlorobenzoic acid and arochlor, and the EC_{50}'s with the grasses were consistent with other assays. Plant toxicity assays have been used as *in situ* biomonitors to assess the extent of metal remediation from soils (Chang *et al.* 1997). For Pb contaminated soils, *A. cepa* germination and growth were used to determine if removal of the metal was effective. Three herbaceous species were developed for assays of bioavailability of metals using Cd, Cr, Ni and Va (Martin *et al.* 1996). This is an important assay when plants are potentially sequestering metals into the food chain. An analogous assay was used for assessment of remediation of crude oil contaminated soils (Salanitro *et al.* 1997).

A life-cycle toxicity test using *Arabidopsis thaliana* has great promise (Kapustka 1997, ASTM 1998).*A. thaliana* is a weed species with a very short life-cycle (35 days from sowing to seed production). It is also a small plant, so intricate experiments can be performed in small areas. Finally, it is part of the international genome projects and a wide variety of mutants have been generated for this species (Somerville and Somerville 1996). Some of the mutants show elevated tolerance or sensitivity to stress.

8.2.4 AQUACULTURE

Aquaculture (hydroponics) simplifies application of test compounds to plants. However, there is debate whether such processes are as environmentally relevant as soil based assays. Toxicity of contaminants to plants is generally greater in aquaculture than in soil (Adema and Henzen 1989, Hulzebos *et al.* 1993). This is probably due to greater bioavailability of the chemical. However, the sensitivity of terrestrial plants in aquaculture is similar to the sensitivity of the aquatic plant *Lemna* (Wang 1991a,b, Huang *et al.* 1993, Ren *et al.* 1996). Thus, for toxicity ranking purposes and determination of toxicity equivalents (TEQs), hydroponic tests with terrestrial plants and *Lemna* are environmentally relevant. *Lemna* is of course fully relevant for toxicity assessment of aquatic environments.

Fresh water algae and macrophytes have been used extensively for toxicity assessment (Lewis 1995). Because of its environmental relevance to terrestrial and aquatic plants, *Lemna* has been used extensively (ASTM 1991, Wang 1991a). Also, laboratory assays of sediment toxicity using rooted submerged macrophytes have been developed. They allow for testing of contaminants in sediments. Species used include *Vallisneria* (Biernacki *et al.* 1997), *Myriophyllum* (ASTM 1997) and Sago pondweed (Fleming *et al.* 1991). In an interesting study, toxicities of *Selenestrum capricornutum* and *Lemna minor* were compared for 16 herbicides that had wide differences in mechanism (Fairchild *et al.* 1997). They included photosystem II (PSII) inhibitors (atrazine) and auxins (2,4-D). There were differences in sensitivity between the two organisms for nine classes of herbicides. Neither species was uniformly more sensitive. This highlights how toxicity depends on the receptor organism, and consequently why a single plant toxicity test for all chemicals has inherent flaws.

In summary, there are a large number of plant toxicity assays. Appropriate plant tests should be employed for comprehensive environmental assessment. They should be selected based on the environmental occurrence of contaminant. For instance, if the contaminant is primarily found in soils then a terrestrial plant grown in soil should be employed. Furthermore, the endpoint can dictate the outcome of the test. In most cases plant growth has been used (e.g., shoot growth or fresh weight accumulation). It is a reasonable compromise between test duration and sensitivity. For ranking relative hazards of toxicants, an aquatic plant system might be more accurate because chemical delivery can be better controlled.

8.3 PLANT BIOMARKERS

Biomarkers generally detect a biochemical aspect of toxicant action (e.g., membrane damage, enzyme inhibition, DNA damage). They can provide rapid and direct indications of toxicant impact in the environment. The specificity of a biomarker gives it great sensitivity, and they often do not rely on knowledge of the history of chemical exposure. Therefore, if plants are sampled from an impacted site and compared at a biomarker level to plants from a reference site, one can determine if the plants are stressed. However, biomarkers tend to be toxicant specific, because not all compounds will inhibit the same biological processes.

Biomarkers have the important benefit of providing knowledge about the mechanism of toxicity. This can lead to an understanding of the biological receptor targeted by a particular chemical class, and extrapolations to related molecules. For instance, for polycyclic aromatic hydrocarbon (PAH) photoinduced toxicity, the physiology of cellular damage via $^1O_2$ is well understood (Foyer *et al.* 1994, Krylov *et al.* 1997). Non-specific peroxidation of lipids and proteins in membranes occurs in the presence of active oxygen, and these types of effects can be assayed as indicators of toxicity.

Inhibition of photosynthesis is a key mechanism of toxicant action in plants (Gressel 1985, Ormrod and Petite 1992, Huang *et al.* 1997a, Krugh and Miles 1997). At the whole plant level, this can result in impaired plant growth, lower yields and loss of competitive advantage (Jansen *et al.* 1993, Dixon and Kuja 1995). There are several reliable methods for measuring impacts on photosynthesis including chlorosis, carbon fixation and electron transport (Hipkins and Baker 1986, Hall and Rao 1987, Gray *et al.* 1996, Huang *et al.* 1997a). In particular, assays of PSII electron transport can be extremely sensitive measures of damage to almost any point in the photosynthetic apparatus (Powles 1984, Krause and Weis 1991, Gray *et al.* 1996, Huang *et al.* 1997a). This is because PSII is the first step in photosynthesis, and inhibition of most metabolic activities in the chloroplast downstream from PSII will lead to a block in PSII electron transport. When PSII does not utilize the light it absorbs, this energy instead results in oxidative damage (Powles 1984, Gray *et al.* 1996, Huang *et al.* 1997a). Photosynthetic electron transport can be measured by a few techniques (Hipkins and Baker 1986, Krause and Weis 1991, Huang *et al.* 1997a). *In vivo*, PSII can be measured by chlorophyll *a* (Chl *a*) fluorescence or O_2 evolution, and primary productivity can be assessed by CO_2 fixation. *In vitro*, PSII activity can be quantified with the aid of electron acceptors.

Gas exchange measurements (e.g., O_2 evolution and CO_2 fixation) are well established techniques used in environmental assessment. For instance, periphyto-photosynthesis was measured using CO_2 fixation as an indicator of effluent toxicity (Lewis 1992). It was found to relate well to animal biomarkers of toxicity. The effects of pentachlorophenol to *L. minor* were measured via O_2 evolution (Huber *et al.* 1982). Also, the phytotoxic effects of industrial effluents and sewage were measured based on transpiration (Gadallah 1996).

Chl *a* fluorescence is a powerful biomarker and has been used for toxicant assessment (Samson and Popovic 1988, Huang *et al.* 1997a, Krugh and Miles 1997). It was used as a toxicity endpoint for metals and pesticides, with most of the impact being observed at PSII (Samson and Popovic 1988, Krugh and Miles 1997). Chl *a* fluorescence was used as a biomarker of photoinduced toxicity of PAHs to plants (Huang *et al.* 1997a, Gensemer *et al.* 1999). The primary site of action was found to be inhibition of electron transport downstream of PSII (Huang *et al.* 1997a). This was followed by inhibition of PSII, probably due to excitation pressure on PSII once the downstream electron transport was blocked. A linkage between inhibition of photosynthesis and inhibition of plant growth was established (Huang *et al.* 1997a).

Tradescantia plants exposed to diesel fuel at 0.1 to 100 mg kg^{-1} in soil were assayed for toxicity via three biomarkers (Green *et al.* 1996): plant morphology, photosynthesis (Chl *a* fluorescence) and a micronucleus assay (DNA synthesis). There was a good correlation between inhibition of growth and photosynthesis. This, along with the above PAH work on photosynthesis, shows the validity of the Chl *a* fluorescence biomarker of toxic action.

Photosynthesis is an enzymatic assay of contaminant impact. Not surprisingly, other enzymatic assays have been used as biomarkers. Many proteins are affected by contaminants. Alterations in the levels of several proteins by NaCl and heavy metals were examined with maize (Ramanathan *et al.* 1996). Peroxidase activity and a few other oxygen detoxification enzymes were elevated in leaves and roots. However, glucose-6-phosphate dehydrogenase and isocitrate dehydrogenase were inhibited at lower Cu^{2+} concentrations than those that affected growth (0.3 μM vs. 3 μM). Peroxidase activity was also employed to examine Cu^{2+} toxicity in the aquatic plant *Hydrilla verticillata* (Byl and Klaine 1991).

Ozone lowered root starch concentrations (Renaud *et al.* 1997). This inhibition could have been due to impairment of starch storage enzymes and/or photosynthesis. Photosynthesis is the likely target, because thylakoid membranes are sensitive to oxygen radical damage (Halliwell and Gutteridge 1985, Gray *et al.* 1996). The starch content assay was validated by comparison to plant growth. This assay was found to be useful for predicting growth and longevity of perennials in the presence of O_3. This assay has great potential because perennials are so difficult to assess at the whole organism level.

Effects of contaminants can be sensitively and specifically assayed by gene expression. A functional homologue of the yeast Cu^{2+} homeostasis ATX1 gene from *Arabidopsis* has been identified (Himelblau *et al.* 1998). It is an antioxidant gene that could be used as an acclimation biomarker. Aflatoxin and benzo(*a*)pyrene activated cytochrome P450 expression in plant cells (Gentile *et al.* 1991, Plewa 1991). Chaperonins and heat shock proteins are up-regulated in response to chemical stress (Ramanathan *et al.* 1996, Schoffl *et al.* 1998). Their function is to stabilize proteins when denaturing conditions exist. It is not surprising they are activated, because many contaminants elevate the levels of oxygen radicals that in turn cause protein denaturation.

Clearly, plant biomarkers have great potential to test the impacts of environmental contaminants on plants. They are rapid, selective and sensitive. Also, they can be used without prior knowledge of the history of exposure. Biomarkers can reveal a great deal about the mechanism of toxicant action, which allows extrapolation to related contaminants.

8.4 IMPACTS OF DIFFERENT GROUPS OF CONTAMINANTS ON PLANTS

A large number of toxicants have been tested on plants using whole organism measures of toxicity. Metals, organic chemicals and non-metal inorganic chemicals can all be potent phytotoxicants. Although these are difficult to summarize because of the number of species tested, the results can be examined based on chemical class.

8.4.1 METAL CONTAMINANTS

Like all living organisms, plants are sensitive to heavy metals in the environment (Fargosova 1994, Wang 1994, Martin *et al.* 1996). Also, metal accumulation by plants is a food chain/biomagnification issue, because there are legislated limits on the content of certain metals in grains (Choudhary *et al.* 1995, Archambault *et al.* 1996). Metal toxicity has been observed in both terrestrial and aquatic plants in the nanomole to micromole range (Wang 1986, Dirilgen and Inel 1994, Suszcynsky and Shann 1995, Das *et al.* 1997). In rice, the order of decreasing toxicity is Cu>Ag>Ni>Cd>Cr>Pb>Zn>Mn (Wang 1994). Toxicity and internal distribution of Hg in tobacco was evaluated following shoot and root exposure (Suszcynsky and Shann 1995). The effects were at the micromole level with little transport in the plant, indicating that the impacts of some metals can be limited to the exposed tissue. Toxicity of Ur in soils has been investigated with three grass species (Buffalo grass, Little blue stem and *Aristida purpurea*) (Meyer and McLendon 1997). Plant emergence, survival and biomass were measured. Toxicity was only observed at high concentrations. This means that plants might be able to survive on metal contaminated soil. On the one hand this has promise for bioremediation of metals (Salt *et al.* 1998), on the other hand it also means that biomagnification of metals in the food chain is a potential problem. Indeed, some metals (e.g., Cd) accumulate in seed grains following root exposure (Choudhary *et al.* 1995, Archambault *et al.* 1996). A real problem is that many soils have naturally high levels of Cd and other metals, so the source may not be anthropogenic.

Toxicity tests with three metals to the aquatic plant *Lemna gibba* (duckweed) growth revealed the order of impact to be $Cu^{2+}>Co^{2+}>Zn^{2+}$, with EC_{50}'s in the micromole range (Dirilgen and Inel 1994). Cd toxicity to *Lemna trisculca* was also examined and an EC_{50} of 6.4 $\mu g\ ml^{-1}$ was reported (Huebert and Shay 1993). Thus, metals are somewhat toxic to aquatic plants. Indeed, these EC_{50}'s may be falsely high, because *Lemna* growth medium has chelating agents that substantially lower metal bioavailability.

Metal bioavailability and soil binding are extremely important in their toxicity to plants. Complexation to soil can be very strong and is pH dependent. Wheat seed germination was affected by metals in acidic soils more than in alkaline soils (Shende *et al.* 1993). In a more applied case, sewage sludge inhibited corn, sorghum and soybean yields due to metals present in the mixture (Berti and Jacobs 1996). Toxicity was thought to be due to Zn and Ni. Metal bioavailability is also important in aquatic systems due to the chelating capacity of humic substances in the water column. Thus, metals often show low toxicity to aquatic plants when there is high binding capacity in the water column (Wang 1986, Dirilgen and Inel 1994).

A potential mechanism of toxicity of metals is redox cycling. This can abstract electrons from bioenergetic electron transport chains and catalytically pass the electrons to O_2 (Halliwell and Gutteridge 1985, Wu *et al.* 1997), forming active oxygen species such as superoxide and hydrogen peroxide. An example of this is the response of higher plants to Pb (Singh *et al.* 1997). Photosynthesis, nitrate reduction and nodulation in N_2 fixing plants were inhibited. It is likely that Pb was affecting the redox status of the plant. In plants, electrons can be abstracted by metals from either the mitochondrial or the chloroplast electron transport chains. Also, metals can inhibit numerous ion pumps in cellular membranes (Das *et al.* 1997).

8.4.2 ORGANIC CONTAMINANTS

There is a wide range of plant responses to organic contaminants, reflecting broad mechanisms of toxicity. Exposure of plants to organic contaminants is common, which is not surprising since organic contaminants are persistent in soils (Jones *et al.* 1989, Tam *et al.* 1996, Salanitro *et al.* 1997). Herbicides are usually organic compounds and obviously were selected for their toxicity to plants (Duke 1990). Also, plants make allelopathic herbicides, specifically evolved to impede the growth of competing plants (Duke 1990).

Plants can assimilate organics from soil, water or air. This is because roots have a high capacity for hydrophobic compounds (Tam *et al.* 1996, Salanitro *et al.* 1997). In aquatic systems, assimilation of contaminants by plants is rapid and efficient, even from non-dissolved phases such as sediment. Because plant tissues have a higher affinity for organics than the aqueous phase, the bioconcentration factors (BCFs) can be very high (Duxbury *et al.* 1997, Thomas *et al.* 1998). Plants can also assimilate organics following aerial deposition on the leaves (Thomas *et al.* 1998). Contaminants received in this manner can be highly toxic, and represent an important entry point of organic compounds into the food chain.

PAHs are contaminants of concern in many industrialized areas. Because PAHs are lipophilic, they tend to accumulate in plants, especially in membrane bilayers (Duxbury *et al.* 1997, Thomas *et al.* 1998). They are highly persistent in soils and sediments, and thus are a risk to terrestrial and aquatic plants. Indeed, plants grown in areas with high PAH loads in the soil or air have high bioconcentrations of PAHs (Jones *et al.* 1989). PAHs without being activated are generally only phytotoxic at high concentrations (Huang *et al.* 1993). For instance, in a study by Chaineau *et al.* (1997), soils contaminated with fuel oils inhibited seed germination and growth in lettuce, barley, clover and maize at high concentrations in soil (0.3 to 4 g kg^{-1}). For bean, wheat and sunflower even higher concentrations were required (4 to 9 g kg^{-1}). The smaller PAHs seemed to be the most toxic.

PAHs absorb sunlight strongly, and can act as photosensitizers (Krylov *et al.* 1997). If sunlight is present, PAH toxicity to aquatic and terrestrial plants increases dramatically (Huang *et al.* 1993, Ren *et al.* 1996, Krylov *et al.* 1997). Obviously, plants provide a relevant model for the study of toxicity of photoactive contaminants. Photoinduced toxicity of PAHs is due to two processes: photosensitization and photomodification reactions (Krylov *et al.* 1997). Photosensitization reactions usually proceed via formation of highly reactive singlet-state oxygen ($^1O_2$). $^1O_2$ formed within a biological organism is highly damaging (Halliwell and Gutteridge 1985). Photomodification of PAHs, which generally occurs via oxidation of the parent compound, results in a mixture of compounds with high toxicity (McConkey *et al.* 1997, Mallakin *et al.* 1999). This is similar to increased PAH hazards following cytochrome P450 activation (Plewa 1991). Interestingly, it was found that humic acids ameliorate PAH toxicity to *L. gibba*, probably due to binding of the PAHs (Gensemer *et al.* 1999). Therefore, PAHs are an example of an environmental toxicant where one environmental factor (light) can enhance risk, while another (binding) can lower risk.

Chlorinated aromatics are broadly toxic to plants (Hulzebos *et al.* 1991, Kapustka 1997, Siciliano *et al.* 1997). However, they are less toxic than PAHs, especially if light is a factor, because chlorinated aromatics are not highly photoactive. A group of chlorophenols was tested for toxicity to millet (Wang 1985). At moderate concentrations they inhibited shoot growth, with the order of toxicity being trichlorophenol>dichlorophenol>chlorophenol>phenol. Pentachlorophenol toxicity to *Lemna* has been examined (Huber *et al.* 1982), and was found to be less toxic than PAHs (cf. Huang *et al.* 1993). Toxicity was associated with inhibited photosynthetic O_2 production and decreased chlorophyll content. Furthermore, low concentrations of pentachlorophenol inhibited glutamate dehydrogenase activity. Structural damage to membranes by chlorophenols has been observed (Hulzebos *et al.* 1993). Interestingly, stimulated growth was observed at low pentachlorophenol concentrations, showing that hormesis is possible (Wang 1985).

Polychlorinated biphenyl (PCB) phytotoxicity has been examined (Mahanty 1986, Mayer *et al.* 1998). They are generally not as toxic to plants as chlorophenols, and thus have received less attention. Algae seem to be more sensitive than higher plants. This may result from different uptake efficiencies. PCBs generally affect membranes and can induce cytochrome P450 activity in plants (Borlakoglu and John 1989, Plewa 1991, Wilken *et al.* 1995). Because plants are relatively insensitive to PCBs, they are a possible route of biomagnification into various food chains. Indeed, plants have thus been used to examine PCB and dioxin transfer to the biosphere (Sinkkonen *et al.* 1995, Thomas *et al.* 1998).

Numerous herbicides have been developed (Gressel 1985, Duke 1990). There is a great deal of data on their phytotoxicity (Wang 1991a). Toxicity of herbicides is often observed at the micromole level (0.1 to 1 $\mu g\ ml^{-1}$). This is in the same range as PAHs in the presence of light or metals, showing contaminants can be as toxic as herbicides. Herbicides are varied compounds and have a number of modes of action. Indeed, they are often designed around specific targets within the plant cell (Duke 1990). For instance, the triazine herbicides (e.g., atrazine) inhibit photosynthesis at PSII (Jansen *et al.* 1993). Plants have been selected that are tolerant to atrazine. Maize can degrade atrazine and thus it has been used extensively on corn fields (Gressel 1985). Other plants have been selected with a mutated protein in PSII, which does not bind atrazine (Duke 1990, Jansen *et al.* 1993). Another herbicide, glyphosate, inhibits the shikimate pathway, preventing aromatic amino acid synthesis (Duke 1990). Crop plants have been selected which are tolerant of glyphosate (Shah *et al.* 1986). Fortunately, glyphosate has a short half-life in soil, so it is not persistent in the environment. This work shows the importance of understanding mechanisms of toxicity as more efficient and safer herbicides can be developed.

Trichloroacetic acid (TCA) and trifluoroacetic acid (TFA) are important phytotoxic air pollutants (Norokorpi and Frank 1995). They are by-products of the new generation of refrigerants (replacing CFCs). Thus, the atmospheric loads of these are predicted to be very high in the future. TCA at 30 $ng\ g^{-1}$ caused defoliation of Scots pine.

Synergistic toxicity of organic compounds with other classes of contaminants has been investigated. The combined effects of fly ash and SO_2 on cucumber were examined (Tung *et al.* 1995). At concentrations where neither fly ash nor SO_2 alone had an effect, the compounds combined caused significant chlorosis. The active organic compounds in the fly ash were probably PAHs. Recently, we found that a metal (Cu^{2+}) combined with an oxygenated PAH (1,2-dihydroxyanthroquinone) had synergistic toxicity (Babu and Greenberg, unpublished observations). This toxicity was due to Cu^{2+} catalysed transfer of electrons to O_2 to form superoxide and hydrogen peroxide.

Higher plants have been used as toxicity biomarkers of effluents that contain organic contaminants. Using a sewage effluent rich in organics, *Lemna*, lettuce and rice were used as test species (Wang 1991b). The endpoints were germination and growth. *Lemna* was the least sensitive plant. Paper mill effluents were also tested using sunflower (Das and Behera 1993). The effects were mostly due to chlorinated aromatics (e.g., PCBs) in the effluents.

8.4.3 INORGANIC CONTAMINANTS

Several typical (non-metal) inorganic contaminants are found in soils, aquatic systems and the atmosphere. The responses of plants to this group of compounds are more varied than to metals. Ozone (O_3), carbon monoxide (CO), salt (NaCl), boric acid (H_3BO_3), nitric acid (HNO_3) and nitric oxide (NO) are common inorganic contaminants. Ozone, NO and CO are primarily atmospheric problems. Acid and salt are common to aquatic systems and soils.

In general, O_3 is the most important phytotoxic air pollutant (Krupa and Legge 1995). O_3 in air was toxic to grasses at 180 $\mu g\ m^{-3}$ (Gruenhage and Jaeger 1994). The effects of O_3 on alfalfa (tolerant and sensitive cultivars) were examined (Renaud *et al.* 1997). Concentrations of 20 to 40 $nl\ l^{-1}$ in air inhibited plant growth, and the effects were more pronounced to the sensitive cultivar. This showed that plant lines sensitive and resistant to contamination could be identified. Interestingly, the mechanism of toxicity is likely to be an active oxygen mechanism because superoxide, hydroxyl radical and singlet-oxygen are readily formed from ozone.

Nitric acid (HNO_3), along with O_3, is important in formation of photochemical smog and acid rain. In a study with oak and pine in southern California forests, HNO_3 is transported into leaves after cuticular deposition (Bytnerowicz *et al.* 1998). In the light, a concentration of 500 ppb in air for 12 h resulted in destruction of the cuticle. However, in the dark 2000 ppb was required for a similar effect. This is another example of photoinduced toxicity. In related work, the impact of acid rain on growth, nutrition, pigment content and photosynthesis in maple and spruce was examined (Dixon and Kuja 1995). Various pH values from 3.2 to 5.6 were tested, with the finding that lower pH had a more negative effect on spruce than poplar.

The effects of CO_2 and CO on plants have been the subject of several studies (Woodrow and Berry 1988, Grodzinski 1992). They are important greenhouse gases, and their anticipated elevated levels could have significant impacts. CO_2 causes plant stomates to close. Also, the efficiency with which plants can consume CO_2 is important for remediation of the gas. CO is an inhibitor of photosynthesis; thus, it will inhibit primary productivity consumption of another greenhouse gas (CO_2).

Boric acid is an important soil phytotoxicant (Kapustka 1997, ASTM 1998). It is broadly phytotoxic and does not bind strongly to soils. Because of these properties it has been employed routinely as a reference toxicant. Similarly, NaCl is an environmental problem in soils, especially where crops are irrigated and/or fertilized (Ramanathan *et al.* 1996, Winicov 1998). It is also broadly toxic to plants and can be used as a reference toxicant.

8.5 QSAR IN PHYTOTOXICOLOGY

Quantitative structure activity relationships (QSARs) correlate the physicochemical properties of molecules to observable biological responses. They are useful for understanding the mechanisms of action of groups of related chemicals and for predicting the environmental risks associated with those chemicals. In developing a QSAR model, especially for plants, it is essential to consider the attributes of the environmental compartment in which the contaminant of interest resides, as this will dictate which physicochemical properties are likely to be most influential in toxicity. For instance, because solar radiation is ubiquitous in the environment and can enhance phytotoxicity of certain chemicals, it represents a modifying factor that can be used in QSAR modeling (see Krylov *et al.* 1997).

QSARs for herbicides have been used to develop more selective and efficient herbicides (Duke 1990). Generally, for herbicides the precise site of action is known. One can then design chemicals that will bind more efficiently to the receptor. One example where this has been performed is the triazine herbicides, which act at PSII (Jansen *et al.* 1993). In this case the binding niche in PSII is well understood and the binding capacity of different compounds to this site has been investigated and modeled. This shows what can be done with toxicity modeling if the mechanism of action is well understood. Unfortunately, for most classes of environmental contaminants, the site of action is generally not characterized, making QSAR modeling less accurate.

QSARs for toxicity of chloroaromatic compounds to plants have been described. With no modifying environmental factors, toxicity was related to lipophillicity, determined as log K_{ow} (Hulzebos *et al.* 1991, 1993). A QSAR was generated for *Latuca sativa* with soil and hydroponic application of the chemicals. Seventy-six different compounds were used, including chloroethylphenols and chloronitrophenols. The EC_{50}'s in soil were very high because the chemicals bind to the soil and were degradable in soil (EC_{50}'s in the 1000 $\mu g\ g^{-1}$ range were observed). Thus, it was difficult to arrive at a reliable QSAR. However, in aquaculture toxicity was much greater and could be related to log K_{ow} (Hulzebos *et al.* 1993). Thus, as with animals, water solubility and lipophillicity (as described by K_{ow}) are a factor that can be used in QSAR modeling. This work also showed that environmental conditions could influence the utility of a QSAR.

A QSAR encompassing modifying factors for environmental perturbations has been generated. It was demonstrated that light could be successfully incorporated as a modifying factor in QSARs (Huang *et al.* 1997b, Krylov *et al.* 1997). Photoinduced toxicity of PAHs occurs via photosensitization reactions (e.g., generation of $^1O_2$) and by photomodification (photooxidation and/or photolysis) of the chemicals to more toxic species. A QSAR developed for toxicity of 16 PAHs to *L. gibba* showed that photosensitization and photomodification additively contribute to toxicity (Krylov *et al.* 1997). This work was extended to a QSAR based on computer modeled structures of PAHs (Mezey *et al.* 1998).

The structures were generated employing *ab initio* quantum mechanical calculations, and it was demonstrated that the shape of the electron density clouds of PAHs fully described photoinduced toxicity.

8.6 CONCLUDING REMARKS

Plant toxicity tests are widespread. A large number of toxicants have been tested on plants using whole organism measures of toxicity and remediation. Germination is often used as a simple and fast assay, however, it lacks sensitivity. Although plant growth and reproduction are usually better toxicity assays, they are more time-consuming. Like all living organisms, plants are sensitive to heavy metals in the environment. Also, metal accumulation by plants is a food chain/biomagnification issue. There is a wide range of plant responses to organic contaminants, reflecting broad mechanisms of toxicity. Ozone (O_3), carbon monoxide (CO), low pH, salt (NaCl), boric acid (H_3BO_3), nitric acid (HNO_3) and nitric oxide (NO) are common phytotoxic inorganic contaminants. In developing QSAR models for plants, it is essential to consider the attributes of the environmental compartment in which the contaminant of interest resides, as this will dictate which physicochemical properties are likely to be most influential in toxicity. Biomarkers can provide a rapid and direct indication of toxicant impact in natural environments. Clearly, plant toxicology is a vital discipline within environmental science. The years ahead will be exciting, especially as the mechanisms of impact become better understood.

8.7 ACKNOWLEDGEMENTS

I want to thank the members of my research group for their assistance in preparing this manuscript. This work was supported by grants from the National Science and Engineering Research Council (NSERC), the Canadian Networks of Toxicology Centres (CNTC) and CRESTech.

REFERENCES

Adema DMM and Henzen L (1989) A comparison of plant toxicities of some industrial chemicals in soil culture and soilless culture. *Ecotoxicology and Environmental Safety*, **18**, 219-229.

Allan AC and Fluhr R (1997) Two distinct sources of elicited reactive oxygen species in tobacco epidermal cells. *The Plant Cell*, **9**, 1559-1572.

Archambault DJ, Zhang G and Taylor GJ (1996) A comparison of the kinetics of aluminum (Al) uptake and distribution in roots of wheat (*Triticum aestivum*) using different aluminum sources. A revision of the operational definition of symplastic Al. *Physioligica Plantarum*, **98**, 578-586.

ASTM (1991) Guide for conducting static toxicity tests with *Lemna gibba* G3. In *Annual Book of ASTM Standards*, Vol. 11.05. E1415-91, American Society for Testing and Materials, West Conshohocken, PA, 10 pp.

ASTM (1994) Standard practice for conducting early seedling growth tests. In *Annual Book of ASTM Standards*, Vol. 11.05. E1598-94, American Society for Testing and Materials, West Conshohocken, PA, 7 pp.

ASTM (1997) Standard guide for conducting static, axenic, 14-day phytotoxicity tests in test tubes with the submersed macrophyte, *Myriophyllum sibiricum* Komarov. In *Annual Book of ASTM Standards*, Vol. 11.05. E1913-97, American Society for Testing and Materials, West Conshohocken, PA, 15 pp.

ASTM (1998) Standard guide for conducting terrestrial plant toxicity tests. In *Annual Book of ASTM Standards*, Vol. 11.05. E1963-98, American Society for Testing and Materials, West Conshohocken, PA, 20 pp.

Berti WR and Jacobs LW (1996) Chemistry and phytotoxicity of soil trace elements from repeated sewage sludge applications. *Journal of Environmental Quality*, **25**, 1025-1032.

Biernacki M, Lovett-Doust J and Lovett-Doust L (1997) Laboratory assay of sediment phytotoxicity using the macrophyte *Vallisneria americana*. *Environmental Toxicology and Chemistry*, **16**, 472-478.

Borlakoglu JT and John P (1989) Cytochrome P-450 dependent metabolism of xenobiotics. A comparative study of rat hepatic and plant microsomal metabolism. *Biochemical Physiology*, **94C**, 613-617.

Byl TD and Klaine SJ (1991) Peroxidase activity as an indicator of sublethal stress in the aquatic plant *Hydrilla verticillata* (royal). In *Plants for Toxicity Assessment: Second Volume*, Gorsuch JW, Lower WR, Wang W and Lewis MA (eds). ASTM STP 1115, American Society for Testing and Materials, West Conshohocken, PA, pp. 101-106.

Bytnerowicz A, Percy K, Riechers G, Padgett P and Krywult M (1998) Nitric acid vapor effects on forest trees — deposition and cuticular changes. *Chemosphere*, **36**, 697-702.

Chaineau CH, Morel JL and Oudot J (1997) Phytotoxicity and plant uptake of fuel oil hydrocarbons. *Journal of Environmental Quality*, **26**, 1478-1483.

Chang LW, Meier JR and Smith MK (1997) Application of plant and earthworm bioassays to evaluate remediation of a lead-contaminated soil. *Archives of Environmental Contamination and Toxicology*, **32**, 166-171.

Choudhary M, Bailey LD, Grant CA and Leisle D (1995) Effect of Zn on the concentration of Cd and Zn in plant tissue of two durum wheat lines. *Canadian Journal of Plant Science*, **75**, 445-448.

Das SS and Behera PK (1993) Physiological effects of sunflower crops grown in paper mill effluents. *Pollution Research*, **12**, 97-100.

Das P, Samantaray S and Rout GR (1997) Studies on cadmium toxicity in plants: a review. *Environmental Pollution*, **98**, 29-36.

Dirilgen N and Inel Y (1994) Effects of zinc and copper on growth and metal accumulation in duckweed, *Lemna minor*. *Bulletin of Environmental Contamination and Toxicology*, **53**, 442-449.

Dixon MJ and Kuja AL (1995) Effects of simulated acid rain on the growth, nutrition, foliar pigments and photosynthetic rates of sugar maple and white spruce seedlings. *Water, Air and Soil Pollution*, **83**, 219-236.

Duke SO (1990) Overview of herbicide mechanisms of action. *Environmental Health Perspectives*, **87**, 263-271.

Duxbury CL, Dixon DG and Greenberg BM (1997) The effects of simulated solar radiation on the bioaccumulation of polycyclic aromatic hydrocarbons by the duckweed *Lemna gibba*. *Environmental Toxicology and Chemistry*, **16**, 1739-1748.

Fairchild JF, Ruessler DS, Haverland PS and Carlson AR (1997) Comparative sensitivity of *Selenastrum capricornutum* and *Lemna minor* to sixteen herbicides. *Archives of Environmental Contamination and Toxicology*, **32**, 353-357.

Fargasova A (1994) Effect of Pb, Cd, Hg, As, and Cr on germination and root growth of *Sinapis alba* seeds. *Bulletin of Environmental Contamination and Toxicology*, **52**, 452-456.

Fiskesjoe G (1993) *Allium* test I: a 2-3 day plant test for toxicity assessment by measuring the mean root growth of onions (*Allium cepa* L.). *Environmental Toxicology and Water Quality*, **8**, 461-470.

Fleming WJ, Ailstock MS, Momot JJ and Norman CM (1991) Response of Sago pondweed, a submerged aquatic macrophyte, to herbicides in three laboratory culture systems. In *Plants for Toxicity Assessment: Second Volume*, Gorsuch JW, Lower WR, Wang W and Lewis MA (eds). ASTM STP 1115, American Society for Testing and Materials, West Conshohocken, PA, pp. 267-275.

Fletcher J (1991) A brief overview of plant toxicity testing. In *Plants for Toxicity Assessment: Second Volume*, Gorsuch JW, Lower WR, Wang W and Lewis MA (eds). ASTM STP 1115, American Society for Testing and Materials, West Conshohocken, PA, pp. 5-11.

Foyer CH, Lelandais M and Kunert KJ (1994) Photooxidative stress in plants. *Physiologia Plantarum*, **92**, 696-717.

Gadallah MAA (1996) Phytotoxic effects of industrial and sewage waste waters on growth, chlorophyll content, transpiration rate and relative water content of potted sunflower plants. *Water, Air and Soil Pollution*, **89**, 33-47.

Gensemer RW, Dixon DG and Greenberg BM (1999) Using chlorophyll a fluorescence induction to detect the onset of anthracene photoinduced toxicity in *Lemna gibba*, and the mitigating effects of humic acid. *Limnology and Oceanography*, **44**, 878-888.

Gentile JM, Johnson P and Robbins S (1991) Activation of aflatoxin and benzo(*a*)pyrene by tobacco cells in the plant cell/microbe coincubation assay. In *Plants for Toxicity Assessment: Second Volume*, Gorsuch JW, Lower WR, Wang W and Lewis MA (eds). ASTM STP 1115, American Society for Testing and Materials, West Conshohocken, PA, pp. 318-325.

Gray GR, Savitch LV, Ivanov AG and Huner NPA (1996) Photosystem II excitation pressure and development of resistance to photoinhibition: II. Adjustment of photosynthetic capacity in winter wheat and winter rye. *Plant Physiology*, **110**, 61-71.

Green BT, Wiberg CT, Woodruff JL, Miller EW, Poage VL, Childress DM, Feulner JA, Prosch SA, Runkel JA, Wanderscheid RL, Wierma MD, Yang X, Choe HT and Mercurio SD (1996) Phytotoxicity observed in *Tradescantia* correlates with diesel fuel contamination in soil. *Environmental and Experimental Botany*, **36**, 313-321.

Gressel J (1985) Herbicide tolerance and resistance: alteration of site of activity. In *Weed Physiology*, Vol. II, Duke SO (ed), CRC Press, Boca Raton, FL, pp. 159-189.

Grodzinski B (1992) Plant nutrition and growth regulation by CO_2 enrichment. *Bioscience*, **42**, 517-524.

Gruenhage L and Jaeger HJ (1994) Influence of the atmospheric conductivity on the ozone exposure of plants under ambient conditions: considerations for establishing ozone standards to protect vegetation. *Environmental Pollution*, **85**, 128-129.

Hall DO and Rao KK (1987) *Photosynthesis*, Edward Arnold, London.

Halliwell B and Gutteridge JMC (1985) *Free Radicals in Biology and Medicine*, Clarendon Press, Oxford.

Himelblau E, Mira H, Lin SJ, Culotta VC, Penarrubia L and Amasino RM (1998) Identification of a functional homolog of the yeast copper homeostasis gene ATX1 from *Arabidopsis*. *Plant Physiology*, **117**, 1227-1234.

Hipkins MF and Baker NR (1986) Spectroscopy. In *Photosynthesis: Energy Transduction, A Practical Approach*, Hipkins MF and Baker NR (eds), IRL Press, Oxford, pp. 51-101.

Huang XD, Dixon DG and Greenberg BM (1993) Impacts of UV irradiation and photooxidation on the toxicity of polycyclic aromatic hydrocarbons to the higher plant *Lemna gibba* L. G3 (duckweed). *Environmental Toxicology and Chemistry*, **12**, 1067-1077.

Huang XD, McConkey BJ, Babu TS and Greenberg BM (1997a) Mechanisms of photoinduced toxicity of anthracene to plants: inhibition of photosynthesis in the aquatic higher plant *Lemna gibba* (duckweed). *Environmental Toxicology and Chemistry*, **16**, 1707-1715.

Huang XD, Krylov SN, Ren L, McConkey BJ, Dixon DG and Greenberg BM (1997b) Mechanistic QSAR model for the photoinduced toxicity of polycyclic aromatic hydrocarbons: II. An empirical model for the toxicity of 16 PAHs to the duckweed *Lemna gibba* L. G-3. *Environmental Toxicology and Chemistry*, **16**, 2296-2303.

Huber W, Schubert V and Sautter C (1982) Effects of pentachlorophenol on the metabolism of the aquatic macrophyte *Lemna minor* L. *Environmental Pollution Series A*, **29**, 215-224.

Huebert DB and Shay JM (1993) The response of *Lemna trisculca* L. to cadmium. *Environmental Pollution*, **80**, 247-253.

Hulzebos EM, Adema DMM, Dirven-Van Breemen EM, Henzen L and Van Gestel CAM (1991) QSARs in phytotoxicity. *The Science of the Total Environment*, **109**, 493-497.

Hulzebos EM, Adema DMM, Dirven-van Breemen EM, Henzen L and van Dis WA (1993) Phytotoxicity studies with *Lactuca sativa* in soil and nutrient solution. *Environmental Toxicology and Chemistry*, **12**, 1079-1094.

Jansen MAK, Mattoo AK, Malkin S and Edelman M (1993) Direct demonstration of binding site competition between photosystem II inhibitors at the Qb niche of the D1 protein. *Pesticide Biochemistry and Physiology*, **46**, 78-83.

Jones KC, Stratford JA, Tidridge P, Waterhouse KS and Johnston AE (1989) Polynuclear aromatic hydrocarbons in an agricultural soil: long-term changes in profile distribution. *Journal of Environmental Pollution*, **56**, 337-351.

Kapustka LA (1997) Selection of phytotoxicity tests for use in ecological risk assessments. In *Plants for Environmental Studies*, Wang W, Gorsuch JW and Hughes JS (eds), Lewis/CRC Press, Boca Raton, FL, pp. 515-548.

Krause GH and Weis E (1991) Chlorophyll fluorescence and photosynthesis: the basics. *Annual Review of Plant Physiology and Plant Molecular Biology*, **42**, 313-349.

Krugh BW and Miles D (1997) Monitoring the effects of five 'nonherbicidal' pesticide chemicals on terrestrial plants using chlorophyll fluorescence. *Environmental Toxicology and Chemistry*, **15**, 495-500.

Krupa SV and Legge AH (1995) Air quality and its possible impacts on the terrestrial ecosystems of the North American Great Plains: an overview. *Environmental Pollution*, **88**, 1-11.

Krylov SN, Huang XD, Zeiler LF, Dixon DG and Greenberg BM (1997) Mechanistic QSAR model for the photoinduced toxicity of polycyclic aromatic hydrocarbons. I. Physical model based on chemical kinetics in a two compartment system. *Environmental Toxicology and Chemistry*, **16**, 2283-2295.

Lewis MA (1992) Periphyton photosynthesis as an indicator of effluent toxicity: relationship to effects on animal test species. *Aquatic Toxicology*, **23**, 279-288.

Lewis MA (1995) Use of freshwater plants for phytotoxicity testing: a review. *Environmental Pollution*, **87**, 319-336.

Mahanty HK (1986) Polychlorinated biphenyls: accumulation and effects upon plants. In *PCBs and the Environment*, Vol. II, CRC Press, Boca Raton, FL, pp. 1-8.

Mallakin A, McConkey BJ, Miao G, McKibben B, Sneikus V, Dixon DG and Greenberg BM (1999) Impacts of structural photomodification on the toxicity of environmental contaminants: anthracene photooxidation products. *Ecotoxicology and Environmental Safety*, **43**, 204-212.

Martin H, Young TR, Kaplan DI, Simon L and Adriano DC (1996) Evaluation of three herbaceous index plant species for bioavailability of soil cadmium, chromium, nickel, vanadium. *Plant and Soil*, **182**, 199-207.

Mayer P, Halling-Sorensen B, Sijm DTHM and Nyholm N (1998) Toxic cell concentrations of three polychlorinated biphenyl congeners in the green alga *Selenastrum capricornutum*. *Environmental Toxicology and Chemistry*, **17**, 1848-1851.

McConkey BJ, Duxbury CL, Dixon DG and Greenberg BM (1997) Toxicity of a PAH photooxidation product to the bacteria *Photobacterium phosphoreum* and the duckweed *Lemna gibba*: effects of phenanthrene and its primary photoproduct, phenanthrenequinone. *Environmental Toxicology and Chemistry*, **16**, 892-899.

Meyer MC and McLendon T (1997) Phytotoxicity of depleted uranium on three grasses characteristic of different successional stages. *Journal of Environmental Quality*, **26**, 748-752.

Mezey PG, Zimpel Z, Warburton P, Walker PD, Irvine DG, Huang XD, Dixon DG and Greenberg BM (1998) Use of QShAR to model the photoinduced toxicity of PAHs: electron density shape features accurately predict toxicity. *Environmental Toxicology and Chemistry*, **17**, 1207-1215.

Mwosu JU, Ratsch HC and Kapustka L (1991) A method for on-site evaluation of phytotoxicity at hazardous waste sites. In *Plants for Toxicity Assessment: Second Volume*, Gorsuch JW,

Lower WR, Wang W and Lewis MA (eds). ASTM STP 1115, American Society for Testing and Materials, West Conshohocken, PA, pp. 333-340.

Norokorpi Y and Frank H (1995) Trichloroacetic acid as a phytotoxic air pollutant and the dose response relationship for defoliation of scots pine. *The Science of the Total Environment*, **161**, 459-463.

Ormrod DP and Petite JM (1992) Sulphur dioxide and nitrogen dioxide affect growth, gas exchange and water relations of potato plants. *Journal of the American Society of Horticultural Science*, **117**, 146-153.

Pfleeger T, McFarlane C, Sherman R and Volk G (1991) A short-term bioassay for whole plant toxicity. In *Plants for Toxicity Assessment: Second Volume*, Gorsuch JW, Lower WR, Wang W and Lewis MA (eds). ASTM STP 1115, American Society for Testing and Materials, West Conshohocken, PA, pp. 355-364.

Plewa MJ (1991) The biochemical basis of the activation of promutagens by plant cell systems. In *Plants for Toxicity Assessment: Second Volume*, Gorsuch JW, Lower WR, Wang W and Lewis MA (eds). ASTM STP 1115, American Society for Testing and Materials, West Conshohocken, PA, pp. 287-296.

Powles SB (1984) Photoinhibition of photosynthesis induced by visible light. *Annual Review of Plant Physiology*, **35**, 15-44.

Ramanathan A, Ownby JD and Burks SL (1996) Protein biomarkers of phytotoxicity in hazard evaluation. *Bulletin of Environmental Contamination and Toxicology*, **56**, 926-937.

Ren L, Zeiler LF, Dixon DG and Greenberg BM (1996) Photoinduced effects of polycyclic aromatic hydrocarbons on *Brassica napus* (Canola) during germination and early seedling development. *Ecotoxicology and Environmental Safety*, **33**, 73-80.

Renaud JP, Allard G and Mauffette Y (1997) Effects of ozone on yield, growth, and root starch concentrations of two alfalfa (*Medicago sativa* L.) cultivars. *Environmental Pollution*, **95**, 273-281.

Ristic Z, Gifford DJ and Cass DD (1991) Heat shock proteins in two lines of *Zea mays* L. that differ in drought and heat resistance. *Plant Physiology*, **93**, 3164-3166.

Salanitro JP, Dorn PB, Huesemann MH, Moore KO, Rhodes IA, Rice-Jackson LM, Vipond TE, Western MM and Wisniewski HL (1997) Crude oil hydrocarbon bioremediation and soil ecotoxicity assessment. *Environmental Science and Technology*, **31**, 1769-1776.

Salt DE, Smith RD and Raskin I (1998) Phytoremediation. *Annual Review of Plant Physiology and Plant Molecular Biology*, **49**, 643-668.

Samson G and Popovic R (1988) Use of algal fluorescence for determination of phytotoxicity of heavy metals and pesticides as environmental pollutants. *Ecotoxicology and Environmental Safety*, **16**, 272-278.

Schoffl F, Prandl R and Reindl A (1998) Regulation of the heat shock response. *Plant Physiology*, **117**, 1135-1141.

Shah DM, Horsch RB, Klee HJ, Kishore GM, Winter JA, Tumer NE, Hironaka CM, Sanders PR, Gasser CS, Aykent S, Siegel NR, Rogers SG and Fraley RT (1986) Engineering herbicide resistance in transgenic plants. *Science*, **233**, 478-481.

Shende A, Juwarkar AS and Dara SS (1993) Phytotoxic effects of heavy metals as influenced by soil properties. *Pollution Research*, **12**, 75-83.

Siciliano SD, Germida JJ and Headley JV (1997) Evaluation of prairie grass species as bioindicators of halogenated aromatics in soil. *Environmental Toxicology and Chemistry*, **16**, 521-527.

Singh RP, Tripathi RD, Sinha SK, Maheshwari R and Srivastava HS (1997) Response of higher plants to lead contaminated environment. *Chemosphere*, **34**, 2467-2493.

Sinkkonen S, Rantio T, Vattulainen A, Aittola JP, Paasivirta J and Lahtipera M (1995) Chlorohydrocarbons, PCB congeners, polychlorodioxins, furans and dibenzothiophenes in pine needles in the vicinity of a metal reclamation plant. *Chemosphere*, **30**, 2227-2239.

Somerville S and Somerville C (1996) *Arabidopsis* at 7: still growing like a weed. *The Plant Cell*, **8**, 1917-1933.

Stephenson GL, Solomon KR, Hale B, Greenberg BM and Scroggins RP (1997) Development of a suitable test method for evaluating the toxicity of contaminated soils to a battery of plant species relevant to soil environments in Canada. In *Environmental Toxicology and Risk Assessment: Sixth Volume*, Dwyer FJ, Doane TR and Hinman ML (eds). ASTM STP 1317, American Society for Testing and Materials, West Conshohocken, PA, pp. 474-489.

Suszcynsky EM and Shann JR (1995) Phytotoxicity and accumulation of mercury in tobacco subjected to different exposure routes. *Environmental Toxicology and Chemistry*, **14**, 61-67.

Tam DD, Shiu WY, Qiang K and MacKay D (1996) Uptake of chlorobenzenes by tissues of the soybean plant: equilibria and kinetics. *Environmental Toxicology and Chemistry*, **15**, 489-494.

Thomas G, Sweetman AJ, Ockenden WA, Mackay D and Jones KC (1998) Air-pasture transfer of PCBs. *Environmental Science and Technology*, **32**, 936-942.

Tung G, McIlveen WD and Jones RD (1995) Synergistic effect of flyash and SO2 on development of cucumber (*Cucumis sativus* L.) leaf injury. *Environmental Toxicology and Chemistry*, **14**, 1701-1710.

Wang W (1985) Use of millet root elongation for toxicity tests of phenolic compounds. *Environment International*, **11**, 95-98.

Wang W (1986) The effect of river water on phytotoxicity of Ba, Cd, and Cr. *Environmental Pollution*, **11**, 193-204.

Wang W (1991a) Literature review on higher plants for toxicity testing. *Water, Air and Soil Pollution*, **59**, 381-400.

Wang W (1991b) Higher plants (common duckweed, lettuce, and rice) for effluent toxicity assessment. In *Plants for Toxicity Assessment: Second Volume*, Gorsuch JW, Lower WR, Wang W and Lewis MA (eds). ASTM STP 1115, American Society for Testing and Materials, West Conshohocken, PA, pp. 68-76.

Wang W (1994) Rice seed toxicity tests for organic and inorganic substances. *Environmental Monitoring*, **29**, 101-107.

Wilken A, Bock C, Bokern M and Harms H (1995) Metabolism of different PCB congeners in plant cell cultures. *Environmental Toxicology and Chemistry*, **14**, 2017-2022.

Winicov I (1998) New molecular approaches to improving salt tolerance in crop plants. *Annals of Botany*, **82**, 703-710.

Woodrow IE and Berry JA (1988) Enzymatic regulation of photosynthetic CO_2 fixation in C_3 plants. *Annual Review of Plant Physiology and Plant Molecular Biology*, **39**, 533-594.

Wu G, Shortt BJ, Lawrence EB, Leon J, Fitzsimmons KC, Levine EB, Raskin I and Shah DM (1997) Activation of host defense mechanisms by elevated production of H_2O_2 in transgenic plants. *Plant Physiology*, **115**, 427-435.

9

Use of Biomarkers of Exposure and Vertebrate Tissue Residues in the Hazard Characterization of PCBs at Contaminated Sites — Application to Birds and Mammals

ALAN L. BLANKENSHIP[1,2] AND JOHN P. GIESY[2–4]

[1] *ENTRIX Inc., East Lansing, MI, USA*
[2] *National Food Safety and Toxicology Center, Michigan State University, East Lansing, MI, USA*
[3] *Institute of Environmental Toxicology, Michigan State University, East Lansing, MI, USA*
[4] *Department of Zoology, Michigan State University, East Lansing, MI, USA*

9.1 INTRODUCTION

9.1.1 NOMENCLATURE AND SOURCES OF POLYCHLORINATED BIPHENYLS

Polychlorinated biphenyls (PCBs) are ubiquitous environmental pollutants that are members of the polychlorinated diaromatic hydrocarbon (PCDH) class of compounds. PCBs are a complex mixture of individual compounds, which are chlorinated with between one and 10 chlorines in various combinations of positions to create a total of 209 possible congeners. The 10 positions are numbered 2–6 on one ring and 2′–6′ on the other ring. Positions 2, 2′ and 6, and 6′, adjacent to the biphenyl bond, are called *ortho* positions; 3, 3′ and 5, 5′, *meta* positions; 4, 4′, *para* positions. PCBs are termed coplanar if the two phenyl rings can attain an alignment that is in the same plane. Chlorines in the *ortho* positions tend to prevent a coplanar alignment. Some of these compounds (especially the coplanar PCBs) are structurally similar to polychlorinated dibenzo-*p*-dioxins (PCDDs), dibenzofurans (PCDFs) and naphthalenes (PCNs). Due to their persistence and widespread use, PCBs are common contaminants at hazardous waste sites in the US and throughout the world. PCBs were manufactured as technical mixtures and sold under various trade names such as Aroclors (e.g., 1242, 1248,

Environmental Analysis of Contaminated Sites. Edited by G. I. Sunahara, A. Y. Renoux, C. Thellen, C. L. Gaudet and A. Pilon

1254, 1260, etc.), clophens, phenclors, kanechlors and sovol for commercial applications such as dielectric fluids in capacitors and transformers, waxes for metal castings, heat transfer agents, plasticizers in paints, coatings and carbonless copy paper, cutting oils, sealants, caulking compounds, pesticide stabilizers and dust control (Eisler 1986, Giesy and Kannan 1998). Sources of PCBs to the environment are most commonly from disposal of, and/or leakage from, capacitors and transformers, industrial and municipal waste discharges, spills, recycling of carbonless copy paper, chloralkali plants, pesticide manufacturing and formulating plants (Eisler 1986, Giesy and Kannan 1998).

9.1.2 WEATHERING OF PCBS

The compositions of PCB congener mixtures that occur in the environment differ substantially from those of the original, technical Aroclor mixtures released to the environment. This is because the composition of PCB mixtures changes over time after release into the environment due to several processes collectively referred to as 'environmental weathering'. This weathering is a result of the combined effects of processes such as differential volatilization, solubility, sorption, anaerobic dechlorination and metabolism and results in changes in the composition of the PCB mixture over time and between trophic levels (Froese *et al.* 1998). Less chlorinated PCBs are often lost most rapidly due to volatilization and metabolism, while more highly chlorinated PCBs are often more resistant to degradation and volatilization and sorb more strongly to particulate matter. More highly chlorinated PCBs tend to bioaccumulate to higher concentrations than low molecular weight PCBs in tissues of animals, and may biomagnify in food webs.

Because of these changes in the relative proportions of individual congeners in the PCB mixture, toxicity reference values (TRVs) derived from laboratory studies using technical Aroclor mixtures may not provide an appropriate point of comparison for PCB mixtures found in environmental samples. The environmental weathering of PCB mixtures may result in a reduction or enrichment of toxic potency of the mixture over time (Williams *et al.* 1992, Corsolini *et al.* 1995, Quensen *et al.* 1998). For example, microbially mediated anaerobic reductive dechlorination is common in sediments and can eliminate the most toxic coplanar congeners (Zwiernik *et al.* 1998), reducing the toxic potency of the resulting PCB mixture by as much as 98% (Quensen *et al.* 1998). In contrast, certain aquatic animals selectively bioaccumulate the coplanar dioxin-like PCB congeners, which can result in a greater relative proportion of these toxic congeners in tissues than in technical mixtures (Williams *et al.* 1992, Corsolini *et al.* 1995).

9.1.3 TOXICITY OF PCBs AND TCDD TOXIC EQUIVALENTS

The toxicity of PCBs has been well established in laboratory and field studies (Giesy *et al.* 1994a,b). Chronic toxicity has been observed in fish, birds and mammals and includes developmental effects, total reproductive failure, liver damage, cancer, wasting syndrome and death (Metcalfe and Haffner 1995). Of particular toxicological significance are the non-*ortho* substituted PCBs (e.g., PCBs 77, 81, 126, 169) and the mono-*ortho* substituted PCBs because these congeners can assume a planar configuration (referred to as coplanar PCBs) similar to 2,3,7,8-tetrachlorodibenzo-*p*-dioxin (TCDD; Safe 1990, Giesy *et al.* 1994a, Metcalfe and Haffner 1995). The mechanism of toxicity of non-*ortho* and mono-*ortho* PCBs is similar to that of TCDD and other PCDHs. The major mechanism of action for the toxicity of PCBs and related chemicals is related to their ability to bind to and activate the aryl hydrocarbon receptor (AhR), which is a cytosolic, ligand-activated transcription factor (Poland and Knutson 1982, Gasiewicz 1991, Blankenship *et al.* 2000). Each of the PCBs, PCDDs and PCDFs binds with different affinity to this receptor and exhibits different biological activities (Safe 1990, Ahlborg *et al.* 1994, Van den Berg *et al.* 1998). Thus, the toxic potency for each of these chemicals varies but can be normalized to TCDD as a toxic equivalent factor (TEF; Van den Berg *et al.* 1998). The concentration of a chemical multiplied by its TEF yields a TCDD toxic equivalent (TEQ) concentration. Because organisms are exposed to environmentally weathered mixtures that contain not only PCB congeners but also PCDD, PCDF, PCN and possibly other AhR-active compounds, the potential effects of PCBs must be interpreted in the context of these other compounds (Giesy *et al.* 1994b). TEFs have been determined for most of the dioxin-like chemicals including PCDDs, PCDFs, PCBs and recently PCNs (Van den Berg *et al.* 1998, Blankenship *et al.* 2000). Class-specific TEFs have been developed for mammals, birds and fish (Van den Berg *et al.* 1998). However, there is an inherent uncertainty in the TEQ approach due to potential species-specific differences in the relative toxicity of PCBs and TCDD and deviations from a simple additivity approach (Villeneuve *et al.* 1999). Despite these acknowledged limitations, the TEQ approach has proved to be an effective way to assess the overall toxicity of a complex mixture of TCDD-related chemicals. TEQs can be determined for any sample for which there are both congener-specific concentrations and either an endpoint- and species-specific relative potency (REP) which is specific to an experiment, or a TCDD equivalency factor (TEF) which is consensus-derived from many relative potency values (Van den Berg *et al.* 1998). TEQs contributed from PCDH congeners can be calculated by solving Equation (9.1):

$$\text{ng TEQ kg}^{-1}\text{sample (lipid)} = \sum_{i=1}^{n} [\text{total conc. PCDH } \#n \text{ in sample (ng kg}^{-1}\text{, lipid)} \times \text{TEF for PCDH } \#n] \quad (9.1)$$

where n is any PCDH congener.

Studies on wildlife in the Great Lakes and elsewhere have found a causal link between adverse health effects and PCBs (Giesy *et al.* 1994a, Bowerman *et al.*

1995, Kennedy *et al.* 1996). However, the observed toxicity usually correlates better to TEQs than to total PCBs (Giesy *et al.* 1994a, Leonards *et al.* 1995). PCBs (expressed as TEQs) are thought to be the most critical contaminants in the Great Lakes due to their persistence (long residence time in the Great Lakes), bioaccumulation and biomagnification potential, human health concerns and wildlife health concerns (especially fish-eating wildlife such as bald eagles, colonial fish-eating water birds, mink and otters; Giesy *et al.* 1994c,d, 1995). Due to their bioaccumulation potential, a food web analysis is often the most appropriate approach for ecological risk assessments for PCBs. At most sites, PCBs are predominantly bound to particles (usually associated with an organic carbon fraction). Within an aquatic system, organisms are exposed to a combination of dissolved PCBs and sediment-associated PCBs. Within a terrestrial system, lower trophic level organisms are exposed to PCBs primarily through ingestion of soil and prey, although for certain species dermal absorption and inhalation could be important routes of exposure. At each subsequent trophic level, certain PCB congeners are selectively enriched or bioaccumulated through mechanisms discussed previously. Generally, those organisms at the top of the food chain are most at risk, depending upon species-specific sensitivities to PCBs.

9.1.4 MECHANISMS OF ACTION FOR PCBs

While the critical mechanism of toxic action of complex mixtures of PCBs is due to the dioxin-like congeners acting through an AhR-mediated mechanism of action (reviewed in Metcalfe and Haffner 1995), the di-*ortho* substituted congeners, which constitute the greatest mass of the total PCBs, have been reported to cause non-AhR-mediated effects (reviewed in Giesy and Kannan 1998). While these effects, which are primarily neurobehavioral in nature, have been observed, they generally occur at much greater concentrations than do the AhR-mediated effects. When hazard quotients (HQ; concentration in a tissue or environmental compartment divided by a species-specific and endpoint-specific reference concentration) are calculated, the greater HQ values are due to the AhR-mediated effects (Giesy and Kannan 1998). Said another way, the AhR-active congeners are the critical toxicants. That is, in the weathered complex mixture of PCBs, the ratio of total dioxin-like activity is greater than the non-dioxin-like effects. The consequences of this are that predicting the toxic effects of total PCB mixtures by their TEQ is more accurate and related to the critical effects (Giesy *et al.* 1994b, Giesy and Kannan 1998). In a risk assessment scenario, protecting from the AhR-active fraction of a weathered environmental mixture of PCBs will most likely be protective of the effects of the non-*ortho* substituted constituents of the mixture. A TEQ/TEF approach can be applied to endpoints other than AhR-mediated effects such as dopaminergic effects on neurobehavioral endpoints (Giesy and Kannan 1998).

9.1.5 EXPOSURE CONSIDERATIONS AND USE OF BIOMARKERS

A fundamental concept in toxicology is the dose–response relationship, which assumes that: (1) a given response is proportional to the concentration of a chemical at a site of action (e.g., receptor) and (2) the concentration at the site of action is proportional to the dose. Environmental monitoring of PCBs at contaminated sites, on the other hand, is typically accomplished by measuring ambient levels of PCBs in water, soil and/or sediment. Estimates of bioavailability, or the amount of a compound that is internally absorbed, are fraught with uncertainty. A more accurate and less uncertain prediction of effects can be made when the internal dose of an individual is related to both the environmental concentration and the biological effects observed (Rangan *et al.* 1997). This is a more proximal measure of effects and does not require predicting tissue concentrations from models that may require the use of conservative or uncertain assumptions and parameters.

Even when an internal dose is available, it is difficult to relate it to the potential for adverse effects. The response of the organism to the presence of a toxicant is a function of the rate of effect relative to the rate of repair. Furthermore, the effect of contaminants is a function of both **toxicokinetics** (which determines the available fraction of a toxicant within an animal through processes such as absorption, distribution, metabolism and excretion) and **toxicodynamics** (which involves the biochemical and physiologic effects of toxicants and their mechanisms of action). Only a portion of the total body burden is available to interact with key biomolecules. Much of the lipophilic compounds are bound in fat, while other compounds are bound to proteins such that they are unavailable to interact with key biomolecules. For instance, when mink were exposed to PCBs and then exposure was terminated, the total concentration in the liver did not decline. However, the activity of the cytochrome P450 IA1 (CYPIA1) monooxygenase enzyme, which is induced by the interaction of PCBs with the AhR, returned to the uninduced level within a few weeks (Shipp *et al.* 1998). This indicates that total internal dose does not relate directly to response. In the case of PCBs, if the rate of exposure is slow such that PCBs are partitioned into fat or bound to proteins such as CYPIA2, rendering the PCBs unavailable, the potential for adverse effects is less than if the rate of response is greater such that a greater proportion of the PCBs is available to interact with critical biomolecules and cause toxicity.

Thus, the concept of biomarkers has been developed. Biomarkers are measurable indicators that the chemical is or was present and that it has caused some measurable effect (see Chapters 7 and 8). Biomarkers of wildlife exposure to PCBs are an active area of research and there is considerable interest in the use of biomarkers for monitoring remediation of PCB contaminated sites. Biomarkers provide a functional measure of exposure that integrates the rate of exposure and bioavailability of the toxicant of interest. The biomarkers may be directly related to an adverse outcome or only an indicator of exposure. Some biomarkers measure adaptive responses to exposures, and if the organism is not forced out of its normal homeostatic range or required to use energy that would normally go into growth or reproduction, there may be no adverse outcome.

9.1.6 PURPOSE OF THIS CHAPTER

The purpose of this chapter is to present currently available approaches to measure effectiveness of remedial actions at sites contaminated with PCBs from a risk-based or hazard-based perspective and to identify data needs. Specifically, approaches that utilize exposure models, biomarkers of exposure and vertebrate tissue residues will be evaluated and limitations of these approaches will be discussed.

9.2 APPROACHES FOR ECOLOGICAL HAZARD ASSESSMENT AND MONITORING REMEDIATION PCB CONTAMINATED SITES – ADVANTAGES AND DISADVANTAGES

9.2.1 INSTRUMENTAL QUANTIFICATION

Analytical chemistry results for PCBs can be reported either as total PCBs, Aroclor mixtures or individual congeners. Historically, only total PCB data were reported. Recent advances (over the last 10–15 years) in analytical chemistry separation techniques have been developed that allow resolution of the most toxic non-*ortho* PCB congeners at low enough detection limits to be useful for risk assessment purposes. Quantification of PCBs can be achieved by comparison to single or combined technical mixture standards, peak summing or pattern recognition for marker PCBs to match a parent Aroclor mixture. PCBs in environmental samples, however, rarely resemble a parent technical mixture due to 'weathering' of PCB mixtures. The most toxic PCBs (those that exhibit the greatest HQ in environmental mixtures) are those with four to six chlorines (these tend to be enriched over time and biomagnify). The most commonly used methods for PCB analysis are USEPA's Solid Waste-846 Methods 8080 and 8081. Both methods, however, warn users that the methods may not be applicable when environmental samples contain weathered Aroclors or complex mixtures of Aroclors. Thus, USEPA Region 9 has recently recommended a congener-specific analytical approach to help ensure that data quality objectives are met for ecological risk assessments (USEPA 1998b). In response to concerns that commercial Aroclors are not representative of PCB mixtures in environmental samples, there is increasing agreement among scientists that toxicity assessments for environmental PCB samples and quantification of total PCBs would be more accurate if performed on an individual congener basis (USEPA 1998b).

9.2.2 *IN VITRO* BIOASSAY

While risk assessments based on the TEQs in complex mixtures are most appropriate in assessing the risk posed by dioxin-like compounds including

PCBs, which are present as complex environmental mixtures, this requires quantitative instrumental analysis of complex mixtures of these compounds that are both difficult and expensive. Furthermore, demonstrating the presence of one or many of these compounds in samples provides only limited information on their biological potency, particularly when present in a complex mixture with many potential interactions. Development of mechanism-based cell bioassays (or bioanalytical assays) to detect specific classes of compounds has enhanced the ability to screen for dioxin-like activity in extracts of environmental compartments such as soil, water and biota. The H4IIE rat hepatoma cell bioassay (Tillitt *et al.* 1991) has been widely used for this purpose. In this assay, ethoxyresorufin-*o*-deethylase (EROD)-inducing potencies of single compounds and environmental samples are determined from complete dose–response curves and compared to that of TCDD in order to express the biological potency of the tested samples in TCDD-equivalents (TCDD-EQs). The bioassay integrates potential non-additive interactions among AhR agonists and other compounds by measuring a final receptor-mediated response (Giesy *et al.* 1994a). A similar but more sensitive recombinant cell line has been developed. H4IIE-luciferase (H4IIE-Luc) cells, which are stably transfected with an AhR-controlled luciferase reporter gene construct, respond specifically to AhR agonists (Sanderson *et al.* 1996). The assay is also referred to as the chemical activated luciferase gene expression (CALUX) system (Murk *et al.* 1996). These cells express firefly luciferase in response to AhR agonists. Luciferase activity is measured conveniently and with high sensitivity as light emission using a plate-scanning luminometer. Luciferase induction potential is assessed by comparison of the response to that of TCDD, the most potent agonist for the mammalian Ah receptor. In addition to H4IIE-Luc cells, which are a mammalian cell line, similar cell lines (using either reporter genes or endogenous genes) have been developed for fish (Richter *et al.* 1997, Villeneuve *et al.* 1999) and birds (Kennedy *et al.* 1996). This is important because fish and birds have been found to respond differently to some PCB congeners than other vertebrates (Van den Berg *et al.* 1998).

Often it is desirable to compare bioassay-derived TCDD-EQs to TEQs derived from instrumental analysis (i.e., mass–balance approaches, tests for mixture interactions, etc.). In these situations, it is important to use a different method to calculate the TEQs that is directly comparable to the bioassay (Tillitt *et al.* 1996). Rather than using World Health Organization (WHO) consensus TEFs (Van den Berg *et al.* 1998), bioassay-specific relative potencies (determined for as many individual compounds as practical that have been detected by instrumental analysis) should be multiplied by the concentration of each congener similarly to equation (9.1).

While *in vitro* bioassays, such as H4IIE-luc cells, are useful to assess TCDD-EQs and to predict the toxicological contribution of PCBs within a complex mixture, additional information is necessary: (1) uptake from various environmental matrices and (2) toxicokinetics in an *in vivo* situation. For example, the bioavailability of PCBs may be limited in sediments or soils that are rich in organic carbon because PCBs bind strongly to sediment organic carbon (Kannan *et al.* 1998). Absorption, body distribution and metabolism are important factors that contribute to the *in vivo* potency of an individual PCDD or PCDF (Van den Berg *et al.* 1994). Similar effects are likely for PCBs. Thus, extracts of the tissues most affected by the toxic effects of PCBs are the most proximal and, thus, most appropriate to be used in bioassays. Since TEQs or TCDD-EQs are an aggregate property of a complex mixture of congeners that separate individually, they cannot be biomagnified (Froese *et al.* 1998). Individual congener movement between trophic levels or environmental matrices must be predicted and then TEFs can be applied to calculate concentrations of TEQs in tissues.

9.3 MONITORING BIOREMEDIATION – METHODS FOR EXPOSURE ASSESSMENT

Before a site can be monitored for the effectiveness of a remedial action, it is important to determine and understand the overall objective of the remedial action. Is the objective to decrease PCB concentrations in fish 10 km downstream? Or is it to increase bald eagle productivity? Prior to the environmental management decision to remediate a site, it is assumed that a proper ecological risk assessment has been conducted in which the nature, extent and causal relationships of ecological effects are established (USEPA 1998a, see Chapter 16). In this process, ecological receptors are identified by specifying assessment endpoints (environmental values to be protected) and measurement endpoints (representative, quantifiable expressions of the assessment endpoints). It is only after such a process has been conducted that an understanding of the key parameters to monitor can be achieved. As mentioned earlier, a food web analysis should be well understood for the receptor(s) of concern. Ideally, the key information to be obtained should include: (1) a spatial understanding of the PCB distribution in water, sediment and/or soil including potential influx of contaminants to the site from other sources; (2) a temporal understanding of the PCB distribution in the same matrices (is there a decreasing trend naturally? — this important point will be discussed in more detail in a later section); (3) data on the distribution of AhR-active PCB congeners in receptor organisms and their food; (4) data on the distribution of receptor organisms in space and time at the site; (5) an understanding of the toxicity of TEQs for specific endpoints and receptors; and (6) matching the exposure distributions to the toxicity to quantify risks. Obviously, only certain sites will warrant this level of detail due to the expense of studies to collect such information. However, a thorough understanding of the aforementioned data should allow for both an appropriate environmental management decision and an appropriate basis for biomonitoring to assess the effectiveness of a remedial action. While the application of TEQ methods can be justified by providing the most appropriate and protective assessment, in some situations the use of a total PCB approach will, due to environmental fate and bioaccumulation processes, overestimate the potential risk to ecological receptors.

In the USA and elsewhere, there have been attempts at setting criteria and/or guidelines for PCBs. For sediments, threshold concentration levels such as effects range low (ERL) and effects range median (ERM) have associated poorly with effects and may be especially inaccurate in predicting effects on upper trophic levels. The Great Lakes Water Quality Initiative set recommended concentrations in water that are meant to be protective of top level predators despite the predominance of a dietary exposure pathway that is most often linked to sediments (reference GLWQI). Recently, the Canadian Council of Ministers of the Environment (CCME) issued guidelines entitled 'Canadian Tissue Residue Guidelines (TRG) for Polychlorinated Biphenyls for the Protection of Wildlife Consumers of Aquatic Biota' (CCME 1998). This TRG is appropriate and we endorse its use because it has more toxicological significance and eliminates much of the uncertainty that can arise from exposure modeling. The disadvantages of such an approach are that source allocation issues are not addressed without further data and it is difficult to establish a cleanup goal (e.g., what level in soil would result in a PCB concentration in bird eggs that is below the TRG?). As for aquatic organisms themselves, including piscivorous fishes, it is stated in the guidelines that they are assumed to be protected by the Canadian Water Quality Guideline for Aquatic Life (CCREM 1987, Moore and Walker 1991, CCME 1998).

9.3.1 WHAT DATA ARE AVAILABLE AND/OR CAN BE COLLECTED?

Sediment and/or soil data are usually available and are important to understand the spatial extent and temporal aspects (e.g., increase or decrease over time) of contamination. Using the Lake Järnsjön, Sweden remediation project as an example, it is important to measure sediment and biota concentrations before, during and after the remedial action and at a nearby reference location (e.g., upstream). Using such an approach, Bremle and Larsson (1998a) reported that 50% decreases in concentrations of PCBs in water and fish after a remedial action were attributable to natural attenuation (i.e., the same changes were observed upstream) despite the success of the remedial effort to remove approximately 95% of the mass of PCBs in the sediments. Moreover, if there are only sediment and/or soil data (i.e., no data available for biota), there will be considerably more uncertainty in the exposure estimates. The uncertainty that is present in such exposure modeling almost always overpredicts actual exposure due to:

- Limited bioavailability (e.g., certain PCB congeners bind strongly to sediments);
- Spatial biases of sampling (e.g., a greater emphasis is usually placed on sampling suspected hot spot locations);
- Foraging behavior (e.g., many of the top level predators have a relatively great home range and also have migratory behavior).

In addition to deciding on the matrices to sample, another concern is how to analyze and quantify the PCB concentrations. Ideally, for the toxicity assessment, congener-specific data should be available as discussed previously. However, congener-specific analyses (including coplanar PCBs and sufficiently low detection limits) are expensive. To be cost-effective, it may be possible to develop a strategy in which a portion of the total number of samples is analyzed both on a congener-specific basis and on an Aroclor basis in order to develop a correlation that can be applied towards other samples from the same site and matrix (USEPA 1998b). Another cost-effective strategy to obtain the TCDD-EQ in a sample is to use a cell bioassay, such as H4IIE-luc cells discussed previously. The advantage of the latter approach is that the response observed will be due to all of the chemicals in the sample (not just the ones for which an analysis was done) and will include interactions between the chemicals in that sample.

Depending on the site, species of concern, exposure pathway of the species of concern and type of remedial action, other items (beyond those discussed in the next section) may need to be sampled for the purposes of monitoring a remediated site or refinement of an exposure model. For example, in an aquatic environment, it may be important to monitor for PCBs in invertebrates (e.g., crayfish), semi-permeable membrane devices (SPMDs) or water (in most cases, however, the contribution of PCBs dissolved in water to fish is negligible compared to dietary exposure to PCBs). In a terrestrial environment, it may be important to measure concentrations of PCBs in invertebrates (e.g., earthworms, crickets, etc.) and plants (as a measure of the volatilization and transport of PCBs).

9.3.2 BIOMONITORING

To address the above issues, studies have reported on the uptake of PCBs from sediments and soils into crayfish, fish, earthworms, crickets and shrews (Hendriks *et al.* 1995, Iannuzzi *et al.* 1997). However, due to site-specific parameters, these studies may or may not relate to other sites. Thus, it is advisable to measure concentrations of PCBs in biota resident at the study site. Few models can accurately predict uptake in a field situation (Belfroid *et al.* 1996) much less in a post-remediation setting. Limited evidence suggests that bioavailability decreases over time as chemicals age in soil (Alexander 1995). This is due to diffusion into soil lattice and Donnan spaces, which is a slow process. However, if contaminated sediment or soil is removed, buried sediments will be exposed which may enhance the bioavailability of contaminants. The resulting contaminant levels in organisms will be hard to predict and should therefore be monitored whenever practical. Finally, because biomonitoring integrates the biologically available PCBs over space and time, biomonitoring can be more cost-effective than analyses of a large number of soil, water and vegetation samples, help establish priorities for site cleanup, and be used to monitor effectiveness of remediation (Talmage and Walton 1991). To be effective biomonitors, the organisms selected should be abundant, easily caught, have a relatively small home range, widespread distribution and generalized food habits (Beardsley *et al.* 1978). While the effects of PCBs on endangered species may be of concern, and since it is inappropriate and/or problematic to monitor these organisms, it is important to monitor the appropriate food items. Examples of extensive national biomonitoring programs include the National Mussel Watch (USA), National Pesticide Monitoring Programs (USA) which has monitored estuarine organisms, starlings, ducks, freshwater fish, etc., the Swedish Dioxin Survey and the Canadian Wildlife Service and Northern Contaminants Program (Canada).

As to which species to use as monitors, the selection should be chosen on a site-specific basis and ideally linked to the assessment endpoint. A few examples of species selected as biomonitors (Table 9.1) in aquatic environments are crayfish, mussels, turtles or fish, and for terrestrial environments white-footed mice, shrews, red-winged blackbirds and tree swallows. A more extensive discussion of tissue residues and effect levels can be found elsewhere (Beyer *et al.* 1996). Some advantages of biomonitoring for PCBs are that:

- Bioavailability is directly addressed;
- Biota typically integrate spatial variation in PCB concentrations in the abiotic environment in which they live;
- Complex assumptions of the foraging behavior and exposure pathways of these biota are unnecessary (assuming they have a limited home range);
- These data can be used in the exposure modeling of top level predators.

Specifically, a biomonitoring approach can be used as a tool to assess contamination, identify sources and assess the successfulness of a remedial activity. Approaches often include capture of feral wildlife or sometimes the release of sentinel animals for bioaccumulation studies (clams, fish, mink, etc.). Animals can be caged at a site, or restricted in movement such as by pinioning, or fitted with radio transmitters so they can be recovered. Examples of biomonitoring to assess the remediation of a PCB-contaminated aquatic ecosystem include studies by Rice and White (1987) using data on water, sediments, caged clams and caged fathead minnows in the Shiawassee River, Michigan; Brown *et al.* (1985) using data on water, sediments, invertebrates and fish in the Hudson River, New York; Bremle and Larsson (1998a,b) using data on water, sediments and perch in the area surrounding Lake Järnsjön, Sweden. Recently, an entire issue of the journal *AMBIO* (Volume 27, Number 5) was dedicated to the remediation project at Lake Järnsjön, Sweden, which is one of the most well documented PCB remediation efforts.

Enzyme induction, such as EROD, is a commonly used biomarker technique to assess exposure to PCBs and other AhR-active chemicals (Stegeman and Lech 1991, Kennedy *et al.* 1996). However, there are some significant limitations to this approach. Dose–response curves can be problematic to interpret due to inhibition of EROD activity, which is observed at great concentrations of an

Table 9.1 Biomarkers for exposure and effects from PCBs

Common name	Scientific name	Endpoint[a]	Reference
Mammals			
Harvest mouse	*Reithrodontomys fulvescens*	Hepatic CYP1A	Lubet *et al.* (1992)
Short-tailed shrew	*Blarina brevicauda*	body burden [PCB]	Watson *et al.* (1985)
Vole	*Microtus pennsylvanicus*	body burden [PCB]	Watson *et al.* (1985)
White-footed mouse	*Peromyscus leucopus*	body burden [PCB]	Watson *et al.* (1985)
		body burden [PCB] and liver weight	Batty *et al.* (1990)
Ground squirrel	*Spermophilus sp.*	[PCB] in liver and muscle	Greichus and Dohman (1980)
Common shrew	*Sorex araneus*	body burden [PCB congeners]	Johnson *et al.* (1996)
Wood mouse	*Apodemus sylvaticus*	body burden [PCB congeners]	Johnson *et al.* (1996)
Field vole	*Microtus agrestis*	body burden [PCB congeners]	Johnson *et al.* (1996)
Sprague–Dawley rats		EROD, BROD, PROD activities from exposure to soil	Fouchecourt *et al.* (1998)
Mink	*Mustela vison*	hepatic burden of PCBs and TEQs, decreased reproduction	Tillitt *et al.* (1996)
		EROD activity	Shipp *et al.* (1998)
European otter	*Lutra lutra*	hepatic [vitamin A] $\propto$ TEQs^{-1}	Murk *et al.* (1998)
Pinnipeds	*Otaria flavescens*	porphyrins, EROD, PROD, BROD and PCBs	Fossi *et al.* (1997b)
Harbor seal	*Phoca vitulina*	P450, P420, and MFO	Troisi and Mason (1997)
Birds			
Common tern	*Sterna hirundu*	multiple responses	Bosveld *et al.* (1995)
Japanese quail	*Coturnix coturnix japonica*	porphyrins in excreta	Fossi *et al.* (1996)
		MFO induction and blood levels of PCBs	Marsili *et al.* (1996)
Great blue heron	*Ardea herodias*	multiple responses	Custer *et al.* (1997)
Tree swallow	*Tachycineta bicolor*	PCB body burdens, reproductive effects, growth	Yorks *et al.* (1998)
Fish			
Fish (general)		P450 induction	Stegeman and Lech (1991)
Amphibians/turtles			
Snapping turtles	*Chelydra serpentina serpentina*	PCB uptake, EROD, developmental effects	Bishop *et al.* (1998)
Leopard frogs	*Rana pipiens*	P450 induction	Huang *et al.* (1998)
Invertebrates			
Crab	*Carcinus aestuarii*	multiple responses	Fossi *et al.* (1997a)
Clams		PCB uptake	Rice and White (1987)

[a]Some of the endpoints listed were assessed but either were not observed or there was no correlation between the endpoint and the concentration of PCBs.

EROD inducer (Kennedy *et al.* 1996). For example, if only a single concentration of an extract is tested in a cell bioassay system, it may be possible that no EROD induction is observed despite the presence of a great concentration of PCBs. To overcome this problem, researchers have measured actual protein levels (rather than activity). Also, the use of other enzyme systems or cell lines (such as H4IIE-luc discussed previously) has proven to be useful and does not exhibit inhibition of activity at great concentrations of PCBs. Another problem observed with the use of EROD as a biomarker is a lack of EROD induction that has been reported in fish at a PCB-contaminated site in New Bedford Harbor due to a mutant protein involved in the AhR mechanism of action (Celander *et al.* 1996). Finally, in a study in which mink were exposed to PCBs in the diet, hepatic EROD induction returned to near control levels after exposure to PCBs was terminated, despite significant liver burdens of PCBs (Shipp *et al.* 1998). In addition to the above concerns, there is no clear link between enzyme induction, such as EROD, and the toxic effects of PCBs and other AhR-active chemicals (Blankenship and Matsumura 1994). Thus, although widely utilized and generally useful to assess whether or not exposure has occurred, there are significant limitations to its use to predict effects, such as differences due to species, sex, season, reproductive status, nutrition and co-contaminants.

9.3.3 AN EXAMPLE USING BALD EAGLE PRODUCTIVITY AS THE ASSESSMENT ENDPOINT

The following is an example of how biomarkers of exposure and vertebrate tissue residues can be used in the hazard characterization of PCBs at contaminated sites. While the example presented applies to an aquatic exposure pathway using bald eagles as the species of interest, a similar approach can be used for other fish-eating birds (e.g., great blue herons, belted kingfishers, etc.) and mammals (mink and otters). Similarly, these approaches can be applied to terrestrial exposure pathways using birds (robins, owls, hawks, etc.) and mammals (shrews, foxes, etc.). To apply such an approach, a thorough evaluation of the most recent literature is necessary due to significant advances in the understanding of biomarkers of exposure and the relationship of tissue residues to effect levels.

The example that is presented is a dietary exposure model that should ideally be only one part of a weight-of-evidence approach. Often data can readily be collected for representative prey species, but may be less available, or not appropriate to collect, for the predator species of interest. If tissue residue data are available for the predator of interest, these data are important for comparison to the model and to compare to literature-derived effect levels.

9.3.3.1 TEFs and Dietary Assumptions

The TEFs used in this analysis are the 1998 WHO TEFs for birds (Van den Berg *et al.* 1998). Note that these bird-specific TEFs differ considerably from previous TEFs that were primarily used for mammals and humans (Ahlborg *et al.* 1994). Most notably for PCB #77, the TEF is 100-fold greater for birds than for mammals. Based on the average value for eight pairs of bald eagles studied on Lake Superior, it was assumed that fish and gulls contribute 94.4% and 5.6% to bald eagle diets, respectively (Kozie 1986). This dietary composition has also been adopted by the authors of the Great Lakes Water Quality Initiative Criteria Documents for the Protection of Wildlife (USEPA 1995).

9.3.3.2 Exposure Approaches

Exposure models can range from simplistic to complex. For the purposes of this example, two different simplistic approaches, described below, can be used to assess potential exposure to either total PCBs or TEQs (Figure 9.1). For these models, it is assumed that data are available for a variety of fish species

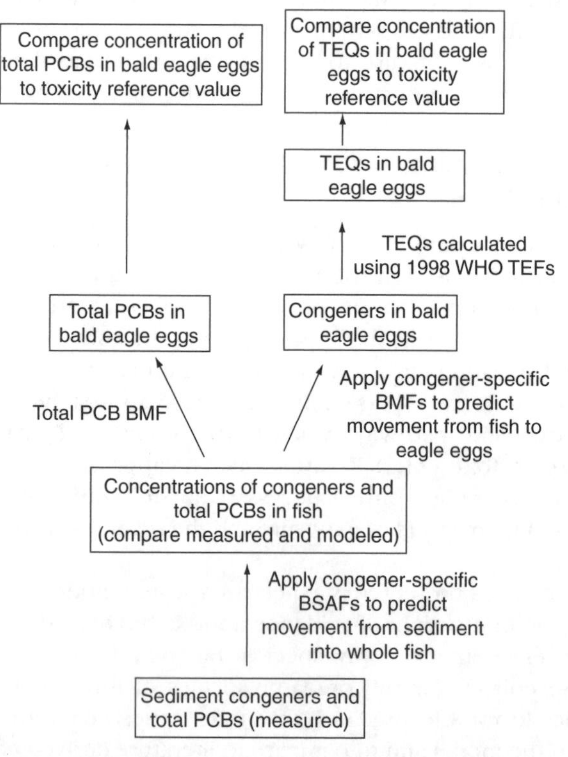

Figure 9.1 Conceptual model for congener-specific uptake of PCBs into bald eagle egg from contaminated fish and sediments. Abbreviations are as follows: TEQs, TCDD-equivalents; WHO, World Health Organization; TEFs, toxic equivalency factors; BMFs, biomagnification factors; BSAFs, biota-to-sediment accumulation factors. In this example (see text), it is assumed that concentrations of PCB congeners have been determined for fish species in the diet of the bald eagle.

in the diet of bald eagles and that this site is sufficiently large to include the entire foraging range (this assumption will overestimate exposure at most sites). While it is possible to estimate exposure from sediment (in the absence of fish data), there is often unacceptable uncertainty in such an approach as discussed previously. Both of these exposure models were compared to corresponding reference doses for calculation of risk quotients.

1. Dietary intake of total PCBs — total PCBs in fish (wet weight basis) was compared to a dietary toxicity reference value for bald eagles proposed by Giesy *et al.* (1995).
2. Congener-specific biomagnification from fish to bald eagle eggs — a congener-specific model was applied to available data in which lipid-normalized concentrations in fish were biomagnified to bald eagle eggs by the use of congener-specific biomagnification factors (BMFs).

Selection and Derivation of BMF Values

Few studies are available for which congener-specific concentrations are measured for both piscivorous bird eggs and representative fish in the diet. Congener-specific data for bald eagles (*Haliaeetus leucocephalus*) are especially rare. At this time, only data from British Columbia are available for congener-specific concentrations in bald eagle eggs (Elliott *et al.* 1996). There were no matching data for fish in the diets of these eagles. One of the most studied regions in the world for which congener-specific data in wildlife such as fish, birds and seals are available is the area surrounding the Baltic Sea, including the countries of Finland, Poland and Sweden. This region has been similarly contaminated as some areas of the Great Lakes, including releases of PCBs from self-copying paper processed at paper mills (Jarnberg *et al.* 1993). In Finland, several studies have been conducted on white-tailed sea eagles (*Haliaeetus albicilla*), which have similar dietary consumption patterns to bald eagles (e.g., mostly fish and some birds; Helander *et al.* 1982). In these studies, congener-specific information is available for fish and eagle eggs from the same geographical area and time period (Koistinen 1990, Koistinen *et al.* 1995, 1997, Vuorinen *et al.* 1997).

Since BMF values depend primarily on dietary habits, metabolic capacity of the organisms of interest and whether or not equilibrium has been reached, BMF values are not generally site-specific, unlike biota to sediment accumulation factors (BSAFs) which can vary depending on site-specific characteristics. Congener-specific BMFs derived from field studies reported in the literature are presented for both fish to bird egg and fish to bird muscle/liver (Table 9.2). Note that these are lipid-normalized BMFs. Since BMFs for white-tailed sea eagles could be calculated for only congeners #77, 105, 126 and 169, BMF values were also derived for fish-eating birds other than eagles for comparison. In general, there was good agreement between fish to bird egg BMFs for congeners #77 and #126, the two most important contributors to TEQs in birds of the Great

Table 9.2 Congener-specific and species-specific biomagnification factors from fish to piscivorous birds

	Lipid-normalized BMFs										
	77	101	105	110	118	126	138	153	156	169	180
Salmon to eagle egg[a]	0.80		30.60			45.00				42.00	
Herring to guillemot egg[b]	0.80					26.50				36.60	
Carp to gull egg[c]	1.32	3.39	7.45	4.90	12.9	20.54	12.8	15.4		15.00	
Shad to gull egg[c]	2.65	10.0	12.5	9.06	50.5	31.94	87.4	75.5		120.00	
Silver bass to gull egg[c]	0.19	2.61	7.90	3.96	24.0		21.7	24.5		12.63	
Smallmouth to gull egg[c]	1.17	4.71	1.41	4.83	29.9		24.4	39.5		26.67	
Trout to grey heron muscle[d]	3.00	0.50	9.90		5.40	11.00	4.60	5.00	7.90	15.00	6.20
Trout to cormorant muscle[d]	1.40	0.43	4.60		2.70	3.10	3.70	4.70	3.50	4.00	6.60
Fish to tern liver[e]		0.50	9.44	0.64	0.80	14.00	0.62	1.10	5.20	9.00	0.79
STATISTICS											
Fish to egg BMFs											
Arithmetic mean	1.16	5.18	11.97	5.69	29.33	31.00	36.58	38.73		42.15	
Geometric mean	**0.89**	**4.52**	**7.95**	**5.40**	**26.15**	**29.74**	**27.74**	**32.57**		**31.25**	
Median	0.99	4.05	7.90	4.87	26.95	29.22	23.05	32.00		31.64	
Std. dev.	0.83	3.33	11.13	2.29	15.78	10.43	34.24	26.45		39.85	
Fish to bird tissue BMFs (muscle and liver)											
Arithmetic mean	2.20	0.48	7.98	0.64	2.97	9.37	2.97	3.60	5.53	9.33	
Geometric mean	**2.05**	**0.48**	**7.55**	**0.64**	**2.27**	**7.82**	**2.19**	**2.96**	**5.24**	**8.14**	
Median	2.20	0.50	9.44	0.64	2.70	11.00	3.70	4.70	5.20	9.00	
Std. dev.	1.13	0.04	2.94		2.31	5.63	2.09	2.17	2.22	5.51	

[a]Koistinen (1990), Koistinen *et al.* (1997).
[b]Koistinen *et al.* (1995).
[c]Koslowski *et al.* (1994).
[d]Zimmermann *et al.* (1997).
[e]Bosveld *et al.* (1995).

Lakes ecosystem (Ankley *et al.* 1993). The BMF values for congeners #105 and #169 were more variable, but these congeners contribute little to TEQs for birds of this area (Ankley *et al.* 1993).

In selecting the most appropriate BMF value for the purposes of modeling exposure from the fish portion of the eagle diet, fish to bird egg BMFs were utilized. In selecting the most appropriate BMF value for the purposes of modeling exposure from the gull portion of the eagle diet, fish to bird tissue BMFs were utilized to transfer congeners to gull and then fish to bird egg BMFs were used to transfer congeners from gull tissue to eagle egg. Multiplication of these lipid-normalized BMFs by lipid-normalized concentrations in fish and gulls resulted in lipid-normalized concentrations in bald eagle eggs which were then multiplied by their respective TEFs and then summed to yield bald eagle egg TEQs as shown by equations (9.2) to (9.6):

$$\text{conc. PCB } \#n \text{ in gull (ng kg}^{-1}\text{, lipid)} = [\text{conc. PCB } \#n \text{ in fish (ng kg}^{-1}\text{, lipid)} \times \text{BMF for PCB } \#n] \quad (9.2)$$

$$\text{conc. PCB \#}n\text{ in eagle egg from gull (ng kg}^{-1}\text{, lipid)}$$
$$= [\text{conc. PCB \#}n\text{ in gull (ng kg}^{-1}\text{, lipid)} \times \text{BMF for PCB \#}n \times \text{diet fraction (0.056)}] \quad (9.3)$$

$$\text{conc. PCB \#}n\text{ in eagle egg from fish (ng kg}^{-1}\text{, lipid)}$$
$$= [\text{conc. PCB \#}n\text{ in fish(ng kg}^{-1}\text{, lipid)} \times \text{BMF for PCB \#}n \times \text{diet fraction (0.944)}] \quad (9.4)$$

$$\text{total conc. PCB \#}n\text{ in eagle egg (ng kg}^{-1}\text{, lipid)}$$
$$= [\text{sum of results of equations 9.2 and 9.3}] \quad (9.5)$$

$$\text{ng TEQ kg}^{-1}\text{ bald eagle egg (lipid)}$$
$$= \sum [\text{total conc. PCB \#}n\text{ in bald eagle egg (ng kg}^{-1}\text{, lipid)} \times \text{TEF for PCB \#}n] \quad (9.6)$$

where n is any PCB congener.

Special Consideration of PCB #77

PCB #77, one of the more potent non-*ortho* PCB congeners, requires special consideration due to its greater susceptibility to metabolism and/or degradation. While originally considered to be resistant to degradation and readily biomagnified due to its lack of *ortho* substitution (Tanabe *et al.* 1988), PCB #77 behaves differently from other non-*ortho* PCBs and differently from other members of the tetrachlorinated homolog group (Willman *et al.* 1997, Froese *et al.* 1998) Therefore, it is *not* appropriate to assume that PCB #77 or other individual congeners have similar properties within a homolog group.

Based on BMFs derived from field studies (Table 9.2), it is unlikely that PCB #77 would biomagnify from fish to eagle eggs to the same extent as congeners #126 and #169. In a study of bird eggs and hatchlings from the lower Fox River and Green Bay, it was observed that PCB #77 decreased when hatchlings reached an age of approximately 25-day old, whereas all other congeners continued to accumulate (Ankley *et al.* 1993). Those authors suggested that this might be due to metabolism of PCB #77. This observation is in agreement with other studies that have observed metabolic degradation and/or depletion of PCB #77 in bird microsomes (Murk *et al.* 1994) marine mammals (Boon *et al.* 1997) and mink (Tillitt *et al.* 1996) relative to other congeners. Similarly, Bright *et al.* (1995) reported that PCB #77 was diminished rather than enriched with increasing trophic status (going from sediment to sea urchins to four-horn sculpins). In a study of Green Bay, Willman *et al.* (1997) found that PCB #77 decreased in relative abundance both within total PCBs and within its homolog group with increasing trophic level. Due to the predominance of PCB #77's contribution to total TEQs (using avian TEFs) in PCB-contaminated sediments, PCB #77 is a critical congener to model when characterizing risk to avian species. Thus, modeling of congener #77 must take the above factors into account.

9.3.3.3 Toxicity Reference Values

TRVs are ideally derived from chronic toxicity studies in which an ecologically relevant endpoint was assessed in the species of concern (or a closely related species). TRVs are usually either defined as NOAELs (no observed adverse effect levels) or LOAELs (lowest observed adverse effect levels), but can also be expressed as the geometric mean of the NOAEL and LOAEL as a conservative estimate of a threshold of effect (Tillitt *et al.* 1996). There are two major problems with the extrapolation of laboratory toxicity data to wild species. The first is the wide range of species sensitivity to AhR-active chemicals, such as PCBs (Gasiewicz 1991). The second is somewhat unique to PCBs, in that most laboratory toxicity studies are primarily based on exposure to parent Aroclors, whereas wild species are likely exposed to congener patterns that are very different from the parent Aroclor due to environmental weathering (refer to the previous discussion).

For dietary intake of total PCBs, TRVs were those of Giesy *et al.* (1995) based on reproductive productivity in bald eagles. The TRV value for NOAEC (no observed adverse effect concentration) was 0.14 mg total PCBs kg^{-1} fish.

For bald eagles and other bird species, the most sensitive dose metric for dioxin-like chemicals are egg concentrations rather than adult tissue concentrations (Giesy *et al.* 1994a). This is due in part to the stage-specific sensitivity of many species (birds, fish and mammals) during development and relative tolerance of adults to the effects of dioxin-like chemicals. In other words, effects can be seen at lesser doses for young, developing animals than for adults. For example, the LD_{50} (the dose that is lethal to 50% of a test population) for chicken eggs (Henshel *et al.* 1993) is 200-fold less than the LD_{50} for an adult chicken (on a wet weight basis; Greig *et al.* 1973). Furthermore, LOAEL values for developmental toxicity occur at doses that are approximately 10-fold lower than LD_{50} endpoints. Thus, when trying to characterize risk to avian species for dioxin-like compounds, the most sensitive endpoint is developmental toxicity.

There are no TRVs for bald eagles based on developmental toxicity. However, Elliott *et al.* (1996) calculated bald eagle TRVs, based on hepatic cytochrome P4501A enzyme induction, of 100 and 210 ng TEQ kg^{-1} egg (ww) for the NOAEC and LOAEC (lowest observed adverse effect concentration), respectively. No significant concentration-related effects for morphological, physiological or histological parameters, such as chick growth, edema or density of thymic lymphocytes, were observed at these concentrations (Elliott *et al.* 1996). Thus, hepatic CYP1A induction appears to be more sensitive than these developmental endpoints. Therefore, these TRVs would be expected to be conservative and protective of eagle chicks from the effects of AhR-active PCB congeners to cause embryo lethality or deformities.

The TEQ-based TRVs from Elliott *et al.* (1996) were calculated from a mammal-based set of TEFs. Since exposure calculations in this study are based on 1998 WHO TEFs for birds, the TRV should be based on the same set of TEFs. After modifying TRVs from Elliott *et al.* (1996) by use of the most recent and bird-specific TEFs, TRVs were 134 and 400 ng TEQ kg^{-1} on a whole egg (wet weight basis) for the NOAEC and LOAEC, respectively. Based on a mean lipid concentration in bald eagle eggs of 7.5% (standard deviation 2.9%), the lipid-normalized TRVs were 1786 and 5329 ng TEQ kg^{-1} for the NOAEC and LOAEC, respectively.

Since these values are suitable for comparison in the current analyses (i.e., same species, same exposure route, similar mixture, etc.), no uncertainty factors have been applied to the TRV in the risk characterization. When TRVs are not available for the species of concern, it may be necessary to modify the TRV (USEPA 1997). Uncertainty concerning interpretation of the toxicity test information among different species, different laboratory endpoints and differences in experimental design (age of test animals, duration of test, etc.) are typically addressed by applying uncertainty factors (UFs) to literature-based toxicity data to calculate the final TRV. Methods for applying UFs have been published (Opresko *et al.* 1994, USEPA 1995). Here we have used the method recently published by USEPA Region 8 for the Rocky Mountain Arsenal (RMA) (USEPA 1997). The RMA procedure uses three uncertainty factors:

1. Intertaxon variability extrapolation, where values range from 1 to 5 (Category A);
2. Exposure duration extrapolation, where values range from 0.75 to 15 (Category B);
3. Toxicologic endpoint extrapolation, where values range from 1 to 15 (Category C).

Also modifying factor (Category D), which incorporates other sources of uncertainty, including:

- Threatened, or listed, and endangered species, where values range from 0 to 2 (d_1);
- Relevance of endpoint to ecological health, where values range from 0 to 2 (d_2);
- Extrapolation from laboratory to field, where values range from -1 to 2 (d_3);
- Study conducted with relevant co-contaminants, where values range from -1 to 2 (d_4);
- Endpoint is mechanistically unclear (vs. clear), where values range from 0 to 2 (d_5);

- Study species is either highly sensitive or highly resistant, where values range from −1 to 2 (d_6);
- Ratios used to estimate whole body burden from tissue or egg, where values range from 0 to 2 (d_7);
- Intraspecific variability, where values range from 0 to 2 (d_8);
- Other applicable modifiers, where values range from −1 to 2 ($d_{9...n}$).

The TRV is calculated using equation (9.7):

$$\text{TRV} = \frac{\text{study dose}}{(A \times B \times C \times D)} \tag{9.7}$$

where $D = (d_1 + d_2 + d_3 + \cdots + d_n)$.

The assignment of uncertainty factors to the NOEL (no observed effect level) for the calculation of the eagle egg TRV is as follows:

- **Intertaxon variability** (A) — this factor is used to estimate an effects concentration or dose for a receptor from laboratory test data with surrogate species. The field study that provided the NOEL used the bald eagle (*H. leucocephalus*) as the test organism (Elliott *et al.* 1996), so there is no extrapolation among species, and the value for $A = 1$.
- **Exposure duration** (B) — this factor is used to estimate the lowest observed effect concentration or dose of a chemical when only acute (short-term) toxicity test data are available. The field study that provided the NOEL (Elliott *et al.* 1996) used eagle eggs collected from the wild, so they reflect actual chronic exposures, hence $B = 1$.
- **Toxicologic endpoint** (C) — this factor is used to estimate NOAEL and/or LOAEL values from studies that report other endpoints. The eagle egg study calculated both NOEL and LOEL (lowest observed effect level) doses (Elliott *et al.* 1996) for induction of hepatic cytochrome P4501A, as expressed by EROD, which is more sensitive than the developmental toxicity endpoints of chick growth, edema or density of thymic lymphocytes. The importance of enzyme induction at the population level is unknown, but there was no extrapolation from other endpoints, so $C = 1$.
- **Modifying factors** (D) — these factors are used to address other aspects of uncertainty:
  - Threatened, or listed, and endangered species — the bald eagle is a Federally Endangered Species, however, the toxicity data are for this species, so $d_1 = 0$;
  - Relevance of endpoint to ecological health — the test endpoint in the field study was EROD induction in bald eagle chicks. The relevance of this endpoint to population sustainability is unknown, so we have conservatively set $d_2 = 1$.

- Extrapolation from laboratory to field — the study by Elliott *et al.* (1996) used bald eagle eggs collected from the wild, and incubated them in the laboratory. Since the exposure doses were actual environmental (field) exposures, $d_3 = 0$.
- Study conducted with relevant co-contaminants — the eagle eggs contained a variety of environmental contaminants, including PCDDs and PCDFs, so $d_4 = 0$.
- Endpoint is mechanistically unclear (vs. clear) — the induction of EROD as an expression of P4501A induction, the mechanism of dioxin toxicity, is clear, so $d_5 = 0$.
- Study species is either highly sensitive or highly resistant - the bald eagle is both the study species and the ecological receptor, so $d_6 = 0$.
- Ratios used to estimate whole body burden from tissue or egg — Elliott *et al.* (1996) measured TEQ concentrations in eagle eggs, which is the endpoint for this shadow risk assessment, so $d_7 = 0$.
- Intraspecific variability — the field study investigated adverse effects on bald eagle hatchlings, the most sensitive life stage, so $d_8 = 0$.

The TRV calculation for eagle eggs is therefore represented by equation (9.8):

$$\mathrm{TRV} = \frac{134 \text{ pg TEQ g}^{-1}\text{egg (ww)}}{(1 \times 1 \times 1 \times (0 + 1 + 0 + 0 + 0 + 0 + 0 + 0))} \tag{9.8}$$

9.3.3.4 Hazard Characterization

HQs are the numerical expression of relative hazard, which are determined by dividing an exposure concentration by an appropriate TRV. A HQ less than 1.0 is considered to represent little environmental risk to bald eagles, whereas a HQ greater than 1.0 represents a level of concern for potential risk such that a more detailed analysis or perhaps monitoring would be warranted. Calculated hazard quotients in this hypothetical example were greater for dietary intake of total PCBs than eagle egg TEQ concentration (Figure 9.2). Ideally, risk-based decision making should be based on a probabilistic approach in which probabilities of effects are characterized (Moore *et al.* 1999; Figure 9.3).

9.4 UNCERTAINTIES ASSOCIATED WITH EACH EXPOSURE ASSESSMENT APPROACH

An uncertainty analysis is required for ecological risk assessments under USEPA guidance and should be performed for the quantitative and qualitative parameters that are included in the risk assessment. While such a requirement may not be explicitly stated during the evaluation process of a remedial action, it is important to be aware of the inherent uncertainties involved. Probabilistic approaches can be useful to identify sources of uncertainty and variability. Some

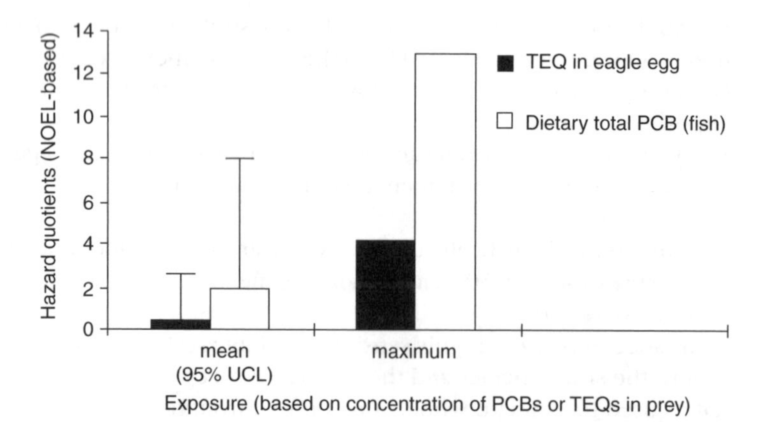

Figure 9.2 Bald eagle hazard quotients using two different approaches: (1) a dietary total PCB model (based on concentration of total PCBs in prey); and (2) a biomagnification model (BMF) in which individual congeners are moved into eagle eggs through modeling and then converted to TEQs.

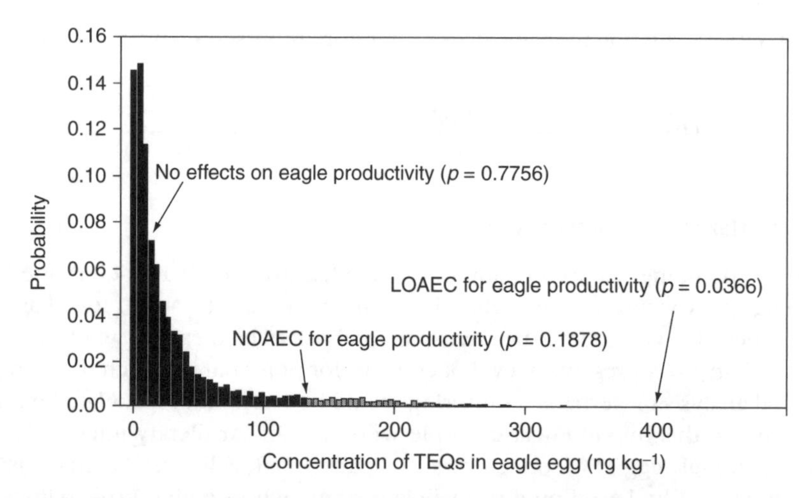

Figure 9.3 Similar to the biomagnification model in Figure 9.2 except that a probabilistic analysis has been performed on the exposure estimation based on the full data set, rather than selecting a central tendency value and high-end estimate (95% upper confidence limit). With this approach, probabilities of effects can be characterized.

common factors that contribute to uncertainty include assumptions relating to exposure models and toxicity thresholds. Additional parameters that contribute to uncertainty and their associated concerns include:

1. The relative impact of the remedial action itself compared to the contaminants left in place (will the remedial action lead to significant habitat alteration?).

2. Spatial distribution of contaminants (by removing hot spots only, what is the exposure from the remaining more diffuse contaminants?).
3. Temporal (including seasonal) distribution of contaminants (in what time frame will exposures and/or body burdens be reduced below a critical threshold and will this time frame be significantly different from that predicted to occur under natural attenuating conditions?).
4. Bioavailability issues (did the remedial action increase the bioavailability of PCBs? If so, for what time period?).
5. Co-contaminants and unknown interactions (if all the PCBs were successfully removed, is there another contaminant or non-contaminant stressor that prevents recovery of the receptor(s) of concern?).
6. Species-specific differences in TEFs.

9.5 RECOMMENDATIONS AND FUTURE DIRECTIONS

Currently there is a need for approaches to measure the effectiveness of a remedial action from a risk-based or hazard-based perspective. In this chapter, several approaches were discussed that utilize exposure models, biomarkers of exposure and vertebrate tissue residues. The methods described here demonstrate how such models can be developed and applied to monitoring programs at contaminated sites. Rather than specifying a target concentration of PCBs in sediment or soil, site-specific considerations of exposure to upper trophic level receptors, including concentrations of contaminants in prey, should be developed and utilized, as recommended in the 'Canadian Tissue Residue Guidelines (TRG) for Polychlorinated Biphenyls for the Protection of Wildlife Consumers of Aquatic Biota' (CCME 1998). Additionally, much more work is needed to reduce uncertainty in post-remediation exposure evaluations and in the characterization of tissue residue effect levels.

REFERENCES

Ahlborg UG, Becking GC, Birnbaum LS, Brouwer A, Derks HJGM, Feeley M, Golor G, Hanberg A, Larsen JC, Liem AKD, Safe SH, Schlatter C, Waern F, Younes M and Yrjanheikki E (1994) Toxic Equivalency Factors for dioxin-like PCBs — Report on a WHO-ECEH and IPCS Consultation, September 1993. *Chemosphere*, **28**, 1049-1067.

Alexander M (1995) How toxic are toxic chemicals in soil. *Environmental Science and Technology*, **29**, 2713-2717.

Ankley GT, Niemi GJ, Lodge KB, Harris HJ and Beaver DL (1993) Uptake of planar polychlorinated biphenyls and 2,3,7,8-substituted polychlorinated dibenzofurans and dibenzo-*p*-dioxins by birds nesting in the lower Fox River and Green Bay, Wisconsin, USA. *Archives of Environmental Contamination and Toxicology*, **24**, 332-344.

Batty J, Leavitt RA, Biondo N and Polin D (1990) An ecotoxicological study of a population of the white footed mouse (*Peromyscus leucopus*) inhabiting a polychlorinated biphenyls-contaminated area. *Archives of Environmental Contamination and Toxicology*, **19**, 283-290.

Beardsley A, Vagg MJ, Beckett PHT and Sansom BF (1978) Use of the field vole (*M. agrestis*) for monitoring potentially harmful elements in the environment. *Environmental Pollution*, **16**, 65-71.

Belfroid AC, Sijm DTHM and Van Gestel CAM (1996) Bioavailability and toxicokinetics of hydrophobic aromatic compounds in benthic and terrestrial invertebrates. *Environmental Reviews*, **4**, 276–299.

Beyer WN, Heinz GH and Redmon-Norwood AW (1996) *Environmental Contaminants in Wildlife, Interpreting Tissue Concentrations*. (A SETAC special publication), CRC Press, New York.

Bishop CA, Ng P, Pettit KE, Kennedy SW, Stegeman JJ, Norstrom RJ and Brooks RJ (1998) Environmental contamination and developmental abnormalities in eggs and hatchlings of the common snapping turtle (*Chelydra serpentina serpentina*) from the Great Lakes–St. Lawrence River basin (1989–1991). *Environmental Pollution*, **101**, 143–156.

Blankenship A and Matsumura F (1994) Changes in biochemical and molecular biological parameters induced by exposure to dioxin-type chemicals. *Biomarkers of Human Exposure to Pesticides*, **542**, 37–50.

Blankenship A, Kannan K, Villalobos SA, Falandysz J and Giesy JP (2000) Relative potencies of individual polychlorinated naphthalenes and halowax mixtures to induce Ah receptor-mediated responses. *Environmental Science and Technology*, **34**, 3153–3158.

Boon JP, van der Meer J, Allchin CR, Law RJ, Klungsoyr J, Leonards PEG, Spliid H, Storr-Hansen E, McKenzie C and Wells DE (1997) Concentration-dependent changes of PCB patterns in fish-eating mammals: structural evidence for induction of cytochrome P450. *Archives of Environmental Contamination and Toxicology*, **33**, 298–311.

Bosveld ATC, de Bont A, Mennen J, Evers EHG, Brouwer A and Van den Berg M (1995) RITOX Report. *Responses to 3,3′,4,4′,5-PeCB and 2,2′,4,4′,5,5′-HxCB in common terns (Sterna hirundu): a three-week feeding study with laboratory hatched chicks*.

Bowerman WW, Giesy JP, Best DA and Kramer VJ (1995) A review of factors affecting productivity of bald eagles in the Great Lakes region: implications for recovery. *Environmental Health Perspectives*, **103**(Suppl. 4), 51–59.

Bremle G and Larsson P (1998a) PCB concentration in fish in a river system after remediation of contaminated sediment. *Environmental Science and Technology*, **32**, 3491–3495.

Bremle G and Larsson P (1998b) PCB in Eman river ecosystem. *AMBIO*, **27**, 384–392.

Bright DA, Dushenko WT, Grundy SL and Reimer KJ (1995) Effects of local and distant contaminant sources: polychlorinated biphenyls and other organochlorines in bottom-dwelling animals from an Arctic estuary. *The Science of the Total Environment*, **160/161**, 265–283.

Brown MP, Werner MB and Sloan RJ (1985) Polychlorinated biphenyls in the Hudson River — recent trends in the distribution of PCBs in water, sediment, and fish. *Environmental Science and Technology*, **19**, 656–661.

CCME (1998) *Canadian Tissue Residue Guidelines for Polychlorinated Biphenyls for the Protection of Wildlife Consumers of Biota*. Canadian Council of Ministers of the Environment. Prepared by the Guidelines and Standards Division, Science Policy and Environmental Quality Branch, Environment Canada, Hull, Quebec.

CCREM (1987) *Canadian Water Quality Guidelines*. Prepared by the Task Force on Water Quality Guidelines of the Canadian Council of Resource and Environment Ministers.

Celander M, Stegeman JJ and Forlin L (1996) CY1P1-, CYP2B- and CYP3A-like proteins in rainbow trout (*Oncorhynchus mykiss*) liver: CYP1A1-specific down-regulation after prolonged exposure to PCB. *Marine Environmental Research*, **42**, 283–286.

Corsolini S, Focardi S, Kannan k, Tanabe S, Borrell A and Tatsukawa R (1995) Congener profile and toxicity assessment of polychlorinated biphenyls in dolphins, sharks and tuna collected from Italian coastal waters. *Marine Environmental Research*, **40**, 33–40.

Custer TW, Hines RK, Melancon MJ, Hoffman DJ, Wickliffe JK, Bickham JW, Martin JW and Henshel DS (1997) Contaminant concentrations and biomarker response in great blue heron eggs from 10 colonies on the upper Mississippi River, USA. *Environmental Toxicology and Chemistry*, **16**, 260–271

Eisler R (1986) *Polychlorinated biphenyl hazards to fish, wildlife, and invertebrates: a synoptic review*. US Fish Wildlife Service Biological Report 85, pp. 1–72.

Elliot JE, Norstrom RJ, Lorenzen A, Hart LE, Philibert H, Kennedy SW, Stegeman JJ, Bellward G and Cheng KM (1996) Biological effects of polychlorinated dibenzo-*p*-dioxins, dibenzofurans, and biphenyls, in bald eagle (*Haliaeetus leucocephalus*) chicks. *Environmental Toxicology and Chemistry*, **15**, 782-793.

Fossi MC, Casini S and Marsili L (1996) Porphyrins in excreta: a non-destructive biomarker for the hazard assessment of birds contaminated with PCBs. *Chemosphere*, **33**, 29-42.

Fossi MC, Savelli C, Casini S, Franchi E, Mattei N and Corsi I (1997a) Multi-response biomarker approach in the crab *Carcinus aestuarii* experimentally exposed to benzo(*a*)pyrene, polychlorobiphenyls and methyl-mercury. *Biomarkers*, **2**, 311-319.

Fossi MC, Savelli C, Marsili L, Casini S, Jimenez B, Junin M, Castello H and Lorenzani JA (1997b) Skin biopsy as a non-destructive tool for the toxicological assessment of endangered populations of pinnipeds: preliminary results on mixed function oxidase in *Otaria flavescens*. *Chemosphere*, **35**, 1623-1635.

Fouchecourt MO, Berny P and Riviere JL (1998) Bioavailability of PCBs to male laboratory rats maintained on litters of contaminated soils: PCB burden and induction of alkoxyresorufin-*O*-dealkylase activities in liver and lung. *Archives of Environmental Contamination and Toxicology*, **35**, 680-687.

Froese KL, Verbrugge DA, Ankley GT, Niemi GJ, Larsen CP and Giesy JP (1998) Bioaccumulation of polychlorinated biphenyls from sediments to aquatic insects and tree swallow eggs and nestlings in Saginaw Bay, Michigan, USA. *Environmental Toxicology and Chemistry*, **17**, 484-492.

Gasiewicz TA (1991) Nitro compounds and related phenolic pesticides. In *Handbook of Pesticide Toxicology* Hayes WJ and Laws ER (eds), Academic Press, San Diego, pp. 1191-1270.

Giesy JP and Kannan K (1998) Dioxin-like and non-dioxin-like toxic effects of polychlorinated biphenyls (PCBs): implications for risk assessment. *Critical Reviews in Toxicology*, **28**, 511-569.

Giesy JP, Ludwig JP and Tillitt DE (1994a) Dioxins, dibenzofurans, PCBs and colonial, fish-eating water birds In *Dioxins and Health*, Schecter A (ed), Plenum Press, New York.

Giesy JP, Ludwig JP and Tillitt DE (1994b) Deformities in birds of the Great-Lakes Region assigning causality. *Environmental Science and Technology*, **28**, A128-A135.

Giesy JP, Verbrugge DA, Othout RA, Bowerman WW, Mora MA, Jones PD, Newsted JL, Vandervoort C, Heaton SN, Aulerich RJ, Bursian SJ, Ludwig JP, Dawson GA, Kubiak TJ, Best DA and Tillitt DE (1994c) Contaminants in fishes from Great Lakes-influenced sections and above dams of three Michigan rivers. I: Concentrations of organo chlorine insecticides, polychlorinated biphenyls, dioxin equivalents, and mercury. *Archives of Environmental Contamination and Toxicology*, **27**: 202-212.

Giesy JP, Verbrugge DA, Othout RA, Bowerman WW, Mora MA, Jones PD, Newsted JL, Vandervoort C, Heaton SN, Aulerich RJ, Bursian SJ, Ludwig JP, Dawson GA, Kubiak TJ, Best DA and Tillitt DE (1994d) Contaminants in fishes from Great Lakes-influenced sections and above dams of three Michigan rivers. II: Implications for the health of mink. *Archives of Environmental Contamination and Toxicology*, **27**, 213-223.

Giesy JP, Bowerman WW, Mora MA, Verbrugge DA, Othoudt RA, Newsted JL, Summer CL, Aulerich RJ, Bursian SJ, Ludwig JP, Dawson GA, Kubiak TJ, Best DA and Tillitt DE (1995) Contaminants in fishes from Great Lakes-influenced sections and above dams of three Michigan rivers. III: Implications for health of Bald Eagles. *Archives of Environmental Contamination and Toxicology*, **29**, 309-321.

Griechus YA and Dohman BA (1980) Polychlorinated biphenyl contamination of areas surrounding two transformer salvage companies, Colman, South Dakota, September 1977. *Pesticide Monitoring Journal*, **14**, 26-30.

Grieg JB, Jones G, Butler WH and Barnes JM (1973). Toxic effects of 2,3,7,8-tetrachlorodibenzo-*p*-dioxin. *Food and Cosmetics Toxicology*, **11**, 585-595.

Helander B, Olsson M and Reutergardh L (1982) Residue levels of organochlorine and mercury compounds in unhatched eggs and the relationships to breeding success in white-tailed sea eagles *Haliaeetus albicilla* in Sweden. *Holarctic Ecology*, **5**, 349-366.

Hendriks AJ, Ma WC, Brouns JJ, Deruiterdijkman EM and Gast R (1995) Modelling and monitoring organochlorine and heavy metal accumulation in soils, earthworms, and shrews in Rhine-delta floodplains. *Archives of Environmental Contamination and Toxicology*, **29**, 115-127.

Henshel DS, Hehn BM, Vo HT and Steeves JD (1993) A short term test for dioxin teratogenicity using chicken embryos. In *Environmental Toxicology and Risk Assessment*, Gorsuch JW, Dwyer FJ, Ingersoll CG and La Point TW (eds), American Society for Testing and Materials, Philadelphia, PA, pp. 159-174.

Huang YW, Melancon MJ, Jung RE and Karasov WH (1998) Induction of cytochrome P450-associated monooxygenases in northern leopard frogs, *Rana pipiens*, by 3,3′,4,4′,5-pentachlorobiphenyl. *Environmental Toxicology and Chemistry*, **17**, 1564-1569.

Iannuzzi TJ, Truchon SP, Keenan RE and Cepko RP (1997) Calculation of hypothetical risks to wildlife receptors associated with polychlorinated biphenyls (PCBs) in the Clear Creek watershed, Bloomington, Indiana. *Organohalogen Compounds*, **33**, 347-352.

Jarnberg U, Asplund L, de Wit C, Grafsrom A-K, Haglund P, Jansson B, Lexen K, Strandell M, Olsson M and Jonsson B (1993) Polychlorinated biphenyls and polychlorinated naphthalenes in Swedish sediment and biota: levels, patterns, and time trends. *Environmental Science and Technology*, **27**, 1364-1374.

Johnson MS, Leah RT, Connor L, Rae C and Saunders S (1996) Polychlorinated biphenyls in small mammals from contaminated landfill sites. *Environmental Pollution*, **92**, 185-191.

Kannan K, Imagawa T, Blankenship AL and Giesy JP (1998) Isomer-specific analysis and toxic evaluation of polychlorinated naphthalenes in soil, sediment, and biota collected near the site of a former chlor-alkali plant. *Environmental Science and Technology*, **32**, 2507-2514.

Kennedy SW, Lorenzen A, Jones SP, Hahn ME and Stegeman JJ (1996) Cytochrome P4501A induction in avian hepatocyte cultures: a promising approach for predicting the sensitivity of avian species to toxic effects of halogenated aromatic hydrocarbons. *Toxicology and Applied Pharmacology*, **141**, 214-230.

Koistinen J (1990) Residues of planar polychloroaromatic compounds in Baltic fish and seal. *Chemosphere*, **20**, 1043-1048.

Koistinen J, Koivusaari J, Nuuja I and Paasivirta J (1995) PCDEs, PCBs, PCDDs, and PCDFs in black guillemots and white-tailed sea eagles from the Baltic Sea. *Chemosphere*, **30**, 1671-1684.

Koistinen J, Koivusaari J, Nuuja I, Vuorinen PJ, Paasivirta J and Giesy JP (1997) 2,3,7,8-Tetrachlorodibenzo-*p*-dioxin equivalents in extracts of Baltic white-tailed sea eagles. *Environmental Toxicology and Chemistry*, **16**, 1533-1544.

Koslowski SE, Metcalfe CD, Lazar R and Haffner GD (1994) The distribution of 42 PCBs, including three coplanar congeners, in the food web of the western basin of Lake Erie. *Journal of Great Lakes Research*, **20**, 260-270.

Kozie KD (1986) *Breeding and feeding ecology of bald eagles nesting in the Apostle Islands National Lakeshore*. MS Thesis, University of Wisconsin, Stevens Point, WI.

Leonards PEG, Devries TH, Minnaard W, Stuijfzand S, Devoogt P, Cofino WP, Vanstraalen NM and van Hattum B (1995) Assessment of experimental data on PCB-induced reproduction inhibition in mink, based on an isomer-specific and congener-specific approach using 2,3,7,8-tetrachlorodibenzo-*p*-Dioxin Toxic Equivalency. *Environmental Toxicology and Chemistry*, **14**, 639-652.

Lubet RA, Nims RW, Beebe LE, Fox SD, Issaq HJ and McBee K (1992) Induction of hepatic CYP1A activity as a biomarker for environmental exposure to Aroclor-1254 in feral rodents. *Archives of Environmental Contamination and Toxicology*, **22**, 339-344.

Marsili L, Fossi MC, Casini S and Focardi S (1996) PCB levels in bird blood and relationship to MFO responses. *Chemosphere*, **33**, 699-710.

Metcalfe CD and Haffner GD (1995) The ecotoxicology of coplanar polychlorinated biphenyls. *Environmental Reviews*, **3**, 171-190.

Moore DR and Walker SL (1991) *Canadian Water Quality Guidelines for Polychlorinated Biphenyls in Coastal and Estuarine Waters*. Inland Waters Directorate, Water Quality Branch, Ottawa, Ontario.

Moore DR, Sample B, Suter G, Parkhurst B and Teed RS (1999) Risk-based decision-making: a case study for piscivores exposed to contaminants at East Fork Poplar Creek, Oak Ridge, Tennessee. *Environmental Toxicology and Chemistry*, submitted.

Murk A, Morse D, Boon J and Brouwer A (1994) In vitro metabolism of 3,3′,4,4′-tetrachlorobiphenyl in relation to ethoxyresorufin-*O*-deethylase activity in liver microsomes of some wildlife species and rat. *European Journal of Pharmacology*, **270**, 253-261.

Murk AJ, Legler J, Denison MS, Giesy JP, van de Guchte C and Brouwer A (1996) Chemical-activated luciferase gene expression (CALUX): a novel in vitro bioassay for Ah receptor active compounds in sediments and pore water. *Fundamental and Applied Toxicology*, **33**, 149-160.

Murk AJ, Leonards PEG, van Hattum B, Luit R, van der Weiden MEJ and Smit M (1998) Application of biomarkers for exposure and effect of polyhalogenated aromatic hydrocarbons in naturally exposed European otters (*Lutra lutra*). *Environmental Toxicology and Pharmacology*, **6**, 91-102.

Opresko DM, Sample BE and Suter II GW (1994) *Toxicological benchmarks for wildlife: 1994 revision*. ES/ER/TM-86/R1, Oak Ridge National Laboratory, Health Sciences Research Division, Oak Ridge, TN.

Poland A and Knutson JC (1982) 2,3,7,8-Tetrachlorodibenzo-*p*-dioxin and related halogenated aromatic hydrocarbons: examination of the mechanism of toxicity. *Annual Review of Pharmacology and Toxicology*, **22**, 517-554.

Quensen JF, Mousa MA, Boyd SA, Sanderson JT, Froese KL and Giesy JP (1998). Reduction of aryl hydrocarbon receptor-mediated activity of polychlorinated biphenyl mixtures due to anaerobic microbial dechlorination. *Environmental Toxicology and Chemistry*, **17**, 806-813.

Rangan U, Hedli C, Gallo M, Lioy P and Snyder R (1997) Exposure and risk assessment with respect to contaminated soil: significance of biomarkers and bioavailability. *International Journal of Toxicology*, **16**, 419-432.

Rice CP and White DS (1987) PCB availability assessment of river dredging using caged clams and fish. *Environmental Toxicology and Chemistry*, **6**, 259-274.

Richter CA, Tieber VL, Denison MS and Giesy JP (1997) An *in vitro* rainbow trout cell bioassay for aryl hydrocarbon receptor-mediated toxins. *Environmental Toxicology and Chemistry*, **16**, 543-550.

Safe S (1990) Polychlorinated-biphenyls (PCBs), dibenzo-para-dioxins (PCDDs), dibenzofurans (PCDFs), and related-compounds — environmental and mechanistic considerations that support the development of toxic equivalency factors (TEFs). *Critical Reviews in Toxicology*, **21**, 51-88.

Sanderson JT, Aarts JMMJG, Brouwer A, Froese KL, Denison MS and Giesy JP (1996). Comparison of Ah receptor-mediated luciferase and ethoxyresorufin-*o*-deethylase induction in H4IIE cells: implications for their use as bioanalytical tools for the detection of polyhalogenated aromatic hydrocarbons. *Toxicology and Applied Pharmacology*, **137**, 316-325.

Shipp EB, Restum JC, Giesy JP, Bursian SJ, Aulerich RJ and Helferich WC (1998) Multigenerational study of the effects of consumption of PCB-contaminated carp from Saginaw Bay, Lake Huron, on mink. 2. Liver PCB concentration and induction of hepatic cytochrome P-450 activity as a potential biomarker for PCB exposure. *Journal of Toxicology and Environmental Health*, **54A**, 377-401.

Stegeman JJ and Lech JJ (1991) Cytochrome-P-450 monooxygenase systems in aquatic species — carcinogen metabolism and biomarkers for carcinogen and pollutant exposure. *Environmental Health Perspectives*, **90**, 101-109.

Talmage SS and Walton BT (1991) Small mammals as monitors of environmental contaminants. *Reviews of Environmental Contamination and Toxicology*, **119**, 47-108.

Tanabe S, Watanabe S, Kan H and Tatsukawa R (1988) Capacity and mode of PCB metabolism in small cetaceans. *Marine Mammal Science*, **4**, 103-124.

Tillitt DE, Giesy JP and Ankley GT (1991) Characterization of the H4IIE rat hepatoma-cell bioassay as a tool for assessing toxic potency of planar halogenated hydrocarbons in environmental samples. *Environmental Science and Technology*, **25**, 87-92.

Tillitt DE, Gale RW, Meadows JC, Zajicek JL, Peterman PH, Heaton SN, Jones PD, Bursian SJ, Kubiak TJ, Giesy JP and Aulerich RJ (1996) Dietary exposure of mink to carp from Saginaw Bay. 3. Characterization of dietary exposure to planar halogenated hydrocarbons, dioxin equivalents, and biomagnification. *Environmental Science and Technology*, **30**, 283-291.

Troisi GM and Mason CF (1997) Cytochromes P450, P420 & mixed-function oxidases as biomarkers of polychlorinated biphenyl (PCB) exposure in harbour seals (*Phoca vitulina*). *Chemosphere*, **35**, 1933-1946.

USEPA (1995). *Great Lakes Water Quality Initiative Criteria Documents for the Protection of Wildlife — DDT, Mercury, 2,3,7,8-TCDD, and PCBs*. EPA-820-8-95-008, US Environmental Protection Agency.

USEPA (1997) *Uncertainty Factor Protocol for Ecological Risk Assessment — Toxicological Extrapolations to Wildlife Receptors*, Region 8 Ecosystems Protection and Remediation Division, US Environmental Protection Agency.

USEPA (1998a). *Guidelines for Ecological Risk Assessment*. EPA/630/R-95/002F, US Environmental Protection Agency.

USEPA (1998b). *Use of PCB Congener and Homologue Analysis in Ecological Risk Assessments*, Region 9 Biological Technical Advisory Group, US Environmental Protection Agency.

Van den Berg M, De Jongh J, Poiger H and Olson JR (1994) The toxicokinetics and metabolism of polychlorinated dibenzo-*p*-dioxins (PCDDs) and dibenzofurans (PCDFs) and their relevance for toxicity. *Critical Reviews in Toxicology*, **24**, 1-74

Van den Berg M, Birnbaum L, Bosveld ATC, Brunstrom B, Cook P, Feeley M, Giesy JP, Hanberg A, Hasegawa R, Kennedy SW, Kubiak T, Larsen JC, van Leeuwen FXR, Liem AKD, Nolt C, Peterson RE, Poellinger L, Safe S, Schrenk D, Tillitt D, Tysklind M, Younes M, Waern F and Zacharewski T (1998) Toxic equivalency factors (TEFs) for PCBs, PCDDs, PCDFs for humans and wildlife. *Environmental Health Perspectives*, **106**, 775-792.

Villeneuve DL, Richter CA, Blankenship AL and Giesy JP (1999) Rainbow trout cell bioassay-derived relative potencies for halogenated aromatic hydrocarbons: comparison and sensitivity analysis. *Environmental Toxicology and Chemistry*, in press.

Vuorinen PJ, Paasivirta J, Keinanen M, Koistinen J, Rantio T, Hyotylainen T and Welling L (1997) The M74 syndrome of baltic salmon (Salmo salar) and organochlorine concentrations in the muscle of female salmon. *Chemosphere*, **34**, 1151-1166.

Watson MR, Stone WB, Okoniewski JC and Smith LM (1985) Wildlife as monitors of the movement of polychlorinated biphenyls and other organochlorine compounds from a hazardous waste site. *Transactions of the Northeast Fish and Wildlife Conference*, pp. 91-104.

Williams LL, Giesy JP, DeGalan N, Verbrugge DA, Tillitt DE, Ankley GT and Welch RL (1992) Prediction of concentrations of 2,3,7,8-tetrachlorodibenzo-*p*-dioxin equivalents from total concentrations of polychlorinated biphenyls in fish fillets. *Environmental Science and Technology*, **26**, 1151-1159.

Willman EJ, Manchester-Neesvig IB and Armstrong DE (1997) Influence of *ortho*-substitution on patterns of PCB accumulation in sediment, plankton, and fish in a freshwater estuary. *Environmental Science and Technology*, **31**, 3712-3718.

Yorks AL, Melancon MJ, Hoffman DJ, Henshel DS and Sparks DW (1998) Nestling tree swallow (*Tachycineta bicolor*) PCB body burdens and their effects on reproduction and growth. *Society of Environmental Toxicology and Chemistry Annual Meeting Abstract Book*, p.164.

Zimmermann G, Dietrich DR, Schmid P and Schlatter C (1997) Congener-specific bioaccumulation of PCBs in different water bird species. *Chemosphere*, **34**, 1379-1388.

Zwiernik MJ, Quenson III JF and Boyd SA (1998) Sulfate stimulates extensive dechlorination of PCBs. *Environmental Science and Technology*, **32**, 3360-3365.

10

Ecotoxicological Approaches in the Field

JÖRG RÖMBKE[1] AND JOS NOTENBOOM[2]

[1]ECT Oekotoxikologie GmbH, Flörsheim am Main, Germany
[2]RIVM, National Institute of Public Health and Environment, Bilthoven, The Netherlands

10.1 INTRODUCTION AND BACKGROUND

The ecotoxicological risk of chemicals and contaminants is usually determined in a tiered system of rising complexity as described in the present book (see Chapters 3 and 4). Proceeding from laboratory tests to semi-field and field studies, ecological reality increases. However, the expense of these studies also increases. Therefore, despite the fact that the ecosystem in its natural context is the most important level for ecotoxicological assessments, standardized field studies have rarely been performed. Before describing the few approaches available for ecosystem level testing, important issues for understanding the current situation in field ecotoxicology should be highlighted. It should be mentioned that this chapter covers mainly the soil compartment, focusing especially on soil invertebrates.

10.1.1 ECOSYSTEM COMPLEXITY

The ecosystem in its natural context is the most important level for ecotoxicological issues (Kimball and Levin 1985). The problem is that every ecosystem is extremely complex, with multifarious reciprocal relationships that are not fully understood, even in the absence of pollution. This is true not only for regions of the world which lack detailed ecological research like the tropics, but also for Northern Hemisphere soils which have been investigated for several decades. A site such as a deciduous beech forest that seems to be very simple at first glance harbors more than 1000 different species, while an anthropogenically impoverished agricultural soil food web can be very complex as well (Moore and De Ruiter 1991). It is not possible to predict exactly how an ecosystem will behave. Therefore, absolute threshold values for pollution

Environmental Analysis of Contaminated Sites. Edited by G. I. Sunahara, A. Y. Renoux, C. Thellen, C. L. Gaudet and A. Pilon

in ecosystems have been attempted (see Chapters 12 and 13) but their establishment is difficult, since an ecosystem's sensitivity to chemicals can change dramatically with time. For this reason, the number of standardized ecotoxicological test procedures is few, while the number of field studies examining the fate or effect of chemicals based upon individual methods and priorities is very high.

There is a clear lack of ecological base data for most soils, since soil science has focused on the physical and chemical properties of soils. In addition, soil ecology is limited by the large heterogeneity of natural soils; whether classified according to their developmental history (soil type) or their physical properties (soil class, mainly grain size distribution). In any case, many different soils have to be considered. In addition the variable time scale has to be taken into consideration; important processes can range between hundreds of years (e.g., soil formation), several months (e.g., litter degradation) or a few hours to days (e.g., leaching or run-off of chemicals to ground- or surface water). However, despite all these difficulties and partly due to long-term international research programs like 'acid rain' in the eighties, our understanding of soils as a living entity has recently increased strongly.

10.1.2 STRUCTURAL AND FUNCTIONAL APPROACH

The lack of an ecological database makes it difficult to demonstrate the range of natural variation in a soil ecosystem. Since analyzing soil ecosystems in their entirety is not possible, certain parameters (physiological, chemical, biotic structures and related interactions) must be selected from the ecosystem and defined. These parameters are then used to describe the normal state of an ecosystem and any potential deviations from it. Such a normal state can serve as a standard of comparison to polluted sites with the same type of ecosystem. Recently, the use of biological soil classification concepts has been discussed in this context in various European countries (Römbke *et al.* 1997). Only chemical effects which exhibit distinct variations in terms of their intensity compared to natural variation (e.g., in relation to an unpolluted control site or a median value over several years) can also be recognized as such and, in turn, evaluated. This is particularly true for microbes, whose density may fluctuate by 90% due to natural stress factors (Domsch *et al.* 1983).

The lack of ecological databases for ecosystems can be attributed to the fact that only a few terrestrial field projects have studied entire ecosystems in quantitative terms and over a period of time (e.g., forests: Ellenberg *et al.* 1986, agricultural sites: Hardy 1986). Furthermore, most ecosystems have been strongly influenced anthropogenically (especially in industrialized countries) or have not yet been examined. The latter is especially true of tropical habitats. For example, not one factor but the interaction of multiple anthropogenic influences (air pollution, land zoning, less crop rotation) has caused species depletion in central European farming regions.

Another approach for the assessment of the status of a soil biocoenosis at a given site is to study whether it is fulfilling its main functions (e.g., decomposition of organic matter). In many natural and managed ecosystems, the bulk of the annual net primary productivity is transferred directly to the decomposer subsystem. Decomposition and mineralization of nutrients contained in this organic material are critical processes affecting soil fertility and primary productivity of ecosystems. The distribution and turnover of organic matter contribute to ecosystem functioning by forming soil organic matter pools and nutrient exchange sites for root uptake. Organic matter decomposition is one of the most integrated processes within soil ecosystems because it involves complex interactions between soil microbial and faunal activities with the soil chemical environment. Species from different taxonomic groups contribute to the breakdown of plant and animal debris. Disturbances which (directly or indirectly) alter organic matter decomposition can result in nutrient losses and declined soil fertility. Therefore, an assessment of altered rates of organic matter decomposition and nutrient retention and release is critical to understand the impact of chemical contamination on overall ecosystem structure and function. However, functional redundancy in the soil ecosystems does not show whether the function is completely fulfilled or whether additional stress would cause a collapse in an ecosystem already stressed.

10.1.3 CHEMICAL TESTING VERSUS SOIL QUALITY ASSESSMENT

In ecotoxicology, methods for the assessment of potentially contaminated soils were originally developed for the testing of individual chemicals (especially pesticides). Field testing in the latter case is relatively simple, acquiring an uncontaminated soil to be spiked with different concentrations of the test compound. The measured effects are then compared to results from untreated controls. The assessment of soil quality is more difficult since a control is often not readily available (cf. Section 10.1.4). The differences between chemical testing and soil quality assessment are summarized briefly in Table 10.1.

Standardized methods for the testing of individual chemicals in field studies can be used as a starting point for the modification of methods to study contaminated soils. Ecotoxicological field studies can have different objectives:

- Validation or calibration of results from laboratory or semi-field tests (e.g., risk limits) as part of an ERA (environmental risk assessment; see Chapters 15 and 19) process;
- Investigation of the impact of (complex) contamination on entire ecosystems;
- Evaluation of ecological risk assessment methodologies;
- Identification and characterization of (potentially) contaminated sites;
- Assessment of the long-term suitability of remediated soils for soil organisms.

Table 10.1 Description of characteristics of field-level standard testing of individual chemicals and quality assessment of (potentially) contaminated soils

Attribute	Testing of single chemicals	Soil quality assessment
Background	part of a prospective registration or notification process (usually at the highest tier)	often retrospective investigation of a site with unknown contamination (mixtures)
Methods	use of internationally standardized tests with structural or functional endpoints (sometimes aiming on single (key) species)	sampling of various organism groups or investigation of ecosystem functions, often using standardized methods known from soil ecology
Interaction	active manipulation of the site (e.g., by spraying a pesticide)	no manipulation
Assessment	comparison of results from treated versus control plots	comparison of results with data from a similar but uncontaminated site or, if such a site is not available, comparison with 'mean' values gained from other field studies
Legal base	chemical, especially pesticide laws (e.g., EU 1991)	soil protection laws (e.g., the German BBodSchG 1998)

Some of these objectives overlap with approaches usually defined as monitoring. For instance, measurements taken to record data at a specific time (survey), repeated measurements taken to record changes over a period of time (surveillance) or ongoing measurements for comparisons between a current state and a previously determined standard (controls or monitoring *per se*). However, details of monitoring approaches and methods will not be presented here.

Neither biological methods like residue analysis for studying the bioavailability of chemicals under real field conditions (see Chapter 2) and characterizing complex contaminant mixtures nor extraction methods will be discussed. Determining the bioavailability of a chemical is an important complement of laboratory studies, which are usually performed assuming 'worst-case' conditions (e.g., no ageing of the chemical).

10.1.4 THE PROBLEM OF CONTROL

When evaluating the ecotoxicological effects of contaminated soils, the lack of a readily available uncontaminated control site becomes problematic. An ideal control site would be close to the contaminated site, and have the same soil properties (ISO 1999a). If such a site is unavailable (as is often the case), a similar soil type should be used. Another option is the use of so-called standard soils; i.e., a limited number of soils with specific properties (e.g., the five EURO-Soils used in standard fate tests — these soils were selected as being representative for European soils in general; Kuhnt and Muntau 1994). Currently, there is no final agreement on how the problem of control could be handled in a practical way. So, a case-by-case decision based on expert knowledge is most probably the best solution for the time being.

10.2 LEGISLATIVE REQUIREMENTS IN FIELD ECOTOXICOLOGY

Field ecotoxicological studies are not mandatory by government agencies responsible for chemical registration or soil protection. The European Union requires testing of the effects of pesticides on earthworms, beneficial arthropods (e.g., carabid beetles) and litter degradation as part of the chemical registration process (EU 1991). USEPA requires terrestrial field studies performed under actual pesticide use (Fite *et al.* 1988). These studies address the acute, subacute and/or long-term effects of pesticide residues on non-target species.

However, for the investigation of the soil of contaminated sites, only very recently governmental requirements have been taken into consideration. For example, the new German Soil Protection Law (BBodSchG 1998) requires ecotoxicological testing to protect natural soil functions. Unfortunately, the enactment lacks direction in describing how these functions should be protected, and how to evaluate the effects of contaminated soils on soil organisms (probably since no standardized methods were available when this enactment was passed). In North American countries the situation is comparable, whereby the assessment of soils (e.g., definition of soil quality criteria or efficacy investigations of remediation efforts) is strongly based on single-species laboratory tests (see Chapter 3; Sheppard *et al.* 1992).

10.3 ECOTOXICOLOGICAL FIELD METHODS

Field studies using different methods have an important role in ecotoxicology and will continue to do so despite the difficulties involved. In the next section, examples are given for the testing of individual chemicals and the investigation of contaminated sites. Section 10.3.3 will describe the testing of individual chemicals and soil quality assessment using functional methods.

10.3.1 FIELD TESTS FOR INDIVIDUAL CHEMICALS

There is a lack of standardized test methods for assessing the ecotoxicological risk of individual chemicals in terrestrial field ecosystems. The earthworm field test, in which the long-term effects of a pesticide on the lumbricid community of an arable or meadow ecosystem are examined, is one exception (ISO 1999b; Figure 10.1). A British guideline comparable in design, albeit extremely general, examining the effects of insecticides on beneficial arthropods (various groups of beetles and spiders) in summer grain fields has been proposed (MAFF 1993). This guideline resembles later proposals, where Collembola are used as test organisms (Wiles and Frampton 1996). Otherwise, there are only methods for evaluating the effect of pesticides on beneficial arthropods in the vegetation layer, e.g., of apple plantations or vineyards (Hassan 1992). These tests are conducted by applying one or more concentrations of a pesticide under conditions comparable to normal agricultural practice. After a specific period of time abundance, biomass and species dominance are used as measurement

Figure 10.1 Hand-sorting and formalin extraction site for the sampling of earthworms at a meadow site in central Europe. Figure shows rectangular plot (50 × 50 cm) for digging out the worms and a can filled with 5 L of formalin solution to be applied into the plot.

endpoints. Well-known methods from soil ecology (e.g., formalin extraction, hand sorting in the case of the earthworm test) are used in these tests (Dunger and Fiedler 1997). Recently, testing the effects of persistent pesticides on soil functions, especially organic matter decomposition, are being standardized (cf. Section 10.3.3).

10.3.2 USE OF FIELD METHODS IN SOIL QUALITY ASSESSMENT

Standardization of field tests is difficult due to the variability of natural conditions. Therefore, field studies can only be conducted on a case-by-case basis following some kind of general criteria and guidance. In other words, the scope of the examination and the details of the performance in each case have to be agreed to by the test performer (i.e., the site owner) and the responsible governmental authority.

Some examples for investigating the effects of (potentially) contaminated soils on soil organisms are given. These investigations can be done actively, using primarily ecotoxicological methods (e.g., exposing laboratory reared earthworms to contaminated soil under field conditions) or passively, using ecological methods (i.e., monitoring the presence of certain key species at a given site).

10.3.2.1 Ecotoxicological Methods

The use of the earthworm as a test organism in field mesocosm studies (or, depending on definition, as combined reaction and accumulation indicators) to assess soil quality is best demonstrated by a study performed at a Superfund Site (Holbrook, USA; Callahan *et al.* 1991). Briefly, field test containers were placed on transects located across the impacted areas and in reference sites located nearby. Plastic test containers were filled with excavated soil, and five *Lumbricus terrestris* were placed on the soil surface. After 7 days, mortality and morbidity endpoints (e.g., coiling, cutaneous lesions) were evaluated. Soil samples and earthworm tissue samples from survivors were randomly selected for chemical analysis (see Chapter 19). This method appears to be sensitive for assessing the potential impact of soil contaminants at field sites (especially when combined with laboratory tests and/or the investigation of native species). The same approach was recently used to assess the effects of soils contaminated by acids or fertilizers on the population dynamics of two springtail species (Kopeszki 1999). Various other methods using introduced or native plants have been suggested in the literature (e.g., Pakeman *et al.* 1998). When selecting organisms for on-site tests the same criteria should be used as when selecting species for a laboratory test.

10.3.2.2 Methodological Considerations

In this section, the question of the selection of test species is discussed in more detail. Soil quality assessments should be aimed at evaluating the ecosystem and not individual species as is often the case for environmental studies (e.g., protection of endangered avian or mammal species). It is necessary to select several organisms from various size classes, functional and trophic groups. The assessment should evaluate either the effects of potential contaminants on the organisms (reaction indicators) or whether the accumulation of a chemical in soil organisms and possible secondary poisoning effects (e.g., for birds of prey) should be identified (accumulation indicators). In addition to the structural endpoints mentioned, functional methods should also be considered (cf. Section 10.3.3).

Invertebrates to be selected as test species in field studies (including monitoring programs) should fulfil the following criteria:

- Clearly identifiable (especially the active life stages);
- Sensitive against a wide range of stress factors;
- Close interaction with soil properties (natural factors as well as contaminants);
- A rapid life-cycle (i.e., one to four generations per year);
- Preferably a good correlation with microbiological activities;
- Easy sampling, e.g., high density;
- Not migratory (so effects can be attributed to a certain site).

Koehler (1996) goes one step further when requiring that 'neither a single, perhaps mythical "most sensitive" species nor the whole community is assessed, but a set of populations with relatively clear and similar ecological requirements (a guild in a broader sense), composed of an array of species, sufficiently diverse to allow the detection of changes'.

The selection of 'ecosystem engineers' or 'key species' is recommended (Jones *et al.* 1994). These organisms directly or indirectly affect the availability of resources to other organisms through modifications of the physical environment. In the soil ecosystems, earthworms and termites (tropical regions) are organisms most often identified as principal engineers (Lavelle *et al.* 1997). Other groups may also be important in this context (e.g., millipedes, ants or, more rarely, springtails or enchytraeids). Ecosystem engineers have sufficient numerical and biomass densities to exert a predominant influence in the formation and maintenance of soil structure, making them ideal species in standardized test methods and assessment studies. Testing or monitoring ecosystem engineers ensures that any changes in their densities or species structure will also influence soil structure and ecosystem processes. The following organism groups fulfil the criteria listed above and have been used successfully in field studies (cf. Koehler 1996).

1. Reaction indicators: earthworms (Edwards and Bohlen 1992), springtails (Hopkin 1997), nematodes (Bongers 1990), gamasid mites (Ruf 1998);
2. Accumulation indicators: earthworms (Edwards and Bohlen 1992), Isopoda (Hopkin 1989).

Methods for sampling some of these groups in the field are presented in Table 10.2.

10.3.3 MEASUREMENT OF SOIL PROCESSES

Two processes (litter decomposition and feeding rate) are described since they are easily measurable (Kula and Römbke 1998). Organic matter decomposition rates may be quantified using a variety of methods, including a cotton strip and a litterbag technique. Additionally, in the case of litterbags, the accessibility of the bag content to soil biota can be controlled by the mesh size.

The cotton strip assay (Howard 1988) uses a physical parameter (loss of tensile strength). Cotton strips are dug into soil and left for a certain period; afterwards, the strength of the material is measured. An advantage of this method is the availability of a well-standardized cotton material. However, to measure loss of tensile strength, special equipment is needed. Another disadvantage is that assessment of degradation of pure cotton instead of natural litter will result in a simplification of this complex soil function.

The litterbag method (MAFF 1986, Paulus *et al.* 1999) determines the effects of chemicals on both mass loss and patterns of nutrient dynamics in decomposing litter (Figure 10.2). It requires more effort than the cotton strip method,

Table 10.2 Overview on methods for the sampling of soil invertebrates in field studies (Dunger and Fiedler 1997; recommendation: +suitable; −not suitable)

Animal group and method	Advantage	Disadvantage	Recommendation
Nematodes			
Sieving	useful for large soil samples	high losses possible	+
Modified funnel method	well suitable for the extraction of small worms (<1 mm)	high losses of large nematodes (>1 mm)	−
Centrifugation	approved, quick	high expenditure of work with large soil samples (100–500 cm^3)	+
Earthworms			
Electrical sampling	simple, 'high-tech' image, 'clean'	expensive equipment, selective assessment, results strongly dependent on soil moisture	−
Hand sorting	simple, accurate	high expenditure of work, 'dirty'	+
Formol extraction	simple, accurate, quick	toxic chemical, high demand of water, only useful in combination with hand-sorting	+
Enchytraeids			
Wet extraction	simple, approved	high expenditure of work	+
Oribatid and gamasid mites			
Dry extraction	approved, accurate	special equipment needed	+
Macrofauna (isopodes, myriapods)			
Hand sorting	simple, approved	high expenditure of work	+

but provides ecologically relevant information and is preferable. The disadvantages of this method are: (i) substrate packed within a litterbag may create a microclimate different from bulk soil; (ii) substrate does not come into contact with contaminated soil; and (iii) litter in the bag may attract soil organisms which in turn may lead to increased biological activity and faster decomposition.

The bait-lamina test (Von Törne 1990, Kratz 1998, Paulus *et al.* 1999) measures the feeding rate of soil invertebrates. Small plastic sticks that contain holes filled with a glucose mixture are put into the soil for several days (Figure 10.3). The number of fed holes is the measurement endpoint. It is both simple and practical with a 'yes' or 'no' answer. The test is standardized, but it does not determine the contribution of different groups of soil biota to the

Figure 10.2 Litterbags (mesh size: 1 cm) were buried at a depth of 5 cm in the soil of a meadow test site containing a test chemical.

Figure 10.3 Sticks of the bait-lamina test at a forest site in winter; the upper part of each plastic stick is visible whereas the lower part, containing the holes with the glucose mixture, is buried in the soil.

decomposition process, nor quantify the decomposition rate or nutrient mineralization dynamics in decomposing material. Therefore, it is usually recommended as a screening test when a control site is available close to the tested area.

10.4 EXAMPLES OF CASE STUDIES FOR STUDYING SOIL ORGANISMS AT CONTAMINATED SITES

10.4.1 OVERVIEW

Sampling programs for the chemical characterization of contaminated sites are routinely performed. However, the number of studies, where at one field site exposure and biological endpoints were measured, is low. Noteworthy exemptions are some studies in which the potential impact of heavy metals on soil organisms at sites close to factories was investigated. In a recent compilation (Pokarzhevskii *et al.* 1998) the following European case studies are listed:

1. Avonmouth, UK, zinc smelter (Hopkin 1989);
2. Merseyside, UK, copper refinery (Hunter *et al.* 1987);
3. Gusum, Sweden, brass mill (Bengtsson and Tranvik 1989);
4. Budel, The Netherlands, zinc smelter (Posthuma *et al.* 1998; see also Section 10.4.2);
5. Ijmuiden, The Netherlands, steel factory (Joosse and Van Vliet 1982);
6. Berlin, Germany, sewage-irrigated field (Kratz and Brose 1998).

In addition, several ecotoxicological field studies (usually for a short period of time) took place on sites contaminated by mining activities (e.g., France: Abdul Rida and Bouché 1994), emissions from highways (e.g., Butovsky 1990) or orchards (highly contaminated by pesticides; e.g., Paoletti *et al.* 1988). This list is not complete by far.

The results of these studies are difficult to interpret (Pokarzhevskii *et al.* 1998):

- Usually chemical and biological studies were not performed at the same place or time;
- Some organisms were assessed on the species, some on the group level — many organisms were not considered at all;
- Mainly direct toxicological effects were taken into account, indirect influences were rarely investigated.

Despite these difficulties, however, some important results from these studies can be summarized as follows (modified according to Pokarzhevskii *et al.* 1998): (1) a decrease of total density and biodiversity is usually observed in the direction of the emission source; (2) the sensitivity of species to the same contaminant found in laboratory experiments may not coincide with effects observed in field studies; (3) different species from the same group of animals may have different patterns of density change when approaching a source of emissions; and (4) differences in the composition of contaminant mixtures may cause the effects in food webs to be very variable.

10.4.2 BUDEL (THE NETHERLANDS)

The project 'Validation of toxicity data and risk limits for soils' studied three aspects (Posthuma *et al.* 1998):

1. The relevance of laboratory toxicity tests for the prediction of effects in the field;
2. The ecological relevance of ecotoxicological risk limits (derived from statistical extrapolation methods);
3. The identification of which factors introduce uncertainties.

Toxic effects of zinc on various organisms (e.g., nematodes, enchytraeids, earthworms, springtails and plants) were studied in the laboratory, in an experimental field and at the field site (a former zinc smelter close to Budel, The Netherlands). Abundance and species diversity of nematodes and enchytraeids and microbial degradation capacity (especially the development of tolerance towards zinc) were the main measurement parameters used to evaluate the ecological risk on the community level (Posthuma *et al.* 1998).

Bioavailability appeared to be the most important factor for all types of soil organisms when comparing the effects of zinc in different soil types. Due to the heterogeneous zinc gradient at Budel, clear relationships between the level of contamination and the observed changes in abundance and species diversity could not be established. However, tolerance of the microbial community was a sensitive parameter even in samples with zinc concentrations slightly above control levels. In general, ecotoxicological risk limits based on the results of laboratory tests gave a good indication for the presence or absence of serious effects. Therefore, the results support the Dutch two-tiered approach for soil ecotoxicological risk assessment.

10.5 OUTLOOK

In the past, the evaluation of ecotoxicological field investigations often suffered from the lack of essential data measured at the same time and place. In any case, a detailed characterization of the field soil to be tested is essential for any assessment of a potential contamination. At least the pH, amount of organic matter (or carbon), C/N ratio, grain size distribution and maximum water holding capacity should be determined according to internationally standardized methods (e.g., ISO guidelines). In addition, it is necessary to measure climatic factors and to know the history of land use of the test site.

In a recent compilation entitled 'Risk Assessment for Contaminated Sites in Europe', Ferguson *et al.* (1998) tried to summarize the status of field testing of contaminated soils. They conclude in their chapter on biological receptors, 'on-site monitoring of changes in community structure may provide a low-cost predictive tool for assessing effects at a high level of biological organization. For retrospective ecological risk assessments, in which the objective is to measure the ecological effects of historical contamination, the identification of species diversity and abundance over time and space seems to offer a logical, straightforward and unambiguous approach'.

One concept aiming in the same direction is the comparison between the biocoenosis actually found at a 'potentially contaminated' site (characterized using mainly qualitative parameters like dominance spectrum but also abundance) and the community which should live there if the soil were not contaminated or otherwise (anthropogenically) affected. Examples are the SOILPACS concept in England (Spurgeon *et al* 1996) or the BBSK concept in Germany (Römbke *et al* 1997). These reference community-defined approaches are based on a correlation between the most important soil properties and the occurrence of organisms at uncontaminated sites deduced from the literature. Currently, the biology of soil organisms and their dependency on certain soil properties are not very well studied, but experiences from aquatic ecosystems using the same concepts are very promising (e.g., BEAST, Canada; Reynoldson *et al.* 1995). However, in addition to the structural assessment, functional methods (especially on litter decomposition) should be used at a field test site in order to get a complete overview of the soil quality at that site.

REFERENCES

Abdul Rida AMM and Bouché MB (1994) A method to assess chemical biorisks in terrestrial ecosystems. In *Ecotoxicology of Soil Organisms*, Donker MH, Eijsackers H and Heimbach F (eds), Lewis, Boca Raton, FL, pp. 383-394.

BBodSchG (1998) *Gesetz zum Schutz des Bodens* [*Federal Soil Protection Act*], Bundesgesetzblatt 1998, Teil I, Nr. **16**, pp. 502-510.

Bengtsson G and Tranvik L (1989) Critical metal concentrations for forest soil invertebrates. *Water, Air and Soil Pollution*, **49**, 381-417.

Bongers T (1990) The maturity index: an ecological measure of environmental disturbance based on nematode species composition. *Oecologia*, **83**, 14-19.

Butovsky RO (1990) Traffic contamination and entomofauna. *Agrokhimija*, **4**, 9-150.

Callahan CA, Menzie CA, Burmaster DE, Wilborn DC and Errst T (1991) On-site methods for assessing chemical impact on the soil environment using earthworms: a case study at the Baird and McGuire Superfund Site, Holbrook, Massachusetts. *Environmental Toxicology and Chemistry*, **10**, 17-826.

Domsch KH, Jagnow G and Anderson TM (1983) An ecological concept for the assessment of side effects of agro-chemicals on soil microorganisms. *Residue Review*, **86**, 65-105.

Dunger W and Fiedler HJ (1997) *Methoden der Bodenbiologie*, Gustav Fischer Verlag, Stuttgart, 539 pp.

Edwards CA and Bohlen PJ (1992) The effects of toxic chemicals on earthworms. *Reviews of Environmental Contamination and Toxicology*, **125**, 23-99.

Ellenberg H, Mayer R and Schauermann J (1986) *Ökosystemforschung. Ergebnisse des Solling-projektes*, Ulmer Verlag, Stuttgart, 507 pp.

EU (1991) Council Directive of 15 July 1991 concerning the placing of plant protection products on the market. *Official Journal of the European Communities*, **34**, 1-23.

Ferguson C, Darmendrail D, Freier K, Jensen BK, Jensen J, Kasamas H, Urzelai A and Vegter J (1998) *Risk Assessment for Contaminated Sites in Europe, Vol. 1, Scientific Basis*, LQM Press, Nottingham, 165 pp.

Fite EC, Turner LW, Cook NJ and Stunkard C (1988) *Guidance Document for Conducting Terrestrial Field Studies*, US Environmental Protection Agency, Washington, DC, 67 pp.

Hardy AR (1986) The Boxworth project — a progress report. *Brighton Crop Protection Conference*, The British Crop Protection Council, Farnham, pp. 1215-1224.

Hassan SH (1992) Guidelines for testing the effects of pesticides on beneficial organisms: description of test methods. *IOBC/WPRS Bulletin*, **XV**, 1-186.

Hopkin SP (1989) *Ecophysiology of Metals in Terrestrial Invertebrates*, Elsevier Applied Science, London, 366 pp.

Hopkin SP (1997) *Biology of the Springtails*, Oxford University Press, Oxford, 330 pp.

Howard PJA (1988) A critical evaluation of the cotton strip assay. In *Cotton Strip Assay — An Index of Decomposition in Soils*, Harrison AF, Latter PM and Walton DWH (eds), ITE Symposium No. 24, Institute of Terrestrial Ecology, Grange-over-Sands, Cumbria, pp. 34-42.

Hunter BA, Johnson MS and Thompson DJ (1987) Ecotoxicology of copper and cadmium in a contaminated ecosystem. II. Invertebrates. *Journal of Applied Ecology*, **24**, 587-599.

ISO (1999a) *Soil Quality — Guidance on the Ecotoxicological Characterization of Soils and Soil Materials*. ISO/DIS 15799 (10th draft), International Standards Organization, Paris.

ISO (1999b) *Soil Quality — Effects of Pollutants on Earthworms. Part 3: Guidance on the Determination of Effects in Field Situations*. ISO 11268-3, International Standards Organization, Paris.

Jones CG, Lawton JH and Shachak M (1994) Organisms as ecosystem engineers. *Oikos*, **69**, 373-386.

Joosse ENG and Van Vliet LHH (1982) Impact of blast-furnace plant emissions in a dune ecosystem. *Bulletin of Environmental Contamination and Toxicology*, **29**, 279-284.

Kimball KD and Levin SA (1985) Limitations of laboratory bioassays: the need for ecosystem-level testing. *BioScience*, **35**, 165-171.

Koehler HH (1996) Soil animals and bioindication. In *Bioindicator Systems for Soil Pollution*, Van Straalen NM and Krivolutsky DA (eds), Kluwer Academic Publishers, Dordrecht, pp. 179-188.

Kopeszki H (1999) Die aktive Bioindikationsmethode mit Collembolen. Möglichkeit der Diagnose von Bodenzustand und-belastung. *UWSF — Zeitschrift für Umweltchemikalien und Ökotoxikologie*, **11**, 201-206.

Kratz W (1998) The bait-lamina test — general aspects, applications and perspectives. *ESPR-Environmental Science & Pollution Research*, **5**, 94-96.

Kratz W and Brose A (1998) *Bodenökologische Untersuchungen zur Wirkung und Verteilung von organischen Stoffgruppen (PAK, PCB) in ballungsraumtypischen Ökosystemen*. Report 1/98, GSF, München, 165 pp.+Annex.

Kuhnt G and Muntau H (1994) *EURO-SOILS. Identification, Collection, Treatment, Characterization*. Special Publication No. 1.94.60, Environment Institute, European Commission, Ispra, 154 pp.

Kula C and Römbke J (1998) Testing organic matter decomposition within risk assessment of plant protection products. *ESPR-Environmental Science & Pollution Research*, **5**, 55-60.

Lavelle P, Bignell D, Lepage M, Wolters V, Roger P, Ineson P, Heal OW and Dhillion S (1997) Soil function in a changing world: the role of invertebrate ecosystem engineers. *European Journal of Soil Biology*, **33**, 159-193.

MAFF (1986) *Laboratory and Field Testing of Pesticide Products for Effects on Soil Macroorganisms. 2. Tests on Organic Matter Breakdown*. Working Document 7/6, Ministry of Agriculture, Forestry and Fisheries, London, pp. 103-104.

MAFF (1993) *Guideline to Study the Within-Season Effects of Insecticides on Beneficial Arthropods in Cereals in Summer*. Working Document 7/7, Ministry of Agriculture, Fisheries and Forestry, London, pp. 105-106.

Moore, JC and De Ruiter PC (1991) Temporal and spatial heterogeneity of trophic interactions within below-ground food webs. *Agriculture, Ecosystems and Environment*, **34**, 371-397.

Pakeman RJ, Hanlard PK and Osborn D (1998) Plants as biomonitors of atmospheric pollution: their potential for use in pollution regulation. *Reviews in Environmental Contamination and Toxicology*, **157**, 1-23.

Paoletti MG, Iovane E and Cortese M (1988) Pedofauna bioindicators and heavy metals in five agroecosystems in north-east Italy. *Revue d'Ecologie Biologie de Sol*, **25**, 33-58.

Paulus R, Römbke J, Ruf A and Beck L (1999) A comparison of the litterbag, minicontainer and bait-lamina methods in an ecotoxicological field experiment with diflubenzuron and btk. *Pedobiologia*, **43**, 120-133.

Pokarzhevskii AD, Van Straalen NM, Butovsky RO, Verhoef SC and Filimonova ZV (1998) The use of detrital food-webs to predict ecotoxicological effects of heavy metals. In *Pollution-Induced Changes in Soil Invertebrate Food-Webs*, Butovsky RO and Van Straalen NM (eds), Vrije Universiteit, Amsterdam, pp. 9-30.

Posthuma L, Van Gestel CAM, Smit CE, Bakker DJ and Vonk JW (1998) *Validation of Toxicity Data and Risk Limits for Soils: Final Report*. Report No. 607505004, National Institute of Public Health and the Environment (RIVM), Bilthoven, 230 pp.

Reynoldson TB, Railey RC, Day KE and Norris RH (1995) Biological guidelines for freshwater sediment based on BEnthic Assessment of SedimenT (the BEAST) using a multivariate approach for predicting biological state. *Australian Journal of Ecology*, **20**, 198-219.

Römbke J, Beck L, Förster B, Fründ C-H, Horak F, Ruf A, Rosciczewski K, Scheurig M and Woas S (1997) *Boden als Lebensraum für Bodenorganismen und bodenbiologische Standortklassifikation — Literaturstudie [literature review report]*. Texte und Berichte zum Bodenschutz 4/97, Landesanstalt für Umweltschutz Baden-Württemberg (Karlsruhe), 390 pp.+Annex.

Ruf A (1998) A maturity index for predatory soil mites (Mesostigmata: Gamasina) as an indicator of environmental impacts of pollution on forest soils. *Applied Soil Ecology*, **9**, 447-452.

Sheppard SC, Gaudet C, Sheppard MI, Cureton PM and Wong MP (1992) The development of assessment and remediation guidelines for contaminated soils, a review of the science. *Canadian Journal of Soil Science*, **72**, 359-394.

Spurgeon DJ, Sandifer RD and Hopkin SP (1996) The use of macro-invertebrates for population and community monitoring of metal contamination — indicator taxa, effect parameters and the need for a soil invertebrate prediction and classification scheme. In *Bioindicator Systems for Soil Pollution*, Van Straalen NM and Krivolutsky DA (eds), Kluwer Academic Publishers, Dordrecht, pp. 95-109.

Von Törne E (1990) Assessing feeding activities of soil-living animals. I. Bait-lamina tests. *Pedobiologia*, **34**, 89-101.

Wiles JA and Frampton GK (1996) A field bioassay approach to assess the toxicity of insecticide residues on soil to Collembola. *Pesticide Science*, **47**, 273-285.

PART TWO

Risk Assessment Approaches for Contaminated Sites

11

Introduction

CLAUDE THELLEN[1] AND CONNIE L. GAUDET[2]

[1]Centre d'expertise en analyse environnementale, Ministère de l'Environnement du Québec, Québec, Canada
[2]National Guidelines and Standards Office, Environmental Quality Branch, Environment Canada, Ottawa, Ontario, Canada

11.1 BACKGROUND

During the past 30 years, contaminated site problems have become a major environmental issue among industrial countries. Corporate and governmental authorities rapidly recognized the necessity to implement appropriate site remediation and protection programs to reduce risk of human and ecological exposure to this contamination. It was also recognized that there were significant challenges in putting together scientifically sound information, appropriate management processes and cost-effective remediation action.

Part One of this book emphasized the many laboratory and field ecotoxicity tools that are now available to study contaminated sites and the potential effects on receptors. These tools complement chemical analyses and ensure more complete information on site contamination. Next, this information has to be assembled to respond to the importance and acceptability of contamination in terms of risk to human health and the environment. At the beginning, environmental studies related to these problems were perceived as 'scientific practice' for decision making. The first numerical soil quality criteria (the 'Dutch list') were proposed in 1983 as a management approach for standardization of requirements in soil protection and remedial actions. This approach was progressively adapted by many jurisdictions.

In parallel to generic approaches, the bases of site-specific concepts were developed from the 'NRC red book paradigm' (1983), which was first designed for human concerns. According to the NRC, risk assessment is the 'characterization of potential adverse effects of exposure to hazards, including estimates of risk and of uncertainties in measurements, and use of analytical techniques and interpretative models'. But this concept was not really adapted for ecological concerns and appeared disconnected from risk management. Different

Environmental Analysis of Contaminated Sites. Edited by G. I. Sunahara, A. Y. Renoux, C. Thellen, C. L. Gaudet and A. Pilon

technical protocols and methodologies were then proposed and tested for applicability in real situations. These have permitted development of national frameworks (for example USEPA 1992, CCME 1996), using common conceptual bases. An update of risk assessment concepts and issues is discussed in depth by Suter (1993).

Part Two of this book introduces the central issues and approaches in assessing risk at contaminated sites, from the use of generic soil guideline values to site-specific ecological risk assessment.

11.2 GENERIC AND SITE-SPECIFIC APPROACHES

Basically, assessment of contamination at a site deals with the problem of defining the level of contamination that poses no appreciable risk to the receptors at the site. This level then serves as the remediation goal or objective for a site, whether in terms of 'clean-up' to these levels or in terms of providing a basis for managing risks at the site. A number of tools have now been proposed to assess these risks. Two major categories are implicitly accepted among environmental professionals and jurisdictions: generic approaches based on benchmark values that describe an 'acceptable' or 'safe' level of contamination that is consistently applied across different sites, and site-specific risk assessment approaches that evaluate risk with respect to the specific receptor and exposure-related conditions at a site. In Chapter 12, Gaudet *et al.* discuss these two approaches, the relative strengths and limitations, and provide examples of how they are used within Canadian and American contaminated site assessment frameworks.

11.3 GENERIC SOIL GUIDELINES

In many countries around the world, there has been a growing reliance on use of generic soil guidelines (variously known as risk limits, intervention or action values, criteria, screening levels and standards) to assess contamination and recommend remediation objectives. The scientific basis of these generic guidelines, as well as the risk management frameworks within which they are applied, have evolved considerably to enable site managers much more flexibility than in the past where 'one number' approaches were the norm. Canada, for example, recommends numbers for four different land uses, and for a number of distinct exposure pathways within these land uses, while The Netherlands recommend a series of values that incorporate, for example, different levels of protection, background levels for naturally occurring substances, or economic and technological considerations. Expanding on the introduction to generic soil guidelines in Chapter 12, Ouellet *et al.* (Chapter 13) provide an in-depth discussion of the development, ecotoxicological basis and future directions for soil quality guidelines using the Canadian and Netherlands approaches as examples.

11.4 RISK-BASED REMEDIATION

The relative merits of generic guidelines-based and site-specific risk assessment-based approaches depend often on the situation at hand. In many cases, a detailed site-specific risk assessment is needed to delineate the receptors of concern at a site, and the nature and extent of exposure to contamination as a basis for risk-based remediation. Generic guidelines reduce receptor and exposure characteristics to a single uniform value of general applicability across a range of sites, and consequently lead to similar remediation effort across sites, whereas site-specific risk assessment enables the use of risk-based remediation strategies targeted at reducing the actual quantifiable risk at that site. As discussed in Chapter 14 (Trépanier *et al.*), for large or complex sites this is important in ensuring effective allocation of limited resources in protecting human health and the environment. As described in this chapter, risk-based remediation may include, for example, capping contaminated soils to limit exposure and reduce risk to acceptable levels rather than more expensive excavation or treatment of soils. Chapter 14 also provides a comprehensive overview not only of the scientific basis for risk-based remediation but also of the key issues and challenges in applying this approach, and key methods to improve the process.

11.5 SITE-SPECIFIC ECOLOGICAL RISK ASSESSMENT

In the 1980s, site-specific risk assessment tools were perceived as a 'black box' for managers and stakeholders involved with contaminated sites problems. It was evident that these approaches had to be more clearly linked to the 'non-scientific' issues and concerns. In response, risk assessment approaches have evolved greatly in the last five years. It is now becoming a 'multi-partner' process.

The challenges of assessing and managing ecological risk at contaminated sites have scientific, socioeconomic and political dimensions. The scientific dimension, usually considered as the ecological risk assessment (ERA) phase of a risk management approach, is in itself daunting: characterizing contamination, describing the array of receptors at a site, the ways that they may be exposed to contamination at a site and their different sensitivities to this contamination are all essential parts of the ERA. Final decisions will also be affected by the relative importance or value placed on species at risk and the level to which they will be protected. Martel *et al.* (Chapter 15) describe the fundamental approaches, steps, tools and challenges in assessing and communicating ecological risk at a site, dealing with uncertainty and the importance that ecological risk assessment plays in providing an objective basis for managing risk.

In the concluding chapter of Part Two, Menzie (Chapter 16) takes the reader through the evolution of ecological risk assessment from its beginnings over a decade ago to its emergence as a mainstream tool for environmental decision making. Following on the preceding chapter that emphasizes the link of risk assessment as a technical approach with the more non-technical aspects of risk communication and decision making, Menzie moves us through the changing social and political environment that has influenced the evolution of ecological risk assessment, emphasizing the context within which risk assessment is applied. He reminds us that risk assessment is a tool, and not an end in itself. Despite major advances in the 'science' of risk assessment, only by understanding the linkages of risk assessment within a broader risk management framework will its role as a key decision support tool be fully effective. Harmonization is still necessary to reduce sources of inconsistencies at the technical, legislative and communication levels, as described by Kamrin (1997). There are also needs for coordination between jurisdictions and between contaminated site studies. The case studies in Part Three provide many examples of the overlapping scientific, social, economic and political dimensions of managing risk at contaminated sites.

REFERENCES

CCME (1996) *An assessment of the risk assessment paradigm for ecological risk assessment*. Commission on Risk Assessment and Risk Management. Prepared by Menzie-Cura & Associates Inc., 50 pp.

Kamrin MA (1997) *Environmental Risk Harmonization: Federal and State Approaches to Environmental Hazards in the USA*. Ecological and Environmental Toxicology Series, Kamrin MA (ed), John Wiley & Sons, New York, 308 pp.

NRC (1983) *Risk Assessment in the Federal Government: Managing the Process*. Committee on the Institutional Means for Assessment of Risks to Public Health, Commission on Life Sciences, National Academy Press, Washington, DC, 191 pp.

Suter GW (1993) *Ecological Risk Assessment*, Lewis, Boca Raton, FL, 538 pp.

USEPA (1992) *Framework for Ecological Risk Assessment*. EPA/630/R-92/001, United States Environmental Protection Agency, Washington, DC.

12

Risk-based Assessment of Soil Contamination: Generic versus Site-specific Approaches

CONNIE L. GAUDET[1], CHARLES MENZIE[2] AND SYLVAIN OUELLET[3]

[1]National Guidelines and Standards Office, Environmental Quality Branch, Environment Canada, Ottawa, Ontario, Canada
[2]Menzie-Cura & Associates, Chelmsford, MA, USA
[3]Strategic Planning and Policy Branch, Environment Canada, Ottawa, Ontario, Canada

12.1 INTRODUCTION

As the 'how clean is clean' controversy erupted in the 1980s, the world was divided into whether to use site-specific risk assessment or numerical guidelines in assessing risk of contaminants in soil and setting remediation targets (see, for example, discussion in Gaudet *et al.* 1992). A decade ago, the dichotomy seemed irreconcilable; precision versus practicality; consistency versus accuracy; regulatory efficiency versus detailed scientific advice; clean-up dollars versus investigation dollars. Even today, with enormous advances in scientific development in establishing contaminated soil remediation objectives, it seems there is still no clear answer as to the 'best' approach. Jurisdictions around the world vary from the use of generic guidelines, to merged approaches, to reliance on comprehensive site-specific risk-based approaches.

Generic or absolute values are usually promulgated by regulatory agencies to promote consistency in, and improve efficiency of, decision making within jurisdictions. Because guidelines embody fundamental principles such as recommended levels of protection and recognized uses of soil, they promote consistency of decisions and provide a level playing field for industry and the public. They are also easily understood and articulated as a basis for decisions (see, for example, Gaudet *et al.* 1995). Recognized limitations include the inability of single-substance guidelines to deal with complex mixtures of substances or to address unique site-specific circumstances such as the presence

Environmental Analysis of Contaminated Sites. Edited by G. I. Sunahara, A. Y. Renoux, C. Thellen, C. L. Gaudet and A. Pilon

of sensitive species or biophysical characteristics of the site that would modify bioavailability of contaminants (see, for example, discussion in Environment Canada 1991; also Hrudey and Pollard 1993). Use of generic criteria may also lead to limitations in dealing with contamination using targeted risk management as discussed later in Chapter 14.

Site-specific risk assessment, though tailored to a site, can be expensive and time-consuming, may involve the adoption of different principles and endpoints for protection across sites within the same jurisdiction, and may be poorly understood by the public and industry alike. On the other hand, site-specific risk assessment is normally considered more realistic and scientifically defensible and can avoid costs and liabilities associated with both over and undermanaging contamination at a site.

Having said this, is the picture really as confused as it sometimes seems? Yes, science is still evolving and there is no single universally endorsed approach to assessing risk of contaminants in soil and in recommending 'clean-up' or remediation goals. But this does not mean that the regulatory and scientific community is in chaos. Far from it. Though many different approaches do exist, guiding principles are converging to recognize key principles for protecting soil as a multi-functional and complex resource that poses risk to both human and ecological health and the major beneficial uses of soil (see, for example, discussion in Sheppard *et al.* 1992). Perhaps most importantly, the risk assessment–risk management paradigm is becoming a uniting scientific force behind virtually all approaches.

At a recent workshop addressing the role of generic benchmarks in assessing environmental risk (Fisheries and Oceans Canada 1999), it was generally agreed that different science-based 'tools', from guidelines to site-specific risk assessment, may be appropriate given particular environmental, economic, technological or socio/political considerations. Though proponents for both site-specific and generic approaches strongly supported their respective approaches, it was clear that it was not the tools themselves but the inappropriate use and understanding of these tools that was problematic. What is necessary is that practitioners and the public understand the inherent scientific basis, as well as the strengths and weaknesses of each approach, in reaching an effective and environmentally supportable decision. What this means in practice is that as a result of the rapid evolution in risk assessment, including the development and application of generic soil values, the toolbox for assessment of contaminated sites now offers an impressive arsenal of effective tools that can be used flexibly to assess and mitigate environmental risk. Even where site-specific risk assessment is strongly promoted, as through the USEPA Superfund Program, the value of generic soil screening values (EcoSSLs) is now recognized as important to streamlining and increasing consistency of the assessment process (USEPA 2000).

The Canadian Council of Ministers of the Environment (CCME) framework for contaminated site assessment and remediation is discussed briefly in this chapter, as one model for risk-based soil assessment that integrates the strengths of both generic and site-specific approaches. Later chapters highlight the generic guidelines and site-specific risk-based approaches respectively.

12.2 SOME BASIC DEFINITIONS

12.2.1 SOIL QUALITY CRITERIA AND GUIDELINES

Generic soil quality values may be variously referred to as guidelines, criteria, standards or benchmarks dependent on the jurisdiction. Ecologically based guidelines normally specify levels of a substance in soil that is considered to present no appreciable risk to soil-associated biota. They are considered 'generic' because a single value, or set of values, based on generic exposure scenarios and default assumptions is recommended as applicable across different sites and soil conditions. Due to differences across jurisdictions in terms of receptors to be protected, ecologically relevant endpoints, level of protection to be afforded, available data and the uses of soil to be considered, soil quality guidelines can vary broadly. Despite these differences, most soil quality values now incorporate rigorous data quality and methodological standards, and consider a range of receptors and exposure pathways. Prior to the mid-1970s, the few criteria that had been established usually addressed simple, long-recognized effects on plants or grazing animals. For example, maximum concentrations were set by some jurisdictions to prevent adverse effects of acute selenosis (blind staggers) in grazing animals or copper toxicity to plants (Environment Canada 1991). By the 1980s many regulatory agencies were faced with the need to establish decommissioning or clean-up criteria, and began to develop more comprehensive soil criteria that considered a broader range of factors. Chapter 13 describes two of these approaches as examples of the scientific rigor now characteristic of the development of generic soil quality values.

More recently, a range of values known as toxicity reference values (TRVs) or screening benchmarks have been developed in some jurisdictions. These values provide a standard benchmark for the toxicity of substances to specified receptors and lend consistency in early assessment of the potential risk at a site. They differ from the guideline or soil quality values described briefly above and in detail in Chapter 13 in that they are not normally promulgated as generic standards or guidelines by a regulatory body, but are recommended as scientific benchmarks for use in site-specific risk assessment. Confusion arises because the generic guidelines are sometimes used in screening level risk assessments interchangeably with the TRVs. In fact, the toxicological basis for TRVs and guidelines can often be similar. The difference may be seen more as one of political endorsement of both the scientific data and underlying assumptions (e.g., level of protection, soil uses to be protected) rather than their scientific basis.

12.2.2 SITE-SPECIFIC OBJECTIVES

Recently, some generic approaches have expanded to include comprehensive guidance on how to incorporate site-specific factors in modifying generic soil values as remediation objectives. This flexibility enables managers to maintain the efficiency and transparency of generic guideline values while adapting to the unique characteristics of a site. The following section discusses this type of framework in more detail.

12.2.3 SITE-SPECIFIC RISK ASSESSMENT

Risk assessment traditionally refers to a range of approaches that are based on a unique characterization of ecological and/or human health risks at a particular site, dependent not only on scientific factors (e.g., receptors, exposure pathways) but also often on management factors related to the issues, public concerns and management endpoints relevant to a particular site. Though the use of generic guidelines can be seen as part of a 'risk-based approach', as described following the CCME framework, normally when risk assessment is referred to it implies detailed site-specific assessment of the risks at a site as described , for example, in Environment Canada (1991) and Hrudey and Pollard (1993).

Though there is no recipe book for risk assessment, there are several guiding frameworks with a common set of features, including, typically, problem formulation, exposure, hazard assessment and risk characterization (see Chapters 14 and 15). More recently, what might be called 'generic' risk-based approaches have evolved such as the ASTM's Risk-based Corrective Action (RBCA) for petroleum contaminated sites (ASTM 1998). RBCA defines basic models and assumptions to guide the risk assessment through three successive tiers of increasing site specificity. Interestingly, RBCA tiers closely mirror the tiers of the CCME framework discussed in the following section. The key difference being that the CCME recommends numerical guidelines in Tier 1 based on generic default assumptions and a standardized exposure model, whereas RBCA does not, in its current form, provide actual numerical guidelines at Tier 1. Current work on petroleum hydrocarbon standards for use in Canada, to be published by the CCME in 2001, has merged features of the RBCA and the CCME approach. The Canadian framework for petroleum hydrocarbons incorporates many of the features of RBCA in terms of the models used to calculate exposure to humans, but provides Tier 1 'look-up tables' (i.e., generic guideline values) and incorporates four distinct land uses (agricultural, residential, commercial and industrial). The CCME model also incorporates values for ecological receptors, a consideration not yet incorporated into RBCA.

12.3 CANADIAN CONTAMINATED SITES FRAMEWORK

The application of generic guidelines and/or risk assessment is normally based on frameworks promulgated within jurisdictions (see, for example, Visser

1993, Nason *et al.* 2001). Increasingly, frameworks are characterized by a tiered approach[1] that incorporates, to varying degrees, both site-specific risk assessment and guidelines. Such tiered frameworks have grown in popularity because they offer a suite of tools that can be flexibly applied in reaching decisions. The CCME framework is one of the earliest examples of such a framework.

The CCME framework provides for a phased acquisition and assessment of site information along either a guidelines-based or site-specific risk assessment-based process (see Figure 12.1). In the general case, the site evaluation procedure begins with consideration of a number of sites that may be candidates for remediation. Using the site CCME National Classification System (CCME 1992), the site manager can quickly sort sites into priority classes and turn her/his attention to the most urgent group. An appropriate sampling and analysis program is then carried out to generate a site characterization data set. These data can then be compared to soil quality guidelines (as described in Chapter 13) to decide if further action might be necessary. Should this be the case, a decision is then made to proceed with a guidelines-based approach or a site-specific risk assessment approach. Details of these options are provided below.

12.3.1 GUIDELINES-BASED APPROACH

In the development of CCME soil quality guidelines, great potential exists for incorporation of site-specific information in the development of site-specific remediation objectives. The CCME process of developing site-specific objectives is, in fact, opening middle ground between the traditional guidelines-based and site-specific risk-based approaches and allowing the development of efficient regulatory decision support frameworks that incorporate the strengths of both approaches while avoiding many of their weaknesses. The 'Guidance Manual for Developing Site-Specific Soil Quality Remediation Objectives in Canada' (CCME 1996b) was written to provide site managers and regulators with consistent direction on application of generic guidelines and other scientific tools of the NCSRP. The guidance manual provides direction on how to choose among the three methods of developing site-specific objectives (SSOs): (1) adoption of guidelines as objectives; (2) adaptation of guidelines as objectives; and (3) development of remediation objectives using site-specific risk assessment techniques. Most of the guidance manual deals with collection and evaluation of information necessary to make a consistent and defensible choice between guidelines-based and risk-based approaches, along with adaptation procedures within the guidelines-based approach.

[1] Risk assessment frameworks may also be tiered, moving from screening level assessments to detailed quantitative assessment (e.g., CCME 1996c). The tiered framework referred to here is a broader framework for site assessment and management that takes into account both generic guidelines and site-specific risk assessment approaches.

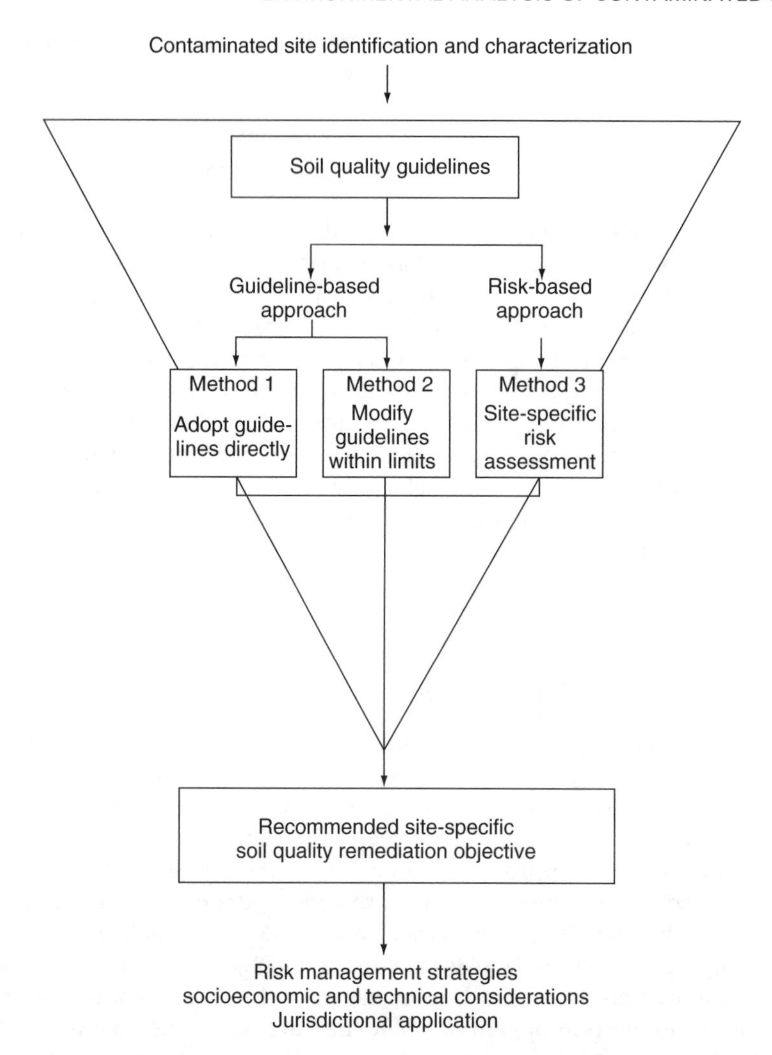

Figure 12.1 Canadian framework for assessment and remediation of contaminated sites (from CCME 1999).

The site-specific objective development process begins with a consideration of the site characterization data set. Experience has shown that it is most efficient to compare these data straight away to any background data that might exist nationally or for the land use or region under consideration. The purpose of this step is to ensure that contaminant levels at a site are not simply a reflection of local or regional background levels of contaminants. If contaminant concentrations do not exceed benchmarks established as indicative of background levels, the investigation can stop. Alternatively, if land use-based backgrounds are exceeded, the proponent may elect to remediate to this background. This is not often done in well-characterized landscapes, but has been used in geologically unusual or ecologically sensitive environments. More often, evaluation proceeds through a review of site, receptor and exposure conditions to assess whether there are any intrinsic conditions that would trigger a more comprehensive site-specific risk assessment. Five such triggers have been identified by the CCME:

- Unusual or sensitive receptors (i.e., endangered or threatened species or people in disease states);
- Unusual or sensitive site conditions (i.e., critical habitat or special land status);
- Unusual or sensitive exposure conditions (i.e., situations where landscapes are unstable or receptor use patterns differ significantly from generic concepts used to develop guidelines);
- Unacceptable data or knowledge gaps (i.e., especially uncharacterized toxic mixtures or chemicals whose fate, behavior or toxicology is poorly understood);
- Guidelines do not exist or complex mixtures must be addressed.

Under the CCME framework, assuming none of these conditions triggering a detailed risk assessment are present, assessment proceeds to use of the soil quality guidelines. Where recommended CCME guidelines exist, they may be adopted directly or modified in certain ways to produce remediation objectives. These limited modifications are based on differences between the generic exposure scenario assumed during derivation of the guidelines (see for example Chapter 13) and the site-specific conditions. Because guidelines are constructed in stepwise fashion, using building blocks or discrete calculations for each of the direct and indirect exposure pathways by which a receptor may be exposed to a soil contaminant (e.g., direct soil contact, soil ingestion, inhalation of indoor air, food ingestion), it is possible to identify the principal factors that affect the final recommended soil guideline value and compare these to actual site conditions. Where conditions at the site vary significantly from the generic default conditions used in deriving the guideline value, the models may be recalculated to more accurately reflect site conditions. For example, the distance to the groundwater table is a key parameter in modeling the exposure of humans to soil-associated contaminants via drinking water.

In developing a generic soil guideline, a default value for distance to an aquifer is used to estimate the extent of groundwater contamination likely to be associated with a level of a contaminant in soil. In setting site-specific objectives, the distance to groundwater may be reset to reflect actual site conditions, such that the final site-specific objective may be lower or higher than the generic soil guideline value. Other modifications may be based on differences in receptors (particularly for ecological receptors) and physical/chemical characteristics of soil that would alter the movement and/or availability of a contaminant (e.g., for pentachlorophenol (PCP), the pH of the soil is a major predictor of movement to groundwater). Such modifications are considered science-based in that they are intended to lend further realism to the estimate of an acceptable level of soil contamination. In more extreme cases, certain exposure pathways or receptors may be removed entirely in calculating a site-specific objective. Without careful consideration of site conditions including present and future uses of the land, such modifications can move into the realm of management-based rather than science-based decisions. The key consideration is whether such conditions are both verifiable and expected not to change in light of the long-term uses of the land.

For example, soil organic matter levels, which strongly influence movement of organic contaminants to groundwater or into the food chain through plants, can be verified and adjustments made through the simplified models that describe cross-media transfers in order to recalculate a site-specific objective. Other conditions such as absence of potable groundwater, actual or potential species present are also verifiable, and to some degree predictable, and can form a basis for a regulatory variance under the guidelines-based approach.

CCME does not recommend adjustment of guidelines on the basis of difficult-to-observe or unstable conditions. For example, restricting the ecotoxicological database to one or two plant species presently of interest might appear practical and desirable. However, changes in economic or social conditions could lead to the need to support other species of higher sensitivity, thereby invalidating the objectives-setting decision.

The proponent can benefit from investigating modification of guidelines only where (s)he expects to gain some economic or technical advantage. If no technical or economic means exist to achieve even an adjusted guidelines-based objective, a risk assessment must be done. In this case risk assessment is said to be triggered by extrinsic factors. Often this will lead to nomination of a risk-based remediation objective that entails long-term monitoring and management of the site. Despite the fact that long-term monitoring and liability are disincentives to risk assessment/risk management, many of the worst sites in Canada are risk-managed because appropriate remedial technologies have not yet been developed.

CCME Soil Quality Guidelines are recommended for protection of both human health and the environment. As the models used in development of these guidelines are very different, the types of modifications that may be appropriate in setting site-specific objectives will vary significantly. CCME guidance has been developed to assist in this interpretation (CCME 1996a,b). Also, a matrix of values derived for the different exposure routes evaluated during guideline development is provided with each recommended guideline (CCME 1999). For example, in developing a soil quality guideline for human health protection, soil ingestion and inhalation of indoor air, drinking water and food ingestion pathways may all be evaluated. Though the recommended CCME guideline is based on the lowest of these values (i.e., the limiting pathway), all of the values are published to enable more flexibility in application of the guidelines-based approach.

12.3.2 SITE-SPECIFIC RISK ASSESSMENT

The CCME framework permits the use of risk assessment in the development of site-specific objectives (CCME 1996b) where triggered by intrinsic or extrinsic factors as discussed in the preceding section. If these triggers are not present, CCME recommends use of the generic, guidelines-based approach. Guidance for ecological risk assessment was prepared specifically for Canada (CCME 1996c). The framework consists of four modules: receptor characterization, hazard (or toxicity) assessment, exposure assessment and risk characterization. With planning, this framework can be applied at any of three levels of effort depending on site requirements and conditions. The framework is intended to provide the information necessary to make a good site management decision.

12.4 UNITED STATES RISK ASSESSMENT FRAMEWORKS

Although the use of generic guidelines is growing, especially as part of integrated frameworks such as that proposed by the CCME above, the use of site-specific risk assessment is also evolving. Whether viewed from the perspective of a site-specific risk-based approach that may incorporate generic guidelines as part of screening level assessment, or from the perspective of a generic guidelines-based approach that may incorporate risk assessment in the final stages, the increasing reliance on both of these tools as part of a coherent decision-making framework is apparent from the following overview of evolution of risk assessment guidance in the United States.

It is worth noting that the use of ecological and human health risk assessment in contaminated site assessment is more actively pursued in the United States than in Canada. Though frameworks are considered site-specific risk-based frameworks, they often incorporate guidelines or other effects-based benchmarks, such as the TRVs discussed earlier in this chapter, in early screening stages of the risk assessment. The following discussion of US risk-based approaches provides an important balance to the Canadian and Netherlands guidelines-driven approaches outlined in Chapter 13, and the CCME site assessment framework discussed in Section 12.3.

12.4.1 US VARIATIONS ON THE RISK ASSESSMENT APPROACH

Guidance for assessing ecological risks at hazardous waste sites and for other environmental problems has been developed by various agencies and departments of the US government and by numerous states. In addition, the ASTM

RBCA Workgroup for ecological risks has completed a review of much of the state guidance. We draw upon these sources to identify the most important guidance and to provide a summary of the major trends taken in approaching ecological risk assessment in the United States.

Primary guidance documents. The two most widely referred to documents when conducting ecological risk assessments at hazardous waste sites in the United States are the Ecological Risk Assessment Guidance for Superfund (USEPA 1997) and Guidelines for Ecological Risk Assessment (USEPA 1998). These two documents are interrelated in that Guidance for Superfund is structured to be consistent with the US Environmental Protection Agency's Framework for Ecological Risk Assessment (USEPA 1992). The Guidelines for Ecological Risk Assessment expand upon the framework and describe the logic and thought needed to apply it to ecological problems. Menzie and Freshman (1997) have discussed how the framework document has influenced the development of other national and state risk guidelines. While the risk assessment processes are becoming more harmonized, the details of the application can vary within and among EPA regions.

Trends in guidance. Our review of trends is drawn largely from the work of the ASTM RBCA Workgroup on Ecological Risk Assessment. In addition to the US national guidance, the workgroup also reviewed state-specific guidance for 17 states. Most US state guidelines are based on either the Ecological Risk Assessment Guidance for Superfund or the Framework for Ecological Risk Assessment and subsequent Guidelines for Ecological Risk Assessment. Some states have adopted features developed in the ASTM RBCA framework.

Screening criteria. Screening procedures are a common element in existing guidance, though state guidance tends to have more explicit early screening level procedures than does the federal guidance. This is probably due to the fact that federal sites and those designated as Superfund sites tend to be large and problematic. As a result, less attention has been given to the use of screening criteria that could exclude a site from further consideration. On the other hand, states need to address many small sites including those located in urban areas. Screening criteria become very important for addressing small and/or urban sites.

While state screening procedures vary, the following seem common:

1. Land use considerations;
2. Determining if receptors and pathways are present;
3. Screening against benchmarks.

Interestingly, in contrast to the CCME framework that starts with generic guidelines and moves to site-specific risk assessment as necessary, the US frameworks largely start with a site-specific risk assessment focus but may use generic guidelines as screening benchmarks as in point 3 above.

Protected receptors. Many states have provided guidance on 'what is important to protect'. Most commonly guidance identifies: (a) rare, threatened or endangered species; (b) special habitats including wetlands or areas of critical environmental concern; (c) receptors recognized as commercially or recreationally important; and (d) ecologically important species. States have developed procedures for determining if ecological receptors and habitats are located at or near the site.

12.4.1.1 Tiered Assessment Approaches

Most guidance supports a tiered assessment approach (see Table 12.1). The number of tiers ranges from two to five. Two tiers are most common with three tiers being the next most common. Though Table 12.1 shows that there is a broad range in approaches, most have an initial screening phase similar to that of the CCME framework described in Section 12.3, though the choice of screening benchmarks varies.

Tier 1 screening benchmarks. Most states and national guidance employs screening benchmarks (i.e., chemical concentrations in environmental media that are considered to pose little or no significant risk to receptors of concern) in

Table 12.1 The use of tiered assessment procedures by US state agencies

ALASKA	**Two tiers**: (1) a conservative screening assessment is performed during scoping using existing documentation and literature values for estimating hazard/effects; (2) a second more detailed assessment is performed if risks are indicated or if bioaccumulative substances are COCs
CALIFORNIA	**Four tiers**: (1) scoping assessment; (2) predictive assessment; (3) validation study; (4) impact assessment
ILLINOIS	**Three tiers**: Tier I is a look-up table (i.e., generic benchmark values); Tier II allows the user to define site-specific inputs and to recalculate objectives by incorporating them into agency-provided equations; Tier III allows the use of site-specific information and alternative models for estimating exposure
INDIANA	Not specified although human health is three tiers
LOUISIANA	**Three tiers**: Tier 1 ERA checklist; Tier 2 screening level ERA; Tier 3 baseline ERA.
MASSACHUSETTS	**Two tiers**: the guidance is structured around a Stage 1 and Stage 2 process
NEW JERSEY	**Two tiers**: Tier 1 is the baseline ecological evaluation (BEE) which is the screening process of the ERA and is required for all SRP sites; Tier II level is required when the results of the BEE indicate the realistic potential for ecological risk at the site
NEW YORK	**Five tiers**
OHIO	**Two tiers**: the Tier I assessment evaluates whether the concentrations of chemicals of concern are equal to or less than the applicable standards for surface water, sediment, soil, groundwater (standards are similar in intent to CCME guidelines) or background levels and whether there are any complete exposure pathways for receptors; the Tier II assessment conducts additional sampling and analysis if contaminant levels are above applicable standards or background levels. An exposure and risk assessment may be performed
OREGON	**Four tiers**

(*continued overleaf*)

Table 12.1 (*continued*)

PENNSYLVANIA	**Three tiers**: (1) basic screening; (2) a more detailed step requiring a site visit; and (3) a site-specific evaluation to determine impact to receptors
TEXAS	**Three tiers**: **Tier 1**, ecological assessment checklist to characterize the ecological setting and identify potentially complete exposure pathways; **Tier 2**, screening level assessment intended to identify any contaminants at a site which pose an ecological risk and to eliminate those that do not (includes three levels of assessment); **Tier 3**, quantitative ecological risk assessment to develop scientifically defensible, site-specific clean-up levels that are protective of ecological receptors. This tier is modeled after the USEPA (1998) document with the exception that in the EPA document risk management can be applied throughout the ERA process
WASHINGTON	**Three tiers**: a checklist is given in each tier for the purpose of an environmental evaluation. (1) Tier I, the initial stage in an environmental evaluation for soil contamination. In this stage, a site is screened to determine whether it qualifies for an off-ramp that requires relatively little information. (2) Tier II, the second stage in an environmental evaluation for soil contamination for use at sites that do not meet any of the Tier I off-ramps. In this stage, the site is screened to determine whether it qualifies for any of the additional off-ramps provided. Tier II off-ramp criteria require more information than those in Tier I. For sites that do not qualify for any of the off-ramps, acceptable methods for evaluating the ecological risks to terrestrial species are provided, or the evaluation can proceed to Tier III. The Tier II evaluation stage is not intended for use at some sites, where the evaluation should instead proceed from Tier I to Tier III. (3) Tier III, the third stage in an environmental evaluation for soil contamination. In this stage, a conventional site-specific ecological risk assessment is conducted. For sites under a Consent Decree or Agreed Order, the risk assessment should be planned and conducted in collaboration with an ecology risk assessor
WEST VIRGINIA	**Three tiers**: (1) *de minimis* screening; (2) uniform risk-based screening; and (3) site-specific ecological risk assessment
WISCONSIN	A tiered approach is presented for the effects assessment. Tiered testing is recommended for toxicity testing and biological community evaluations

Tier 1 assessments. Table 12.2 summarizes the basis for screening benchmarks. Dependent on the media, these screening level benchmarks may be considered more or less acceptable as follows.

- **Water** — benchmarks have high state and federal regulatory acceptance. Benchmarks are either state or federal water quality criteria[2] or are derived using similar methods. The scientific basis for deriving these criteria are well established and recognized. They are derived to be protective of most kinds of aquatic biota including plants, invertebrates and fish.
- **Sediments** — benchmarks have moderate state and federal regulatory acceptance. Benchmarks are derived from a variety of methods, each of which have strengths and limitations. Most available benchmarks are designed to be protective of invertebrate animals living on or within the

[2] Standards, criteria and guidelines are used interchangeably in this chapter. In Canada, generic benchmark values for soil quality, water quality and sediment quality are termed 'guidelines'; in the US they may be referred to as criteria or standards.

Table 12.2 Use of benchmarks by various US state agencies

ALASKA	NOAELs (no observed adverse effects levels) are preferred as the basis for screening toxicity reference values. The maximum concentration at the site is used in calculating the estimated environmental concentrations. A hazard quotient approach based on screening exposure assumptions is used to select contaminants of concern
MASSACHUSETTS	A chemical benchmark approach is used in Stage 1 for screening. Benchmarks may also be used in Stage 2. The state water quality criteria are used as explicit benchmarks for judging risk
NEW JERSEY	During Tier I, maximum measured contaminant concentrations are to be compared to based benchmarks or screening values, using a 'weight-of-evidence' approach
NEW YORK	Technical Guidance for Screening Contaminated Sediments and Division of Water Technical and Operational Guidance Series 1.1.1 Ambient Water Quality Standards and Guidance Values contain sediment and surface water quality criteria which have been established as screening tools for the identification of contaminated sediments and surface water. These quality criteria are intended to serve as tools in a preliminary assessment of the toxicity of contaminants
OHIO	A chemical benchmark approach (applicable standards, background and results of site-specific sediment and surface water testing) is included in the Phase II assessment
OREGON	During Level II, maximum detected concentrations compared with applicable criteria and background concentrations
PENNSYLVANIA	1/10 of standards for sediment, surface water and groundwater as criteria. They are not species-specific
TEXAS	At Tier 2 (Levels A & B) site values are compared to established state standards, federal guidelines and other benchmarks to identify COPCs (chemicals of potential concern)
WEST VIRGINIA	Appendix E of the guidance contains several *uniform benchmark standards*. The risk assessor may also choose other benchmarks using a weight-of-evidence approach

sediments. Some are based on state or federal water quality criteria modified to account for partitioning to sediments, while others are based directly on effects on sediment-associated biota.

- **Soils** — benchmarks have low state and federal regulatory acceptance. While some states have adopted soil benchmarks, most recognize the difficulty of deriving benchmarks designed to be protective of 'valued ecological entities'. The difficulties stem from the heterogeneity of soils, lack of a generally accepted method and differences of opinion over what receptors to protect with the benchmarks. However, there is a current federal initiative to develop ecologically-based soil screening levels for use federally (EcoSSLs). The initiative was based on the stated need (USEPA 2000) for peer-reviewed benchmarks against which site data could be screened in order to streamline the risk assessment process. Contaminants found at concentrations below these accepted benchmarks could be removed from further evaluation, whereas chemical concentrations above these accepted exposure levels could be evaluated further in the baseline ecological risk assessment (USEPA 2000). The USEPA expects to develop a consensus-based guidance that will present ecological soil screening levels (EcoSSLs) for 20–25 of the chemicals that are often found in soil at hazardous waste sites at levels that could cause an ecological risk. A technical supporting document will also be prepared presenting the data, models, equations, etc. and the rationale used to estimate these EcoSSLs.

Toxicity reference values. Ecological risk assessments often depend on TRVs to estimate risks to wildlife. TRVs are typically derived from the literature. Uncertainty factors or scaling factors are used to extrapolate these data to various wildlife species. A number of the guidelines include discussions concerning how to select toxicity values and how to apply uncertainty factors. Examples include Alaska, California, Illinois and New Jersey. A few states allow for the use of other methods and seem to encourage innovation.

Contaminant exposure. About one-half the guidance documents include information on how to consider exposure. While some of the guidance documents support the use of food chain models, very little specific guidance is given. Use of fate and transport models is frequently included in the guidance documents. Although exposure involves both spatial and temporal dimensions, it appears that existing guidance provides little useful information on this.

Level of risk. State guidance is generally vague on what constitutes a 'significant ecological risk'. Some states consider risks to be present when a hazard quotient exceeds '1'. A few use numerical standards as a basis for determining risk. Most guidance documents support either comparison to benchmarks or simple hazard quotient approaches. Some of the guidance documents go further and include weight-of-evidence approaches within higher tiers. Examples of how states evaluate risk are given in Table 12.3.

12.4.2 AREAS FOR FUTURE GUIDANCE DEVELOPMENT

Based on a review of available US guidance, there are a few areas where additional guidance is needed and where there are efforts currently underway to address these needs. Several questions that new guidance could address include:

- What constitutes a 'significant ecological risk' at the individual and population level? For the purpose of site management, how should these be characterized and presented to the risk manager?
- How should ecological information be integrated with other elements of site management including (1) human health considerations and (2) potential risks (impacts) associated with remedial decisions?

Table 12.3 Examples of how states evaluate risk

ALASKA	Only the quotient/benchmark approach. A weight-of-evidence approach is not supported in this manual
CALIFORNIA	In Phase I, hazard quotients and indices are calculated. Detailed assessment techniques are allowed beyond the screening level assessment (Phase I) in Phase II
ILLINOIS	The guidance generally relies on the quotient method and benchmark approaches in Tier I and Tier II. Field assessment is supported in Tier II and weight-of-evidence and other methods appear to be supported in Tier III
INDIANA	Use of a professional experienced in performing ecological risk assessments
MASSACHUSETTS	Exceedance of water quality criterion is considered to reflect risk of harm. Quotient methods are commonly used. Finally, a procedure is presented on the use of a weight-of-evidence method
NEW JERSEY	Maximum measured contaminant concentrations are compared to ecotoxicological-based benchmarks or screening values during Tier I. If the measured concentration exceeds the benchmark, Tier II assessment may be warranted. Tier II must be conducted in accordance with EPA guidance. One can assume that the use of quotient method, field assessment and weight-of-evidence would be supported
NEW YORK	Contaminant-specific and site-specific criteria are used to determine the impact on fish and wildlife resources. Criteria-specific analysis uses numeric criteria for contaminants of concern that have been established for specific media or biota
NORTH CAROLINA	Risks exist if numerical standards are exceeded. Specific methods are not specified
OHIO	The guidance generally relies on the quotient method and benchmark approach. Field assessment is supported for assessing risks to aquatic organisms. Qualitative assessments are acceptable for aquatic habitats that do not lend themselves to quantitative evaluation (shallow water bodies)
OREGON	Uses maximum detected concentrations compared to ecotoxicological-based benchmarks or screening values during Level II
WASHINGTON	Utilizes benchmarks and is flexible on the application of other methods
WEST VIRGINIA	The guidance relies on the benchmark approach as well as the weight-of-evidence and site-specific approaches
WISCONSIN	The guidance recognizes that a wide variety of approaches can be used to develop the weight of evidence needed to allow risk managers to make informed decisions

- How should 'recovery' be characterized under no action, limited action or following remediation? How does the consideration of temporal scales associated with recovery impact site management decisions?
- What tools can be identified for communicating information to managers and remedial engineers?

12.5 SUMMARY

As the array of site assessment frameworks from Canada and the United States attests, there is no single universally accepted approach to assessing the risk of contamination at a site. There are however some striking consistencies across the approaches.

- Most approaches are based on tiered risk assessment that begins with generic screening level assessment and proceeds to more detailed site-specific assessment where required. This tiered approach is increasingly seen as maximizing the effectiveness and efficiency of assessments, while merging the strengths of both generic and detailed site-specific approaches.
- Though the use may vary, screening level benchmarks are now a consistent feature of most contaminated site assessment frameworks. These benchmarks, whether referred to as guidelines, TRVs, literature values, criteria or standards, all have a common purpose to provide an estimate of the concentration of a chemical in soil, or other media, that poses no appreciable or unacceptable effects on receptors of concern (ecological or human). Where the concentrations of a chemical at a site exceed these benchmarks, there is a potential risk and further action is warranted. These generic, effects-based values lend efficiency, consistency, transparency and predictability to the process.
- At the other extreme, almost all frameworks accommodate site-specific risk assessment approaches. These approaches lend increased realism and precision to estimates of risk at a site.
- The Canadian (CCME) approach is more overtly 'guideline-based' than most US approaches in that the guidelines can be used as the basis for remediation objectives for a site. In the US, the guidelines, or equivalent values, are used for initial screening of potential risk, primarily to determine the need for further action.
- Site-specific risk assessments can be technically demanding and streamlined risk assessment approaches, such as ASTM's RBCA, are being developed to facilitate consistent assessment at sites. RBCA reflects the tendency towards tiered approaches, and is technically and philosophically consistent with the CCME framework for contaminated site assessment (with the exception that CCME provides Tier 1 'look-up values' whereas RBCA provides simplified models for calculating these values for a site).

REFERENCES

ASTM (1998) *ASTM Risk-based Corrective Action*, American Society for Testing and Materials, Philadelphia, PA. http://www.ucop.edu/facil/eps/astm.html.

CCME (1992) *National Classification System for Contaminated Sites*. Report No. EPC-CS39E.

CCME (1996a) *A Protocol for the Derivation of Environmental and Human Health Soil Quality Guidelines*. CCME Subcommittee on Environmental Quality Criteria for Contaminated Sites.

CCME (1996b) *Guidance Manual for Developing Site-Specific Soil Quality Remediation Objectives for Contaminated Sites in Canada*. CCME Subcommittee on Environmental Quality Criteria for Contaminated Sites.

CCME (1996c) *A Framework for Ecological Risk Assessment: General Guidance*. CCME Subcommittee on Environmental Quality Criteria for Contaminated Sites.

CCME (1999). *Canadian Environmental Quality Guidelines.* Canadian Council of Ministers of the Environment, Winnipeg, Manitoba.

Environment Canada (1991) *Review and Recommendations for Canadian Interim Environmental Quality Criteria for Contaminated Sites*. Ecosystem Science Directorate Scientific Series No. 197.

Fisheries and Oceans Canada (1999) *Proceedings of the 26th Annual Aquatic Toxicity Workshop: 4-6 October 1999, Edmonton, Alberta*. Canadian Technical Report of Fisheries and Aquatic Sciences 2293, Baddaloo EG, Mah-Paulson MH, Verbeek AG and Nimi AJ (eds).

Gaudet CL, Brady A, Bonnell M and Wong MP (1992) Canadian approach to establishing cleanup levels for contaminated sites. In *Hydrocarbon Contaminated Soils and Groundwater*, Vol. 2, Calabrese EJ and Kostecki PT (eds), Lewis, Chelsea, MI, pp. 49-65.

Gaudet CL, Keenleyside KA, Kent RA, Smith SL and Wong MP (1995) How should numerical criteria be used? The Canadian approach. *Human and Ecological Risk Assessment*, **1**(1), 19-28.

Hrudey SE and Pollard SJ (1993) The challenge of contaminated sites: remediation approaches in North America. *Environmental Research*, **1**, 55-72.

Menzie CA and Freshman JS (1997) An assessment of the risk assessment paradigm for ecological risk assessment. *Human and Ecological Risk Assessment*, **3**, 853-892.

Nason *et al.* (2001) Soil quality guidelines. In *Assessing Contaminated Soils: From Soil-Contaminant Interactions to Ecosystem Management*, Fairbrother *et al.* (eds), SETAC Press, Pensacola, FL, in press.

Sheppard SC, Gaudet CL, Sheppard ML, Cureton PM and Wong MP (1992). The development of assessment and remediation guidelines for contaminated soils, a review of the science. *Canadian Journal of Soil Science*, **72**, 359-394.

USEPA (1992) *Framework for Ecological Risk Assessment*. EPA 630/R-92/001, Risk Assessment Forum, U.S. Environmental Protection Agency, Washington, DC.

USEPA (1997) *Ecological Risk Management Guidance for Superfund: Process for designing and conducting ecological risk assessments*. Interim Final EPA-540-R-97-006, Environmental Response Team, US Environmental Protection Agency, Edison, NJ.

USEPA (1998) *Guidelines for Ecological Risk Assessment*. EPA/630/R-95/002F, Risk Assessment Forum, US Environmental Protection Agency, Washington, DC.

USEPA (2000) *Ecological Soil Screening Levels* (web-site). US Environmental Protection Agency. http://www.epa.gov./oerrpage/superfund/programs/ecossls/index.htm.

Visser WJF (1993) *Contaminated Land Policies in Some Industrialized Countries*. TCB R02, (1993), Technical Soil Protection Committee, The Hague, The Netherlands.

13

Ecological Effects-based Soil Guidelines: Case Studies from The Netherlands and Canada

SYLVAIN OUELLET[1], TRUDIE CROMMENTUIJN[2] AND CONNIE L. GAUDET[3]

[1]Policy and Communications, Environment Canada, Ottawa, Ontario, Canada
[2]National Institute for Public Health and Environmental Protection, Bilthoven, The Netherlands
[3]National Guidelines and Standards Office, Environmental Quality Branch, Environment Canada, Ottawa, Ontario, Canada

13.1 INTRODUCTION

Since the first introduction of numerical soil quality criteria in 1983 (the 'Dutch list'), numerical soil criteria have steadily gained momentum as key tools in the assessment and remediation of contaminated sites. Within a decade, several countries and jurisdictions were developing and using soil quality criteria (also known variously as guidelines, intervention, action and alert levels) for specific chemical substances with the intent to protect human health and or ecological receptors when evaluating and managing soil contamination (Visser 1993). While soil quality guidelines were evolving, it seemed the world remained divided on whether to use a site-specific risk assessment approach or a numerical guidelines-based approach in assessing and remediating contaminated soils (see, for example, discussion in Gaudet *et al.* 1991). In response, many jurisdictions are moving towards practical, risk-based frameworks for site assessment and management that marry the strengths of both approaches to varying degrees. The focus of the present chapter is on the development and use of generic or numerical soil quality guidelines for protection of ecological receptors. We present here two specific examples of ecological effects-based soil quality guidelines that have served as a template for many other existing efforts worldwide: The Netherlands and Canada.

Although many early guidelines were based on professional judgment or simply adopted from other jurisdictions, by the mid to late 1990s the scientific

Environmental Analysis of Contaminated Sites. Edited by G. I. Sunahara, A. Y. Renoux, C. Thellen, C. L. Gaudet and A. Pilon

basis and frameworks within which soil guidelines were used had evolved considerably. Most soil quality guidelines in use today are considered 'risk-based' in that they consider receptors to be protected, likely exposure scenarios (where a land use-based system is used) with level of protection required to sustain the 'quality of soil' as defined in jurisdictional policy. For example, guidelines may be targeted at protecting certain land uses (e.g., CCME 1997), or may be based on implications for action (VROM 1999). These management decisions will, to a large extent, dictate the form of the final guideline values. The review of Netherlands and Canadian approaches below illustrates these differences in management frameworks, and the ways in which ecotoxicological data is used to provide values that support the stated protection goals.

Using slightly different procedures, both countries generally derive their guidelines on a substance-by-substance basis, with the exception of groups of substances with similar properties like dioxins and furans. The initial step is a literature search and evaluation on physicochemical properties, toxicity, fate, behavior and persistence. Following this step, toxicological data are selected in order to proceed with the calculation of the soil quality guidelines using risk-based approaches.

Though most soil quality guidelines worldwide have both a human health and an ecological/environmental protection component, this chapter focuses on the ecotoxicologically-based guidelines. Information on human health-based guidelines can be found elsewhere (e.g., Visser 1993, CCME 1996, VROM 1999).

13.2 THE NETHERLANDS APPROACH

The Netherlands (or Dutch) approach differentiates between ecotoxicological risk limits (ERLs) and environmental quality criteria as illustrated in Table 13.1. Ecotoxicological and physicochemical data are used to derive these ERLs which are solely based on scientific considerations. On the other hand, environmental quality criteria are policy benchmarks developed using the scientific basis of

Table 13.1 Ecotoxicological risk limits for soil and soil quality criteria in the project 'Setting Integrated Quality Objectives' and 'Intervention values' (VROM 1994)

Ecotoxicological risk limits	Soil quality criteria
ECOTOX SCC[a] and HUM-TOX SCC[b]	
	intervention value for soil clean-up
MPC[c]	
	target value
NC[d]	

[a]ECOTOX SCC, ecotoxicological serious soil contamination concentration.
[b]HUM-TOX SCC, human-toxicological serious soil contamination concentration, based on human toxicological information, not dealt with in this paper.
[c]MPC, maximum permissible concentration.
[d]NC, negligible concentration.

ERLs and may also include other considerations (e.g., economic and social) in addition to the risk limit. The different risk limits: ECOTOX SCC, MPC and NC (see Table 13.1) and the different quality criteria: intervention value and target value represent a different level of protection for ecological receptors. The concentration in the environment is compared to the soil quality criteria or ERLs (in ascending order: target value, MPC, intervention value) to identify the type of actions required, once a particular benchmark is exceeded (VROM 1999).

13.2.1 THE DERIVATION OF RISK LIMITS USING ECOTOXICITY TEST RESULTS

13.2.1.1 Soil Contact

Toxicological data is retrieved from the existing scientific literature for soil associated ecological receptors including soil plants (grain species, trees, etc.) and invertebrates (e.g., worms, arthropods, etc.). Toxicological endpoints related to potential effects at the population level, such as survival, growth and reproduction and commonly expressed as an $L(E)C_{50}$ or NOEC (no observed effect concentration), are preferred for development of values. Effect data on microbiological processes and enzymatic activity, commonly expressed as a NOEC or EC_x value, are also considered. In addition, other toxicological data are taken into account when the data in question is considered ecologically relevant, e.g., histopathological effects on reproductive organs of a species.

13.2.1.2 Bioaccumulation in the Food Chain

Contaminants accumulating through the food chain may exert toxic effects on birds and mammals. From physical chemical parameters like $\log K_{ow}$ and water solubility an indication can be obtained for the bioaccumulative potential of the substance in question. If there is a positive indication, data on the sensitivity of birds and mammals and bioconcentration factors (BCFs) from diets (e.g., worms, fish, mussels) of wildlife consumers is searched for. The substances for which this step is considered are organic substances with $\log K_{ow} > 5$ and molecular weight < 600. For metals this is considered case-by-case.

13.2.1.3 Normalization

Not all of the tests described in the literature are performed under the same conditions. Therefore normalization of terrestrial test results was proposed by Denneman and Van Gestel (1990). This proposal was studied by Van Gestel *et al.* (1995) who reviewed literature on the influence of pH and organic matter on toxicity for metals. Other studies described by Van Gestel (1990) showed the influence of organic matter content on the toxicity of organic chemicals. Based on these studies, currently all data on the sensitivity of species are recalculated for a standard soil containing 10% organic matter and 25% clay.

13.2.1.4 Reference Values

Because metals may occur naturally at levels in the environment that exceed effects concentrations from laboratory studies, reference values are developed to ensure that recommended soil values serve as realistic benchmarks. Reference values for metals in soil are based on 'reference lines' derived by correlating measured ambient background concentrations (total concentrations in the soil matrix) at a series of remote rural sites in The Netherlands to the percentage clay and organic matter content of these soils (see Edelman 1984, De Bruijn and Denneman 1992 and CSR 1996 for calculating the reference values). A study has recently been initiated where background concentrations in relatively unpolluted locations in The Netherlands are measured for all substances for which the target and intervention values are available.

13.2.1.5 Dealing with Data Limitations

Where no data on terrestrial/sediment species are available, the equilibrium partition (EP) method is applied to derive ERLs for soil. In addition, the EP method is used for harmonization (see Section 13.2.3) of MPCs/NCs. To be able to apply the EP method, data on partition coefficients are required. For deriving K_p values, only studies in which the humus or organic matter content or organic carbon content is reported are taken into account. Organic carbon content is derived from the organic matter content by dividing it by 1.7.

When none or only few experimental data are available, the organic carbon partition coefficient (K_{oc}) can be estimated using the regression equations described by Sabljic *et al.* (1995) and DiToro *et al.* (1991). Both give empirical formulas from which $\log K_{oc}$ can be derived using $\log K_{ow}$.

13.2.1.6 Data Selection

The aim of selecting toxicity data is firstly to derive one single toxicity datum for each compound and species, and second to select reliable toxicity data. One parameter per species is necessary as input in the extrapolation methods (Section 13.2.2). Therefore chronic as well as acute toxicity data are weighted as follows (Slooff 1992):

- If, for one species, several toxicity data based on the same toxicological endpoint are available, these values are averaged by calculating the geometric mean;
- If, for one species, several toxicity data based on different toxicological endpoints are available, the lowest value is selected. The lowest value is

determined on the basis of the geometric mean, if more than one value for the same parameter is available (see above);

- In some cases, data for effects of different life-stages are available. If from these data it becomes evident that a distinct life-stage is more sensitive, this result may be used in the extrapolation.

13.2.2 CALCULATING ECOTOXICOLOGICAL RISK LIMITS

In The Netherlands, five different extrapolation methods are used for the calculation of ERLs. The *Refined Effects Assessment* is applied when the chronic data for more than four different taxonomic groups are available. The *Preliminary Effects Assessment* is used only if chronic data for less than four species of different taxonomic groups or less than four different microbiological processes or only acute data are available. For parameters, like metals, having a natural background concentration, the *Added Risk Approach* is applied. For substances with potential for bioaccumulation, an ERL for secondary poisoning is also derived applying the *Secondary Poisoning Approach*. Where no soil toxicity data are available, ERLs are derived on the basis of aquatic toxicity data and applying the *EP Method*. Each of these methods is briefly described below.

13.2.2.1 Refined Effects Assessment

The refined effects assessment or statistical extrapolation method is based on the assumption that the sensitivities of species in an ecosystem can be explained by a statistical frequency distribution that describes the relationship between the concentration of the substance in a compartment and a certain percentage of species that will be unprotected. The method of Aldenberg and Slob (1993) is used for criteria if NOECs for four or more different taxonomic groups are available as follows (this method assumes that the NOECs used for estimating the distribution fit the log-logistic distribution):

$$\mathrm{PAF}(x) = 1/[1 + \exp((a - x)/\beta)] \qquad (13.1)$$

in which $\mathrm{PAF}(x)$ is the potentially affected fraction of all possible species at concentration x, a is the mean value of the log-logistic distribution, β is the scale parameter of the log-logistic distribution and $x = \log(c)$ is the logarithm of the concentration. $\mathrm{PAF}(x)$ has a value between 0 and 1 and is the fraction of species that have log(NOEC) values smaller than x. The concentration corresponding to a 50% protection level (potentially affected fraction of all species of 50% or $\mathrm{PAF} = 0.5$) is the ECOTOX SCC.

For the maximum permissible concentration (MPC), intended to protect all species in an ecosystem, a 95% protection level is chosen as cut-off value (VROM 1989). The concentration corresponding to a 95% protection level (potentially affected fraction of all species of 5% or PAF = 0.05) can be derived using statistical extrapolation methods. The negligible concentration (NC), in contrast to the MPC, is not based on a fraction of species protected and is derived by dividing the MPC by a factor of 100. This factor is applied to take into account combination toxicity (VROM 1989).

13.2.2.2 Preliminary Effects Assessment

In the preliminary effects assessment method, assessment factors are applied to toxicity data. The size of this factor depends on the number and kind of these toxicity data. The factors and conditions used for deriving MPCs for soil are shown in Tables 13.2 and 13.3 for terrestrial MPCs and secondary poisoning MPCs respectively. For deriving MPCs the method is often referred to as the modified EPA method (Van de Meent *et al.* 1990). The factors and conditions used for deriving ECOTOX SCCs for soil are shown in Table 13.4.

Table 13.2 Modified EPA method for terrestrial ecosystems to derive the MPC

Available information	Assessment factor
Lowest acute $L(E)C_{50}$ or QSAR[a] estimate for acute toxicity	1000
Lowest acute $L(E)C_{50}$ or QSAR estimate for acute toxicity of microbe-mediated processes, earthworms or arthropods and plants	100
Lowest NOEC or QSAR estimate for chronic toxicity[b]	10
Lowest NOEC or QSAR estimate for chronic toxicity for three representatives of microbe-mediated processes or of earthworms, arthropods and plants	10

[a] QSAR, quantitative structure activity relationship.
[b] This value is compared to the extrapolated value based on acute $L(E)C_{50}$ toxicity values. The lowest one is selected.

Table 13.3 Modified EPA method for birds and mammals to derive the MPC

Available information	Assessment factor
Less than three acute LC_{50}'s and no chronic NOECs	1000
At least three acute LC_{50}'s and no chronic NOECs	100
Less than three chronic NOECs[a]	10
Three chronic NOECs	10

[a] This value is compared to the extrapolated value based on acute $L(E)C_{50}$ toxicity values. The lowest one is selected.

Table 13.4 Assessment factors used to derive the ECOTOX SCC

Available information	Assessment factor
Only acute LC_{50}'s and no chronic NOECs	10
Geometric mean of chronic NOECs	1

13.2.2.3 Added Risk Approach

The added risk approach, which is modified from Struijs *et al.* (1997), is used to calculate risk limits for metals in the different environmental media. The approach starts with calculating a maximum permissible addition (MPA) on the basis of available data from laboratory toxicity tests in the same way that the MPC is calculated (see Section 13.2.2.1). This MPA, which may be related to anthropogenic activities, is considered to be the maximum concentration to be added to the background concentration (C_b). The MPC is then the sum of the C_b and the MPA. The background concentration is based on data available for concentrations in non-polluted sites. Conventionally, the MPC for metals was calculated using either fixed assessment factors or by applying a statistical extrapolation method (Van de Meent *et al.* 1990). The MPA is calculated using a similar approach as the MPC for substances having no natural background concentration. The NC is defined as the background concentration (C_b) plus the negligible addition (NA): $NC = C_b + NA$, where $NA = MPA/100$. The factor 100 is a safety factor, to take into account combination toxicity (VROM 1989). For practical reasons, the background concentration is added to the MPA or NA, since it can thus be compared to monitoring data. It must be noted that the background concentration and the MPA are independently derived values.

The theoretical description of the added risk approach as described by Struijs *et al.* (1997) includes bioavailable fractions of the background concentrations that can vary between 0% and 100%. At the moment the approach has been worked out for MPCs for metals by assuming 100% bioavailability (Crommentuijn *et al.* 1997). This policy decision was made based on the fact that the effects of natural background concentrations of metals may be neglected or even considered desirable, because of their contribution to species biodiversity. In addition, not enough information is currently available to evaluate the bioavailability of the background concentrations for metals. Furthermore, Crommentuijn *et al.* (2000) have shown that assuming different bioavailabilities for background concentrations for most of the metals considered had little influence on the MPC. With regard to the bioavailable fraction of the metals and metalloids in laboratory tests, it is here assumed that the metals and metalloids added to the test medium are completely bioavailable, i.e., the bioavailable fraction of the added metal and metalloid in the laboratory tests is 100%.

13.2.2.4 Secondary Poisoning Approach

Most NOECs for substances are based on environmental concentrations in water, air, sediment or soil. Some species, which are in general those high in the food chain, are additionally or mainly exposed to substances via their food and may accumulate these substances to high concentrations. Several food-chain models have been developed for estimating the risk for secondary poisoning. All these models propose a more or less simple food chain in which the effect concentrations of birds and mammals are divided by the BCF to obtain a concentration in the soil. The following two simple food chains, taken as representative examples, are currently being used in the Dutch approach. The first one is a simple aquatic food chain: water → fish or mussel → fish-eating bird or mammal while the second one represents a simple terrestrial food chain: soil → worm → worm-eating bird or mammal (Romijn *et al.* 1993, 1994). The MPC related to accumulation through the food chain is defined as:

$$\mathrm{MPC}_{\mathrm{water/soil}} = \frac{\mathrm{NEC}_{\mathrm{bird/mammal}} \times f}{\mathrm{BCF}_{\mathrm{fish/worm/mussel}}} \qquad (13.2)$$

in which $\mathrm{MPC}_{\mathrm{water/soil}}$ is the maximum permissible concentration for water or soil; $\mathrm{NEC}_{\mathrm{bird/mammal}}$ is the no effect concentration for birds or mammals, which is obtained from extrapolation of effect data; $\mathrm{BCF}_{\mathrm{fish/worm/mussel}}$ is the bioconcentration factor for fish, worm or mussel; f is a factor which takes into account the caloric content of the food, default values are 0.32 for fish, 0.20 for mussel and 0.23 for worm as food. The factor of 0.23 in the formula is used to correct for differences in calorific content of the food used in the laboratory for birds and mammals and the calorific content of worms (Westerterp *et al.* 1982, as cited by Van de Plassche 1994).

Recently a more elaborate food chain has been proposed by Jongbloed *et al.* (1994). They designed a (still) simplified food web with three trophic levels: plants and invertebrates at the first level, small birds and mammals at the second level, and birds and beasts of prey at the third level. It was concluded however that incorporating more levels has the consequence that only for a few substances is it possible to calculate risk limits for secondary poisoning since data are limited. The MPC, which includes secondary poisoning, is compared to that for direct exposure. Usually, the lowest MPC of the different MPCs is taken as the respective MPC for water or soil. However, the uncertainties in both the MPCs and the BCFs are taken into account.

13.2.2.5 Equilibrium Partitioning Method

The EP method, originally proposed by Pavlou and Weston (1984) to develop sediment quality criteria, has been described in detail by Shea (1988) and DiToro *et al.* (1991) and is based on three underlying assumptions. Firstly, the approach assumes that bioavailability, bioaccumulation and toxicity are closely related to the pore water concentrations. Secondly, it is assumed that sensitivities of aquatic organisms are comparable with sensitivities of organisms living in the sediment. The third assumption dictates that equilibrium exists between the chemical sorbed to the particulate sediment organic carbon and the pore water and that these concentrations are related by a partition coefficient (K_{oc}).

It can be reasoned that the EP method is valid for species that are exposed mainly through the pore water. For instance soft-bodied organisms like earthworms and enchytraeids will be mainly exposed via the pore water. The amount of metals available in the pore water depends strongly on soil characteristics, for instance pH and organic matter content. Relationships between the accumulation of metals by invertebrates and soil characteristics have been found (reviewed in Van Gestel *et al.* 1995). Also some relationships between toxicity and soil characteristics have been found, for instance for cadmium and earthworms (Van Gestel and Van Dis 1988) and between chlorobenzenes and earthworms (Belfroid *et al.* 1994). However, it is important to keep in mind that, for many species living in or on top of the soil, exposure occurs through other routes as well, for instance through food. For these species, the equilibrium partition method may not be representative of actual exposure conditions.

13.2.3 HARMONIZING INDEPENDENTLY DERIVED MPCs

When independently derived MPCs for water and sediment/soil are available, MPCs for water and sediment/soil have to be harmonized. This is done by calculating the ERL for sediment or soil from the ERL for water and applying the EP method as described earlier. In principle, the lowest value of the independently derived MPC (equation (13.2)) and the MPC resulting from the equilibrium partitioning approach is then taken as the harmonized MPC. However, the uncertainties in both the MPCs and the partition coefficient are taken into consideration. Further information on harmonization can be found in Van de Plassche and Bockting (1993).

13.3 THE CCME APPROACH

In the early 1990s, the Canadian Council of Ministers of the Environment (CCME), with members of all Canadian provincial, territorial and federal governments, mandated the development of a new approach that would produce scientifically defensible soil guidelines by a transparent method. Simultaneously, the CCME mandated the development of guidance on application of the guidelines and alternative risk assessment techniques. Experience has shown that these three tasks are highly interdependent and share many properties and steps. In this section, we first discuss the derivation of the CCME soil guidelines as a major model for soil guideline development that differs from the Netherlands approach (above).

13.3.1 GENERAL CONSIDERATIONS

The Canadian soil quality guideline development protocol marries a multimedia approach for human health protection to biological effects-based procedures for soil organisms, plants, livestock and wildlife. Some provinces or territories may have separate protocols for developing soil quality guidelines that may different from the CCME protocol (e.g., Ontario, Quebec and British Columbia). As some of these other protocols were developed during the development of the national CCME protocol, there can be some similarities between the various procedures used for deriving soil quality guidelines.

Under the CCME protocol, soil quality guidelines protective of human health are first nominated on the basis of direct exposure, then evaluated and adjusted as required by groundwater and food production checks. Environmental quality guidelines are similarly developed on the basis of direct exposures and later adjusted as required to take account of nutrient cycling effects, bioaccumulation and plant nutrition requirements where appropriate. The lower of the human health and environmental quality guidelines becomes the recommended soil quality guideline for four defined land uses based on the Canadian land zoning system (agricultural, residential/parkland, commercial and industrial). Guidelines nominated on these toxicological bases are then evaluated by 'check' mechanisms for plant nutritional requirements, geochemical background and analytical detection limits. The final guidelines for sensitive land uses are intended to represent levels of contaminants in soil which present no appreciable risk to humans and no observable adverse effects to ecological receptors. For commercial and industrial guidelines, the goals are protection of children and adults, respectively, and minimal adverse effects to ecological receptors.

13.3.1.1 Guidelines Derivation Protocol

The national protocol for development of soil guidelines (CCME 1996) was designed to generate effects-based guidelines for defined land uses, based on development of a credible exposure scenario for each land use category. Plausible receptor arrays, exposure scenarios and management objectives (i.e., levels of risk) have been defined and assumed for each of the four defined land uses. Clear statements on these objectives provide direction for the use of three domains of information required for guideline derivation: contaminant fate and behavior, biological effects and exposure routes. The 1996 version of CCME protocol addresses protection of both human health and the environment.

13.3.1.2 Human Health Soil Quality Guidelines

The human health soil quality guidelines recommended by the CCME are derived using calculation procedures similar to those used in site-specific risk assessment discussed earlier, with the difference that generic conservative values (e.g., weight of an adult, exposure duration, exposure pathways), assumed to be representative of typical Canadian conditions, replace site-specific information. The resulting human health soil quality guidelines provide concentrations of contaminants at or below which no appreciable risk to human health is expected. In this section, we address only the environmental component. For further information on the derivation of human health soil quality guidelines see the CCME protocol (1996).

13.3.1.3 Land Use-based Exposure Scenarios

For each land use, identification of receptors of concern was regarded as the first step in defining the exposure scenario with supported dependence on ecological receptors considered to decrease in the order agriculture>residential/parkland>commercial>industrial. Agricultural land use requires that the full spectrum of living creatures be protected by targeting indicator plants, invertebrates, grazing animals and soil microbial processes (see Table 13.5). Less sensitive land uses are accommodated by working with subsets of these receptors. For example, residential uses of soil require protection for a similar suite of biota except that livestock need not be considered. At the other end of the sensitivity scale, industrial lands may be considered adequately protected provided plants, invertebrates and soil nutrient and energy cycling are not seriously impaired. A brief description of the derivation process for environmental guidelines follows.

13.3.1.4 Toxicological Data Considered

Protection of the environment is complicated because soil-associated biota are members of a complex soil ecosystem where individuals may experience different exposures depending on spatial and temporal considerations. Like other ecosystems, the soil ecosystem may be examined at the level of the individual, population or community and within each of these, different periods of observation yield information on different processes. Traditionally, toxicologists have focused on the growth, reproduction and mortality of individual species.

Table 13.5 Receptors and exposure pathways for the land use categories considered in the derivation of Canadian soil quality guidelines

Route of exposure	Agriculture	Residential/ parkland	Commercial	Industrial
Soil contact	soil nutrient cycling processes, soil invertebrates, crops/plants, livestock/wildlife	soil nutrient cycling processes, soil invertebrates, plants, wildlife	soil nutrient cycling processes, soil invertebrates, plants, wildlife	soil nutrient cycling processes, soil invertebrates, plants, wildlife
Soil and food ingestion	herbivores	not applicable	not applicable	not applicable

Accordingly, the greatest abundance of ecotoxicological information for soil is from single-species bioassay (soil contact pathway), and this is used as a primary source in the CCME protocol (CCME 1996).

13.3.2 CCME APPROACHES FOR DERIVATION OF GUIDELINES

13.3.2.1 Derivation of Preliminary Soil Guideline

A preliminary effects-based guideline (threshold effects concentration or TEC) is derived based on the available toxicological data using methods as described below. The final derivation of soil quality guidelines also considers what are referred to as 'checks' (similar to the Dutch system).

13.3.2.2 Direct Soil Contact

Derivation of the preliminary environmental soil quality guidelines for soil contact is intended to protect plants and invertebrates and is done using three different methods depending on the availability of data as well as on the level of confidence in those data: the weight-of-evidence method, the LOEC method and the median effect method (Figure 13.1). Minimum data requirements have been set for each of these three methods so that when requirements for the preferred method (weight-of-evidence) cannot be met, the second alternative method (LOEC) is used or the third method (median effects) if requirements are also not met for the second method. The weight-of-evidence approach is a modification of the approach used for calculating sediment quality guidelines for the National Status and Trends Program (Long and Morgan 1990) that used a percentile of the effects data set or the combined effects and no effects data set to estimate a concentration expected to cause no, or negligible, adverse effects. For agricultural land use and residential and parkland land use, the 25th percentile of the effects and no effects data distribution is chosen as the 'no potential effects range' (NPER). For commercial and industrial land uses, the 25th percentile of the effects data distribution only (i.e., no effects data excluded) is selected as the effects range low (ERL). If the minimum data requirements for the weight-of-evidence approach cannot be met, the preliminary guideline is derived using the lowest observed effect concentration (LOEC) for agricultural land use and residential and parkland land use, and the geometric mean of the LOECs for commercial and industrial land uses.

13.3.2.3 Food Chain Effects

Another important exposure pathway is ingestion of contaminated soil and food by grazing animals and wildlife. Models have been developed for the protection of grazing animals ingesting both soil and plants grown on contaminated soil, and have recently been expanded to account for predators consuming prey from contaminated sites (CCME 1996, 1997). These models consider uptake and accumulation of substances through the food chain and, despite

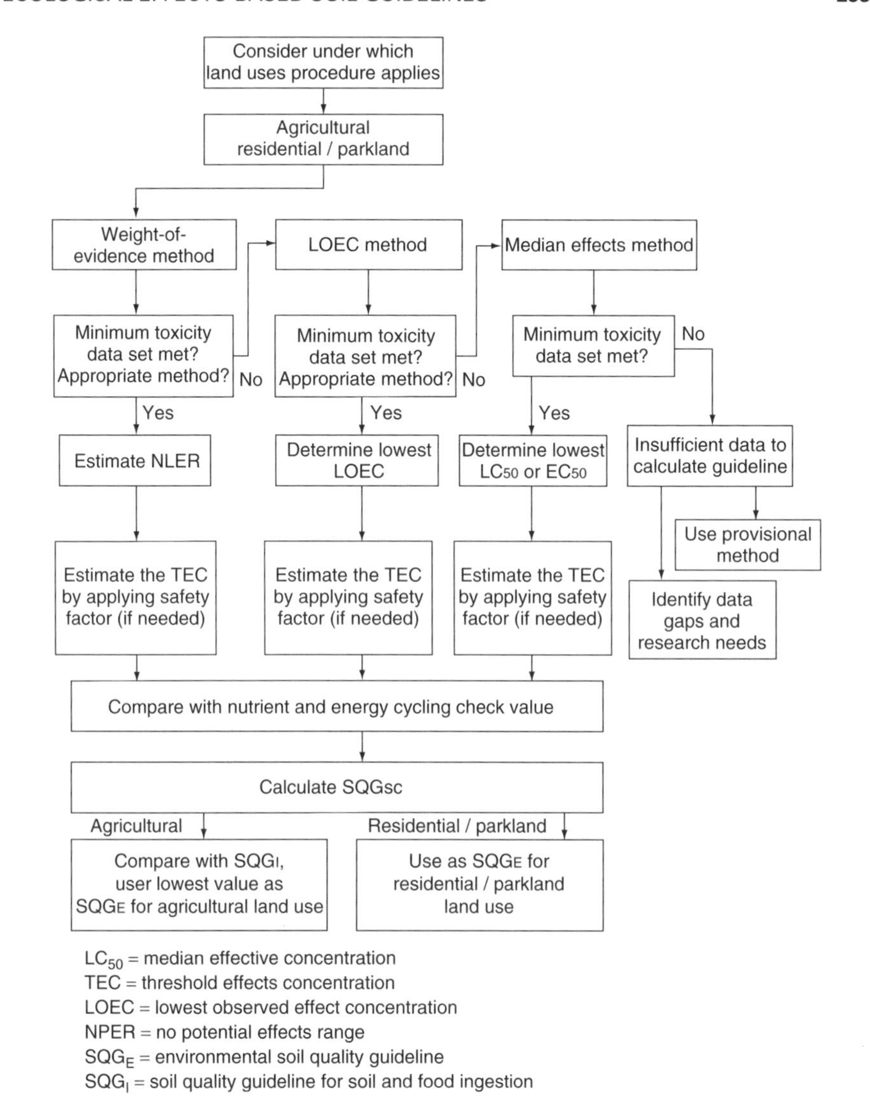

Figure 13.1 CCME soil contact guidelines derivation procedure for agricultural land use (from CCME 1996).

differences in approach, are analogous to the secondary poisoning approach in The Netherlands described above.

13.3.2.4 Guidelines for Soil and Food Ingestion

Where data are adequate, a soil and food ingestion guideline is calculated for this pathway using the most sensitive mammalian or avian species to be protected.

The lower of the soil contact and soil ingestion guidelines is selected as the environmental soil quality guideline for agricultural land use.

13.3.2.5 Ecosystem Protection

Soil contact and ingestion data are augmented with data on microbial processes to ensure protection of community or soil ecosystem levels. To address this concern, information on effects of contaminants on C and N cycling processes is used as an integrator of soil community level health. Further details on these procedures can be found in CCME (1996).

13.3.2.6 Derivation of Final Recommended Soil Quality Guidelines

A recommended soil quality guideline developed on the toxicological bases discussed above must undergo three reality checks as a final step. Effective soil quality guidelines must be evaluated against: plant nutritional requirements, geochemical background and analytical detection limits.

13.3.2.7 Guideline Check Mechanisms

During execution of the above checks, it may be necessary to adjust a soil guideline nominee to generate a final soil quality guideline. Such an adjustment can be necessary to avoid contamination of other media (e.g., groundwater) at the proposed guidelines level. However, any such adjustment is accompanied by full scientific documentation in the guideline document, which summarizes the critical data, and steps in developing the environmental and human health guidelines. The guideline document also provides the site manager with decision support information.

13.4 CONCLUSIONS AND CHALLENGES FOR ECOTOXICOLOGICAL BASIS OF GUIDELINES

The practice of soil quality assessment and remediation is rapidly evolving. Only a little over a decade ago, soil quality guidelines were based largely on professional judgement and no comprehensive guidance on risk assessment existed. This is a significant contrast with the sophisticated development procedures and site assessment procedures outlined in this chapter.

As is evident above, the major difference in approach between The Netherlands and Canada is the initial protection goals. Whereas Canada has adopted a land use-based approach, and developed exposure scenarios and levels of protection relative to these land uses, The Netherlands has adopted an approach based on levels of risk that prescribe different actions (i.e., action or alert levels). Most jurisdictions in the world today use an approach based on one of these two basic models. Despite key differences in approach, the ecological receptors, endpoints (growth, reproduction, survival) and pathways (e.g., direct soil contact, food chain) considered as the scientific basis for guidelines are similar, as is the type of ecotoxicological data considered in the derivation of guidelines, and the additional considerations such as background or reference levels. Most importantly, the derivation of soil quality guidelines across jurisdictions is moving more and more towards transparent, risk-based methods that require credible soil toxicity data to be generated, addressing ecologically significant endpoints. While soil quality guidelines can be used as indicators of soil quality or directly as remediation objectives, some jurisdictions use them as screening criteria within broader assessment and remediation frameworks, as discussed previously.

When using generic soil quality guidelines, it is important to keep in mind several critical outstanding issues and questions related to their development.

Validation of test methods and risk limits. Terrestrial effects assessment is currently based only on the results of laboratory toxicity tests and it is uncertain whether the values calculated from these methods really represent concentrations relevant to ecosystems. In The Netherlands, a project entitled 'Validation of Toxicity Data and Risk Levels for Soils' was executed. In the first place, attention was paid to the question (Posthuma 1997): do species in the field, and in the laboratory, react in the same way if exposed to a comparable level of metal contamination?

Determining the actual risk. When talking about actual risk, a comparable question is addressed as when talking about validation: do the values calculated from laboratory toxicity test-based methods (representing a potential risk) really represent concentrations relevant to a specific site? For example, differences in bioavailability due to differences in physicochemical characteristics of the site and combined toxicity from co-occurring contaminants at a site may affect actual risk. The derived generic risk limit is assumed to be valid for all ecosystems, and may result in an underestimation or overestimation of the risks at a specific site.

Harmonization of risk limits. Environmental quality guidelines are determined for the individual compartment (air, water and soil) without taking into account the transport of chemicals from one medium (air, water and soil) to another and the effect that can result in compartments other than where the chemical was released. Harmonization of these guidelines is needed. Multimedia fate models have been used (Van de Meent and De Bruijn 1995) to harmonize independently derived quality guidelines.

Including natural backgrounds and bioavailability. In the past, during the setting of Dutch environmental quality objectives (EQOs) for metals, some specific problems have become clear. In some cases, the methodology proposed resulted in risk limits lower than the natural background concentrations in the environment. In this case the EQOs are usually set on the level of the natural background concentration. The actual effect of naturally occurring metals may be lower than expected on the basis of laboratory tests, because of the lower bioavailability under field conditions. As emphasized by Allen (1997) and Cook and Hendershot (1996) on developing guidelines for metals, it is preferable to account also for the bioavailability of natural background and anthropogenic concentrations, anticipating that only the bioavailable part of the total concentration may cause an adverse effect. At the moment, guidelines are generally based on total concentrations in the soil. Methods for calculating and measuring potentially and actually occurring (bio)available metal concentrations in soils are being developed and validated to enable a better estimate of the risks imposed by heavy metals in the future (Janssen *et al.* 1997a,b, Peijnenburg *et al.* 1997), but much work is needed as this will vary not only with the substance and exposure pathway but also with the receptor.

Data limitations. Deriving soil quality guidelines is a compilation of methods, knowledge and data available at the moment. It is to be expected that because of ongoing efforts to improve effects assessment, new methods will be incorporated in the future. There are many areas in soil toxicology where knowledge and data are lacking.

- Toxicity data on the sensitivity of terrestrial species and terrestrial processes is still very scarce.
- Soil quality guidelines are derived for single substances. Since many soil contamination problems are the result of *mixture toxicity*, much more research is needed in this area as additive, synergistic or antagonistic effects are known to occur in contaminant mixtures.
- Some metals for which quality guidelines have been developed are essential elements for which ecological receptors may have a minimum requirement to supply their needs, and a maximum exposure level above which the element is toxic (Scheinberg 1991). This minimum requirement is necessary because most trace metals play an essential role in the metabolism of the organism (Rainbow 1993). The range between the minimum and maximum levels is often called the window of essentiality (Hopkin 1993). In ecotoxicological studies with essential elements, it can theoretically occur that the effects observed are caused by element limitation instead of toxic effects. Knowledge on this topic is lacking at this time.

REFERENCES

Aldenberg T and Slob W (1993) Confidence limits for hazardous concentrations based on logistically distributed NOEC toxicity data. *Ecotoxicology and Environmental Safety*, **25**, 48–63.

Allen HE (1997) Standards for metals should not be based on total concentrations. *SETAC-News*, **January**.

Belfroid A, Sikkenk M, Seinen W, Van Gestel K and Hermens J (1994) The toxicokinetic behaviour of chlorobenzenes in earthworm (*Eisenia andrei*) — experiments in soil. *Environmental Toxicology and Chemistry*, **13**, 93–99.

CCME (1996) *A Protocol for the Derivation of Environmental and Human Health Soil Quality Guidelines*. Canadian Council of Ministers of the Environment, Subcommittee on Environmental Quality Criteria for Contaminated Sites.

CCME (1997) *Recommended Canadian Soil Quality Guidelines*. Canadian Council of Ministers of the Environment, Soil Quality Guidelines Task Group.

Cook N and Hendershot WH (1996) The problem of establishing ecologically based soil quality criteria. The case of lead. *Canadian Journal of Soil Science*, **76**, 335-342.

Crommentuijn T, Polder MD and Van de Plassche EJ (1997) *Maximum permissible concentrations and negligible concentrations for metals. Taking background concentrations into account*. RIVM Report no. 601501 001.

Crommentuijn T, Polder M, Sijm D, De Bruijn J, Van de Plassche E (2000) Evaluation of the Dutch environmental risk limits for metals by application of the added risk approach. *Environmental Toxicology and Chemistry*, **19**, 1692-1701.

CSR (1996) *QA-procedures for deriving environmental quality objectives (INS and I-values)*, CSR-KD/003.

De Bruijn JHM and Denneman CAJ (1992) *Achtergrondgehalten van negen sporen-metalen in oppervlaktewater, grondwater en grond van Nederland*. VROM-publikatiereeks Bodembescherming nr. 1992/1 (in Dutch).

Denneman CAJ and Van Gestel CAM (1990) *Bodemverontreiniging en bodemecosystemen: Voorstel voor C-(toetsings)waarden op basis van ecotoxicologische risico's*. RIVM Report nr. 725201 001 (in Dutch).

DiToro DM, Zarba CS, Hansen DJ, Berry WJ, Swartz RC, Cowman CE, Pavlou SP, Allen HE, Thomas NA and Paquin PR (1991) Technical basis for establishing sediment quality criteria for nonionic organic chemicals using equilibrium partitioning. *Environmental Toxicology and Chemistry*, **10**, 1541-1583.

Edelman Th (1984) *Achtergrondgehalten van stoffen in de bodem*. VROM-publikatiereeks Bodembescherming nr. 34, Staatsuitgeverij, Den Haag (in Dutch).

Gaudet CL, Brady A, Bonnell M and Wong MP (1991) Canadian approach to establishing cleanup levels for contaminated sites. In *Hydrocarbon Contaminated Soils and Ground Water*, Calabrese EJ and Kostecki PT (eds), Lewis, Boca Raton, FL, pp. 49-65.

Hopkin SP (1993) Ecological implications of 95% protection levels for metals in soil. *Oikos*, **66**, 137-141.

Janssen RPT, Peijnenburg WJGM, Posthuma L and Van den Hoop MAGT (1997a) Equilibrium partitioning of heavy metals in Dutch field soils. I. Relationships between metal partition coefficients and soil characteristics. *Environmental Toxicology and Chemistry*, **16**, 2470-2478.

Janssen RPT, Posthuma L, Baerselman R, Den Hollander HA, Van Veen RPM and Peijnenburg WJGM (1997b) Equilibrium partitioning of heavy metals in Dutch field soils. II. Prediction of metal accumulation in earthworms. *Environmental Toxicology and Chemistry*, **16**, 2479-2488.

Jongbloed RH, Pijnenburg J, Mensink BJWG, Traas ThP and Luttik R (1994) *A model for environmental risk assessment and standard setting based on biomagnification. Top predators in terrestrial ecosystems*. RIVM Report no. 719101 012.

Long ER and Morgan LG (1990) *The potential for biological effects for sediment-sorbed contaminants tested in the National Status and Trends Program*. NOS OMA 52, National Oceanic and Atmospheric Administration Technical Memorandum, Seattle, WA, 175 pp + appendices.

Pavlou SP and Weston DP (1984) *Initial evaluation of alternatives for development of sediment related criteria for toxic contaminants in marine waters (Phase II)*. Report prepared for US Environmental Protection Agency, Washington, DC.

Peijnenburg WJGM, Posthuma L, Eijsackers HJP and Allen HE (1997) A conceptual framework for implementation of bioavailability for environmental purposes. *Ecotoxicology and Environmental Safety*, **37**, 163-172.

Posthuma L (1997) Gradient studies as a possibility to associate ecotoxicological risk assessment with toxic effects on population and community parameters in the field. In *Ecological Principles for Risk Assessment of Contaminants in Soil*, Van Straalen NM and Løkke H (eds), Chapman & Hall, London.

Rainbow PS (1992) The significance of trace metal concentrations in marine invertebrates. In *Metals in Invertebrates*, Dallinger R and Rainbow PS (eds). A SETAC special publication, Lewis, Boca Raton, FL, pp. 3-24.

Rainbow PS (1993) The significance of trace metal concentrations in marine invertebrates. In: Dallinger R and Rainbow PS (Eds.) *Metals in invertebrates*. SETAC-publication, Lewis-Publishers, pp. 3-24.

Luttik R, Romijn CAFM and Canton JH (1993) Presentation of a general algorithm to include secondary poisoning in effect assessment. *Science of the total environment*, Supplement 1993, 1491-1500.

Romijn CAFM, Luttik R and Canton JH (1994) Presentation of a general algorithm to include secondary poisoning in effect assessment, Part 2. Terrestrial food chains. *Ecotoxicology and Environmental Safety*, **27**, 107-127.

Sablijic A, Gusten H, Verhaar H and Hermens J (1995) QSAR-modelling of soilsorption. Improvements and systematics of log Koc vs. log Kow correlations. *Chemosphere*, **31**, 4489-4514.

Scheinberg H (1991) Copper. In *Metals and Their Compounds in the Environment*, Merian E (ed), VCH, Weinheim, pp. 893-908.

Shea D (1988) Developing national sediment quality criteria. *Environmental Science and Technology*, **22**, 1256-1261.

Slooff W (1992) *RIVM Guidance document. Ecotoxicological effect assessment: deriving maximum tolerable concentrations (MTC) from single-species toxicity data*. RIVM Report no. 719102 018.

Struijs J, Van de Meent D, Peijnenburg WJGM, Van den Hoop MAGT and Crommentuijn T (1997) Added risk approach to derive maximum permissible concentrations for heavy metals: how to take into account the natural background levels? *Ecotoxicology and Environmental Safety*, **37**, 112-118.

Van de Meent D and De Bruijn JHM (1995) A modeling procedure to evaluate the coherence of independently derived environmental quality objectives for air, water and soil. *Environmental Toxicology and Chemistry*, **14**, 177-186.

Van de Meent D, Aldenberg T, Canton JH, Van Gestel CAM and Slooff W (1990) *Desire for levels. Background study for the policy document 'Setting environmental quality standards for water and soil'*. RIVM Report no. 670101 002.

Van de Plassche EJ (1994) *Towards integrated environmental quality objectives for several compounds with a potential for secondary poisoning*. RIVM Report no. 679101 012.

Van de Plassche EJ and Bockting GJM (1993) *Towards environmental quality objectives for several volatile compounds*. RIVM-report no. 679101011.

Van Gestel CAM (1990) *Earthworms in ecotoxicology*. PhD Thesis, State University of Utrecht.

Van Gestel CAM and Van Dis WA (1988) The influence of soil characteristics on the toxicity of four chemicals to the earthworm *Eisenia andrei* (Oligochaeta). *Biology and Fertility of Soils*, **6**, 262-265.

Van Gestel CAM, Rademaker MCJ and Van Straalen NM (1995) Capacity controlling parameters and their impact on metal toxicity in soil invertebrates. In *Biogeodynamics of Pollutants in Soils and Sediments*, Salomons W and Stigliani WM (eds), Springer-Verlag, Berlin.

Visser WJF (1993) *Contaminated land policies in some industrialized countries*. TCB RO2, Technische Commissie Bodembescherning.

VROM (2001). *Environmental Quality Standards*. Ministry of Housing, Physical Planning and the Environment.

VROM (1989) *Premises for risk management. Risk limits in the context of environmental policy*. Second chamber, session 1988-1989, 21137, no 5.

VROM (1999) *Stoffen en normen. Overzicht van belangrijke stoffen en normen in het milieubeleid*. Environmental Quality Objectives in The Netherlands. Ministry of Housing, Physical Planning and Environment (English version in press and available in 2000).

14

Risk-based Remediation Approach: Critical Review and Implications

JEAN-PIERRE TRÉPANIER[1], STAN PAUWELS[2], RENÉE GAUTHIER[3] AND LORRAINE ROUISSE[1]

[1]Risk Analysis Department, SANEXEN Environmental Services, Longueuil, Quebec, Canada
[2]ABT Associates, Cambridge, MA, USA
[3]Service des lieux contaminés, Quebec Ministry of Environment, Quebec, Canada

14.1 WHAT IS A RISK-BASED REMEDIATION APPROACH?

In the past, the assessment of contaminated soils as well as other media has often been based on generic contaminant-specific criteria. These criteria, discussed in more detail in Chapters 12 and 13, represent maximum concentrations which are not expected to result in unacceptable risk to humans or ecological receptors. Remediation may be required if contaminant levels exceed these concentrations. In many cases, the entire volume of soil that exceeds one or more criteria is excavated and disposed of, or treated in such a way that concentrations fall below the criteria. This represents a substance-by-substance concentration-based approach that is easy to understand and implement. It is a useful tool to apply at sites where more complicated evaluations may not be suitable or cost-effective. However, the criteria or guidelines-based approach can be quite limited in defining remediation plans at larger or complex sites because it implicitly assumes that exposure conditions are the same everywhere. Risk-based remediation normally requires more than just the use of generic criteria (see also Chapter 12 for a discussion on the use of generic criteria within a broader risk-based decision framework).

Risk-based remediation simply refers to the idea of managing cleanup issues in such a way that human health and the environment will not experience unacceptable risks. This approach integrates risk assessment practices with more traditional site investigation and remediation activities. Under this integrated approach, sites are characterized not only in terms of contaminant concentrations, but also in terms of potential receptor exposure to these contaminants.

Environmental Analysis of Contaminated Sites. Edited by G. I. Sunahara, A. Y. Renoux, C. Thellen, C. L. Gaudet and A. Pilon

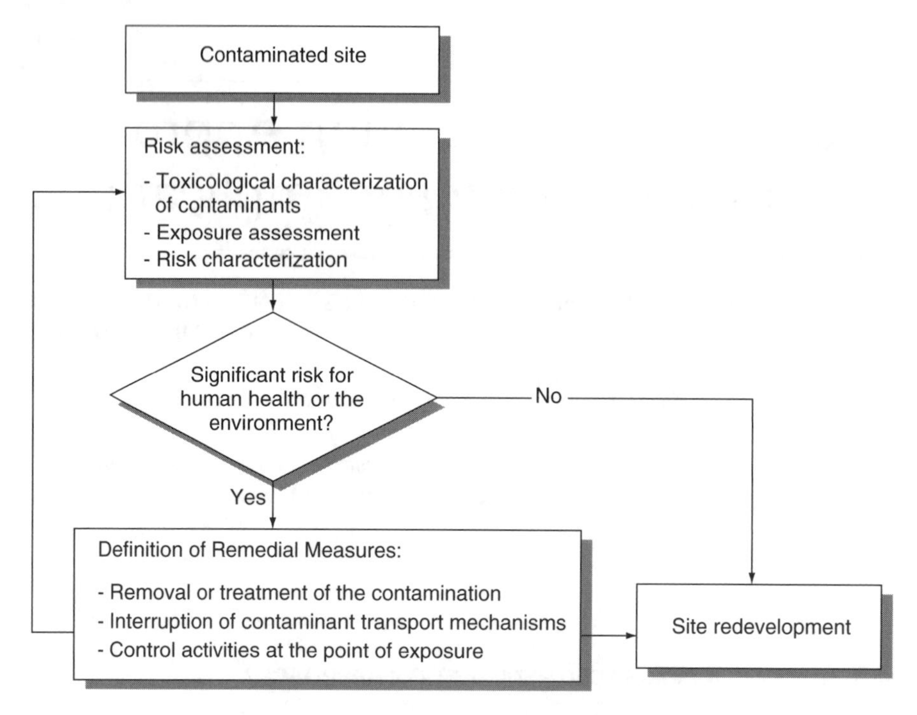

Figure 14.1 Simplified view of the risk assessment and risk management process under a risk-based remediation approach for contaminated sites.

Remedial measures are then defined and applied to limit or prevent exposure to potential harmful levels of contaminants. Figure 14.1 shows an example of the process involved in a risk-based remediation for contaminated soils.

Risks at contaminated sites occur from the interaction between hazardous chemicals and the potential exposure of human or ecological receptors to such chemicals. This definition not only involves concentration, but also exposure, which includes several elements:

- The form and bioavailability of the contaminant in the environment (organic compound, metal complexes, adsorbed or dissolved, etc.);
- The location of the contaminant (surface or subsurface soils, surface water or groundwater, on-site or off-site);
- The presence or absence of receptors (human or ecological);
- The nature, duration and frequency of contact between contaminated soils and potential receptors (short-term, long-term, seasonal).

By including exposure, a risk-based approach may conclude that contaminant levels, which exceed generic criteria, do not necessarily represent a source of significant risk to human health or the environment. It also implies that remedial efforts could focus on minimizing exposure rather than eliminating contaminants. Such a level of flexibility is not available with traditional concentration-based cleanup approaches. It also implies, however, that a remediated site may contain residual contamination. Specific deed restrictions may therefore be required to ensure that responsibilities are properly transferred to a future site owner, in case of a real-estate transaction.

Specific tools are used in risk-based remediation to assess site-related risks. Because risk assessment techniques can be complex, much attention has been focused on their technical limitations rather than on the fundamental implications of risk-based management. One major implication is that risk-based remediation allows for the redevelopment of sites which remain 'contaminated', i.e., sites on which measurable concentrations of contaminant remain after the cleanup is completed. This fact also holds when using generic criteria. In fact, non-risk-based generic criteria would not necessarily pass the test of a good risk assessment. A site, which is remediated based on generic criteria, could also remain contaminated, unless 'contamination' is defined only as contaminant levels which exceed their criteria. Therefore, criteria-based cleanups could pose a risk if they have not been validated by a health or ecological risk assessment. It is probably true that most generic criteria are reasonably protective, since there have been no reports of obvious deleterious effects at sites that were remediated down to criteria.

With a risk-based approach, high concentrations of contaminants may be left behind, as long as exposure can be controlled to limit risks to acceptable levels. Covering contaminated soil with 'clean' material may be sufficient to limit exposure and prevent significant risk to human health. However, such remedial alternatives may leave behind contamination that could last for a long time. Changes in land use — for example, from industrial to residential — or resource use — such as the tapping of contaminated groundwater as a new source of drinking water — may alter exposure such that the initial condition of no significant risk is no longer valid. More important, the extensive use of a risk-based approach could lead to many partially remediated sites, which could become a significant liability to future generations, if land use changes.

Generic risk-based criteria are a reasonable compromise between concentration-based criteria and a cleanup strategy developed from site-specific risk assessment. Such an approach effectively combines the ease of use of generic criteria with a conservative protection of human health and the environment. It also ensures some uniformity in remediation efforts across sites. However, our experience suggests that risk-based criteria lack the flexibility of remediation plans based on site-specific risk assessment. Such an option may not provide cost savings as long as generic criteria are applied as 'concentrations not to be exceeded' on a site. Risk-based concentrations for surface soils may be lower than the current non-risk-based generic values, especially if based on an acceptable additional cancer risk of one in a million (1×10^{-6}). The cost advantage of a risk-based approach relates to the possibility of managing exposure, and not to the fact that it provides less stringent criteria. Validation of generic criteria by risk assessment should be encouraged as a means to ensure adequate protection of health and the environment.

14.2 THE OBJECTIVES OF RISK-BASED REMEDIATION

A key objective of risk-based remediation is the proper allocation of limited resources. This approach has become an essential tool for managing contaminated sites because it offers remedial alternatives in cases where the concentration-based approach would be too expensive or impractical. The objectives of risk-based remediation are not the same for all interested parties. Regulators want to ensure that remediation will protect human health and the environment, whereas owners are concerned with cleanup costs. Remediation may also occur at sites that are being sold or redeveloped. In such cases, the costs of remediation should not exceed the value of the land itself, including the potential benefits associated with its redevelopment.

For both government agencies and stakeholders, there may be a third objective for risk-based remediation, i.e., ensuring that any remediation is based on a consistent and technically defensible approach. Generic criteria have often been criticized in the past because the technical basis for their derivation was unclear or inconsistent. For a risk-based approach, one must be able to explain what the remediation plans are based on. This issue is critical for the credibility of government agencies, which are responsible to the public, as well as for site owners, who may wish to create a climate of confidence in the community where the cleanup project takes place.

The financial benefits of risk-based remediation have been demonstrated at a number of sites. Major cost reductions are generally tied to a drop in the volumes of soil or groundwater to be removed and treated. In cases where only human health is of concern, no excavation or treatment may be required at all. For example, simply covering the contaminated soils with a clean layer may be sufficient to reduce the current risk to negligible levels. In such cases, the risk assessment only has to show that the soil covering will make the site safe for its intended purpose. The presence of hazardous wastes (drums) or the contamination and use of underground water should not be an issue, however.

In other cases, a risk assessment can clearly define the location and volumes of soil to be excavated. The costs associated with such an approach are typically lower than those based on generic criteria which have to be met everywhere on a site, irrespective of exposure. For this reason, sites with subsurface soil contamination can benefit substantially (in terms of costs) from risk-based remediation plans. This is an important incentive for using risk-based remediation. Using partial mitigation measures — such as covering up surface contamination with clean soil-justified by a risk-based remediation plan still entails responsibilities on the part of the site owner. For example, the site owner may have to implement a strict monitoring program to ensure that the integrity of those mitigation measures remains intact over time. If any problems arise, the owner will also intervene to get the site restored to safe conditions. This long-term commitment can have serious financial implications or operational constraints.

In terms of protecting human health and the environment, risk-based approaches can increase our understanding of the advantages and limitations of remediation plans. Concentration-based approaches often lead to the false perception that no contamination remains after a cleanup is completed. In most cases, contaminants do remain on site, but at levels below the criteria. The actual risks resulting from this residual contamination are not known unless a risk assessment is carried out. Again, estimating how efficient a risk-based approach is in protecting health and the environment, as compared to traditional concentration-based approaches, will depend on the quality and representativity of the methods used to carry out the risk assessment. In other words, it is difficult to state that the risk-based approach protects health and the environment in a manner which is better or worse than generic criteria. However, one can estimate 'how safe' the residual contamination is, which represents a major advantage over the non-risk-based criteria approach.

14.3 A CRITICAL LOOK AT RISK-BASED REMEDIATION

The potential cost saving from using a risk-based approach at contaminated sites has made it a useful environmental management tool. This approach, however, also has several shortcomings worth discussing.

14.3.1 STRENGTHS AND WEAKNESSES OF RISK-BASED REMEDIATION

The strengths and weaknesses of risk-based remediation relate to two distinctive, yet interrelated elements:

- The concept of environmental management based on risk;
- The advantages and limitations of risk assessment, at the center of risk-based remediation.

Managing environmental issues at contaminated sites based on risk is a fundamental concept that must be fully understood and accepted by both the government and the public. Implementing this option in regulations or administrative practices should ideally reflect a societal choice, based on a clear, complete and unbiased examination of the implications of risk-based environmental management. Oftentimes, however, risk-based remediation has been implemented before a true debate on its implications had been completed.

The major implications of risk-based remediation are summarized below:

- Risk-based remediation allows for the redevelopment of contaminated sites where traditional concentrations or criteria-based approaches do not prove practical;

- Sites can be redeveloped even with contamination left behind when it is not technically or economically feasible to clean them up. A technical example of such a situation is the presence of contamination beneath an existing building whose excavation could result in structural failure;
- Compared to a criteria-based approach, the large-scale use of risk-based remediation in a given area could result in health liabilities to future generations if those sites were redeveloped so as to increase exposure.

Balancing out these competing considerations is not simple. Considering that non-risk-based environmental assessments could prevent site redevelopment due to unreasonable cleanup costs, such sites would remain more contaminated than after a risk-based cleanup. Fortunately, one can implement a risk-based approach in such a way that the community benefits from its advantages, while minimizing potential adverse effects. Conditions for such an implementation will be discussed later in this chapter.

Other important issues to consider in a critical examination of risk-based remediation relate more specifically to the risk assessment process. Risk assessment has been criticized in the past because of the high uncertainty levels associated with risk numbers. These uncertainties made it a challenge to quantify the true magnitude of actual risks posed by residual contaminants. Several tools are available to address such uncertainties. By definition, the concentration-based approach using criteria not based on risk cannot assess uncertainties. On this issue, the risk-based approach provides more useful answers than the traditional criteria approach.

Two aspects of the risk assessment paradigm are worth examining on this issue. First, risk assessment usually relies on mathematical modeling. This technically complex approach can be difficult to understand by the public and regulatory authorities. Second, as mentioned above, risk assessment can be unreliable because large uncertainties remain on the final estimates, and because some risk estimates have proven to be unrealistic.

These criticisms are interconnected. It is difficult not to use mathematical tools to estimate exposure and risk. Such models include both the mathematical equations and the variables used in the calculations; the two elements are related to the final uncertainty of risk estimates. Even if the reliability of these estimates is not a direct function of the models themselves, simple models cannot assess the full complexity of a system in which exposure occurs via multiple routes, to many different receptors (i.e., people of different age, sex and occupation or ecological receptors of different species), with variations in space and time. It is, therefore, desirable to use models that are sophisticated enough to consider these complexities. The flip side of using sophisticated models is that the technical aspects of the risk assessment become even more difficult to understand by the legislators, the public and the site owner.

14.3.2 THE PROBLEM OF VALIDATION IN RISK ASSESSMENT

Validation is an important risk assessment challenge. Risk assessment is essentially a predictive exercise. Even when examining an existing contaminated site, it is used to predict current or future risks. In theory, epidemiology rather than risk assessment could be used to measure the actual impact of contamination on human health. This is not practical, however, because most exposures are at the low end of the spectrum, the potentially affected populations are usually relatively small and site cleanup decisions cannot wait for the results of such studies. Therefore, risk assessments estimate if an existing situation poses a significant risk now or in the future. Since we tolerate only risks so small that epidemiology cannot detect them, it becomes intrinsically difficult to validate the final risk values.

Risk usually arises from multi-media exposure. Although analysts typically make the assessment less complicated by considering only major exposure routes, such studies simplify the actual exposure and may not represent all the pathways and interactions involved in an actual environmental exposure. In fact, risk analysts often consider *a priori* that certain exposure pathways are negligible when, in fact, they may not be. Figure 14.2 illustrates the multi-media dimension of human exposure to environmental contaminants. Although many pathways shown here may not be involved in a given situation, it would be unusual if most of them were not involved at least to some

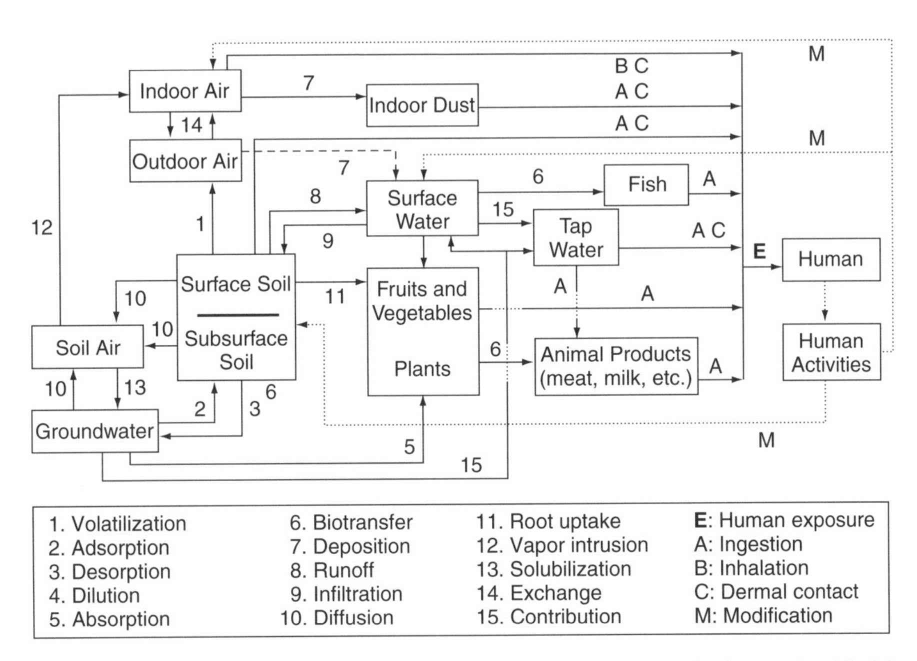

Figure 14.2 Schematic representation of the multi-media exposure for human health risk assessment.

extent. To determine with 'certainty' if a specific pathway is significant or not can only be done by performing the exposure assessment for this pathway. In fact, significance may not be obvious when checking for additional risks below 1×10^{-6}. Experienced risk analysts have learnt that what seems at first insignificant may become more important than what could be expected intuitively. For example, dermal contact has often been omitted from risk assessments because it was considered to be a negligible exposure pathway and because the tools were not available to estimate such exposure. However, under certain circumstances this pathway could contribute significantly to the total absorbed dose, i.e., bathing in contaminated water (USEPA 1992a). Similar difficulties are associated with ecotoxicological risk assessments, which also deal with multi-media systems (see Figure 14.3).

An exposure assessment is usually difficult to validate because of its multi-media dimension. A complete validation should consider all potential pathways, which implies the sampling and analyzing of many environmental media. Moreover, detection and quantification limits need to be low, even well below the detection limits of current analytical methods. Furthermore, because temporal, spatial and individual variations can introduce a large margin of error in measurements, the number of media-specific samples required to conduct a sound validation can be quite large.

Validating dose estimates is also a complex issue. As with environmental concentration measurements, sensitive analytical tools are required. Many

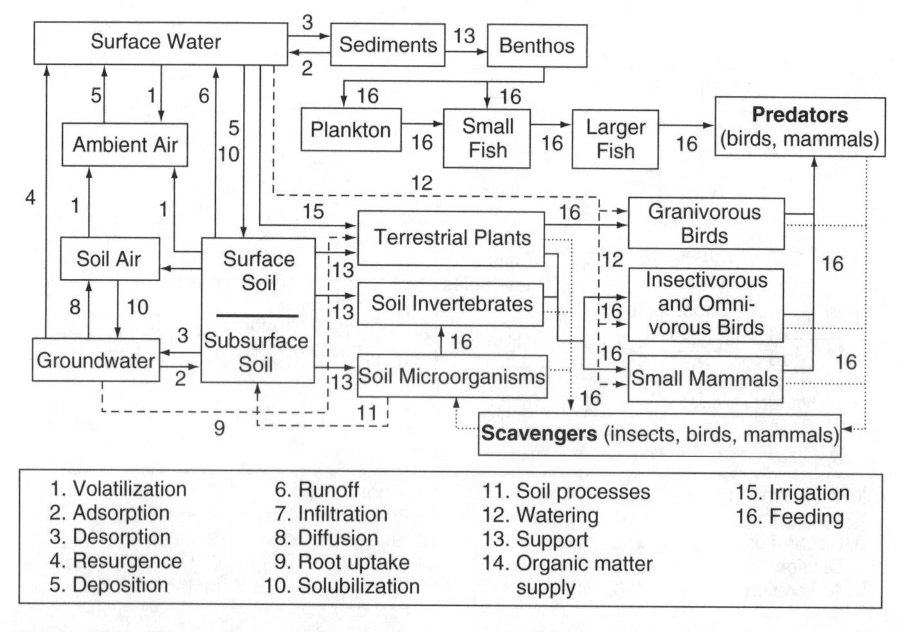

Figure 14.3 Schematic representation of the multi-media exposure from a contaminated soil for ecotoxicological risk assessment (simplified example).

physiological factors, including absorption, distribution, metabolism and elimination, modify contaminant levels in biological tissue. The media being analyzed — biological tissue versus water or soil — can also interfere with the analytical methods. These factors make an exact measurement of exposure quite difficult.

Another complication is our lack of understanding of the relationship between exposure and expected concentrations in tissues. Even when tissue concentrations can be measured with precision, it remains a challenge to quantify the exposure which caused the concentration. Although physiologically-based pharmacokinetic (PBPK) models can help to assess this relationship for certain chemicals, uncertainties remain when using these tools, especially if one considers the actual environmental exposure versus laboratory-controlled conditions. Actual exposures vary in terms of frequency, duration and pattern across three exposure routes (i.e., inhalation, ingestion and dermal contact). For each of these routes, the nature of the media in contact with the receptor, as well as the physicochemical properties of the contaminant, can affect the amounts absorbed. Depending on the route of exposure, the fate of the contaminant in the organism will also vary. The well-known first-pass effect, which occurs in the liver after ingestion, is an example of such a factor. Exposure validation is also complicated by variability in the receptor. Some factors like age, sex and strains may affect contaminant uptake and absorption. Individual variations are also present among similar organisms of the same age, sex and strain.

Another confounding factor is the simultaneous presence of different compounds in the environment and the diet. Some ions are known to increase or reduce the absorption of other ions. For example, cadmium decreases the absorption of zinc and copper, and calcium that of cadmium (Klaassen and Rozman 1995). More complex mixtures may also produce similar effects. For example, milk can increase lead absorption (Kelly and Kostial 1973). But what is known about these interactions represents only an infinitesimal part of the extremely complex world of living organisms. Assessing the interactions between two or three substances is a difficult and complex task, even if one considers a single biological species. In this context, understanding and predicting the interactions between hundreds of contaminants and species appears practically impossible.

These and other factors make it difficult to validate exposure assessments. The last part of the risk assessment process is potentially even more uncertain. With an adequate toxicity database, it is fairly easy to predict the risk of high dose exposure for a single contaminant. The extremely low levels of risk targeted by risk analysts, however, make it difficult to quantify risk with a reasonable level of confidence. This is especially true for cancer risk estimates, which are typically based on high to low dose extrapolation models whose biological basis can be more hypothetical than scientifically proven. For some of these models, the mathematical form does not even allow the additional risk to reach a zero value, even when the dose is zero (i.e., a small but calculable risk remains even if the contaminant of concern is absent). These models were generally developed to describe the behavior of the dose/response curves at high doses, and not to predict the risk at very low doses (especially with a precision of 1×10^{-6}). In all cases, these models are fitted to experimental data in order to estimate chemical-specific parameters. In addition, the experimental data are usually restricted to a few data points (typically three experimental doses, plus one or two control groups). The confidence limits of a given mathematical model adjusted to these points can therefore be quite large. Typically, the 95% confidence intervals on cancer incidence in an experimental assay may be as large as 15 or 20%. Predicting a risk at a level of 1×10^{-6} from these curves is not an exact science. This is why risk assessment usually relies on highly conservative assumptions. It should be recognized that even when trying to estimate the worst-case scenario of risk, significant uncertainties remain.

14.3.3 THE CURRENT PRACTICE OF RISK ASSESSMENT

Risk assessment has been applied to many situations, including the assessment of risk posed by new facilities (e.g., hazardous waste incinerators) as well as risks of existing contamination or risks resulting from new activities or new products. Until the early 1990s, much effort was put into developing dose/response models that could estimate risks at very low doses. Integrated models were also developed to estimate human exposure from environmental media. These models tended to be quite simplistic because the data necessary for such exposure assessments were incomplete.

After the initial tool development stage in the early and mid 1980s, risk assessment was used extensively to estimate risks at contaminated sites. Early assessments were often simplistic or overconservative. For instance, only adults were considered in human health risk assessment, even when children were of concern, or the only exposure route considered was the one that appeared to be the main route or for which data were available.

Even today, the following limitations still exist in the application of risk assessment:

- Age-related variations in the exposure patterns are poorly assessed.
- Background exposure is often omitted from the total exposure.
- Variability among individuals is rarely adequately considered. For instance, when Monte Carlo simulations are used instead of a simple deterministic approach, analysts often do not consider the interdependence between variables. In other cases, it is only applied to a small number of variables, though many others also vary among individuals. Results from such limited simulations cannot be interpreted as statistical distributions of the actual exposure.
- Uncertainty is usually not correctly accounted for, even though some risk assessments claim to consider it. When uncertainty is discussed, it is often

confused with variability (i.e., use of Monte Carlo simulations to 'describe' uncertainty). Sophisticated numerical methods, such as second-order Monte Carlo, may result in impressive-looking curves. Any statistical approach, however, would simply skip the fundamental nature of uncertainty, because it has no statistical distribution. Fuzzy numbers arithmetic can be a useful tool to quantify uncertainty, but is rarely used in risk assessments. Its use would show the high level of uncertainty inherent in risk assessment estimates. This uncertainty, describing what is unknown about a system, should be viewed as a reason for better data and better practices in risk assessment.

Another limitation in current risk assessment practices is the oversimplicity of models used in site-specific studies. Such models often replace actual measurements to help reduce the cost of a study. Sound modeling of environmental transfer of contaminant or multi-media exposure, however, should rely on a detailed and validated characterization of input data. Such a characterization is expensive and may be in conflict with the site owner's objective of cost control. That is why most risk assessments use very simplified models, with little or no calibration and validation.

Government agencies and professional organizations have developed risk assessment guidelines based on a tiered approach, which includes two or three 'levels' of evaluation. The ASTM standard 1739 (ASTM 1995) is an example of this approach. This guideline includes three levels of risk assessment to assess potential human health effects:

- **Tier 1 evaluation** develops non-site-specific soil and groundwater criteria for direct and indirect exposure pathways using conservative exposure factors, simple fate and transport models for potential pathways, and various land use categories.
- **Tier 2 evaluation** applies the direct exposure values established under a Tier 1 evaluation at the point(s) of exposure developed for a specific site and develops values of potential indirect exposure pathways at the point(s) of exposure based on site-specific conditions.
- **Tier 3 evaluation** develops values for potential direct and indirect exposure based on site-specific conditions.

The level of site-specificity and complexity of models used within each tier are closely related. As a general rule, more complex models are used in detailed (Tier 2 or 3) risk assessments. In practice, truly detailed (Tier 3) risk assessments are not frequently performed because Tier 2 evaluations are usually sufficient to address most situations.

Figure 14.4 shows the effect of the level of evaluation on various parameters. As a general rule, the more specific the risk assessment, the higher the complexity of the models used and the lesser the degree of conservatism in risk estimates.

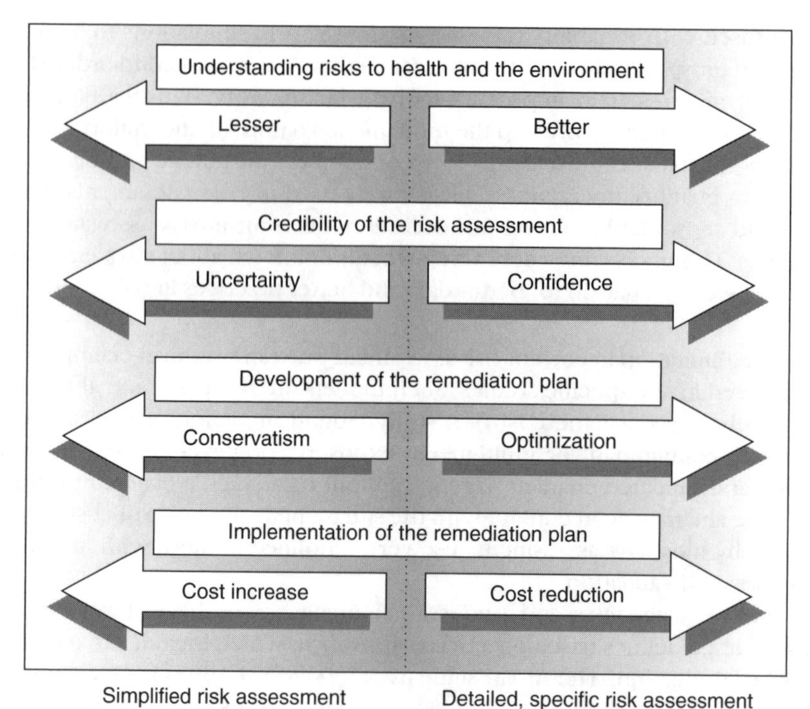

Figure 14.4 Effect of simplified and detailed specific risk assessment on various considerations.

14.4 IMPROVING THE PROCESS BASED ON PAST EXPERIENCES

While legitimate criticisms have been formulated against risk assessment and risk-based environmental management, it is also useful to examine how these processes could be improved.

14.4.1 UNIFORMITY AND QUALITY OF RISK ASSESSMENT METHODS

The first element to consider is the quality of the risk assessment methods used to perform site-specific studies. Because they are at the center of the risk-based remediation approach, these assessments should be reliable and scientifically sound. The credibility of the approach depends on the validity of the risk estimates.

Numerous government agencies at the federal, provincial or state levels have developed and published risk assessment guidelines, or technical guidance documents (USEPA 1992b, 1996a,b, CCME 1996). Although such documents discuss the basic theory of risk assessment, they cannot provide detailed technical guidelines for every potential situation. The expertise of individual analysts remains essential to ensure a quality risk assessment. The following basic principles define the validity and reliability of risk assessments.

- Risk assessments should consider all contaminants detected at the site for which adequate toxicological information is available. Although this statement does not conform with some existing guidelines, it does not appear judicious to perform a risk assessment based on inadequate toxicological information.
- Risk assessments should address all potential exposure pathways. One should not consider *a priori* that a specific pathway is unimportant and should therefore be omitted from the analysis. Carrying out the risk assessment is the only way to ensure that a pathway is indeed negligible, in a specific situation, for all groups potentially exposed and for all contaminants.
- Risk assessments should consider many age groups in the exposed human population to ensure that the exposure for any given age group does not exceed safe levels. Moreover, since a number of age-related physiological factors can alter the absorption, distribution and toxicodynamics of xenobiotics, these factors should be considered along with the variations in the exposure patterns when known.
- Risk assessments should consider all potential adverse toxicological effects, including cancer, chronic effects other than cancer and acute effects.
- If feasible and practical, risk assessments should consider the potential interactions between all contaminants present. This is especially true in human health risk assessments, when several contaminants can have the same toxicological mechanisms or produce similar effects (i.e., cancer); this is a frequent occurrence in occupational exposures.
- For contaminants with toxicological threshold effects (i.e., most effects other than cancer), the risk assessment should consider natural background exposures to chemicals to ensure that the *total* exposure is properly quantified.
- Risk assessments should consider the effect of seasonal variations on exposure and, thus, on risk. The reason is that seasonal variations may significantly modify exposure patterns. This concerns cancer risk estimates (i.e., different weighted long-term average doses) as well as other effects (i.e., seasonal doses that may exceed a safe exposure during a significant period of time).
- Risk assessments should be conducted by, or under the supervision of, well-trained qualified personnel with adequate background and training in toxicology, multi-media environmental exposure assessment and risk characterization.

14.4.2 POTENTIAL CHANGES OF LAND USE

Implementing a risk-based approach to manage contaminated sites typically implies that some level of contamination may remain at the site. Because risk-based remediation considers exposure patterns that are specific to the site under study, its conclusions cannot be extrapolated to other situations. Site remediation based on site-specific exposure scenarios has to be accompanied by some mechanism which ensures that a change of land use will not give rise to unacceptable risks.

A similar constraint should also exist to control changes in the future development of a site, even if site ownership does not change hands. For example, contaminated soil from deep layers may be brought to the surface as a result of excavation work, thereby allowing potential exposure to contaminants that were previously inaccessible. Remediation plans should include provisions to ensure adequate long-term protection of human health and the environment. Examples of such provisions include limitations on land use (i.e., building residential housing on sites cleaned up only to support commercial or industrial activities) or restrictions on the type of work allowed on a site.

Such limitations can be prescribed for a definite time period. For example, residual concentrations of petroleum hydrocarbons in soil can decrease over time due to biotic and abiotic degradation processes. Including this knowledge in the assessments can help define the time period at the end of which the limitations could be modified or cancelled after due verification.

14.5 SPECIAL CONSIDERATIONS FOR ECOTOXICOLOGICAL RISK ASSESSMENT

In the past, contaminated sites were assessed mainly for human health risks. In fact, the first risk assessment guidelines developed in the early 1980s focused only on human health concerns. Since then, ecological risk has received increased attention. The reasons are that: (1) new tools and approaches to screen or quantify ecological effects have been and continue to be developed; (2) the public wants more attention to be paid to the environment; and (3) ecological risks may outweigh human health risks at some sites.

Assessing risks to ecological receptors is challenging because terrestrial and aquatic environments can be more complex than human populations. For example, instead of dealing with a single, well-studied species (i.e., *Homo sapiens*), one must address multiple species and lifestages that interact within specific assemblages and communities. The mix of species and exposure pathways also varies depending on whether contaminants are located in sediment, soil or surface water. Some receptors, such as fish, plants or benthic and soil invertebrates, live within the contaminated matrix and therefore are subject to direct exposure. Others, such as birds, mammals or piscivorous fish, are also exposed to contaminants via the food chain. In addition, the same compound has different fates depending on whether it is found in soil, sediment or surface water.

This inherent complexity creates a unique challenge for the risk assessor. Since it is difficult to quantify risk for all the combinations of receptors, exposure pathways and contaminants, one has to define which components of the ecosystem should be considered to properly evaluate the ecotoxicological risks posed by the site contamination. Upon completion of this step, one should be able to eliminate certain matrices, exposure pathways, contaminants and/or species from further consideration. It then becomes possible to: (1) identify the major risk 'drivers'; (2) develop a conceptual risk model for the site; (3) select sensitive receptors or habitats to be protected; and (4) choose a set of tools to measure and better quantify unacceptable effects.

A variety of tools have been developed to help quantify ecological risks. Examples include the following:

- Literature searches to identify species-specific toxicity values or exposure factors and contaminant-specific fate parameters;
- Biological field surveys to identify sensitive receptors, assess the site-specific potential for exposure by target species and compare site-specific species, compositions or distributions and population levels with those found at reference sites;
- Laboratory or field ecotoxicity testing to measure the toxicity of contaminated soil, sediment or surface water samples to target species (such as plants, invertebrates and fish);
- Laboratory or field bioaccumulation studies to determine how much of the contaminants present in soil or sediment samples collected at the site can be taken up by organisms;
- Food web modeling to calculate the movement of bioaccumulative contaminants into terrestrial or aquatic food webs;
- Wildlife exposure modeling to calculate more realistic contaminant daily doses based on site-specific values for home ranges, migration patterns, concentrations of contaminants in prey items, contaminant bioavailability, bioaccumulation potentials or probability distributions.

There really is no standard way to analyze and interpret such data because the results need to be assessed within a site-specific context. In addition, some of the data (particularly those from toxicity testing or biological field studies) can be quite ambiguous and difficult to interpret. This is another major characteristic of ecotoxicological risk assessments. Although everything is not perfectly clear in human health risk assessment, the large number and diversity of potential receptors in ecotoxicological risk assessment significantly increases the complexity of the study. For example, the apparent toxicity of soil samples may vary based on exposure duration, species used and lifestage tested. In addition, confounding answers that sometimes result from different evaluation approaches (e.g., deterministic evaluation versus biotests) may be quite difficult to interpret.

Ultimately, the results of most ecological risk assessments are interpreted within a semi-quantitative 'weight-of-evidence' context to determine if a significant risk is actually realized at a site. Therein lies probably one of the limitations of most ecological risk assessments as currently applied at contaminated sites.

The final data interpretation usually relies on a healthy dose of professional judgement. Therefore, depending on one's viewpoint — the site owners, regulator, environmental pressure group or the public at large — different interpretations can be provided for the same data set. This injects a measure of subjectivity into the whole process, which can lead to diverging opinions about the level and extent of ecological risk and whether or not one should proceed with remediation.

Even though the risk characterization step can be open to diverging interpretations, the ecological risk assessment process itself has been successfully applied at numerous sites. By proceeding with a study in a tiered fashion, keeping the assessment focused on a limited but clearly-defined set of objectives and maintaining open channels between the site owners, regulators, risk assessors and risk managers, it is possible to limit misunderstandings and minimize differences of opinion.

14.6 A FINAL WORD OF CAUTION

Risk-based remediation refers to 'corrective' actions at a contaminated site based on risk considerations. Such an approach is desirable because past disposal practices, especially in industrial areas, typically did not consider the potential long-term effects to soils and other environmental media. In many cases, the contamination of soil and groundwater resulted from improper practices of landfilling or waste disposal.

Risk assessment and risk management are practical tools offering flexibility within the identification process of the safest and most cost-effective remediation solutions. Although still perfectible, this approach is undergoing constant revision and benefits from the most recent scientific and technical progress. In many cases, risk-based remediation can provide solutions where a more traditional approach would be practically impossible to apply.

As discussed earlier in this chapter, however, the large-scale use of a risk-based corrective approach may result in numerous partially decontaminated sites. Although this cannot be totally avoided in an industrial world, a preventive approach can certainly reduce the long-term degradation of our environment. Past experiences clearly show that cleanup activities can be prohibitively expensive. Soil contamination may also result in unexpected long-term effects on public health as well as on the environment, even when proper risk assessments are carried out. This is particularly true if future land use patterns at partly decontaminated sites change in such a way that exposure to residual contaminants is enhanced. These considerations should help us keep in mind that prevention still remains the best way to avoid adverse effects of site contamination.

REFERENCES

ASTM (1995) *Standard Guide for Risk-based Corrective Action Applied at Petroleum Release Sites*. ASTM Standard E 1739-95, American Society for Testing and Materials, Philadelphia, PA, 51 pp.

CCME (1996) *A Protocol for the Derivation of Environmental and Human Health Soil Quality Guidelines*. Document 108-4/8-1996-E, Canadian Council of Ministers of the Environment, 169 pp.

Kelly D and Kostial K (1973) The effect of milk diet on lead metabolism in rats. *Environment Research*, **6**, 355-360.

Klaassen CD and Rozman K (1995) Absorption, distribution, and excretion of toxicants. In *Casarett and Doull's Toxicology — The Basis Science of Poisons*, fifth edition, Klaassen CD (ed), McGraw-Hill, New York.

USEPA (1992a) *Dermal Exposure Assessment: Principles and Applications*. Interim Report EPA/600/8-91/011B.

USEPA (1992b) *Guidelines for Exposure Assessment*. EPA/600Z-92/001, Federal Register 22888-22938.

USEPA (1996a) *Proposed Guidelines for Carcinogen Risk Assessment*. EPA/600/P-92/003C, Federal Register 17960-18011m.

USEPA (1996b) *Guidelines for Reproductive Toxicity Risk Assessment*. EPA/630/R-96/009.

15

Application Requirements for ERA of Contaminated Sites

LOUIS MARTEL, RAYNALD CHASSÉ, ANNE-MARIE LAFORTUNE AND SYLVIE BISSON

Centre d'expertise en analyse environnementale du Québec, Ministère de l'Environnement du Québec, Québec, Canada

15.1 INTRODUCTION

The decision making process for the management of contaminated sites must examine and integrate both scientific and non-scientific (e.g., economic, social, engineering, legal, administrative) criteria, since there is a direct relationship (and correlation) between the two. The scientific component of the decision making process must be grounded on a sound scientific information base; ecotoxicological risk assessment (ERA) represents a useful process to obtain the scientific information needed to facilitate effective management decisions.

In this chapter, after having introduced the field of ERA for contaminated sites, we will focus on all important concepts and steps for the realization of ERA for contaminated sites and conclude by examining three essential, though often neglected, elements that are critical to the usefulness and credibility of ERA, namely uncertainty analysis, quality assurance and control, and ecotoxicological risk communication.

15.2 ECOTOXICOLOGICAL RISK ASSESSMENT OF CONTAMINATED SITES

ERA can be defined as a rational process, that supports the decision making process by identifying, comparing and analyzing descriptive measures toward formulating a global judgment with respect to the environmental fate of contaminants and their effects on receptors (CEAEQ 1998). It involves the estimation of likelihoods/probabilities and magnitudes of undesired events (Suter 1993) and takes explicitly into consideration uncertainties associated with the assessment. This process can provide useful and accurate information to risk managers about the potential adverse effects of different management options (AIHC 1997), while offering an immense challenge to basic and applied science (Bartell *et al.* 1992).

Environmental Analysis of Contaminated Sites. Edited by G. I. Sunahara, A. Y. Renoux, C. Thellen, C. L. Gaudet and A. Pilon

Initially, ERA frameworks were based on the human health risk assessment framework (NRC 1983). This premise, however, was rapidly recognized as inappropriate, given the complexity of ecological systems. Consequently, in the early 1990s, ERA frameworks evolved to encompass and integrate distinctive characteristics of this discipline (NRC 1992, USEPA 1992).

In the context of contaminated sites, ERA can be termed 'source-driven retrospective assessments', since they are initiated by the observation of a contaminated site (the source) leading to the requirement for an assessment of possible effects associated with it (Suter 1993). The tool box available for this kind of ERA is more complete and diverse (e.g., toxicity tests in laboratory/field, chemical analysis of field samples, biomarkers, biological surveys, etc.) than for conventional, predictive assessments. Furthermore, as argued by Suter (1993), the usual separation between source, exposure and effects assessment is no longer appropriate since each act and occur simultaneously in the field. Therefore, this kind of assessment might benefit from a linear operational process that describes the logical sequence of actions and operations, leading to the estimation of ecotoxicological risk (CEAEQ 1998). More information on such a process can be found in Section 15.3.

There is general agreement that ERAs are best conducted using a tiered or iterative approach (CRAM 1996, 1997a,b). Through an iterative process, only the necessary information pertinent to management needs will be incorporated into the assessment. This process progresses from a qualitative to a quantitative assessment, the latter being more detailed than the former. At each iteration, ERA becomes less conservative, more site-specific and the level of uncertainty decreases. In return, however, ERA becomes more sophisticated and expensive (Belluck *et al.* 1993). Hence, the need exists to balance the tier level with management needs and conclude the assessment where appropriate. At the same time, it is critical to avoid compromising the required depth of the assessment in favor of a less expensive ERA; when the precision level is inadequate to support decision making, there is a real and likely risk that more costly consequences will arise (Powell 1999).

Contaminated site risk management is a complex process involving and balancing many diverse considerations (e.g., regulatory, political, social, economic, technical, health and ecological factors) (CRAM 1997a). ERA is an essential input to this process, and is generally used to ascertain if a site presents a problem and, if so, to determine the acceptable site-specific contamination level. While this is the general application of ERA for contaminated sites, it may also be used to support other activities associated with contaminated site management. Often contaminated site managers are confronted by numerous problematic sites or by a contaminated site of great expanse. Most frequently, in these situations, the manager's first task is to prioritize the intervention activities, with ERA offering practical assistance in this exercise. Generally, in this use, it is not necessary for the ERA to proceed beyond the first assessment tier.

ERA can be used for the selection of the appropriate treatment technology for the contaminated soil or management scenario for the contaminated site. In these cases, ecotoxicological risks associated with each technology or with each scenario are compared; ERA contributes to this process as it can express changes in ecological effects as a function of changes in exposure to contaminants. However, a distinctive feature of this type of application is its prospective nature. Almost any remediation applied to a site will have its own impact on the site's ecosystem. Therefore, these ERAs must be compared to the ERA of the present (no action scenario) site conditions (Hope 1995).

After having implemented a management scenario, it is important to evaluate its effectiveness. Environmental monitoring can be used to determine if the action applied to the site has indeed been effective in reducing the ecotoxicological risk level. The monitoring program should focus on the same aspects (assessment endpoints) as did the ERA leading to the implementation decision, and should be designed following the ERA process (Suter 1993, Bartell 1998).

Regardless of its use, the value of the ERA in the contaminated sites management process is directly linked to its contribution to informed decision making. Toward this end, the realization of an ERA requires that management and assessment-related activities be undertaken. A dialogue between those who are in charge of risk management and those in charge of risk assessment must be initiated at the onset of the ERA. Risk assessors must understand the objectives and needs of risk managers, similarly, risk managers must understand and take into account the results of ERAs and their ecological significance. To summarize, ERAs will contribute effectively to contaminated sites management only if there is a clear connection between assessment and management activities, and a clear distinction between the responsibilities of risk assessors and risk managers (CRAM 1997a,b). As an example, precise definition of the question at hand or the selection of the level of conservatism necessary are risk management duties, whereas exposure estimates of ecosystem components are clearly risk assessment duties (Jasanoff 1993, Fairbrother *et al.* 1994).

15.3 ERA'S CRITICAL STEPS

Numerous jurisdictions have developed their own ERA guidelines, which are generally based on one of the two authoritative frameworks for ERA: the US Environmental Protection Agency Framework for Ecological Risk Assessment (USEPA 1992) and the National Research Council Paradigm for Ecological Risk Assessment (NRC 1992). In this section, we present and describe a linear process designed specifically for the ERA of contaminated sites, which accommodates both frameworks. The two principal advantages associated with this type of linear approach are that:

- It organizes the activities and operations according to a linear work structure that facilitates planning (e.g., time, cost, resources, etc.) and implementation of tasks;

- It can be used in conjunction with any guidelines because it is compatible with and based on both existing ERA frameworks.

The linear ERA process leading to the estimation of ecotoxicological risk associated with contaminated sites is presented in Figure 15.1. It consists of four principal phases. The first phase is planning, which corresponds to management activities and whose principal goal is to explicitly define the ERA's objective. Since the planning is not part of the assessment phases, it is not included in the iteration loop of the process. In the latter three phases of the process, greater emphasis is placed on the formulation phase as it has a marked influence on the description and characterization phases. But it is important to recognize that each phase will be influenced by the others, either prospectively or retrospectively, as indicated by double arrows in Figure 15.1. Note that the process encompasses activities relating to risk managers and assessors. Moreover, it explicitly includes communication between risk assessors, risk managers and stakeholders, as needed.

The process will now be described in detail, with discussion focused on its most critical steps. As appropriate, information pertaining specifically to the first or higher tiers will be identified as such.

15.3.1 PLANNING

The linear ERA process begins with a planning phase whose purpose is to establish the goals, scope and focus of the assessment, such that the results of the ERA will be useful for the risk management of the contaminated site (Moore and Biddinger 1995). Critical to the success of this phase is the dialogue between the risk management and risk assessment teams, which must be initiated at the beginning of the planning phase. As part of the ecotoxicological risk communication process, this dialogue provides the basis for integrating

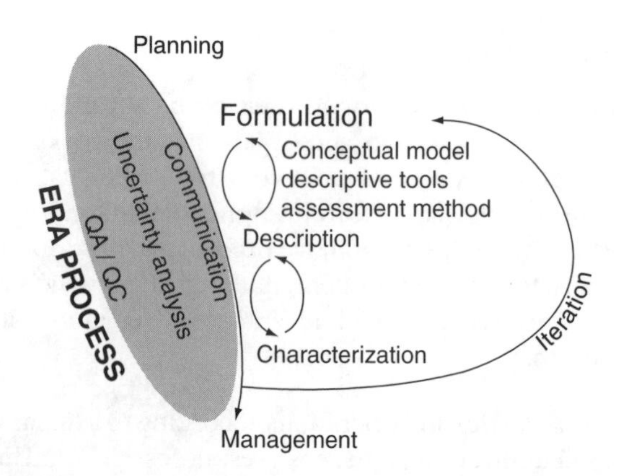

Figure 15.1 Linear ERA process (modified from CEAEQ 1998).

the risk management needs into ERA. It is imperative to remember, however, that planning is a management activity that remains distinct from the scientific conduct of the assessment.

The first aspect of planning is to clearly establish the management problem in an ecotoxicological context, such that the ERA can adequately address the issues as initially conceived by the risk manager and stakeholders. To this end, regulatory, technical and socio-economic issues and implications must be clearly defined, as well as applicable spatio-temporal scales. In addition, time and budget constraints must also be considered.

The second aspect of planning is to define the risk management decision that will be supported by the assessment. This involves the formulation of a decision statement that the study will attempt to resolve. According to USEPA (1994a), a decision statement links the principal study question to actions based on possible outcomes of the assessment. To define a decision statement, the problem should be reviewed and, where possible, broken down into manageable pieces. This break-down can be performed according to a temporal sequence of conceivable actions (e.g., confirming risk → determining the objective of the decontamination → monitoring) or on a spatial base, that is, by dividing the contaminated site into more homogeneous or consistent units.

Next, we must define for each fragment or unit the principal question to which the ERA must respond, the possible responses to these questions, and the actions which must be undertaken based on the responses obtained. All that remains is to combine the question with the conceivable actions toward defining the decision statement, and where the problem has been fragmented, to structure the logical resolution sequence of the decision statements. It is especially important not to underestimate the importance of the formulation of the decision statement, as it is a critical input for the next phase of the linear ERA process.

These two aspects of planning lead to the elaboration of a general objective which must define the expectation of risk managers concerning the ERA. Also, the following questions should be addressed in planning. Why is the ERA needed? What are the ecological concerns associated with the context? What are the questions which will need to be answered? What decisions will ERA influence? What are management concerns (logistic, public or socio-economic considerations)? Ideally, the results of the planning phase should be recorded and summarized in report format, and subsequently transferred to the assessment team who can initiate the next phase of the ERA — problem formulation.

15.3.2 FORMULATION

The foundation for the entire ERA is the ecotoxicological problem formulation. This phase of the ERA involves three major steps: elaboration of the conceptual model, determination of descriptive tools, and definition of the assessment method. The results of this phase lead to the development of an assessment plan, which constitutes the guiding report for the formulation phase. The assessment plan details the technical and scientific elements necessary to the assessment, identifies the biases and potential difficulties, offers a timeline for ERA realization and identifies required resources.

15.3.2.1 Conceptual Model

The ecotoxicological conceptual model is a representation of the environmental system which examines the source and routes of transport of contaminants, contaminated media, routes of exposures and ecological receptors. It is particularly essential to determining the routes of exposure and the potential ecotoxicological effects on receptors (ASTM 1995). Supported by a definition of the ecotoxicological problem, the conceptual model is based on an analysis of available information and a site visit (IEA 1995), as well as an analysis of the source of stress and the implicated ecosystem. It leads to the formulation of hypotheses of perturbations which describe the potential ecological impacts of the contamination.

The conceptual model results from a logical analysis of the situation under study, in terms of contamination and receptors. Toward its construction and refinement, the conceptual model must adopt a systematic and iterative approach that incorporates all information that has been obtained during the assessment. The complexity of a conceptual model should be consistent with the complexity of the site and available data (ASTM 1995). The quality and quantity of information and how well available information — on stressor sources and characteristics, exposure opportunities, characteristics of ecosystems potentially at risk and ecological effects — is integrated and used determine the quality of the assessment and the limits of its conclusions (USEPA 1998).

A conceptual model should be presented in both graphic (flow charts or pictorial representation) and narrative form (Suter 1996). It must attempt to present all pertinent information, aggregate and accessible, on a common spatio-temporal scale. Because it is the conceptualization of an existing or potential situation, the conceptual model cannot, by definition, be exhaustive. Nevertheless, it must be as complete and pertinent as possible because it forms the base for the following assessment. The final model should contain sufficient information to support the development of exposure scenarios under the current and anticipated future site uses and conditions (ASTM 1995).

The conceptual model must present the most exact portrait of the significance of the contamination, identifying the critical contaminants, their nature and form, while estimating the magnitude and distribution of the contamination. Establishment of this portrait demands a good knowledge and understanding of the source and characteristics of stress, the physical and chemical properties of the contaminants, as well as the site characteristics, in order to define the mechanisms of transport and transformation within the different environmental compartments (e.g., air, water, soil and biota).

The construction of the conceptual model also demands a good knowledge and understanding of the relationship between the source of stress and the receptors that could be present in the spatial limits of the assessment. This will assist in determining receptors which are susceptible to being exposed either directly, through contact with a contaminated abiotic environment, or indirectly, through the food chain. It is also important to identify the ecological links of receptors with other biological or ecological entities for which there are no routes of exposure. In addition, it is important to know and understand the biological systems and their ecological links in order to conceptualize the influence of the stress on potentially affected ecosystems.

In ERA, the conceptual model is used to integrate all site information and to determine whether information, including data, is missing and whether additional information needs to be collected at the site. The conceptual model should be acceptable to all parties involved in the problem (Hope 1995) since it helps identify missing data and provides a framework for further data collection (USEPA 1998). Also, uncertainties associated with the conceptual model need to be identified clearly so that efforts can be taken to reduce these uncertainties to acceptable levels (ASTM 1995). More information on uncertainty analysis can be found in Section 15.4.

The conceptual model leads to the formulation of explicit hypotheses of perturbations that clarify and articulate the proposed stressor-effect relationship (USEPA 1998). These hypotheses constitute the point of departure for the second step of the formulation phase — that being the determination of descriptive tools. This involves identifying assessment and measurement endpoints, their inclusion into decision rules and establishing desirable precision levels.

15.3.2.2 Descriptive Tools

The assessment endpoints are used to estimate the risk and must identify what will be evaluated in order to verify the hypothesis of perturbation. They specify, first and foremost, the way in which the results of the ERA guide decisions about the measures to be taken following the assessment. It is therefore crucial that assessment endpoints be defined clearly and precisely and reflect what must be verified (Barnthouse 1995, Hope 1995). Their range must also correspond to the spatial limits and temporal scales established by the conceptual model.

An assessment endpoint is an explicit expression of the ecological value that is to be protected (Suter 1990, USEPA 1992). Valuable ecological resources are those without which ecosystem functions would be significantly impaired (e.g., edaphic systems, marsh), those providing critical resources (e.g., habitat, fisheries) and those perceived as valuable by humans (e.g., endangered species and other issues addressed by legislation) (USEPA 1997a). Assessment endpoints must be specific and clearly defined to provide direction and boundaries for risk assessment and minimize miscommunication (Linthurst *et al.* 1995). Useful assessment endpoints define both the valued ecological resource and characteristics of the resource to be protected (reproductive success, production per unit area, area extent). It must be understood that an ERA which neglects to examine ecological values that are not of social or economic importance may ultimately be ineffective at protecting socially or economically ecological values since they can be linked together (Linthurst *et al.* 1995).

Sometimes the assessment endpoint can be measured directly. Usually, however, it is not amenable to direct measurement and measurement endpoints must be selected for each assessment endpoint (ASTM 1995). A measurement endpoint is a measurable biological or ecological response to a stressor that can be related to an assessment endpoint (Suter 1990, USEPA 1992). It may be useful to use more than one line of evidence to reasonably demonstrate that contaminants from a site are likely to cause adverse effects on the assessment endpoint (USEPA 1997a). Thus, more than one measurement endpoints group may be selected for each assessment endpoint.

Measurement endpoints can involve data or results from a combination of laboratory and field investigations. Thus, some form of relationships based on expert judgments, statistical methods or simulation models is usually required to link measurement endpoints to an assessment endpoint (Barnthouse and Brown 1994). This link must be clearly described and must be based on scientific evidence. Other desirable characteristics of a measurement endpoint include: it must be reliable and scientifically recognized; it must present an adequate sensitivity and specificity and a low natural variability. Also, it must be suitable to the phenomenon and the problem being assessed.

Endpoints are selected using a progressive approach that begins with the definition of assessment endpoints, followed by the selection of measurement endpoint groups and their links to the assessment endpoint. The critical step in selecting endpoints is deciding, among a broad array of possibilities, what ecological characteristics are important to decision making and are scientifically defensible.

At this point of the formulation phase, the risk assessor must demonstrate that the conceptual model and defined assessment endpoints correspond adequately to management needs. This ensures that the eventual ERA results will be useful for the management of contaminated sites (Van Leeuwen *et al.* 1998). Therefore, selection of final assessment and measurement endpoints must be discussed between the risk assessment and risk management teams. During these exchanges, each assessment endpoint is integrated into the decision logic leading to the definition of decision rules. These describe, in a single statement, a logical basis for choosing among alternative actions that will solve the problem. In the first tier of ERA, a single decision rule is usually sufficient to address a decision statement, assuming that the risk is estimated using a quotient method. In subsequent tiers, there is generally a decision rule for each assessment endpoint. The decision rule integrates the assessment endpoints, the statistical parameter (e.g., geometric mean, 95th percentile), the action level (e.g., quotient greater than 1) and the measures to be taken if the action level is reached (e.g., undertake higher tier assessment).

Next, the decision maker's task is to determine the tolerable limits on the decision errors (Moore and Biddinger 1995). In the first tier of ERA, this step involves specifying the degree of conservatism necessary to minimize, to an acceptable level, the possibility of obtaining false negatives. In subsequent tiers, the aim of establishing precision levels is to minimize the false positives as well as the false negatives. This requires the utilization of a systematic approach that permits the quantification of necessary precision levels.

The quantification of precision levels must be conducted for each assessment endpoint. To arrive at such a result, the possible extent of the ecotoxicological response must first of all be determined (e.g., 0% to 100%). Following this, the decision errors (e.g., false positives and false negatives) and the consequences of each must be identified. These consequences may be ecological, economical, social or other. Next, the tolerable levels of probability for the occurrence of decision errors are assigned on the basis of possible consequences.

The determination of the decision rules and precision levels is very important in establishing the plan for collecting data and necessary information for undertaking the assessment. In fact, decision rules and precision levels determine the effort level required to obtain the data and information in order to meet the general objective of the assessment as defined in the planning phase.

Also, it is during the exchange between the risk assessment and risk management teams that uncertainties in the identification and selection of assessment and measurement endpoints should be addressed. Conceptual model development and endpoint selection may be the most important sources of uncertainties in a risk assessment. While some uncertainties cannot be avoided, the risk assessment team should document and justify those which are known and present the nature of these uncertainties. One of the most common criticisms of ERAs is inadequate discussion of associated uncertainties (USEPA 1992). Masking or omission of uncertainty does not lend a higher credibility to the information presented, it simply hampers the subsequent decisions by preventing an informed evaluation of the information (USEPA 1997b).

15.3.2.3 Assessment Method

The preceding discussion between risk assessment and risk management teams not only ensures that the conceptual model and endpoints address managers' concerns, but is a way to assess the need for change before advancing too far in the assessment (USEPA 1998). It leads to the last major step of the formulation phase — the assessment method definition. This step starts with a review of the conceptual model for each assessment endpoint, with consideration given to the selected measurement endpoint groups. For each assessment endpoint, this review leads to the elaboration of a specific scenario that preserves from the conceptual model only the elements that are pertinent to the assessment endpoints. The degree of detail and conservatism in the scenario depends on the tier of the assessment (Suter 1997). In the first tier, the exposure scenario is of the 'plausible worst case' type, that presents a level of conservatism sufficiently elevated in accordance with the defined precision levels and that attributes extreme, but realistic, characteristics to measurement endpoints. At this level of assessment, resorting to the plausible worst case scenario is strongly justified because the uncertainty level can be high. This scenario also permits the assessment to be efficient, less expensive and useful in identifying potential exposures that are so minimal that detailed analysis is not warranted. In subsequent tiers of the ERA, the exposure scenario is of the 'best approximation' type. In contrast to the first tier, measurement endpoints are revised such that they can be described by more realistic quantification of characteristics.

From the conceptual model, each specific exposure scenario must characterize the relationship between the stressor and the assessment endpoint. Ideally, exposure scenarios should be performed using reliable and representative environmental and biological monitoring data, as well as toxicological and ecotoxicological data. Such data, however, are seldom available or do not completely relate to the problem under study. Thus, an exposure scenario is a set of assumptions about the stressor, the pathway of stressor transport (information based on intrinsic physicochemical and environmental fate properties of the substance or group of substances), the routes of contamination and exposure, the trophical and ecological links and the relevant spatial and temporal scales. Each scenario must also specify the way in which the variations of magnitude and spatio-temporal distribution of the contamination, as well as the behavioral characteristics of receptors, are quantified.

If a scenario fails to consider all factors affecting the problem defined by the assessment endpoint, uncertainty can be introduced in the form of basic lack of knowledge. The uncertainty associated with the scenario can be addressed by applying imaginative thinking about all possible factors that come to bear in real-life environmental problems (Cullen and Frey 1999). These sources of uncertainty, like those that have been identified previously, will determine the necessary effort level to be invested in the next phase (description phase) to address data gaps and outstanding questions. Also of interest and importance is to subsequently identify the key contributors to these sources of uncertainty.

The next step of the formulation phase involves the detailed definition of assessment methods, one for each assessment endpoint, that will be used for the estimation of ecotoxicological risks. A formal analytical process (Suter *et al.* 1994) may assist in identifying the different terms for the risk estimation methods. In the first tier of ERA, the quotient method is used to estimate risk. The quotients assist in verifying the absence of significant risk and in qualitatively estimating the risk for all receptors exposed to the stressor.

In subsequent tiers of the ERA, for each assessment endpoint the risk is estimated using a more quantitative method. Different risk estimation methods can be used in the quantitative ERA. These can be stochastic methods based on statistics and probability theory. Their choice depends principally on data that are necessary for their application. In many cases, several risk estimation methods can be applied for each assessment endpoint.

The problem formulation phase concludes with the production of an assessment plan. This plan documents the decisions and evaluations made during problem formulation, and identifies additional investigative tasks needed to complete the assessment (USEPA 1997a). The assessment plan presents the elements of the problem formulation, as well as the planned activities needed to support the assessment. The assessment plan is an important tool, useful in organizing resources in a way that directs efficient data collection and treatment. Sound planning will streamline the study process, increase the likelihood of efficiently collecting appropriate and useful data for the goal of the assessment (PNL 1996), and support defensible decision making in relation to the acceptable level of uncertainties (Hope 1995). The risk assessment and risk management teams should agree that the assessment plan describes a study that will provide the risk manager with the information needed for the site remediation process (USEPA 1997a).

15.3.2.4 Description

The data necessary for estimating the risk for each assessment endpoint must be obtained. These data can be gathered from the analysis of retrospective data (see Lugsdin and Breton 1996 for more details) or by performing analysis, measures and/or tests. The analysis of retrospective data consists of searching, collecting and selecting useful data, among the available data and information that correspond to the measurement endpoints. This review of existing data must be conducted by examining the type and quality of data and their capacity to support management needs, according to the selected decision rules and the required precision levels.

The selection of pertinent data also assists in identifying missing information that will be required to perform the assessment. In most cases, new data will be collected for the purpose of filling gaps in the existing data set (Bilyard *et al.* 1997). These data may be derived from laboratory and field activities with the goal of producing physical, chemical, toxicological and ecological data in response to identified measurement endpoints. The sampling plans should be designed in accordance with the precision levels defined by the management team during the determination of descriptive tools step of the formulation phase.

Whatever the source of data retrieval, the risk assessor must strive for an optimal trade-off between cost and accuracy/precision of the data. A systematic and structured quality assurance and quality control (QA/QC) program may help to do that, as discussed later.

15.3.2.5 Characterization

The selected data is compiled and analyzed systematically to judge if it is of sufficient quantity and quality for the characterization of risk. Characterization is the final phase of the linear ERA process and includes two major components: risk estimation and risk interpretation. The risk estimation attempts to resolve, for each assessment endpoint, the different terms of the equation that forms the risk estimation method. This means, for the first tier of ERA, generating estimated exposure and reference values, and for subsequent tiers, generating exposure and response profiles. Afterwards, the risk is estimated by comparing these values or profiles.

In the first tier of ERA, where an estimated risk presents a quotient greater than 1, it is pertinent to undertake a sensitivity analysis toward identifying the elements of the method that have the greatest influence on this estimation. In subsequent tiers, the conformity of estimated risk to the precision level previously established is evaluated. As needed, the risk may be estimated more accurately by entering the iteration loop to generate the necessary information to achieve the selected precision levels.

The risk interpretation provides information for discussing risk results obtained for assessment endpoints. Risk interpretation should be consistent with the values of 'transparency, clarity, consistency, and reasonableness' (USEPA 1995). Thus, when preparing the risk interpretation, the risk assessment team should make sure that the documentation of risks is easy to follow and understand, with all assumptions, defaults, uncertainties, professional judgments and any other inputs to the risk estimate clearly identified.

Each estimated risk provides a qualitative or quantitative description of the magnitude of the ecotoxicological response linked to an assessment endpoint. Nevertheless, this description does not take into account the considerations related to uncertainty (except in higher tiers for the quantifiable uncertainty already incorporated into the estimated risk), ecological significance and causal evidence. The risk interpretation must therefore specify the manner in which the assessment endpoint was estimated, based on measurement endpoint groups and their links to the assessment endpoint. It must also identify the uncertainty associated with the risk estimates. The risk interpretation, therefore, is principally based on a final analysis of uncertainty involved in the assessment. This necessitates a revision and synthesis of dominant sources of uncertainties, techniques and methods used for taking uncertainties into account, and their impact on the estimated risk. It must also be done on the basis of the QA/QC program associated with the risk estimation method. The final analysis of uncertainty must identify the terms of the risk estimation method for which the uncertainty has the most influence on the result. The impact of these terms on the precision and credibility of the estimated risk must finally be analyzed.

Moreover, besides interpreting the estimated risk in accordance with the inherent uncertainty, this interpretation must also consider the ecological significance and causal evidence of the response. The discussion of ecological significance of risks should include an evaluation of intensity and scale of effects, both spatial and temporal, as well as recovery potential of the ecosystem (Harwell *et al.* 1994). The analysis of causal evidence examines the plausibility of the stressor-effect causal link, and the possibility that the response is associated with another factor besides the stressor.

When more than one risk estimation method is used for an assessment endpoint, a strength-of-evidence approach may be used to integrate the different types of data to support a conclusion. This approach will ensure that risk interpretation is objective and not biased to support a preconceived answer. Furthermore, USEPA (1998) indicates that confidence in the conclusion of a risk assessment may be increased by using several lines of evidence to interpret and compare risk estimates. According to them, there are three principal categories of factors for risk assessors to consider when evaluating lines of evidence:

1. Adequacy and quality of data.
2. Degree and type of uncertainty associated with the evidence.
3. Relationship of the evidence to the risk assessment.

Risk characterization involves integrating and subsequently communicating all information gathered and analysis performed in the course of the assessment to the risk management team in an understandable manner (Wiegert and Bartell 1994). Discussion of risk estimates should identify the strengths and limitations of the risk conclusions in such a way as to provide a 'complete, informative and useful' set of information for decision makers (USEPA 1995). Clarity and completeness are essential. The assessor must be able to communicate the results of the assessment in a user-friendly form (Macler 1998).

15.4 ESSENTIALS FOR THE CREDIBILITY OF AN ERA

In the preceding section, we have defined the critical steps needed to design and carry out a valid and pertinent ERA. In addition, to be a useful input to contaminated site management, ERA should not only be technically valid, but also credible to all parties involved in the management process. As pointed out by Roberts (1987), this is not to say that results and their validity are not of concern, but that correct results are unlikely to be accepted unless there is confidence in the process. Credibility is more closely related to the perception stakeholders have of the assessment process than to the technical tools used in its performance. Therefore, it is easily understood that the way in which expert judgement is used in the risk assessment can have a profound impact on the credibility of the assessment. Note that the usual way to compensate for lack of credibility is to make very conservative choices (Chrostowski *et al.* 1998), which can lead to less than optimal risk management decisions.

Expert judgement is a necessary constituent of any risk assessment. It can be used to define, first and foremost, specifications applicable to all cases pertaining to a specific category. This is what we call the generic use of expert judgement. Expert judgement is also used in a case-specific manner, during the conduct of an ERA. Generally speaking, the smaller the ratio of the generic expert judgement to the case-specific expert judgement, the more likely the ERA is to provide specific insight into the problem under consideration. But there is a corollary to the use of specific expert judgement: the probability that credibility of the assessment will be challenged can be viewed, at least in part, as a function of the perception that it is based on case-specific expert judgement.

Thus, to be effective and useful, the ERA should have gained and maintained an adequate level of credibility. To achieve this level, the ERA should have incorporated the following three essential tools: uncertainty analysis, quality assurance and quality control, and ecotoxicological risk communication. It is important to understand that these tools are closely interrelated. For this reason, they have to be treated as such (i.e., with interactions, feedback, iterations). Obviously, the effort level afforded to each tool must be adjusted to the management context, the complexity of the situation and the particular tier of the assessment. Generally, the required effort level will be minimal for pristine conditions and catastrophic conditions, but will have to increase as the conditions diverge from these two extremes (Chapman 1998).

15.4.1 UNCERTAINTY ANALYSIS

Too often, we are faced with the fact that uncertainty associated with an ERA is in part ignored or hidden and, at best, in part qualitatively described and quantitatively 'integrated' into the risk computation. It is now well recognized that uncertainty must be considered in all steps of ERA (Warren-Hicks and Moore 1998). Uncertainty analysis should be a formal process kept as simple as possible but no simpler. ERAs of contaminated sites present situations when it may be important to address uncertainty, as stated by Morgan and Henrion (1990):

- When one is performing an analysis in which people's attitude toward risk is likely to be important;
- When one is performing an analysis in which uncertain information from different sources must be combined;
- When a decision must be made about whether to expend resources to acquire additional information.

The quantitative part of uncertainty analysis is often considered the main aspect of this process. But in fact it is only the 'tip of the iceberg'. Uncertainty analysis consists of qualitative and quantitative elements that are adequately described for all phases of ERA. To be useful and significant, it must be conducted in relation to the formulation, description and characterization phases of the ERA (Norton 1998).

Hence, there is a necessity to apply a structured uncertainty analysis process throughout the ERA. Such analysis will assist in defining the level of confidence associated with the scientific information by identifying, minimizing and describing the uncertainty associated with the entire ERA process. We have to keep in mind that the goal of uncertainty analysis is to fully describe the strengths and weaknesses of the ERA. Therefore, the ERA must include a description of the assumptions, limitations, information gaps, confidence, variability and sources for all the information used (Warren-Hicks and Moore 1998). Figure 15.2 presents a flowchart that can be used to guide the development of uncertainty analysis. It must be applied for each assessment endpoint.

First, a listing of the sources of uncertainty is made by revisiting all aspects of the ERA that can contribute to uncertainty. Of course, this should encompass the exposure scenario, the risk equations and the measurement endpoints. However, it is also essential to globally take into account the problem formulation phase since a part of uncertainty can be ascribed to ill-defined management needs (Finkel 1990). To be valuable, the list must identify all sources of uncertainty. Since it is the starting point of uncertainty analysis, appropriate effort should be allocated to development of this list.

Second, each element of the listing has to be qualified based on the type of uncertainty implicated (see Box 15.1) and on the direction in which it is expected to influence the risk estimates (positive or negative). From that qualification, the assessor should rank the importance of the elements. For those associated with models, sensitivity analysis can be useful for that ranking.

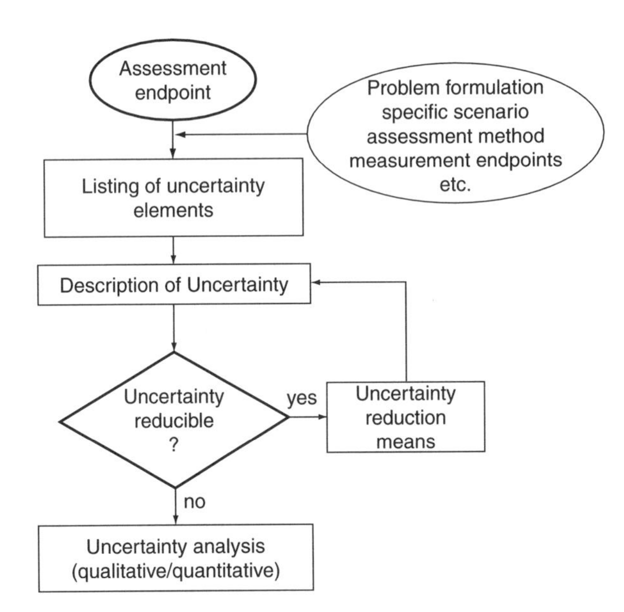

Figure 15.2 Flow chart describing activities associated with uncertainty analysis (modified from CEAEQ 1998).

Note that summarization of this uncertainty description will be a most valuable tool for risk management discussion.

Box 15.1 Categorization of uncertainty.

Elements of uncertainty can be categorized in different ways. For simplicity, a three-category classification is presented (Suter 1993): stochasticity or phenomena-inherent variability/heterogeneity, ignorance or incomplete knowledge, and error or operational mistakes. For some needs, it may be appropriate to use more precise categories (error might be subdivided into measurement error, random error, systematic error, etc.).

- Stochasticity refers to the uncertainty that can be described and estimated, but that cannot be eliminated since it is an inherent characteristic of the system. Abiotic (i.e., rain, wind, seasonal cycles) and biotic (i.e., colonization, competition) factors that are in perpetual modification, as well as some variable characteristics of organisms (i.e., water and food consumption, weight, age, territory area use) are effectively stochastic at the levels of interest in risk assessment.
- Ignorance refers to the incomplete understanding of some system properties. It can be fundamental, as for the influence of an unknown phenomenon. This type of ignorance cannot be described or quantified because it is not identified. Ignorance can also be caused by our inability to precisely measure, determine or describe some pertinent elements for the estimation of risk. As an example, it is not possible to investigate all the toxicological responses for all the potentially exposed species. So, ignorance includes incomplete or inaccurate data that might require approximation, simplification or extrapolation.
- Error refers to human and technological mistakes. It arises principally during sampling, measurements, data compilation and treatment. It can also be caused by poor or incomplete observance of model application limits.

Third, the possibility and necessity of minimizing ERA uncertainty is questioned. For that, the question that needs to be answered is 'for what element should uncertainty be reduced?'. The previous ranking will facilitate the response to this question. It may also be necessary to go into greater depth with sensitivity analysis for certain elements. This type of analysis allows for the assessment of the importance that the element has to the model output. The final decision on the need to minimize uncertainty for an element should be done on a 'cost-benefit' assessment that takes into account both the management and the assessment context. Before moving into qualitative and quantitative uncertainty analysis, it is essential to know what is uncertain and how that uncertainty might influence the management decision (Suter 1998).

A subsidiary question for elements that are retained is 'how to reduce the uncertainty associated with it?'. It is a matter of judging relevance and feasibility of modifying those elements. Often, this will be addressed simultaneously to the preceding cost-benefit assessment.

Lastly, when reducing uncertainty is no longer required or possible, the impact of residual uncertainty on risk estimates must be established. This analysis uses appropriate mixing of qualitative and quantitative methods to describe the uncertainty in the risk estimates and the confidence that can be attached to them. Tools that can be used range from qualitative description to visualization of data and model output to sophisticated mathematical and statistical treatments (Monte Carlo simulation, Bayesian inference, fuzzy logic, etc.). Selected methods must be used in ways that conform to their underlying assumptions. This is especially important for quantitative tools. Also, some tools require choices among different options, producing in this way new sources of uncertainty (i.e., the choice of a specific distribution form for a variable in a Monte Carlo simulation). These choices should also be examined as part of the uncertainty analysis. A description of such tools is beyond the scope of this chapter. However, to obtain more details on these and other tools, readers are referred to more specific works on uncertainty analysis (among others, Finkel 1990, Morgan and Henrion 1990, Bartell *et al.* 1992, Warren-Hicks and Moore 1998).

15.4.2 QUALITY ASSURANCE AND QUALITY CONTROL

The ERA process relies extensively on the use of data and models (Bradbury *et al.* 1998). As indicated in Box 15.1, errors associated with generation, treatment and manipulation of data can be an important source of uncertainty. It is possible to reduce errors in data management by developing an appropriate QA/QC program and applying it throughout the ERA.

The objective of the QA/QC program is to minimize the errors associated with controllable factors in the assessment. In order to be fully effective, this program should be planned and applied at the very beginning of an ERA, and carried through to completion (USEPA, 1988). The planning of a QA/QC program must address the stakeholders' needs in a manner that allows the data quality objectives to be defined quantitatively and qualitatively. This can be done by defining statements describing the quality of data needed to support the risk management decision.

Although it may seem unproductive and costly to spend time on this planning, we should understand that an ineffective plan will result in greater cost and lost time (USEPA 1988, 1994b). The content and level of details of a QA/QC program should be adjusted to the case-specific ERA and should always adhere to the following principle (CEAEQ 1998):

All the data and the information generated or used in an ERA should be of sufficient quality for their intended use, and an adequate documentation of their quality and acceptability should be presented together with this data and information.

What is meant by quality assurance and quality control? QA is the written, detailed procedures that are to be followed for each step of the ERA, while QC can be seen as the activities that occur during the implementation of the QA elements (i.e., duplicate samples, replicate analysis, peer review, etc.). The QA/QC program must encompass all the elements of quality assurance and control, allowing us to assess precision, accuracy, representativeness, completeness and comparability of data and information used. Components of this program include, but are not limited to, the following:

- Description of the quality control activities.
- For each duty (i.e., sampling, laboratory analysis, modeling, etc.), personnel qualification and training requirements.
- Protocols, standard operating procedures (SOPs) and all necessary documentation and records.
- Data selection criteria.
- Procedures to ensure that results are complete, correct and interpreted in a consistent manner, such as performing peer review on the ERA calculations using risk assessors not already involved in the analysis, verifying adherence of analysis to ERA protocol and documenting any deviation from protocol as well as preparing written records of QA/QC procedures and conclusions (Chrostowski *et al.* 1998).

A comprehensive listing of elements that must be addressed in the QA/QC program can be found in Table 15.1. Additionally, to be effective it is essential that all the personnel involved in the ERA have copies of the QA/QC program and understand the requirements prior to the start of data generation activities. Also, changes in the process of data generation can be frequent and should be incorporated into the current version of the program (USEPA 1994b). Other aspects including communication, management support, independence, staff involvement and back-up support should also be taken into account to have a successful QA/QC program (Simes 1995).

15.4.3 ECOTOXICOLOGICAL RISK COMMUNICATION

It is becoming increasingly recognized that ERA should not be conducted in seclusion and that communication among risk assessors, risk managers and stakeholders is an integral part of ERA (Leis and Krewski 1987, Pittinger *et al.* 1998). We define ERA communication as an interactive process of exchanging information and opinions, including information flows between risk assessors, risk managers, academic experts and stakeholders. In particular, stakeholder involvement in an ERA requires that a real discussion be implemented rather than one-way communication. To succeed in significantly addressing the risk managers' and stakeholders' concerns, the communication process must be based on mutual understanding.

Table 15.1 Elements of the QA/QC program

Tasks	Elements
Sampling	• sampling network • sampling frequency • types and number of samples • equipment and methods • sample handling and custody • constraints
Existing data acquisition	• data sources identification • access to the primary literature (original reference) • data acceptance criteria • data usability and limitations
Laboratory analysis and studies	• samples identification, reception and conservation procedures • organisms and chemicals supplies • analytical protocols and SOPs • calculations and statistics methodologies • analytical methodologies validation • quality control requirements
Field studies	• study design • field methods and SOPs • calculations and statistics methodologies • study validation • quality control requirements
Model and data processing	• model description and up-to-date documentation (theory, programming codes, installation and exploitation procedures, etc.) • models assumptions and application domain which agree with the prescribed application • model validation (performance, replication and credibility)
Risk characterization	• uncertainty analysis, integration and propagation methods description • weight-of-evidence approach • assessment of adequacy of ERA's results for the needs of risk managers and stakeholders • quality control requirements (internal or external peer review, etc.)

In practice, several communications will be necessary during an ERA, and the communication process must be adjusted for each of them. Communication is an integral part of the linear ERA process. It can be seen as a valuable source of information and data that can only be generated by the active participation of stakeholders (Grima 1987). Moreover, acceptance of a risk assessment is not only a function of the information it contains, but is also of interest to the parties implicated. For a specific ERA, the detail of what and how risks are communicated can be crucial to the decision making process, especially when the issue is localized (Pittinger *et al.* 1998), as is the case for contaminated sites. Experience shows that adequate consideration of communication during the conduct of an ERA allows for the arrival at a more accurate and reliable assessment and increases the probability that the results, conclusions and recommendations of the ERA will be accepted (Viteri 1996, Hull *et al.* 1998).

It should also be understood that there are areas of possible conflict in the risk communication process associated with the attitudes, beliefs, values and expectations of the involved parties. All these factors should be treated as valid and legitimate, and be taken seriously into account in the risk communication process.

The communication process should be initiated at the beginning of the linear ERA process by meeting with the risk managers to ensure that the problem formulation phase includes a constructive means of achieving both societal and scientific goals. This is also the occasion to agree on a preliminary list of stakeholders. This meeting is the starting point of the general communication process plan that should include definition and scope of the problems, identification of stakeholders and audience, identification and description of the issues in question, determination of how messages will be formulated and delivered, and determination and identification of the ERA communicators (Warren-Hicks and Moore 1998). For each communication, the following principles should be considered: plan and prepare carefully, gauge your audience, tailor your communication tools, and let your audience understand that you are responsive to their concerns (Viteri 1996).

This leads to a set of basic risk communication rules that may seem obvious, but that are frequently violated in practice (Covello 1987, Warren-Hicks and Moore 1998):

- Accept and involve all the stakeholders as legitimate partners.
- Listen to your audience.
- Assess and nurture your credibility.
- Plan carefully before communicating, and evaluate your performance.
- Be honest, frank and open.
- Speak clearly and with compassion.
- Coordinate and collaborate with other credible sources.
- Meet the needs of the news media.
- Follow up on commitments.

Finally, during the presentation of the ERA results, it will be important to provide a focus on important insights derived from the assessment, put those insights into their ecological context and give enough information to allow the audience to evaluate the validity and implications of the assessment.

15.5 CONCLUSION

Attempts to eliminate risks associated with human activities in the face of uncertainties and potentially high costs present a challenge to risk managers (Ruckelsaus 1984, Suter 1993). The principal objective of ERA is precisely to provide appropriate information to risk managers about the potential adverse effects on ecological systems of different potential management decisions. The practical application of ERA in the decision making process not only provides a credible scientific description of the problem, but also considers the specific risk information needs of the managers in the context of the relevant statutory requirements, precedence, available resources, time constraints and social and political situation.

The preceding sections describe a linear ERA process, including the elements critical in performing useful and credible ERAs for the contaminated sites management process. This process establishes the foundation for generating scientific information of adequate quality to support effective decision making. In so doing, ERA also defines the relationship between ecotoxicological science and environmental management.

However, unlike large sites which generally include a pilot study to focus efforts, small sites that are often in urban areas typically lack sufficient resources for problem formulation (Suter 1999). Therefore, the selection of assessment endpoints for these sites constitutes a difficult and time-consuming task, which is too often at the origin of confrontation between the regulated and the regulating risk managers, compromising the utility of ERA for the decision making process. To facilitate the application of ERA for small sites, 'generic' assessment endpoints are now being proposed as a starting point for the assessment endpoints selection step in ERA (Suter 2000). This approach is probably the most promising strategy for the proper use of ERA for small contaminated sites.

REFERENCES

AIHC (1997) *Ecological Risk Assessment: Sound Science Makes Good Business Sense*, Ecological Risk Assessment Committee, American Industrial Health Council, 13 pp.

ASTM (1995) *Standard Guide for Developing Conceptual Site Models for Contaminated Sites*. E 1689-95, American Society for Testing and Materials, Philadelphia, PA, 8 pp.

Barnthouse LW (1995) A framework for ecological risk assessment. In *Methods to Assess the Effects of Chemicals on Ecosystems*, SCOPE 53, IPCS Joint Activity 23, SGOMSEC 10, Linthurst RA, Bourdeau P and Tardiff RG (eds), John Wiley & Sons, Chichester, pp. 367–379.

Barnthouse L and Brown J (1994) Issue paper on conceptual model development. In *Ecological Risk Assessment Issue Papers. Risk Assessment Forum*. EPA/630/R-94/009, US Environmental Protection Agency, Washington, DC, pp. 3.1–3.70.

Bartell SM (1998) *Risk-based monitoring programs for environmental decision-making*. Integrated Approaches for Interpreting Environmental Effects Monitoring Data Workshop, 22 October, 1998, Château Frontenac, Québec City, Canada.

Bartell SM, Gardner RH and O'Neill RV (1992) *Ecological Risk Estimation*, Lewis, Boca Raton, FL, 252 pp.

Belluck DA, Hull RN, Benjamin SL, French RD and O'Connel RM (1993) Defining scientific procedural standards for ecological risk assessment. In *Environmental Toxicology and Risk Assessment*, Vol. 2. STP 1216, Gorsuch JW, Dwyer FJ, Ingersol CG and La Point TW (eds), American Society for Testing and Materials, Philadelphia, PA, pp. 440-450.

Bilyard GR, Beckert H, Bascietto JJ, Abrams CW, Dyer SA and Haselow LA (1997) *Using the Data Quality Objectives Process During the Design and Conduct of Ecological Risk Assessments*, US Department of Energy, Assistant Secretary for Environment, Safety and Health, Office of Environment, Washington, DC, 82 pp.

Bradbury S, Hermens J, Karcher W, Niemi G, Purdy R and Richards C (1998) Obtaining data for ecological risk assessment. In *Ecological Risk Assessment Decision-Support System: A Conceptual Design*, Reinert KH, Bartell SM and Biddinger GR (eds), SETAC Press, Pensacola, FL, pp. 29-37.

CEAEQ (1998) *Procédure d'évaluation du risque écotoxicologique pour les terrains contaminés*, Ministère de l'Environnement et de la Faune, Gouvernement du Québec, 139 pp.

Chapman PM (1998) An ecological perspective of uncertainty. In *Uncertainty Analysis in Ecological Risk Assessment*, Warren-Hicks WJ and Moore DRJ (eds), SETAC Press, Pensacola, FL, pp. 131-139.

Chrostowski P, Foster S, Durda J and Preziosi D (1998) *Good ecological risk assessment practices*. Presented at the 19th SETAC Meeting, Charlotte, NC.

Covello VT (1987) Informing people about risks from chemicals, radiation, and other toxic substances: a review of obstacles to public understanding and effective risk communication. In *Prospects and Problems in Risk Communication*, Leiss W (ed), University of Waterloo Press, pp. 1-49.

CRAM (1996) *An Assessment of the Risk Assessment Paradigm for Ecological Risk Assessment*, Commission on Risk Assessment and Risk Management. Prepared by Menzie-Cura & Associates Inc., 50 pp.

CRAM (1997a) *Framework for Environmental Health Risk Management*. Final Report, Vol. 1, Commission on Risk Assessment and Risk Management, Washington, DC.

CRAM (1997b) *Risk Assessment and Risk Management in Regulatory Decision-Making*. Final Report, Vol. 2, Commission on Risk Assessment and Risk Management, Washington, DC.

Cullen AC and Frey HC (1999) *Probabilistic Techniques in Exposure Assessment. A Handbook for Dealing with Variability and Uncertainty in Models and Inputs*, Plenum Press, New York, 335 pp.

Fairbrother A, Kapustka LA, Williams BA and Glicken J (1994) *Ecological risk assessment benefits environmental management*. Presented at Ecological Risk Assessment: Uses and Abuses Conference, Corvalis, OR, 15-16 November 1994. Report No. SAND-94-3062C; CONF-9411167-1 1994, Department of Energy, Washington, DC, 11 pp.

Finkel AM (1990) *Confronting Uncertainty in Risk Management: A Guide for Decision-Makers*, Center for Risk Management, Resources for the Future, Washington, DC, 68 pp.

Grima APL (1987) Improving risk information transfer: instrumental and integrative approaches. In *Prospects and Problems in Risk Communication*, Leiss W (ed), University of Waterloo Press, pp. 117-134.

Harwell M, Gentile J, Norton B and Cooper W (1994) Issue paper on ecological significance. In *Ecological Risk Assessment Issue Papers: Risk Assessment Forum*. EPA/630/R-94/009, US Environmental Protection Agency, Washington, DC, pp. 2.1-2.49.

Hope BK (1995) Ecological risk assessment in a project management context. *The Environmental Professional*, **17**, 9-19.

Hull RN, McKee PM, Fahey JJ and Benson DC (1998) *Risk communication during the risk assessment process: an ontario case study*. Presented at the 19th SETAC Meeting, Charlotte, NC.

IEA (1995) *Guidelines for Baseline Ecological Assessment*, Institute of Environmental Assessment/E&FN Spon, London, 142 pp.

Jasanoff S (1993) Relating risk assessment and risk management: complete separation of the two processes is a misconception. *EPA Journal*, **19**(1), 35-37.

Leiss W and Krewski D (1987) Risk communication: theory and practice. In *Prospects and Problems in Risk Communication*, Leiss W (ed), University of Waterloo Press, pp. 89-112.

Linthurst RA, Bourdeau P and Tardiff RG (1995) *Methods to Assess the Effects of Chemicals on Ecosystems*. SCOPE 53, IPCS Joint Activity 23, SGOMSEC 10, John Wiley & Sons, Chichester, 416 pp.

Lugsdin T and Breton R (1996) Data collection and generation. In *Ecological Risk Assessment of Priority Substances Under the Canadian Environmental Protection Act*. Resource Document (Draft 1.0), Environment Canada, Chemicals Evaluation Division, Commercial Chemicals Evaluation Branch, pp. 2.1-2.24.

Macler BA (1998) Managing ecological risks: what information do regulators need and want? In *Ecological Risk Assessment: A Meeting of Policy and Science*, de Peyster A and Day K (eds), SETAC Press, Pensacola, FL, pp. 55-63.

Moore DRJ and Biddinger GR (1995) The interaction between risk assessors and risk managers during the problem formulation phase. *Environmental Toxicology and Chemistry*, **14**(12), 2013-2014.

Morgan MG and Henrion M (1990) *Uncertainty: A Guide to Dealing with Uncertainty in Quantitative Risk and Policy Analysis*, Cambridge University Press, 332 pp.

Norton SB (1998) An ecological risk assessor's perspective of uncertainty. In *Uncertainty Analysis in Ecological Risk Assessment*, Warren-Hicks WJ and Moore DRJ (eds), SETAC Press, Pensacola, FL, pp. 186-192.

NRC (1983) *Risk Assessment in the Federal Government: Managing the Process*, Committee on the Institutional Means for Assessment of Risks to Public Health, Commission on Life Sciences, National Academy Press, Washington, DC, 191 pp.

NRC (1992) *A Paradigm for Ecological Risk Assessment*, Committee on Risk Assessment Methodology, Board on Environmental Studies and Toxicology, Commission on Life Sciences, National Academy Press, Washington, DC, 120 pp.

Pittinger CA, Bachman R, Barton AL, Clark JR, deFur PL, Ells SJ, Slimak MW, Stahl RG and Wentsel RS (1998) *A Multi-Stakeholder Framework for Ecological Risk Management*. Summary of a SETAC Technical Workshop; 23-25 June 1997, Williamsburg, VA, SETAC Press, Pensacola, FL, 24 pp.

PNL (1996) *Why Use the DQO Process?* Pacific North West National Laboratory. http://etd.pnl.gov:2080/DQO/why.html.

Powell MR (1999) *Science at EPA. Information in the Regulatory Process*, Resources for the Future, Washington, DC, 433 pp.

Roberts JR (1987) The conundrum of risk communication: error, precision and fear. In *Prospects and Problems in Risk Communication*, Leiss W (ed), University of Waterloo Press, pp. 193-205.

Ruckelshaus WD (1984) *Risk Assessment and Management: Framework for Decision Making*. EPA600/9-85-002.

Simes GF (1995) *Quality Assurance. Organizational-Catalytic-Technical*. EPA/600/A-95/099, US Environmental Protection Agency, Office of Research and Development, National Risk Management Research Laboratory, 19 pp.

Suter II GW (1990) Endpoints for regional ecological risk assessments. *Environmental Management*, **14**(1), 9-23.

Suter II GW (1993) *Ecological Risk Assessment*, Lewis, Boca Raton, FL, 538 pp.

Suter II GW (1996) *Guide for Developing Conceptual Models for Ecological Risk Assessments*. ES/ER/TM-186, Oak Ridge National Laboratory, Oak Ridge, TN, 21 pp.

Suter II GW (1997) *A Framework for Assessing Ecological Risks of Petroleum-Derived Materials in Soil*. ORNL/TM-13408, Oak Ridge National Laboratory, Oak Ridge, TN, 55 pp.

Suter II GW (1998) An overview perspective of uncertainty. In *Uncertainty Analysis in Ecological Risk Assessment*, Warren-Hicks WJ and Moore DRJ (eds), SETAC Press, Pensacola, FL, pp. 121-130.

Suter II GW (1999) Lessons for small sites from assessments of large sites — Editorial. *Environmental Toxicology and Chemistry*, **18**(4), 579-580.

Suter II GW (2000) Generic assessment endpoints are needed for ecological risk assessment. *Risk Analysis*, **20**(2), 173–178.

Suter II GW, Gillett JW and Norton S (1994) Issue paper on characterization of exposure. In *Ecological Risk Assessment Issue Papers. Risk Assessment Forum*. EPA/630/R-94/009, US Environmental Protection Agency, Washington, DC, pp. 4.1–4.64.

USEPA (1988) *You and Quality Assurance in Region 10*. EPA 910/R-88-100, US Environmental Protection Agency, Region 10 Quality Assurance Management Office, 12 pp.

USEPA (1992) *Framework for Ecological Risk Assessment*. EPA/630/R-92/001, US Environmental protection Agency, Washington, DC.

USEPA (1994a) *Guidance for the Data Quality Objectives Process*. EPA QA/G-4, US Environmental Protection Agency, Quality Assurance Management Staff, Washington, DC, 68 pp.

USEPA (1994b) *EPA Requirements for Quality Assurance Project Plans for Environmental Data Operations*. Draft Interim Final, EPA QA/R-5, US Environmental Protection Agency, Quality Assurance Division, Washington, DC, 22 pp.

USEPA (1995) *Elements to Consider When Drafting EPA Risk Characterizations*. US Environmental Protection Agency. http://www.epa.gov/ORD/spc/rcelemen.htm.

USEPA (1997a) *Ecological Risk Assessment Guidance for Superfund: Process for Designing and Conducting Ecological Risk Assessment*. Interim Final, EPA/540/R-97/006, US Office of Solid Waste and Emergency Response, US Environmental Protection Agency, Washington, DC.

USEPA (1997b) *Supplemental Ecological Risk Assessment Guidance for Superfund*. EPA 910-R-97-005, USEPA Region 10, Office of Environmental Assessment, Risk Evaluation Unit, US Environmental Protection Agency, Washington, DC, 65 pp.

USEPA (1998) *Guidelines for Ecological Risk Assessment*. EPA/630/R-95/002F, Risk Assessment Forum, US Environmental Protection Agency, Washington, DC, 153 pp.

Van Leeuwen C, Biddinger G, Gess D, Moore D, Natan T and Winkelmann D (1998) Problem formulation. In *Ecological Risk Assessment Decision-Support System: A Conceptual Design*. Proceedings of the Pellston Workshop on Ecological Risk Assessment Modeling, 23–28 August 1994, Pellston, MI, Reinert KH, Bartell SM and Biddinger GR (eds), SETAC Press, Pensacola, FL, pp. 7–14.

Viteri Jr. APE (1996) *Communicating uncertainties contained in ecological risk assessment*. Presented at the 17th SETAC Meeting, Vancouver, BC.

Warren-Hicks WJ and Moore DRJ (1998) *Uncertainty Analysis in Ecological Risk Assessment*, SETAC Press, Pensacola, FL, 227 pp.

Wiegert RG and Bartell SM (1994) Issue paper on risk integration methods. In *Ecological Risk Assessment Issue Papers. Risk Assessment Forum*. EPA/630/R-94/009, US Environmental Protection Agency, Washington, DC, pp. 9.1–9.66.

16

The Evolution of Ecological Risk Assessment During the 1990s: Challenges and Opportunities

CHARLES A. MENZIE

Menzie-Cura & Associates Inc., Chelmsford, MA, USA

16.1 BEGINNINGS

During the 1990s, ecological risk assessment moved towards center stage as a tool for environmental decision making. The formalized process, which we refer to as ecological risk assessment, is relatively new and the increased use of this tool has raised numerous questions about when it is appropriate to apply it, how to apply it, and how to interpret the results of an assessment. As I discuss in this chapter, these are not simply technical questions but also involve management, social and economic issues.

I consider ecological risk assessment a relatively new process. By this I mean that the language and formalized process are new. In contrast, ecological risk 'thinking' has been a part of human culture throughout recorded history and has evolved along with the evolution of our societies. For example, one of the earliest recorded risk management decisions to preserve biodiversity can be found in the story of Noah's Ark.

Individuals and societies have experienced and related to 'nature' in many different ways over time. Cronon (1995) observes that 'The work of literary scholars, anthropologists, cultural historians, and critical theorists over the past several decades has yielded abundant evidence that "nature" is not nearly so natural as it seems. Instead, it is a profoundly human construction'. This is not to say that the non-human world is somehow unreal or a mere figment of our imaginations — far from it. But the way we describe and understand the world is so entangled with our own values and assumptions that the two can never be fully separated. What we mean when we use the word 'nature' says as much about ourselves as about the things we label with that word. As the British literary critic Raymond Williams once famously remarked, 'The idea of nature contains, though often unnoticed, an extraordinary amount of human history'.

Environmental Analysis of Contaminated Sites. Edited by G. I. Sunahara, A. Y. Renoux, C. Thellen, C. L. Gaudet and A. Pilon

Cronon poses a fascinating question: 'What happens to environmental politics, environmental ethics, and environmentalism in general once we acknowledge the deeply troubling truth that we can never know at first hand the world "out there" — the "nature" we seek to understand and protect — but instead must always encounter that world through the lens of our own ideas and imaginings?'.

The formalized language and process of ecological risk assessment finds its roots during the late 1960s and early 1970s in legislation such as the National Environmental Policy Act (NEPA). Bartell (1998) notes that the NEPA process afforded an opportunity to integrate ecology, social science, policy and economics in a decision-making framework. Ecologists were forced to learn how to practice and apply their professional skills within the legal context of NEPA, often in courtroom situations. The basic language and four-step process for organizing risk assessments were presented in Risk Assessment in the Federal Government: Managing the Process, commonly known as the 'red book' (NRC 1983). This was the starting point for organizing the US Environmental Protection Agency (USEPA) Framework for Ecological Risk Assessment (USEPA 1992). However, the framework (Figure 16.1) expanded on the NRC process in several important ways. I attended many of the technical discussions leading to the development of the framework and chaired the reviews of EPA's ecological risk assessment case studies, as well as several workshops and meetings in the early 1990s related to the framework. In my view the most important developments in the process are:

1. Identifying the importance of 'problem formulation'.
2. Increasing the communication between risk assessors and risk managers.
3. Integrating effects and exposure information into 'analysis'.
4. Incorporating data acquisition, monitoring and verification.

During the early 1990s we were commissioned by the Presidential/Congressional Commission on Risk Assessment and Risk Management to identify outstanding issues related to the conduct of ecological risk assessment (Menzie and Freshman 1997). We conducted a survey of existing and developing approaches and recommended that the USEPA's framework should be accepted as the paradigm for most ecological risk assessments. We further recommended that the framework be augmented to: (a) reflect the importance of communication among stakeholders, risk managers and risk assessors throughout the process and (b) identify the iterative (e.g., tiered) nature of risk assessments. Finally, we recommended that additional guidance be developed for implementing components of the framework through a series of case studies. We suggested that this be undertaken as a collaborative effort involving stakeholders, risk managers and risk assessors. We stressed that guidance was needed for: (a) problem formulation; (b) risk characterization including lines-of-evidence approaches as well as simple and more sophisticated (e.g., probabilistic) quantitative methods; (c) tiered or phased approaches; (d) methods for identifying, quantifying and managing uncertainties in the analyses; (e) communication; (f) stakeholder involvement; and (g) education.

Our findings and recommendations to the Presidential/Congressional Commission on Risk Assessment and Risk Management (CRARM 1997) were echoed by many in academia, government and industry and a number are addressed in the USEPA's guidelines for ecological risk assessment (USEPA 1998a). I will focus on a few that I consider most important for advancing the practice of ecological risk assessment and the value of this process for environmental decision making. I have organized these around the following challenges:

- Understanding the decisions.
- Identifying 'What is important to protect?'.
- Dealing with scales.
- Using the right tools.
- Evaluating and managing uncertainties.
- Communicating about the problem, the analyses and the decisions.

These challenges are interrelated but I have found that they emerge as key discussion topics prior to, during and after the conduct of risk assessments.

16.2 UNDERSTANDING THE DECISIONS

Risk assessment by itself is an orphan. A request — often heard from industrial clients or regulatory agencies — to 'go do a risk assessment' is incomplete if it is separated from the overall risk management context. Risk assessment is a tool, not an end in itself. To understand how, or even when, to use risk assessment, the risk management context must be identified and the questions or information needs must be clearly defined. Good communication is essential during planning and problem formulation (Moore and Biddinger 1996). While this may appear obvious, there are many risk assessors and risk managers who fail to have the interactions necessary to develop and implement effective risk management approaches. To some degree this reflects historical misconceptions concerning the roles of risk assessment and risk management.

There is a perspective that risk assessment is best conducted and 'kept pure' if it is carried out independently from risk management. This perception — that risk assessment and risk management must be kept apart — is a long-standing misinterpretation of the 1983 National Research Council (NRC) red book. North (1996), one of the authors of the red book, notes their report never called for a 'separation' between risk assessment and risk management but only a 'distinction' between the two (EPA Risk Policy Report, 14 June, 1996, Volume 3, Number 8, p. 9). Cooper (1998) also argues for better integration of risk assessment and risk management for ecological issues.

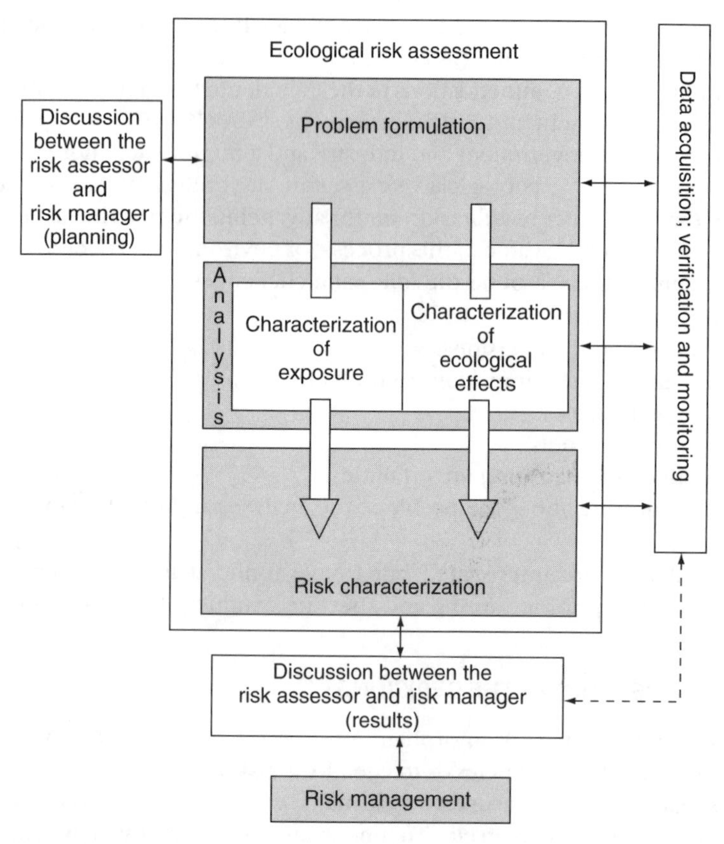

Figure 16.1 USEPA Risk Assessment Framework (source USEPA 1998a).

The emergence of risk management as an organizing framework that incorporates risk assessment is underscored by recent reports from the President/Congressional Commission on Risk Assessment and Risk Management (CRARM 1997) and the NRC (1996). The Commission proposed a conceptual approach that emphasizes solutions to environmental risk reduction through an integrated process involving stakeholders. This theme is also present in the NRC's Understanding Risk: Informing Decisions in a Democratic Society. The report presents seven principles which will influence how risk assessment, risk characterization and risk management are carried out in the future:

1. Risk characterization should be a decision-driven activity, directed towards informing and solving problems.
2. Coping with a risk situation requires a broad understanding of the relevant losses, harms or consequences to the interested and affected parties.
3. Risk characterization is the outcome of an analytic – deliberative process.

4. The analytic - deliberative process leading to a risk characterization should explicitly deal with problem formulation early on in the process.
5. The process should be mutual and recursive.
6. Those responsible for a risk characterization should begin by developing a provisional diagnosis of the decision situation so they can better match the analytic - deliberative process to the needs of the decision, particularly in terms of the level and intensity of effort and representation of parties.
7. Each organization responsible for making risk decisions should work to build organizational capability to conform to the principles of sound risk characterization.

There are a number of examples of risk management approaches that involve explicit decisions. The ASTM Risk-based Corrective Action (RBCA) process involves a tiered assessment strategy wherein risk-related information, decisions and remedial options are considered in a sequential manner (ASTM 1995, 1998). The USEPA Superfund Program (USEPA 1997a) and the Canadian Council of Environmental Ministers (CCME 1996) National Contaminated Sites Remediation Program also illustrate how risk assessment information is integrated into decision making. Further, a number of states and provinces have also moved toward tiered risk management approaches that link risk assessment activities to specific risk-based questions and remedial options.

With the increased involvement of stakeholders in environmental decision making, there will be challenges on how to best relate assessment information to decisions. McDaniels (1998) suggests that a clear framework with well-articulated objectives and alternatives for a specific decision should make it easier for 'lay' participants to understand the focus of the complex issues that are inherent in ecological risk assessments. I found this suggestion useful during a recent facilitation process involving an environmental agency and lay people. The approach included the development of 'management goals' and 'assessment endpoints' which were to be evaluated by 'multiple lines of evidence'. However, the diversity of measures and assessment tools led to confusion among the lay participants until the elements of the analysis and possible outcomes were related to a defined set of conclusions and associated management options. This step — which took time to develop, discuss and modify — enabled all sides to understand the process, the results that might be generated and how they would be used to make decisions. Prior to this exercise, the ecological risk assessment seemed to many lay participants like a 'black box' with unknown outcomes.

Over the next decade, increased reliance will be placed on using risk management approaches for addressing environmental issues. Risk assessments will be used to inform many of these decisions. In order for these to be effective, the risk assessor and risk manager must have a clear understanding of the decisions and what is needed to inform them. Because there will be increased involvement by stakeholders, efforts will also be needed to engage them in a meaningful way.

16.3 IDENTIFYING 'WHAT IS IMPORTANT TO PROTECT?'

This question is being asked in hundreds of meetings, workshops and environmental policy-related forums. Recent peer-review literature and reports from regulatory agencies, industry and academia have focused on this question from various perspectives. It remains one of the key challenges for ecological risk assessments.

Identifying 'What is important to protect?' has been approached from several directions. In the agency's guidelines for ecological risk assessment, the USEPA (1998a) has identified criteria for considering this question as part of selecting assessment endpoints. These include: (a) ecological relevance; (b) susceptibility to known or potential stressors; and (c) relevance to management goals. The USEPA notes that ecological relevance and susceptibility are essential for selecting assessment endpoints that are scientifically defensible; however, to increase the likelihood that the risk assessment will be used in management decisions, assessment endpoints are more effective when they also reflect societal values and management goals. This last statement underscores the importance of stakeholder involvement.

Two USEPA staff members (Barton and Sergeant 1998) recently focused on the question 'How can we decide what we are trying to protect?'. They began by stressing the importance of clearly articulated management objectives. They then identified ecological entities associated with three categories: (a) plants, animals and their habitats; (b) whole ecosystems, their functions and their service; and (c) special places. These are proposed not to classify the entities, but as a checklist to guide the process of setting objectives. Finally, they provided a short list of general criteria that risk managers might use to select entities within these categories: mandated protection, other societal value, rarity or special vulnerability and ecological significance.

State and provincial risk assessment guidance also reflects a common set of criteria or ecological entities that are judged important to protect. This was apparent in a review of approximately 27 risk assessment guidance documents undertaken as part of developing the ASTM RBCA ecological risk assessment standard. My experience indicates that common ground on 'What is important to protect?' is most frequently achieved when the ecological entities are well recognized as 'important' and/or 'valuable' by scientists as well as the public. Examples include rare and endangered species, large wildlife species and well-recognized ecosystems such as wetlands. Common ground is more difficult to reach when the entities are ecologically relevant or significant (as judged by scientists) but not intrinsically valued by the public or many risk managers. Frequently cited examples in this latter category include soil microbes and earthworms.

Another perspective on identifying 'What is important to protect?' is to begin with the stated management goals as the umbrella under which other criteria and guidelines are subsequently considered. A recent Pellston Workshop on Contaminated Soils emphasized the importance of well-defined management goals (Menzie *et al.*, 2001): 'It is clear that contaminated soil issues can be addressed in a number of ways and that these depend not only on the conditions at a site but also on how the site may be used in the future and the overall site management goals. Because these vary, a "one-size fits all" approach for assessing risks at contaminated sites will likely not meet management needs across different scales of site problems. Management goals are formulated either at the policy or at the site-specific level. Policies may involve land use issues, establishing a course of action for particular types of operations or facilities, or determining resources that need to be protected. Actions taken to address these are sometimes referred to as technical policy decisions. Management goals related to specific sites may involve similar issues but applied at a local or site-specific basis. The assessment of risks and decisions concerning what needs to be done flow from stated management goals. Therefore, establishing these goals is critical to both policies and site-specific actions related to management of contaminated soils. [Alternative management goals] raises issues of land ownership, land use, future development plans, and ecological values. A management goal sets a framework for discussing these issues and for identifying an appropriate course of actions. When ecological risk assessment is used as a tool, the risk management goal is translated into assessment endpoints'.

I have suggested elsewhere (Menzie 1995, 1998) that the risk assessor must be an informed and careful listener when considering the question 'What is important to protect?'. Perhaps most difficult for the risk assessor is my suggestion that he or she exercise restraint in putting forward a point of view or assuming that he or she has the answers. The assessor must also be open-minded and recognize that what is important to other people may not overlap squarely with what the assessor believes is important ecologically. Finally, the assessor should listen for what is missed in these discussions that may be ecologically important but not obvious to others. In my view alternative views should be joined in a way that informs, expands and constrains the other. I believe that this can only be accomplished through good communication.

Because the question 'What is important to protect?' has societal as well as technical/ecological components, it is apparent that stakeholders will and should have increased input in addressing this question either at a policy level or for particular environmental problems. Some of the best examples of stakeholder involvement involve watershed management programs. A number of these are being implemented by the USEPA and various state agencies. Stakeholder involvement has been a key aspect of most, and the agencies are gaining insights into this aspect of the process. As part of a USEPA project, I had an opportunity to discuss lessons learned with participants in five watershed programs. With respect to selection of ecological receptors, the major lessons learned were as follows.

1. Stakeholders and ecologists collectively provide valuable contributions toward the identification of ecological receptors. In a few of the watershed case studies, ecological receptors were added to the list prepared by the watershed workgroup based on the input received from stakeholders. These additions reflected what people felt were important to protect and which the workgroup subsequently judged were helpful additions. Presentations of conceptual models and lists of ecological receptors facilitated the identification of such 'gaps' and prompted stakeholders to think about what else should be included.
2. Two useful starting points for identifying ecological receptors potentially at risk include historical information on what people consider important and available biological measures for ecological resources. Many of the participants in the watershed case studies relied upon existing knowledge, historical precedents or biological criteria for selecting ecological receptors. These considerations included an assessment of what people value. Such information was obtained through stakeholder participation or from surveys of the public. To a large degree these defined what was important to protect and were part of a common knowledge base about the system.
3. Ecological receptors potentially at risk can include ecological components other than individuals or populations of species. The Middle Platte case study provides an example where a range of ecological components was utilized to capture various aspects of the system that were judged to be important. These included particular habitat types, mosaics of habitats as well as populations of selected species.
4. Because stakeholders do not always agree on valued ecological entities, education and facilitation may be necessary when selecting ecological receptors. Education may be required to help stakeholders understand components of the system as a basis for selecting appropriate ecological receptors for analysis. For example, the selection of eelgrass in Waquoit Bay is one example where the ecological importance of this receptor was not obvious to the general public. In cases where there are disagreements, facilitation may be helpful. Where consensus cannot be reached, the risk manager will need to make decisions on what to include in consultation with risk assessors and upon considering the bases for including or excluding receptors within the analysis. To some degree differences of opinion or emphasis can be handled by distinguishing between how receptors are evaluated either as assessment endpoints (primary focus) or as measures of effects and exposure (used to judge broader or higher trophic level assessment endpoint).

The question of 'What is important to protect?' will remain a challenge. The question will likely be addressed through a combination of management, scientific and human value considerations. Where land ownership or water rights issues are present, these will also affect the resolution of the question. The question will sometimes be addressed at national or state policy levels and at other times at site or problem-specific levels. Stakeholders will have a voice in the discussion to greater or lesser degrees depending on the nature of the problem. Because of the diversity of value systems and the various ways in which decisions will affect people, it is unlikely that there will be complete agreement on the question. However, addressing the question will require good communication, a search for common ground, recognition that there will be differences of opinion and a fair means of addressing such differences.

16.4 DEALING WITH SCALES

How should ecological risks be measured? Is it enough to answer the question of ecological risk with a simple 'yes' or 'no'? Human health risk assessments are typically conducted at relatively simple scales: one species, sensitive individuals, cancer and non-cancer effects. The assessment of exposure either involves a particular person or a particular area. Both are defined in terms of an individual's or group's activities. Many screening-level ecological risk assessments extend these ideas to ecological systems. In such cases, effects are estimated for particular individuals (in a population) or for particular parts of an ecosystem. Exposure is usually simplified to one or more statistics.

Because ecological risk assessments are intended to inform environmental decision making, conservative screening-level approaches can be useful. (Note: the assessment is only a tool, not an end in itself.) However, screening-level assessments often do not capture the dimensions of risk that are most relevant to ecological entities. In many cases, such information is important for informing decisions, and needs to be considered in the assessment (Clifford *et al.* 1995, Freshman and Menzie 1996, Washington-Allen and Sample 1997, Akçakaya 1998, Applied Biomathematics 1998, Gentile and Harwell 1998, Menzie *et al.* 2001). The three major scales that influence ecological risks are:

- Spatial scales (for distribution of the stressor and for the receptor);
- Temporal scales (for release, transformation, sequestration rates as well as for biological processes including recovery);
- Effects scales (for type of effect and magnitude of the effect).

Spatial scales can range from local situations such as those encountered in the vicinity of a hazardous waste site, to watersheds, regions and ultimately global levels. They may involve the identification of local populations of organisms, the extent of ecosystems and the distribution of mosaics of habitats or landscapes. They may be thought of in terms of foraging areas, breeding areas, migration pathways and other geographic areas occupied by species either permanently or temporarily. The stressor(s) of concern also exhibit spatial characteristics.

They could range from a small 'hot spot' of contamination to a region-wide influence such as changing water elevations associated with global warming. During the 1990s there was increased interest in developing tools that capture the spatial distributions of ecological entities as well as of stressors. Geographic Information Systems (GIS) are the most notable examples (Clifford *et al.* 1995). I expect that there will be increased attention given to utilizing these tools to quantify ecological risks in a way that incorporates spatial scale. Policy decisions still need to be made concerning appropriate spatial scales for assessments. In particular, policy direction is often needed on what constitutes a 'local population' and the spatial scale associated with that population. Relatively few regulatory agencies have provided guidance on how to do this. Oregon is an example of one that has addressed this technical policy decision (Oregon Department of Environmental Quality 1998).

Temporal scales include seasonal patterns of behavior and changes in exposure magnitude with time. Both influence exposure to the stressor. Other aspects of temporal scale include the ability of the species or system to recover from stress as well as the natural temporal variability of the system.

Effects scales refer to the type and magnitude of the effect. These can be thought of in terms of dose–response relationships at the individual, population or system level. The most commonly considered effects include mortality, morbidity and reproductive success. However, other effects such as behavior (e.g., avoidance) can be important. Screening-level assessments typically treat potential effects as 'yes or no' results, i.e., conditions are either above or below a threshold. Actual effects, especially those that influence populations, are more complex. For example, the effect of reproductive impairment on a population of fish will depend on the population characteristics of that fish population and the magnitude of the effect (e.g., on the recruitment rate of young to the population).

Incorporating information on spatial, temporal and effects scales into ecological risk assessments received limited attention in most ecological risk assessments during the 1990s. As assessments attempt to evaluate regional and global influences, explicit consideration of all these scales will be an essential part of conducting scientifically defensible risk assessments. Gentile and Harwell (1998) identify 'scales' as key criteria for judging significance in ecological risk assessments and present decision frameworks for evaluating the importance of changes.

16.5 USING THE RIGHT TOOLS

Within the language of ecological risk assessment the 'right tools' include those that provide 'measures of effect', 'measures of exposure' and 'measures of ecosystem and receptor characteristics' as these are described in the USEPA's guidelines (USEPA 1998a). The question of 'What are the right tools?' for an ecological risk assessment must start with the stated assessment endpoints.

Studies that are conducted without careful consideration of the assessment endpoints might be inappropriate for the eventual risk assessment. During the 1990s, increased emphasis was given to the preparation of analysis plans and ecological risk assessment workplans. These documents are intended to document how the proposed work, including data collection efforts, will be used in the risk assessment, and are important communication tools.

The Massachusetts Weight-of-Evidence Workgroup proposed a set of criteria as an aid for selecting the right tools (Menzie *et al.* 1996). These include: (a) strength of association between the measure and the assessment endpoint; (b) site-specificity; (c) stressor-specificity; (d) quality of data and overall study; (e) availability of an objective measure for judging environmental harm; (f) sensitivity of the measure for detecting change; (g) spatial representativeness; (h) temporal representativeness; (i) quantitative; (j) correlation of stressor to response; and (k) use of a standard method. In the weight-of-evidence approach, these criteria are used by the risk assessor to evaluate the degree of confidence in the measure as a tool for evaluating the stated assessment endpoint.

Over the past few decades a plethora of measurement and modeling tools have been introduced. Investigators are drawn to both simple and sophisticated tools. An attraction of the former is 'cheap and quick', the later 'cutting edge and high tech'. Either may be appropriate. However, the assessor needs to take a broader view when identifying if the candidate method is adequate and appropriate. Three questions that are useful to ask when making such judgement are:

- How well does the method relate to the stated assessment endpoint?
- How much technical support is needed to reach a conclusion adequate for the decision?
- How much uncertainty can be accepted in the analysis and can it be managed adequately for the decision?

16.6 EVALUATING AND MANAGING UNCERTAINTIES

The evaluation — especially quantitative evaluation — of uncertainty has received much attention during the 1990s. SETAC held a Pellston Workshop on the topic with reference to ecological risk assessment (Warren-Hicks and Moore 1998). The USEPA's (1998a) guidelines provide a discussion of how uncertainties should be evaluated. However, actual experience at quantitative uncertainty analyses is limited for ecological risk assessments. This probably reflects the fact that the ecological risk assessment methods are still being developed, as well as a concern by many in the regulatory community that quantitative uncertainty analyses might obscure or complicate interpretation of results. In 1999, concerns regarding the role of quantitative uncertainty analyses were raised during the USEPA's collaborative process to derive ecological soil screening levels. My impression is that such concerns over the use of quantitative uncertainty analysis arise mainly from a perceived 'lack of knowledge' concerning the parameters. In such cases, bounding analyses and professional judgement become important tools.

The concerns over using quantitative uncertainty analyses can be resolved and I gained some insight into this when I chaired the USEPA's Monte Carlo Workshop (USEPA 1996). The workshop report provides a useful framework for considering when and how to use quantitative uncertainty analyses as well as how to express the results. In particular, it suggested a tiered approach to using these analyses at various stages in the risk assessment. It also indicated the value of deterministic analyses and underscored the need to present deterministic results along with probabilistic results. It described the role of professional judgement in uncertainty analyses. The workshop report led to a set of principles (USEPA 1997a,b) which I believe should be considered when conducting quantitative evaluations of uncertainty for ecological risk assessments.

Risk managers often think about uncertainty in terms of how it can be reduced and/or managed within the context of a decision that will be protective of the environment. A key issue that arises is how to insure that the decision is environmentally protective; for example, overall protection of the environment is the first decision criterion for remedial decision making in the Superfund Program (USEPA 1998b). The issue has often been addressed by relying on deterministic estimates of risk that are based on assumptions or estimates known to be conservative. The resulting risk estimate is presumed to be conservative and therefore environmentally protective. For the risk manager, uncertainties have been managed by erring on the side of protectiveness. While this strategy for managing uncertainty can work in many instances, it can also lead to environmental decisions that are unbalanced and not cost-effective. In such cases, it is useful to make a more in-depth examination of the sources and magnitudes of uncertainty.

The desire to manage uncertainty and insure that decisions are environmentally protective has led to the development of tiered approaches for evaluating ecological risks (CCME 1996, USEPA 1997a,b, 1998a, ASTM 1998). These begin with simple conservative screening-level assessments and proceed to more sophisticated analyses as needed in order to make a decision. With each tier, knowledge is increased and uncertainty reduced: sources and magnitudes of uncertainty are also better known and characterized to insure that they are properly managed. Quantitative uncertainty analysis is one of the tools that can be used at later tiers to provide this additional insight. Thus, the tiered assessment strategy is itself a tool for managing uncertainty. A tiered strategy also guides risk assessors and risk managers in allocating the resources needed for assessing risks at a level adequate for the decision.

As increased emphasis is given to estimating population-level risks, probabilistic analysis will need to be relied upon to estimate the fractions of population that may be affected by various stressors. From a population biology standpoint, the death or reproductive impairment of individuals becomes important when it is large enough to influence the characteristics of the population. The tools for making these evaluations are available. However, additional experience is needed in applying them to ecological risk assessments. I expect that this experience will identify areas where there is a lack of knowledge concerning how populations respond to stressors. However, understanding where we lack information provides a starting point for identifying where uncertainties can be reduced.

16.7 COMMUNICATING THE PROBLEM, THE ANALYSES AND THE DECISIONS

Communicating about ecological risks is a challenge and experience indicates that it needs to begin at the planning and problem formulation stages of an assessment and continue throughout the process. There are at least two important interrelated channels of communication: (1) between the risk manager and the risk assessor and (2) with stakeholders. The former is more operational and directed toward insuring that the management goals are translated appropriately into assessment endpoints and an appropriate analysis to support decision making. I have already discussed the challenge of achieving these operational aspects. Communication with stakeholders can be an important part of planning as well as the overall risk assessment process. The challenge is knowing when and how to engage stakeholders in this communication. I explore this below.

An over-arching goal for engaging stakeholders in planning and problem formulation as well as the rest of the risk assessment process is to achieve a common understanding for the purpose and direction of the assessment. McDaniels (1998) draws upon his research experience on risk perception to identify blind spots in how lay people perceive ecological risk. He suggests that risk communication on ecological issues is likely to work best if done as one component of a public decision-making process, within a discursive approach to clarifying public preferences, in which there is a clear purpose and task to motivate the participants. Fisher (1998) identifies the development of a 'shared understanding' as the goal of the risk communication process.

When should stakeholders be involved? I don't think there is a simple answer but key criteria seem to include: (1) the potential that they will be affected by the decision; (2) the potential that they have information important for the assessment and the decisions; (3) their level of interest; and (4) the magnitude of the potential problem. Based on my experience, the following reflects the extent to which broad stakeholder involvement is sought.

Stakeholder involvement almost always sought

- Forming local, state/provincial or federal technical policy decisions and laws.
- Developing guidance for implementing technical policy decisions and laws.

- Environmental impact evaluations of new facilities or operations.
- Permitting new major dischargers.
- Use of public lands for harvesting and extraction of natural resources.
- Regional and watershed management plans.
- Unregulated releases of chemicals from large sites that extend beyond the property boundaries of a facility.

Stakeholder involvement sometimes sought

- Forming a technical policy or operational decision for a specific facility (e.g., an industrial practice).
- Periodic reviews of existing discharge or operational permits.
- Development of private lands where abutters may be impacted.
- Unregulated releases of chemicals from large sites that do not extend beyond the property boundaries of a facility.

Stakeholder involvement seldom sought

- Operations and activities conducted in accordance with technical policy decisions and laws that have been arrived at with stakeholder involvement.
- Performance monitoring of various facilities (e.g., discharge permits) by government agencies.
- Unregulated releases of chemicals from small sites that do not extend beyond the property boundaries of a facility.

How should stakeholders be involved? This challenge will depend on the nature of the problem. However, much attention has been given over the past decade to how to reach out to and engage stakeholders in the process. I recently had an opportunity to be involved in several risk assessments that involved stakeholders. I also discussed 'lessons learned' with individuals responsible for the five USEPA watershed case studies and with individuals such as Alvin Chun who do training on stakeholder involvement. I have compiled a set of lessons learned that may be useful to consider for ecological risk assessments. These reflect situations where stakeholder involvement was a major component of the planning process and therefore they may not be applicable for all situations. The lessons learned are as follows.

1. Advance and centralized planning of outreach activities, meeting logistics and meeting structure can provide a well-defined but flexible framework for conducting the stakeholder meetings.
2. Knowledgeable local persons familiar with the various stakeholders and the issues can identify the most effective outreach activities and provide important insight into the issues.
3. Advance notice and publicity, using effective mechanisms for reaching a broad range and large number of stakeholders, is critical to ensure that stakeholders are aware of the opportunity to participate in the meetings.

4. Providing stakeholders with a 'straw man' document in advance of the meetings stimulates and helps focus thinking and discussion.
5. The use of a knowledgeable and neutral third party facilitator contributes to the success of meetings.
6. A clearly structured meeting and ground rules for time and conduct helps stakeholder meetings run smoothly, enables the facilitator to be flexible where appropriate, and helps individual participants feel that their views would be heard and respected.
7. Invited formal presentations by individuals from various stakeholder groups help insure that the views of stakeholders are articulated and provide a basis for others to comment.
8. Encouraging and welcoming a broad range of discussion enables stakeholders to express their views in their own way and to comment on areas that, while not directly related to the issue at hand, could be very important to the overall risk management process.
9. Risk assessment jargon and emphasis on the 'process' should be kept to a minimum when engaging the public in discussions about an environmental issue.
10. Effective stakeholder involvement requires a commitment of people and resources adequate to support the process.
11. Continuity of key contact people is important for building and maintaining trust and supporting the development of institutional knowledge related to an issue.
12. Formal meetings may not be adequate for engaging all stakeholders. In some cases, it may be necessary to meet with people in less formal settings or to seek out their viewpoints.
13. Illustrative examples, tutorials and/or the involvement of a facilitator knowledgeable about the ecological risk process can help stakeholders understand the process and more efficiently examine the issues related to their specific program.
14. A variety of methods have been identified for involving stakeholders. Training on stakeholder involvement can help key contact people develop and maintain effective stakeholder programs.

An additional and especially important 'lesson learned' regarding communication among all parties is the value of developing understandable conceptual models. The development of a conceptual model through an iterative process is probably the most useful technical exercise within problem formulation and should be initiated early in the process. Conceptual models provide useful tools for developing a common understanding of the problem and identifying how possible stressors are related to ecological receptors. Experience gained from the USEPA's watershed case studies and at hazardous waste sites indicates that people with little or no technical background found the process of developing conceptual models very useful. The conceptual model was frequently identified as the strongest component of the problem formulation. Initial models may be overly simple and sometimes wrong, but these early models provide a basis for identifying gaps in knowledge about the system and relationships among valued ecological entities and stressors. The models help individuals understand how they or their organizations may influence or be influenced by processes or activities related to the risk decision. As such, conceptual models — together with management goals — help personalize the problem formulation process.

16.8 CONTINUINGS

During the next decade we will see an increased use of ecological risk assessment as a tool for environmental decision making. I envision the evolution of the field to involve:

1. Integration of the assessment process into explicit decision frameworks;
2. Refinement of tools to account for spatial, temporal and effects scales;
3. Increased emphasis on population-level and ecosystem risks;
4. Better education and communication about environmental issues.

We as a society will also continue to evolve in our understanding of nature and our relationship to it. This is already reflected in our language. Ideas that are being formulated today concerning 'responsible care', 'environmental stewardship', 'sustainable development' as well as the 'precautionary principal' will be further integrated into our policies and practices.

Environmental decision making is a complex, multi-disciplinary, multi-dimensional process that depends on technical information joined with information from disciplines such as economics, sociology and risk perception (Patton 1998). Better communication among risk assessors, risk managers and stakeholders is needed to insure a proper integration of social and scientific issues. Kapustka and Landis (1998) emphasize the importance of preserving science as a methodology for acquiring objective information separated from societal value. Glicken and Fairbrother (1998) observe that the ethical and moral responsibilities we perceive in terms of our relationships with and responsibilities to the environment cause us to begin our ecological assessments with social questions. I agree with these observations, but think that progress needs to be made on integrating science with societal issues. The tug and pull between perceptions and science-based realities (or attempts to represent reality) is a central issue in risk management and communication about risks (Menzie 1998) and I propose a conceptual approach for bringing these together (Figure 16.2). This can only be accomplished through good communication. Because the process of integrating alternative views involves expanding as well as constraining perceptions and science-based perspectives, careful listening is critical, especially on the part of risk scientists.

Finally, many people and organizations are making active contributions to the field of Ecological Risk Assessment. I especially want to acknowledge the

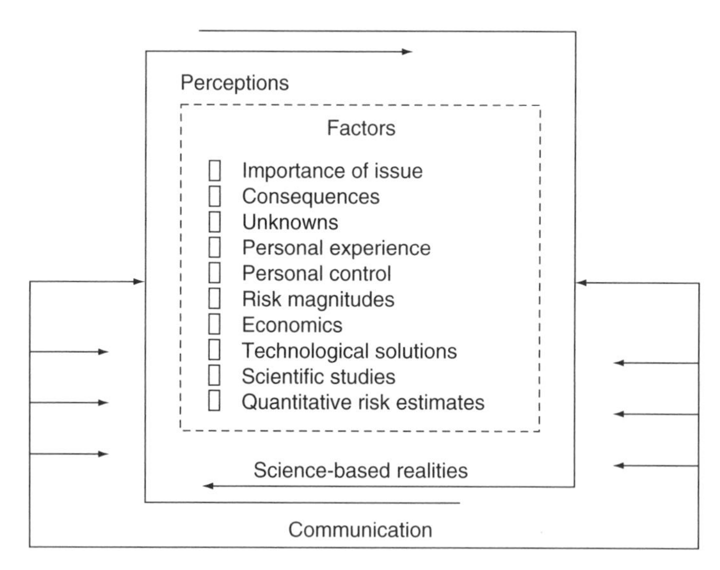

Figure 16.2 A conceptual framework for integrating perceptions and science-based realities about risk.

work carried out at Oak Ridge National Laboratory (ORNL) during the 1980s and 1990s as well as the efforts of the Society of Environmental Toxicology and Chemistry, especially the Ecological Risk Assessment Advisory Group (ERAG). Many of the conceptual ideas and assessment tools that were developed during the past few decades are well presented in Suter (1993) and this serves as a useful text for those entering the field.

REFERENCES

Akçakaya HR (1998) *RAMAS GIS: Linking Landscape Data with Population Viability Analysis*, version 3.0, Applied Biomathematics, Setauket, NY.

Applied Biomathematics (1998) *Integrating Physical-Chemical and Ecological Models*, Applied Biomathematics, Setauket, NY.

ASTM (1995) *Standard Guide for Risk-based Corrective Action Applied at Petroleum Release Sites*. ASTM Standard E 1739-95, American society for Testing and Materials, Philadelphia, PA.

ASTM (1998) *Provisional Standard Guide for Risk-based Corrective Action Applied at Chemical Release Sites*.

Bartell SM (1998) Ecology, environmental impact statements, and ecological risk assessment: a brief historical perspective. *Human and Ecological Risk Assessment*, **4**, 843-851.

Barton A and Sergeant A (1998) Policy before the ecological risk assessment: what are we trying to protect? *Human and Ecological Risk Assessment*, **4**(4), 787-795.

CCME (1996) *A Framework for Ecological Risk Assessment: General Guidance*. En 108-4/10-1996E, The National Contaminated Sites Remediation Program, Canadian Council of Ministers of the Environment, Winnipeg, Manitoba.

Clifford PA, Barchers DE, Ludwig DF, Sielken RL, Klingensmith JS, Graham RV and Banton MI (1995) An approach to quantifying spatial components of exposure for ecological risk assessment. *Environmental Toxicology and Chemistry*, **14**, 895-906.

Cooper WE (1998) Risk assessment and risk management: an essential integration. *Human and Ecological Risk Assessment*, **4**, 931-937.

CRARM (1997) *Framework for Environmental Health Risk Management*. Commission on Risk Assessment and Risk Management, Washington, DC.

Cronon W (1995) Introduction: in search of nature. In *Uncommon Ground*, Cronon W (ed), W.W. Norton & Co., New York, pp. 23-57.

Fisher A (1998) The challenges of communicating about health risks and ecological risks. *Human and Ecological Risk Assessment*, **4**, 623-626.

Freshman JS and Menzie CA (1996) Two wildlife exposure models to assess impacts at the individual and population levels and the efficacy of remedial actions. *Human and Ecological Risk Assessment*, **2**, 481-498.

Gentile JH and Harwell MA (1998) The issue of significance in ecological risk assessments. *Human and Ecological Risk Assessment*, **4**, 815-828.

Glicken J and Fairbrother A (1998) Environment and social values. *Human and Ecological Risk Assessment*, **4**, 779-786.

Kapustka LA and Landis WG (1998) Ecology: the science versus the myth. *Human and Ecological Risk Assessment*, **4**, 829-838.

McDaniels TL (1998) Systemic blind spots: implications for communicating about ecological risk. *Human and Ecological Risk Assessment*, **4**, 633-638.

Menzie CA (1995) The question is essential and ecological risk assessment. *Human and Ecological Risk Assessment*, **1**(3), 159-162.

Menzie CA (1998) Introduction: risk communication and careful listening — resolving alternative world views. *Human and Ecological Risk Assessment*, **4**, 619-622.

Menzie CA and Freshman JS (1997) An assessment of the risk assessment paradigm for ecological risk assessment. *Human and Ecological Risk Assessment*, **3**, 853-892.

Menzie CA, Henning MH, Cura J, Finkelstein K, Gentile J, Maughan J, Mitchell D, Petron S, Potocki B, Svirsky S and Tyler P (1996) Special report of the Massachusetts weight-of-evidence workgroup: a weight-of-evidence approach for evaluating ecological risks. *Human and Ecological Risk Assessment*, **2**(2), 277-304.

Menzie CA, Efroyson R, Ells S, Henningson G, and Hope B (2001) Risk assessment and risk management. In *Assessing Contaminated Soils: From Soil-Contaminant Interactions to Ecosystem Management*, Fairbrother *et al.* (eds), SETAC Press, Pensacola, FL, in press.

Moore DRJ and Biddinger GR (1996) The interaction between risk assessors and risk managers during the problem formulation phase. *Environmental Toxicology and Chemistry*, **14**, 2013-2014.

North W (1996) *EPA Risk Policy Report* (1996), **3**, 8-9.

NRC (1983) *Risk Assessment in the Federal Government: Managing the Process*, National Academy of Sciences Press, Washington, DC.

NRC (1996) *Understanding Risk*, National Academy Press, Washington, DC.

Oregon Department of Environmental Quality (1998) *Guidance for Ecological Risk Assessment, Level III, Appendix A: Procedure for Performing a Population-Level Ecological Risk Assessment*, Waste Management and Cleanup Division, Portland, OR.

Patton DE (1998) Environmental risk assessment tasks and obligations. *Human and Ecological Risk Assessment*, **4**, 657-670.

Suter GW (1993) *Ecological Risk Assessment*, Lewis, Chelsea, MI, 538 pp.

USEPA (1992) *Framework for Ecological Risk Assessment*. EPA 630/R-92/001, Risk Assessment Forum, US Environmental Protection Agency, Washington, DC.

USEPA (1996) *Summary Report for the Workshop on Monte Carlo Analysis*. EPA/630/R-96/010, Risk Assessment Forum, US Environmental Protection Agency, Washington, DC.

USEPA (1997a) *Ecological Risk Management Guidance for Superfund: Process for Designing and Conducting Ecological Risk Assessments*. Interim Final EPA-540-R-97-006, Environmental Response Team, US Environmental Protection Agency, Edison, NJ.

USEPA (1997b) *Guiding Principles for Monte Carlo Analysis*. EPA/630/R-97/001, Risk Assessment Forum, US Environmental Protection Agency, Washington, DC.

USEPA (1998a) *Guidelines for Ecological Risk Assessment*. EPA/630/R-95/002F, Risk Assessment Forum, US Environmental Protection Agency, Washington, DC.

USEPA (1998b) *Ecological Risk Management Principles for Superfund Sites*. Directive 9285. 7-28 P, Office of Solid Waste and Emergency Response, US Environmental Protection Agency, Washington, DC.

Warren-Hicks WJ and Moore DRJ (1998) *Uncertainty Analysis on Ecological Risk Assessment*. Proceedings from the Pellston Workshop on Uncertainty in Ecological Risk Assessment, 23-28 August 1995, Pellston, MI, SETAC Press, Pensacola, FL, 314 p.

Washington-Allen RA and Sample BE (1997) *Determination of the Spatial Risk to Wildlife From Dispersed Contaminants on the Oak Ridge Reservation*. ES/ER/TM-229, Oak Ridge National Laboratory, Oak Ridge, TN.

PART THREE

Case Studies Showing Applications of Toxicological and Ecotoxicological Risk Methods to Contaminated Sites and Remediation Technologies Management

17

Introduction

ADRIEN PILON

Biotechnology Research Institute, National Research Council of Canada, Montreal, Quebec, Canada

17.1 INTRODUCTION

Contaminated site restoration and clean-up have become very challenging areas involving a large range of stakeholders. Over the last decade, a large number of major site restoration projects have led the regulators, the site owners and the environmental experts to question the basis of the decisions involved during this process. Environmental, economical and social issues are interrelated in the decision process leading to the restoration of contaminated sites to a proper level of quality, which satisfies both the environment and human dimensions.

Risk assessments and analysis of impact on human and environmental receptors have become a predominant step to go through during the restoration, and the reuse of contaminated sites. The risk analysis of the human and environmental procedure is generally used for the following purposes.

- Determining if the actual contamination at a given site represents a risk to the environment receptors and humans, and therefore requires that the site be restored.
- Site restoration includes implementing mitigation measures, containment, passive systems, total removal of contaminants and control of emissions. A wide range of solutions depend on the potential risks and exposure to receptors. Risk analysis may also lead to 'no action' based on current land use.
- Risk is associated with the restoration actions as well as with the remediation technologies. This refers mainly to the advantages or disadvantages of implementing remediation solutions with respect to the fate of the contaminants.
- The level of risk needs to be associated with the future land use, which could be residential, recreational, commercial or industrial.
- When the contamination impacts properties or sites outside the boundary of the site.

Environmental Analysis of Contaminated Sites. Edited by G. I. Sunahara, A. Y. Renoux, C. Thellen, C. L. Gaudet and A. Pilon

The procedures and their respective advantages have been presented in Part Two of this book. Gaudet *et al.* (Chapter 12) demonstrated the advantages of risk-based assessments compared to generic criteria. More precisely, Ouellet *et al.* (Chapter 13) and Martel *et al.* (Chapter 15) discussed the use of ecotoxicological risk assessment for site restoration and management. Finally, Trépanier *et al.* (Chapter 14) presented the advantages of the risk-based remediation approach and their potential applications and limitations for site restoration and risk management.

The following series of chapters will present six case studies where human and environmental risk assessments were performed. These cases will highlight very important aspects related to the use of risk assessments. The major issue being addressed is the contaminants, which are being evaluated throughout this process. For instance, many categories of contaminants like hydrocarbons (Chapter 22) or explosives (Chapter 19) have toxic properties, which need to be assessed, and for which no reference doses are known. Even, in some cases, intermediates or metabolites are newly characterized and therefore lacking the basic information for assessing risks.

Risk assessment will help to determine the added value of the selected remediation technology and will show how different treatment impacts on the bioavailability or degradation of contaminants (Chapter 20). Risk assessments will assist in the process of transferring a site for future use after it has been restored using the best available technology (Chapter 18). Finally, risk assessments methodology can lead to different conclusions based upon the model being used. This will be demonstrated in a comparative study (Chapter 21). In many cases presented herein, economical comparisons are made to evaluate different options. As this aspect is becoming increasingly important, due to the high number of contaminated sites discovered as a result of investigations and land use changes, the economical burden of site restoration is increasing. Therefore, risk assessments performed using sound scientific data and technical expertise will definitely help the decision makers in the process of sustainable development of contaminated lands. Finally, Chamberland *et al.* (Chapter 23) will discuss the use of risk-based corrective action at petroleum sites where residual hydrocarbon contamination in groundwater was suggesting a limitation of usage. The following is a brief description of the chapters found in Part Three of this book.

17.2 RISK ASSESSMENTS AND SITE RESTORATION: CASES IN THE NETHERLANDS

Several case studies will show the potential use of the risk assessment and how it leads to new options for site restoration. In Chapter 18, Elkenbracht *et al.* present two cases. The first case discusses a contamination problem of soil and groundwater in an urban area in The Netherlands. The Dutch experience also presents a major change in the Dutch soil protection policy, which is now implementing the approach of site restoration based on its functionality or future use, compared to past practices where sites had to be cleaned to allow multi-functionality. Before implementing the risk assessment approach, this site could not be redeveloped due to high costs, as a result of the analysis of potential institutional use and the implementation of a series of restoration and mitigation measures. The risk analysis, in this case, was focused mainly on human receptors due to the low probability that ecological receptors would be exposed to pollutants.

Groundwater impacted with pesticides (HCH, chlorobenzene, etc.) required the installation of a biological treatment system to prevent the spreading of contaminants. The ecotoxicological risk assessment in this case mainly deals with specific functions within the soil ecosystem once the site is restored to its land use. In this particular case, the ecotoxicology functions refer to nitrogen cycle by microbial activity and to the capacity of soil to sustain plant growth.

Elkenbracht *et al.* present a second case where the ecological receptors are predominant at the site. The site is a natural feature inherited from the glacial period and contains a pond where the aquatic ecosystem has been polluted over the years by metals (i.e., lead). The case describes the application of the different restoration scenarios considered to protect the lake and the forest ecosystem, to remove the sediments without impacting the aquifer underneath. Each case presents the scenarios with consideration given to the environment and the economical value.

17.3 ENVIRONMENTAL RISK ASSESSMENT FOR EXPLOSIVES-CONTAMINATED SITES

Robidoux *et al.* in Chapter 19 discuss a novel approach for explosives-contaminated sites. At first sight, an industrial site located near a major river with agricultural land and wooded lots nearby could represent a major ecological risk. In this case, TNT is the major contaminant found at the site, and its physical and chemical properties suggest that this compound would behave in the environment in such a way that low bioaccumulation and biotransformation are expected. In this study, the potential risk of contamination to receptors (earthworms, birds, deer, plants and soil microbes) present at the site will be assessed. Effects of TNT on different organisms in soils, groundwater and surface water will be evaluated. However, cumulative risk associated with byproducts and TNT degradation products cannot be estimated since scant toxicity data are available for these contaminants. The CalTOX model is used to predict the potential transport and transformation of TNT in groundwater.

17.4 INFLUENCE OF SOIL REMEDIATION TECHNIQUES ON THE BIOAVAILABILITY OF HEAVY METALS

Remediation technologies are developed and applied to reduce the levels of contaminants in soils and groundwater. In Chapter 20, Van Gestel *et al.* discuss the use of ecotoxicological indicators to determine the quality of soils after remediation. Currently, chemical analyses are used to determine the levels of remediation and quality of soils. In addition to current practices, soil remediation technologies can remove or sometimes add 'toxicity' to the environment through its treatment chain. Van Gestel *et al.* discuss a few cases where heavy metals-contaminated soils were remediated using different techniques, and how toxicity is removed or increased at different steps of treatment. A battery of ecotoxicological tests (bioassays) was applied to assess the bioavailability of heavy metals submitted to soil treatment technologies at laboratory and commercial scales. The study also evaluates the effect of residual heavy metal contamination in soils after treatment. Both studies focus on bioaccumulation of metals in test organisms (earthworms and plants).

17.5 CONTAMINATED SITES REDEVELOPMENT USING THE RISK ASSESSMENT/RISK MANAGEMENT APPROACH: THE MONTREAL EXPERIENCE

Like all large cities, Montreal is faced with managing numerous contaminated sites (brownfields) inherited from old industrial activities. In order to implement proper environmental management practices, the City of Montreal conducted a pilot study where both human and ecotoxicological risk assessments would be used and assessed. This work was conducted in collaboration with an environmental firm (DDH) and compared to existing Contaminated Sites Rehabilitation Guidelines in the Province of Quebec, Canada. First, a comparison of two methodologies for human risk assessment was performed. Results from Guay and Barbeau (Chapter 21) show how different methods would lead to major differences in redevelopment practices and thus impact on the overall costs of an environmental restoration project. The 'brownfields' issue in Canada has been studied by several organizations and it is generally recommended that environmental management methods be made available to all stakeholders involved in redevelopment activities.

17.6 ENVIRONMENTAL FATE AND HUMAN EXPOSURE MODELING OF THE RESIDUAL TPH CONTAMINATION IN A BIOREMEDIATED PETROLEUM STORAGE SITE

The case presented in Chapter 22 by Loranger *et al.* is very representative of a major issue at petroleum-contaminated sites in Canada, where remediation and reuse of sites based on risk assessment is not fully accepted due to its complexity. Loranger *et al.* describe how this approach was used during the restoration of a large petroleum storage facility. The remediation was first established on the basis of generic criteria. The complexity of the petroleum contamination is related to the mixture of compounds found in petroleum products and the lack of toxicological data for each individual compound. Different methodologies are currently being developed in Canada and the USA in order to establish representative toxicological values (reference dose or Rfd). In this particular case study, Loranger *et al.* base their risk assessment on extractable petroleum hydrocarbons (EPHs) instead of total petroleum hydrocarbons (TPHs) and use the toxicity data for each subgroup of petroleum compounds instead of the entire range of petroleum hydrocarbons. The CalTOX model was used to assess the potential transfer of contaminants and exposure pathways of selected petroleum subgroups. The risk assessment shows that in some areas of the site mitigation measures should be implemented to prevent exposure of toxicants to humans through ingestion.

17.7 AN APPLICATION OF RISK-BASED CORRECTIVE ACTION (ATLANTIC RBCA) FOR THE MANAGEMENT OF TWO SITES IMPACTED WITH RESIDUAL HYDROCARBONS IN GROUNDWATER

As pointed out in Chapter 22, RBCA is not accepted in all jurisdictions in Canada. In Chapter 23, Chamberland *et al.* describe a very specific use of this approach for petroleum-contaminated sites in Atlantic Canada. However, the range of hydrocarbon compounds described in this study is more limited to the light fractions, C_6-C_{12} (aromatic and aliphatic). The major difficulty is related to low residual hydrocarbon concentrations left in groundwater bodies where remedial actions would be very costly to remove contaminants and would not necessarily be amenable to remediation under normal operations. The same methodology was applied to two sites and investigations included hazard identification, receptors considered, exposure assessment and determination of site-specific target levels. The use of RBCA will help to recommend the use of groundwater in both cases, with some limitations of use only in one case.

18

New Dutch Policy on the Remediation of Soil — Balancing Human Health and Ecological Risks, Land Use and Remediation Costs: Two Case Studies

ERNESTINE ELKENBRACHT[1], PETRA KREULE[2], FERRY VAN DEN OEVER[2] AND JUDITH RAES[2]

[1]*Quintens advies en management, Bunnik, The Netherlands*
[2]*TAUW BV, Deventer, The Netherlands*

18.1 INTRODUCTION

This chapter summarizes the modernization process which the Dutch soil remediation policy is currently undergoing. The aim of this process is to accelerate the national soil remediation programme by better attuning site remediation to the intended land use and level of risk posed by soil contamination with respect to this land use. Weighing the remediation costs of various options is an important aspect. This new approach, still under development, is termed 'functional remediation' in that risk is assessed in light of the desired function of the soil. This approach is considered a practical alternative to the previous 'multifunctional' approach.

Two cases are described that illustrate the consequences of this new policy. The first is a function-oriented remediation project at a site where human health risks exist in relation to the desired land use due to soil contamination, and where further spread of contaminants has to be prevented. The second is a situation where exposure of the ecosystem to soil contamination is intended to be reduced by carrying out a function-oriented remediation project.

18.1.1 MODERNIZATION OF DUTCH SOIL REMEDIATION POLICY

In 1998, a change in the direction of Dutch soil remediation policy occurred as a result of an evaluation of the remediation regulation contained in the 1995

Environmental Analysis of Contaminated Sites. Edited by G. I. Sunahara, A. Y. Renoux, C. Thellen, C. L. Gaudet and A. Pilon

Soil Protection Act (Tweede Kamer der Staten-Generaal 1997). The evaluation showed that soil remediation projects have, in general, been too expensive. In addition, the cost-effectiveness of the national soil remediation operation has been too low because, per unit of time, too few cases have been tackled. In view of the large number of soil contamination problems in The Netherlands, the national soil remediation operation has shown relatively little progress, partly because of the few means available. All this has resulted in the stagnation of both spatial and economic processes. This stagnation has brought about social damage, a narrow basis for soil remediation projects, a low investment level and adverse effects on the quality of the living environment. The change in direction of the Dutch soil remediation policy is necessary to remedy this social stagnation. It also aims to increase the cost-effectiveness of soil remediation projects and to accelerate the national soil remediation operation. It should be noted that the philosophy of the new policy is clear, but not all the details have yet been finalized.

18.1.2 WHAT DOES THE NEW POLICY MEAN?

The new policy addresses the weighing of remediation objectives. This weighing starts from the point that it is clear that the site has to be remediated due to environmental or social reasons. The Dutch policy about determining the *necessity* of remediation is risk-based and is not changing. The big difference between the past and proposed approaches for weighing of remediation objectives is that the multifunctional objective is no longer the basis for soil remediation. Until now, the weighing of remediation objectives started by considering the multifunctional restoration of the soil. Only in case of disproportionately high costs due to site-specific reasons was a variant permitted, characterized by isolation, containment and monitoring (ICM). This often resulted in perpetual aftercare. Therefore, restoration of multifunctionality no longer appears to be realistic. On the other hand, an increase of perpetual aftercare because of the ICM method is not desirable.

The aim of the new policy is to raise both the social and environmental cost-effectiveness of the national soil remediation operation. Social cost-effectiveness can be raised by promoting social activities (the building of new houses, economic development, the development of rural areas) at desired locations whenever such is wished for, and by preventing unnecessary delay. An objective of the renewed soil remediation programme is that the stagnation in social and economic development posed by the existing soil contamination is removed as soon as possible, and that such stagnation is prevented in the future.

The remediation programme's environmental cost-effectiveness can be raised by making soil suitable for the desired use at more contaminated sites per unit of time. As a consequence, on a national level, more contaminated soil will be tackled and the spread of contaminants will be reduced. At individual sites, human health risks will be eliminated, and the spread of contaminants and the exposure of ecosystems to soil contamination will be reduced.

This can be accomplished, in part, by using such social means as increasing market dynamics and broadening social support for the execution of soil remediation projects in order to raise the cost-effectiveness of the national soil remediation operation. This can also be accomplished technically by making remediation projects less expensive by better attuning the remedial measures to the current or future desired land use (i.e., *function-oriented remediation*). Although the new approach is still being worked out, the basic principles have been formulated in an important policy document 'Van Trechter naar Zeef' (Kooper 1999) in which remediation projects should meet the following conditions.

1. The measures taken to remediate a contaminated site should be based on an integrated vision for tackling the *entire* contamination. In other words, the vision should be tackled considering all relevant aspects of soil contamination, including immobile and/or mobile contaminants.
2. The soil should be made suitable for the desired use by eliminating potential routes of exposure and preventing the spread of contaminants. The following two types of measures can be distinguished:
   - Those aimed at reducing undesired human health and ecological risks posed by contaminants located in the contact zone. Contaminants beneath the contact zone will be left behind, when there will be no spreading of these contaminants. However, contaminants left behind have to be registered.
   - Those aimed at the cost-effective prevention of the spreading of mobile contaminants from the subsoil beneath the contact zone (about 1.5 m below ground level). This depth is also the level of the mean phreatic groundwater level in The Netherlands.

   These points are further elaborated upon in Sections 18.1.3 and 18.1.4.
3. The remediation project should result in a situation where as little aftercare as possible will be needed. Such aftercare may vary from mere registration or monitoring to the pumping and treatment of contaminated groundwater. The duration and costs of aftercare should be minimized. During the remediation, it should be checked at regular intervals whether this aim will be attained, so that the project can be adjusted as necessary. This means that an ICM variant — which would entail perpetual, active aftercare — should be applied only if it is the only cost-effective option.

18.1.3 FUNCTION-ORIENTED TACKLING OF THE CONTACT ZONE

Function-oriented tackling of the contact zone may consist of either a standard approach or one of several variants; the latter is indicated if the site concerned needs to be remediated more thoroughly or less thoroughly. The standard approach has four points for attention:

- What is the desired type of land use on the site?
- Which thickness of the clean layer requires this land use type? The contact zone or clean cover should have a certain minimum thickness and is attuned to the desired land use. This thickness may vary from 0 m (if the entire site is paved) to 1.5 m (e.g., in gardens with deeply rooting crops).
- What construction of the clean layer is required to sustain the specified land use type?
- What quality of soil is needed to sustain the specified land use type? The quality of the contact zone is attuned to the desired land use. The quality of the soil in the contact zone should meet the so-called *land use standards,* under which each contaminant has been set for four types of land use.

The following four land use types have been distinguished:

I. Gardens and intensively used areas (relatively much contact with soil possible), such as play and recreational areas.
II. Extensively used areas, like verges, green spaces near industrial areas and undeveloped areas.
III. Buildings and unbroken pavement like asphalt and paving stones that fit well together.
IV. Agricultural and ecological areas.

In agricultural and ecological areas, no standard approach is possible. For each specific case, the best function-oriented quality of the contact zone will be established. The thickness of the contact zone depends on ecological factors.

The land use standards were derived from land use and quality requirements, which include:

- People should be able to safely consume crops raised in their own gardens;
- The unrestricted growth of plants, grassland, shrubs and trees should be possible;
- Macrofauna (e.g., rainworms) should be present at local background values;
- Microbial processes (e.g., nitrification) should function properly.

Which of these conditions are taken into account in the derivation of land use standards depends on the type of land use. These standards were formulated in order to prevent unacceptable risk to humans and (for the functions to which this applies) the ecosystem. They are more stringent than the maximum allowable risk level set for humans and the HC_{50} (i.e., hazardous concentration 50, the limit above which more than 50% of the terrestrial species present will be harmed) that is applied in relation to ecological risks in general risk assessment. Based on these land use standards, it will be attempted to achieve a sustainable soil quality in view of the intended land use by also taking into account sources of contamination other than the soil to which humans will be exposed (e.g., air polluted by distant emission sources).

The principle of land use standards is illustrated in Figure 18.1. For three of the four land use types, a standard approach is possible. For type III (an unbroken pavement covering the entire site), no remediation is needed, as long as humans cannot come into contact with the contamination and no leaching or spreading can occur. For type I, the quality of the contact zone has to be improved until it meets the required quality of the land use type. This function-oriented remediation of the site will be registered and the restrictions in land use will be set down. When changes in land use occur on-site, it will be necessary to check whether the soil quality permits this new type of land use or the site has to be remediated again.

The above standard approach may be deviated in some individual cases and in special areas. These solutions are called tailor-made. The tailor-made solutions for the individual cases have not yet been elaborated, though some principles for tailor-made solutions for special areas are already recorded in the policy.

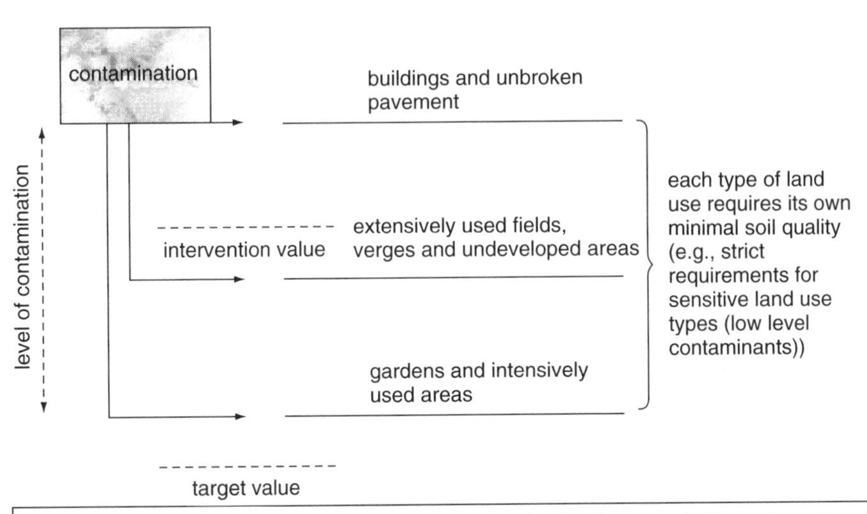

Explanation: It is necessary to remediate contaminated soil with concentrations above the intervention value. Depending on the land use the soil quality in the contact zone has to be improved and meet the land use standard values. When no contact with the contaminated soil is possible (due to unbroken pavement) and there is no spreading of contaminants, the contaminated soil will be allowed to remain provided that the contamination and the land use restrictions are registered.

Figure 18.1 The contact zone quality to be attained depends on the type of land use and the accompanying land use standard.

There can be special characteristics in an area which could be a reason to remediate the site more or less thoroughly. When the contaminated site is located, for example, in a vulnerable area (e.g., groundwater protection area), it may be necessary to remediate more thoroughly. Less thorough remediation is only permitted when it is not possible to attain the standard land use level. This is only allowed by exception, and the standard land use quality will remain the future objective.

18.1.4 FUNCTION-ORIENTED APPROACH TO SUBSOIL/MOBILE CONTAMINANTS

The best way of preventing contaminants from spreading is to remove the source of contamination as well as the contaminant plume. However, total removal of the contamination is not always possible due to technical as well as financial reasons. In the new policy in The Netherlands, the extent of removal is determined by the concept of cost-effectiveness. In addition, for the remediation of the subsoil, a standard approach has been formulated. At this moment, the approach is still being elaborated in practice.

For the standard approach, the next points of departure are important:

- The source of the contamination has to be removed as much as possible, in order to attain as little active aftercare as possible;
- The contaminant plume has to be removed as much as possible;
- The objective of the remediation is to attain a stable final situation. This stable final situation depends on the characteristics of the soil and the contaminants in the specific situation. In this situation no further spreading of contaminants takes place. The scope of the stable final situation is the whole contaminated area, recognizing that small differences in concentrations are possible due to differences in soil conditions.

The objective has to be attained within 30 years and it is obligatory to measure progress. Therefore, it will be possible to adjust the measurements when necessary and determine the stable final situation. It is permitted to deviate from the above aims if such is desirable from a financial viewpoint (i.e., such a deviation may consist of remediating the site more thoroughly or less thoroughly). The cost-effectiveness of various options can be determined by means of the RMK (risk reduction, environmental merit and cost) model (Table 18.1) (Tauw Milieu bv. 1998a). This also applies to the weighing of measures to be taken in the contact zone.

The aim of the above approach is to attain a situation in which no aftercare will be needed. However, the achievement of this aim would be extremely expensive in exceptional situations. Then an ICM variant could be chosen, which usually involves perpetual aftercare.

Table 18.1 Tool for the weighing of options for remediation

RMK: risk reduction, environmental merit and cost *(M. Nijboer et al. Tauw Milieu bv. 1998a)*
Each part of the RMK model (risk reduction, environmental merit and cost) contains indices which can be useful in the weighing of options for remediation. For each option, these indices clarify its main consequences by making the available data manageable.

Risk reduction deals with exposure because of soil pollution on and around a specific location. First, all exposure scenarios, the number of exposed people and objects are documented. Then, the exposure is quantified with physicochemical exposure models (e.g., CSOIL). Finally, the risk reduction is calculated by dividing the risk at the beginning by the expected risk after remediation.

Environmental merit concentrates on potential influence on the surroundings of the pollution or the remediation. Negative environmental effects should be as small as possible. To illustrate the elements that play a role when employing this method, the factors included in the indices for environmental merit are detailed below. The part of the RMK model dealing with environmental merit can be seen as a simplified LCA (life-cycle analysis) tailored to soil remediation. This means that the 'proceeds' (positive effects) are compared with the 'sacrifices' (negative effects). In this, the following effects are distinguished.

Positive environmental effects:
1. *higher soil quality;*
2. *higher groundwater quality;*
3. *the prevention of future groundwater contamination.*

Negative environmental effects:
1. *use of uncontaminated soil;*
2. *use of uncontaminated groundwater;*
3. *energy consumption;*
4. *surface water contamination;*
5. *air pollution;*
6. *production of waste, space is required for remediation activities.*

For every factor, a score is calculated by means of an RMK spreadsheet with formulas. All scores are summarized in a table, which then gives an understanding of the strong and weak sides of the remediation variant. Finally, the environmental merit index is calculated. When calculating the costs of a variant an expected highest and lowest value for the different costs is taken into account. In addition, different time schedules of the remediation are considered.

18.2 CASE I: FUNCTION-ORIENTED REMEDIATION OF A SITE WITH HUMAN HEALTH RISKS AND RISKS OF SPREADING

18.2.1 INTRODUCTION

18.2.1.1 General

This is one of the first cases in The Netherlands where the function-oriented approach was followed. For a long time, the site — located on Handelskade, which is strategically situated at the entrance of the Hansa town of Deventer — was unsuitable for further development due to soil and groundwater contamination. This contamination, which started in 1946, was caused by the production of pesticides at this site. Complete remediation of the site (i.e., recovery of multifunctionality) was not feasible. If an isolation option had been applied, the Municipality of Deventer would not have been able to fully develop the site (i.e., very few uses would have been possible).

In collaboration with the authorities, a remediation plan was drawn up by applying detailed risk assessment models. The remediation plan was carried out, and a new high school building could be constructed on the site. This project was one of the first large-scale remediation projects carried out in The Netherlands where extensive use was made of exposure models intended for risk assessment.

18.2.1.2 History

A pesticide factory operated at the site from 1946 to mid-1962. The insecticides produced at this factory included the highly toxic hexachlorocyclohexane (HCH), whose active ingredient was gamma-HCH (Lindane). The remaining components are practically unusable and were considered a waste. This waste product was temporarily stored at the site (inside and/or outside the buildings), part was discharged into the municipal sewage system. Pesticides were not only produced but also stored at and sold from the site. The insecticides sold included HCH, DDT, dieldrin and aldrin, and such products as mouse granules, rat poison and Landison (a mercury-containing disinfectant).

18.2.1.3 Soil Structure and Hydrogeology

Deventer is located on the River IJssel in the IJssel valley. The top layer of the soil, which is 6 m thick, is a moderately permeable Holocene covering layer. The first (non-confined) aquifer consists of sandy deposits (Formation of Twente and Kreftenheye). At the site, the aquifer extends to a depth of 30–40 m NAP (normaal Amsterdams peil (Amsterdam ordnance datum)). Below the aquifer, there is a layer of barely permeable sediments belonging to the Formation of Drenthe (Tauw Milieu bv. 1989a).

The regional groundwater flow is mostly towards the River IJssel (in a south-westerly direction), but near the river, the groundwater flows towards the river from both sides. The results of a hydrogeological investigation show that below the river the groundwater seeps upwards as a result of the draining effect of this river.

Running along the southern border of the site is a canal (De Buitengracht). Because the water in this canal causes infiltration, the phreatic groundwater around the canal flows almost vertically downwards.

18.2.2 CONTAMINATION SITUATION

Since 1985, several soil investigations have been carried out at the site (Tauw Milieu bv. 1987, 1989a,b, 1991). Their results regarding soil and groundwater contamination are given below.

18.2.2.1 Soil

- The soil at the former company site was found to be heavily contaminated with HCH and other organochloro-pesticides (especially chlorobenzenes), mainly in the top layer, which extends to 2.0 m below ground surface. The content values of these contaminants ranged between 100 and 1000 mg/kg d.m. (16 and 160 times the intervention value, respectively).
- Because of leakage from the sewer pipes below Handelskade, the soil around this sewer was heavily contaminated with HCH (at approximately 2–3 m beneath ground level). Spreading occurred mainly downwards. Contaminants were encountered to a depth of 10 m below ground surface.
- At the locations in soil where elevated HCH content values were encountered, high content values of chlorobenzene were also found (this substance is a natural degradation product of HCH).

18.2.2.2 Groundwater

- The superficial (phreatic) groundwater at the site was found to be heavily contaminated with HCH (at most $380 \times I$, where I is the intervention value). In addition, high content values of di- and trichlorobenzenes were measured (at most $60 \times I$).
- At the former depot, the first aquifer was found to be heavily contaminated with HCH to 10 m below ground surface (at most $40 \times I$). High content values of HCH ($3 \times I$) were also measured at 15 m beneath ground level.
- High content values of HCH and chlorobenzenes were measured between the canal and the IJssel at approximately 7 m below ground surface. High content values were also measured 100–150 m downstream of the site at the same depth.
- Locally in the lower part of the first aquifer, high content values of benzene ($6 \times I$) were measured.
- In view of the depth at which groundwater was found to be contaminated and the draining effect of the river, the superficial groundwater downstream of the site will probably not be contaminated. The deeper groundwater seeps upwards only at the river (see Figure 18.2).

18.2.2.3 Problem Analysis in Relation to Desired Land Use

- *Risks of exposure posed to humans and the ecosystem.* During further investigation, an extensive risk analysis was carried out. It showed that human health risks existed in the former situation. Humans could have run risks during excavation work, nuisance from odours could have occurred during hot, calm weather, and there was a risk from vapours. In view of the planned construction of a building, vapours entering indoor air could have caused problems. There are no unacceptable ecological risks, because the site is located in an urban area. The ecosystem in urban areas, being mostly

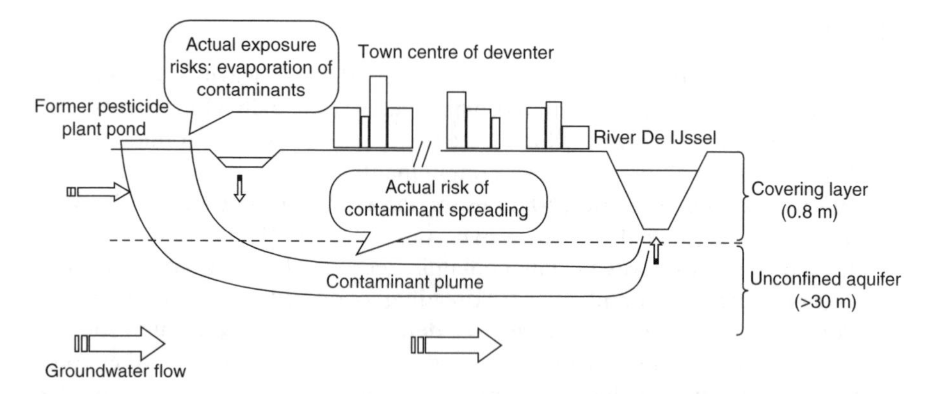

Figure 18.2 Diagram of the contamination and its spreading.

paved, is considered to be less susceptible to environmental hazards than in other forms of land use.

- *Environmental risks.* Environmental risks resulted mainly from the spreading of contaminants via groundwater because the volume of contaminated groundwater increases every year by more than 100 m^3 (Dutch criteria for environmental risks). The deeper groundwater was found to be contaminated to such an extent that the contamination plume had reached the IJssel. Because of the vast water volume of the river 'De IJssel', it is expected that there are no unacceptable risks for the organisms living in the river, due to dilution effects.

18.2.3 REMEDIATION OF THE SITE

18.2.3.1 Options for Remediation

A total of six options for remediation were worked out (Tauw Milieu bv. 1993). The options that are most relevant are:

- Multifunctional variant — EUR36 000 000 (equivalent to US$32 400 000)
- Isolation variant (hydrogeological and/or civil engineering) — EUR12 000 000 (equivalent to US$10 800 000)
- Functional variant EUR14 000 000 (equivalent to US$12 600 000)

The first two variants are traditional ones, which are characterized by the principle of 'multifunctional remediation, unless'. In the present case, the first variant (recovery of multifunctionality) was unfeasible because of high costs. The second variant was financially feasible, but would restrict site use.

The third variant is a precursor of the new type of functional remediation. In consultation with the parties involved (province, municipality, National Institute of Public Health & Environmental Protection and Tauw bv.), this variant was developed and implemented. It involves hydrogeological isolation with optimization of uses. This made the site suitable for the construction of a high school, which had been planned for many years. In this variant, the remediation value is based on the risk approach.

18.2.3.2 Weighing the Measures to be Taken

- *Measures in relation to the contact zone (human health risks).* One of the tools used to work out and substantiate the selected variant was the CSOIL exposure model (see Table 18.2) (Tauw Milieu bv. 1996). The model relates the concentration of a contaminant in soil to intake by humans, based on assumptions regarding the properties of the contaminant, the partition coefficients between the various environmental compartments and human behaviour.

The model can be used in two ways. It can be used to carry out risk assessments for sites. It is used either to identify further investigation needs or to support application of temporary measures. The model can also be used to compare the residual risks of different options for remediation and to calculate risk limits (i.e., remediation values).

Model calculations show that the only relevant route of exposure was the inhalation of air containing toxic vapours, due to migration of soil vapour into

Table 18.2 CSOIL exposure model

The CSOIL model is a model that quantifies the exposure to contaminated substances. It consists of a large number of theories and empirical and mechanical relations. These theories and relations describe:

- *division of a substance over soil phases;*
- *transport processes;*
- *direct exposure (oral intake of soil, water and air, dermal contact with soil, water and air, inhalation of soil, water and air);*
- *indirect exposure (consumption of food contaminated through contaminated soil).*

For the substances, the following data set served as a starting point:

- *molecular weight;*
- *water solubility;*
- *water partition coefficient (expressed as its log value);*
- *Henry constant;*
- *air diffusion coefficient;*
- *permeation coefficient;*
- *bioconcentration factor (for metals).*

Also, some soil characteristics (e.g., soil type), the depth of the water table and information on the contamination situation like the concentration of the substance in ground or groundwater are used as input for the model calculations.

The output of the model calculation is compared to target risk values and a conclusion will be drawn, dependent on the outcome: does the concentration of the compound on the site exceed the intervention value, target value, intermediate value, tolerable concentration in air (TCA) or hazardous concentration for 50% of the terrestrial organisms (HC_{50}) for ecosystems or does the volume of groundwater with a concentration above the intervention value increase by more than 100 m^3 in a year?

either ambient air or indoor air via crawl spaces. To prevent vapour release, it was proposed to construct an uninterrupted pavement on the unbuilt part of the site. This barrier would also prevent contact with contaminated soil. These measurements should be kept intact. Whenever major changes occur, one has to check whether new measurements are necessary.

- *Measures to prevent spreading (risks of spreading).* It was proposed to control groundwater contamination hydrogeologically by means of a groundwater withdrawal system to prevent further spreading downstream of the site. It was decided to withdraw groundwater in the immediate vicinity of the site. This means that the deep groundwater contamination downstream of the site in the first aquifer will not be remediated. The contamination plume will continue to spread in the direction of the river and will take a long time to disappear (by natural discharge). However, because of the remedial measures taken, the plume will no longer be supplemented with more contamination, because the source of contamination has partly been excavated and partly contained.

18.2.3.3 Final Remediation of Soil and Groundwater

In 1994, the planned soil remediation measures and hydrogeological containment measures were taken. At the location where later the new building was constructed, contaminated soil was excavated to a depth of approximately 1 m and transported to a treatment plant. As a result of the excavation, the calculated risk limits for HCH (1.8 mg/kg dry weight) and chlorobenzenes (0.2 mg/kg dry weight) were almost attained. At those locations where excavation was not feasible for technical reasons, the soil in the crawl space of the new building was sealed with HDPE foil and concrete.

To prevent further spreading of contamination from the contained site via groundwater, deep wells were installed at five places downstream of the site. A total of 40 m^3/h groundwater was withdrawn and treated in the groundwater treatment plant.

18.2.3.4 Groundwater Treatment Plant

In view of the nature of the contaminants in the groundwater, it was decided to treat most of the groundwater biologically. Compared to physicochemical treatment techniques, biological treatment has the following advantages:

- mineralization of the contamination
- cost savings
- relatively simple operation
- low production of residues
- simultaneous stripping effect.

However, the following aspects should be taken into account:

- longer start-up time
- aftertreatment (polishing step)
- moderate biodegradability of HCH
- fluctuations in flow rate and concentrations.

Biodegradation can occur in various types of reactor. It was decided to use a biocontactor system. In such a system, the microorganisms important for degradation become attached to the surface of a repeatedly submerged plastic plate. Compared to a bacteria bed, a biocontactor system has the following advantages: its operation requires little labour; less stripping effects; freezes up less easily; and the residence times can be more easily adjusted.

Before the final design of the groundwater treatment plant was made (see Figure 18.3) and the system was selected, an extensive study was conducted to determine the optimum size of the plant. First, the relationships between the residence time in the biocontactor system or treatment efficiency and the accompanying cost were worked out for several variants.

The plant contains the following elements, in the sequence in which they occur:

- biocontactor for biological treatment
- sand filter for trapping coarser components, oxidized iron and biological sludge
- activated carbon filters as a polishing step
- a biobed to treat exhaust air.

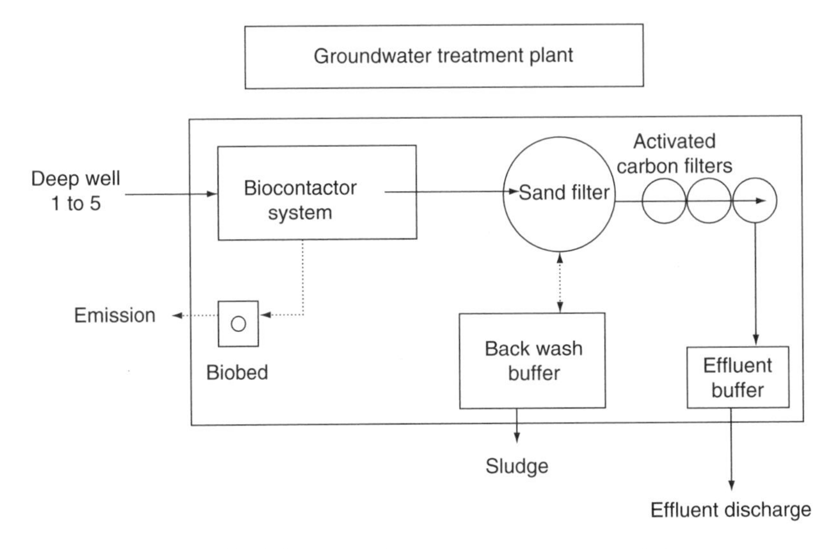

Figure 18.3 Diagram of the groundwater treatment plant.

The water originating from the deep wells is led to the biocontactor system. The system has a nutrient-dosing device, which is used to dose the groundwater with nitrogen. The microorganisms (biofilm) attached to the carrier material of the biocontactor biodegrade aromatics, chlorobenzenes and HCH. Activated carbon filters are expected to adsorb the remaining organic compounds that have not been biodegraded. After passing the activated carbon filters, the air is led to the biofilter/biobed, which contains lava rock. Ventilating air originating from the biocontactor system is also led to the biofilter. In this filter, the organic components of the air are biodegraded.

18.2.4 AFTERCARE

The start-up of the groundwater extraction and treatment process also meant the start-up of the aftercare phase, which in principle will be perpetual. An aftercare plan was drawn up to check the operation of the facilities during this phase (Tauw Milieu bv. 1995). The aim of this plan is to control the situation in the future, which includes checking whether the accompanying functional limitations are maintained. This means that the buildings at the site may be used only as offices or industrial buildings, that various restrictions apply to excavation work, and that the (non-treated) groundwater is unusable and will remain contaminated for a long time. These functional limitations are partly based on the conducted analysis of residual risks. The regional government is responsible for maintaining the aftercare. The costs of the aftercare were capitalized for 30 years and are included in the total costs of the remediation.

The aftercare plan consists of the following:

- Measurement of indoor and crawl-space air (twice a year) by taking samples with activated carbon tubes. These tubes will be analysed for total HCH and chlorobenzenes. If the risk limits are exceeded, the air change rate should be checked.
- Regular checks on the operation of the groundwater withdrawal and treatment systems.

Since the remediation work has been carried out and the aftercare phase begun, there are no actual risks of exposure (Tauw Milieu bv. 1994, 1997b). However, if the containment measures are not taken or continued, there will be potential risks of exposure. The following exposure routes will then be relevant: ingestion and inhalation of soil particles; dermal contact with soil; and inhalation of soil vapour.

18.2.4.1 Results of Checks

The results of soil vapour samples taken at the bottom of the crawl space show no elevated concentrations of contaminants, which is contrary to the potential risks calculated using the model. This confirms that it is important to carry out measurements — especially in relation to risks posed by evaporation — because model results give only a limited and conservative impression of the complex reality; in fact most models are highly conservative. As long as no increased concentrations are encountered in the crawl space, it will not be necessary to increase the air change rate.

18.3 CASE II: FUNCTIONAL REMEDIATION OF A SITE WITH ECOLOGICAL RISKS

18.3.1 INTRODUCTION

18.3.1.1 General

This case shows how in The Netherlands the new policy was applied to sites of high natural value. The site concerned was a former skeet-shooting range, which was located within the ecological main structure and was included as a nature reserve in a designated land use plan. The location was part of a pingo ruin, which had a high nature-historical value. A natural water pool was present at this location.

The results of a soil investigation show that this range was contaminated at a level of lead far above the intervention value. Until now, no effects of the contamination on flora or fauna are to be seen, but effects are predicted on this location with a low pH value and the wet peaty (water) soil. To gain an insight into the human health and ecological risks and the possible need for remediating the site, a further remedial investigation was conducted. To determine the most suitable option for remediation, use was made of the new weighing system and the RMK model (see Section 18.1.4).

18.3.1.2 History and Site Description

Formerly, the skeet-shooting range was used a few days per year, from 1930 to 1990. Due to land consolidation and changes in farm practices, after 1900 most of the varied flora and fauna disappeared from the surroundings of the site. This applies to a lesser extent to the site itself, because it was used sporadically.

The site comprised several types of land use. The western part of the site consisted of pasture. Around the pingo ruin (which is filled with water), lies a zone of woods. The woods comprised about 65% of the site. The landscape at the border of the woods consisted of grasslands and arable land. All together, this was a site with a varied flora and fauna on a relatively small surface area (80 000 m^2). Figure 18.4 gives an impression of the site.

18.3.1.3 Soil Structure and Hydrogeology

An overview of the local and regional soil structure is shown in Table 18.3. Between 0 and 0.5 m beneath ground level, the soil consists of fine humic

Figure 18.4 Photograph of the pool and its surroundings.

Table 18.3 An overview of the local and regional soil structure

Local		Regional			
Depth (m) below ground level	Soil	Depth (m) below ground level	Soil	Formation	Hydrogeological units
0–0.5	fine sand	0–0.5	fine sand	Twente	aquitard
0.5 –1.5	medium fine sand	0.5–5	loamy sand	Drenthe	aquitard
1.5–5	loamy sand	5–12	fine sand	Eindhoven	aquifer
		12–50	very fine sand/clay	Peelo	semi pervious layer
		>50	coarse sand	Urk/Harderwijk	aquifer

sand, and of loam and loamy sand between 1.5 and 5.0 m below ground. The topsoil is barely permeable, and through this a horizontal flow occurs only via thin connected sandy layers. Because these layers are mostly scattered, the horizontal flow does not occur in a uniform direction.

Regionally, there is a precipitation excess, most of the water in the topsoil flows vertically, which means that there is an overall infiltration situation. The groundwater level varies between approximately 1 and 2 m below ground surface.

The area of the pool is about 24 000 m^2. The depth of the surface water varies from 1 to 1.95 m. The thickness of the sediment layer, including peaty material, is 1 m at maximum. Underneath the sediment layer there is the hard loamy underground of the pingo ruin.

18.3.2 CONTAMINATION SITUATION

18.3.2.1 Soil

The soil at the range was strewn with pellets of shot. The lead-content values measured in the topsoil (to a depth of approximately 0.5 m below ground surface) were far above the intervention value (varying between five and 20 times this value with outliers up to 100 times the intervention value) (de Vries *et al.* 1995, Tauw bv. 1998). The intervention value for lead, varied for different soil types, ranges from 300 up to 450 mg/kg dry weight. Locally, lead-content values far above the intervention value were encountered to a depth of approximately 1 m below ground surface (approximately five times the intervention value).

Indicative lead-content values were used in the assessment of the investigation results (i.e., assessment of the contamination situation, determination of urgency and selection of an option for remediation; Tauw Milieu bv. 1997a). The reason for this is that the availability of lead to the environment cannot be easily derived from these extremely high content values (which are partly the result of measuring small lead granules in samples). Indicative content values are those that may exist in the future as a result of oxidation and/or weathering of lead granules.

18.3.2.2 Groundwater

The shallow groundwater samples taken from monitoring wells at the skeet-shooting range showed high concentrations of lead (far above the intervention value of 75 μg/L), and samples taken from monitoring wells from the shallow groundwater at the arable land showed moderate concentrations (just below the intervention value). The high concentrations are only site-specific, and are highly correlated with the presence of lead granules. Furthermore, in an actualization survey, no measured concentrations exceeded the target value, which confirms the assumption that the groundwater contamination is only very local.

The depth and horizontal extent of the groundwater contamination were not delineated. It was assumed that the first aquifer is not contaminated and that the horizontal extent of the groundwater contamination is equal to that of the soil contamination, because of the high adsorption capacity of Pb in combination with the presence of loam which contains sulphates and carbonates that bind with Pb in ionized form, and the limited vertical flow in the topsoil. This assumption approaches the real situation.

18.3.2.3 Surface Water and Sediment

The lead concentration in the water of the pool at the site exceeded the target value for surface water. The lead content in part of the pool sediments was found to exceed the intervention value, whereas in another part of these sediments the lead content was found to exceed only the target value. The thickness of the contaminated sediments was also measured; it varies from 0.3 to 0.6 m. The thickness of the sediment layer, including peaty material, is 1 m at maximum. The loamy soil beneath the sediment layer is not contaminated with lead.

18.3.2.4 Problem Analysis Regarding the Desired Land Use

To determine the most suitable option for remediation, the risks posed by the contamination at the site were assessed using the CSOIL model (de Vries 1995, Tauw Milieu bv. 1996). These are human health and ecological risks posed by contaminants in the contact zone as well as risks of spreading.

Human Health Risks

The human health risks posed by contaminants in the contact zone were determined for the two functional variants most relevant to the site: one for the arable land and one for the nature reserve at the site.

- *Arable land.* Food crops grown in the area include potatoes, beets and maize. The exposure calculations are based on the standard functional variant, and the consumption of food crops grown at the site is taken into account. In this, it is assumed that at most 10% of the food annually consumed by one person consists of food grown at the site.

  Based on the standard functional variant chosen, the risk limit for lifetime exposure to lead is derived using the CSOIL model. At this limit, the total calculated lifetime mean exposure equals the permissible daily intake. If content values in soil exceed this limit, lifetime exposure is assumed to pose actual human health risks. The mean content value measured at the part of the site used as arable land does not exceed the value of 1.48 mg/kg dry weight calculated for lifetime exposure. This means that the lead contained in the arable land does not pose any potential human health risks.
- *Nature reserve.* A separate exposure variant was selected for the nature reserve, because the manner in which this area (around the pool) is used differs greatly from the use of the surrounding arable land. The functional variant chosen for the nature reserve is not based on crop consumption

but on the number of visitors per unit of time. A comparison between the analysis results and the risk limits derived shows that in the nature reserve the risk limit for lifetime exposure to lead is exceeded. Although this potential human health risk is predicted by the model calculation, it is certain that there are no actual human health risks, because compared to the standard functional variant the actual number of visitors is lower. In addition, the nature reserve will be completely closed in the near future (see Section 18.3.3.4).

Ecosystem

In The Netherlands, the standard procedure for the determination of ecological risks consists of comparing the contaminated surface area exceeding the standard set for ecosystems (HC_{50}, see Section 18.1.3) to the critical surface area permitted for the type of area concerned. A distinction is made between three types of area based on their sensitivity to harmful effects. The nature reserve discussed belongs to the most sensitive class. This means that the surface area of the contamination in the contact zone may not be larger than 50 m^2. Because this figure is exceeded to a great extent, it is assumed that there are ecological risks.

Spreading to the Surroundings

During the investigation, some groundwater samples were taken in order to analyse the lead content. In the monitoring wells at the nature reserve, very high concentrations of lead (exceeding the intervention value) were measured. According to the Dutch assessment system, there is a risk of spreading if it is expected that annual spreading will occur over a volume in excess of 100 m^3. Field measurements are taken to obtain a realistic set of input parameters for the CSOIL model. These measured input parameters are, among others, the retardation factor, flowing speed of the groundwater and soil structure. The CSOIL calculations show that, in the current situation, no risks of spreading are to be expected. The outcome of the calculations corresponds to the knowledge that, in general, the spreading of Pb through groundwater is of minor importance. This is a result of the high adsorption capacity of Pb. In this situation, the Pb adsorbs easily to sulphates and carbonates that are present in the loamy soil.

Regarding the contact zone, no human health risks are expected (if the nature reserve is not frequently visited) but the ecosystem is at risk. However, the contamination is not expected to spread to the groundwater and deeper subsoil because of the presence of the almost impermeable clay layer.

18.3.3 REMEDIATION OF THE SITE

18.3.3.1 Options for Remediation

For the site, two traditional variants were compared with the functional one (Tauw Milieu bv. 1997a, Tauw bv. 1998):

- Multifunctional variant — EUR4 100 000 (equivalent to US$3 690 000)
- Containment variant (ICM) — EUR800 000 (equivalent to US$720 000)
- Functional — variant EUR1 000 000 (equivalent to US$900 000)

Multifunctional Variant

This variant consists of excavating all soil containing lead at concentrations above the target value. Contaminated soil would be excavated to 0.5 m below ground surface and locally to 5 m below ground surface. Excavation would necessitate the removal of the existing vegetation.

Most of the ground contamination would be removed by excavation of soil to 5 m below groundwater surface at the locations where the groundwater was found to be contaminated. Groundwater remediation by groundwater withdrawal is not feasible in view of the long remediation period expected (>200 years). The surface water would be remediated by water withdrawal and treatment. The pool sediments would be remediated by removing the layer of contaminated sludge. At the end of the remediation project, the site could be landscaped and, in theory, the site would then be multifunctional. In total it was estimated that approximately 47 000 m^3 of soil had to be excavated.

However, in practice it would not be possible to preserve the natural value of the current specific ecosystem, because an investigation showed that the dry woods are the most sensitive to excavation. To excavate the topsoil, the woods would have to be removed. As a result, the biotope containing most of the animal species would disappear. Complete recovery of woods would take at least 50 to a 100 years. This process could not be accelerated by afforestation, because as a result of the application of new fill, a new state of equilibrium between abiotic conditions would need to be established. Furthermore, as a result of the removal of sludge from the pool, the seed supply in the pool would disappear. The seed supply contains mainly *Juncus effesus*, which is not very rare. However, recovery of the pool would take a long time. From this it can be concluded that, as a result of a multifunctional remediation project, the dry woods area and the pool would have low ecological value for a long time.

Containment Variant (ICM)

The simplest and most efficient ICM variant with which the ecological risks could be reduced consists of removing all contaminated soil and sludge exceeding the intervention value for lead in and around the pool (soil and sludge with concentrations above the target value would be left behind and contained). This would amount to removing approximately 17 000 m^3 of soil. A monitoring plan (aftercare) would be drawn up for monitoring possible spreading of contaminants via groundwater and possible lead intake from crops grown on the arable land. In addition, functional limitations would be put on groundwater withdrawal. The effects of this variant on the natural value of the site would be comparable to those of the multifunctional one, except that in the case of the containment variant the remediated area would be smaller.

Functional Variant

The aim of the functional variant is to preserve the specific natural function of the area and to preserve also the nature-historical value of the site, because the former skeet-shooting range was situated in a pingo ruin. Consequently, the remedial measures focus on this, contrary to the two traditional variants, which are based on general standards that are not specific for the situation. The manner in which the measures of the functional variant are weighed is discussed below. The amount of contaminated soil to be removed is estimated at 27 000 m^3.

18.3.3.2 Weighing Measures for the Functional Variant

Because contamination was not spreading, functional remediation was needed only for the soil contact zone, which posed ecological risks. The risk-reduction measures to be taken at the skeet-shooting range were weighed by taking into account the desired site function, namely the present natural value of the area, a pool surrounded by woods. To gain a clear picture of the effects playing a role, a system was used to determine whether there are possibilities for nature development at the site (see Table 18.4).

18.3.3.3 Final Selection of a Variant

A comparison between the three worked-out variants revealed that the functional variant is to be preferred over the multifunctional and ICM variants, especially in relation to technical feasibility, social effects, functional limitations and uncertain factors. The RMK results show that the functional variant has an average score for risk reduction and higher scores for environmental merit and cost (Tauw Milieu bv. 1998b). The higher costs of the functional variant compared to the ICM variant are due to the fact that this variant is to be executed in several phases over 3 years. Based on these considerations, it was decided to select the functional variant.

18.3.3.4 Remediation of the Contact Zone

The ecological value of the site is related to the presence of the pool and surrounding woods. Based on the results and interpretation of the reference study, remedial measures need to be taken at the pool and its surroundings. However, such measures should cause the least possible harm to the current abiotic conditions, since the current natural value depends on these conditions

Table 18.4 Functional remediation in relation to ecosystems

At the request of the Ministry of Agriculture, Nature Management and Fisheries, the National Institute of Public Health and Environmental Protection wrote a guide to soil assessment, which can be used to assess ecotoxicological risks of contamination at sites. Tauw bv. developed a decision-making system that completely fits in with this guide and can be used to assess the possibilities for nature development at contaminated sites. This system comprises the following three phases:

Phase 1: Are the abiotic preconditions (i.e., conservation of natural values) met?
Phase 2: Does the existing contamination pose risks to the desired nature development?
Phase 3: Should measures be taken to eliminate these risks, and if so, what measures?

The first phase focuses on the abiotic conditions at the site. Field studies showed that the abiotic natures of a site are closely related to the creation or presence of a certain natural value. Contaminations can have a strong effect but are not necessarily the main reason why the desired natural value is lacking. Therefore, first the abiotics are examined and then the effects of contaminations.

The following discusses how this was applied to the skeet-shooting range.

Abiotic value
The pool is probably a pingo ruin. At locations where water could not easily infiltrate during the last ice age, ice lenses formed, which pushed the soil up to form hills (pingos). During the melting of ice, pingos collapsed to form barely permeable pools from which water could not escape. A pingo ruin is characterized by a circular wall of sand and loam. Such a wall is recognizable at the site. The groundwater level at the pool is higher than that in the surroundings. This is indicative of a pseudo-groundwater level. Because the bottom of the pool is barely permeable, the water cannot infiltrate to the actual groundwater. The bottom of such a pool is vulnerable. When the impermeable layer is broken by excavation, the pool can lose its water and become irreparably damaged.

Effect of the contamination on the natural value
In soil, two forms of lead from shot can be a threat: lead granules and ionized lead. The latter has a stronger effect on the ecosystem because it is bioavailable. The current assessment system assesses ecological risks based on the exceedance of so-called HC_{50} values. Such values have been derived from NOEC values (no observed effect concentration) for organisms. NOEC values were determined during laboratory experiments. HC_{50} values indicate the mean concentration at which 50% of the potentially occurring species are assumed to suffer damage or nuisance from the contamination concerned. However, it is not possible to indicate what species would fall victim to them. In other words, it cannot be indicated whether at a certain HC_{50} value those species would be protected that are typical of a certain area or are desired in it. Therefore, to assess properly the actual effects of soil contamination, HC_{50} values should not be considered to be of high utility. Toxic effects of soil contamination can better be determined in the field by checking whether soil contamination causes visible effects in flora and fauna. A proper approach is to conduct a comparative field study, in which the ecological functioning of the contaminated site is compared to a clean site that is as similar as possible. Such a study was conducted in 1995. During this study, the skeet-shooting range was compared with a similar reference site. The results show that existing lead contamination does not pose a threat to the natural functioning of the site, despite the high concentrations of total lead. Actual ecotoxicological risks are posed mainly by the bioavailable ionized lead, which is formed by the weathering of lead shot. As a result of acidic conditions (especially at the bottom of the pool), the amount of ionized lead will gradually increase along with the probability for toxic effects to occur. Ingestion of lead shot pellets can cause the death of (water) birds. Particularly pheasants, pigeons and ducks are at risk of ingesting such pellets. The birds probably mistake them for grit, which they swallow to facilitate the grinding up of food in their gizzard. Once the pellets reach the birds' gizzard, they are ground up to fine particles. The lead then enters the blood stream. It is a known fact that certain species of water birds are particularly prone to lead poisoning. Experiments conducted with ducks have shown that within one month, the ingestion of one single lead shot pellet will result in death in 35% of cases. Mortality will be 100% within five weeks of the ingestion of two lead shot pellets.

(Tauw bv. 2000). Furthermore, such measures should be aimed at preventing an increase in the amount of bioavailable lead.

For remedial measures to be aimed at the conservation and recovery of natural values, they need to be taken in phases focused on subareas based on lead concentrations. Subareas that are most at risk should be remediated first. Therefore, the following phases are proposed:

- Phase I — remediation of the pool.
- Phase II — remediation of soil containing high content values of total lead with a remediation duration of approximately 5 years.
- Phase III — remediation of soil containing lower content values of total lead with a remediation duration of approximately 5 years.

The scheduling of remedial activities considers such factors as the breeding season of birds. The pool would be remediated once by removing only the heavily contaminated sludge. The sediment in the pool is therefore partly remediated. The vegetation present in the pool is *J. effesus* and is not 'rare' in The Netherlands, therefore remediation in phases is not necessary in the pool. Totally, about 8000 m^3 of sediment was removed, including peaty material on the sludge layer.

The soil surrounding the pool will be cleaned in the years 2000, 2001 and 2002. Because of this 'phased' remediation, the flora and fauna on the site have the possibility to mobilize from one spot on the site to another. It is estimated that about 3500 m^3 of soil will be removed. Remediation of soil (phase II) would have to occur in phases by annually remediating a few narrow strips to approximately 0.25 m beneath ground level. The excavated strips would not be filled. Trees would be saved, as far as possible. To promote recolonization, the strips to be remediated in a year should not be adjacent to each other. After phase II, no activities would occur during 4 years for the purpose of recolonization. Phase III would be carried out similarly to phase II. All remedial measures would be attuned to the future management of the site.

18.4 CONCLUSION

The two cases detailed above show how the proposed new policy has been carried out in The Netherlands. In both cases the functional variant turned out to be the best solution. The main feature of this variant is that the measures focus on the risks that need to be reduced to enable the desired land use.

18.4.1 CASE I

Case I concerned the remediation of a heavily contaminated site in the centre of Deventer where human health risks and risks of spreading played a role. The impasse existing at the site could eventually be broken by applying the functional variant. The traditional multifunctional variant was too expensive and the cheaper ICM variant would not make it possible to use the site as desired. Although the functional variant was somewhat more expensive than the ICM one, its advantages outweighed the additional cost, because it made all desired types of land use possible and thus remedied the social stagnation. However, it should be noted that perpetual aftercare is needed, especially to control the groundwater contamination.

18.4.2 CASE II

Case II concerned a variant worked out for a contaminated skeet-shooting range located in a nature reserve, the ecological values of which are threatened by exposure to lead contamination. This area has a special natural value, which should be preserved. Based on the available data and investigation results, three variants were worked out. The ICM and functional variants are similar, except that the latter is focused more on the ecological aspects of the site. The functional variant selected is a creative solution, one which will reduce the risks posed to the ecosystem while preserving the specific natural value. This is a better option than the complete removal of all contamination (multifunctional remediation), which would involve great effort, would not be technically feasible and would result in the disappearance of the specific natural value of the area.

REFERENCES

de Vries HJ, Bokeloh DJ and Siepel H (1995) *Ecologische risico's van de loodverontreiniginging op de kleiduivenschietbaan Loon DR/010/18*, nr. 92384, LB&P ecologisch advies BV, The Netherlands.

Kooper W (1999) *Afwegingsproces saneringsdoelstellingen — van Trechter naar zeef II,SDU uitgevers*, Bunnik, The Netherlands.

Tauw bv. (1998) *Saneringsplan*. KSL Loon, nr. R3698629.N02, Tauw bv. Deventer,The Netherlands.

Tauw bv. (2000) *Studie naar sanering van kleiduivenschietbanen in de provincie Drenthe*, Tauw bv. Deventer, The Netherlands.

Tauw Milieu bv. (1987) *Nader onderzoek fase I en II*. Handelskade Deventer, nr. 52527.09/R0-01, Tauw Milieu bv. Deventer,The Netherlands.

Tauw Milieu bv. (1989a) *Nader onderzoek fase III*. Handelskade Deventer, nr. 60472.45/R0-01, Tauw Milieu bv. Deventer, The Netherlands.

Tauw Milieu bv. (1989b) *Grondwateronderzoek*. Handelskade Deventer, nr. 60472.46/R0-01, Tauw Milieu bv. Deventer, The Netherlands.

Tauw Milieu bv. (1991) *Saneringsonderzoek*. Handelskade Deventer, nr. 3142205.S03/MTB, Tauw Milieu bv. Deventer, The Netherlands.

Tauw Milieu bv. (1993) *Saneringsplan*. Handelskade Deventer, nr. 525795.S05/CHS, Tauw Milieu bv. Deventer, The Netherlands.

Tauw Milieu bv. (1994) *Evaluatie grondsanering en aanleg geohydrologisch beheerssysteem*. Handelskade Deventer, nr. 3293963.S12, Tauw Milieu bv. Deventer, The Netherlands.

Tauw Milieu bv. (1995). *Nazorgplan locatie*. Handelskade Deventer, nr. R3305171.T01/GWO, Tauw Milieu bv. Deventer, The Netherlands.

Tauw Milieu bv. (1996) *CSOIL*, versie 6.3a, Tauw Milieu bv. Deventer, The Netherlands.

Tauw Milieu bv. (1997a) *Beslissystematiek ten behoeve van inschatting ecologische risico's*, Tauw Milieu bv. Deventer, The Netherlands.

Tauw Milieu bv. (1997b) *Verzamelde rapportages met betrekking tot geohydrologische beheersing IBC-locatie.* Handelskade te Deventer, nr. R3466833.T05/GHW, Tauw Milieu bv. Deventer, The Netherlands.

Tauw Milieu bv. (1998a) *Beslisondersteunende systeem RMK voor het boordelen van varianten voor bodemsanering, een methodiek gebaseerd op Risicoreductie, Milieuverdienste en Kosten*, Tauw Milieu bv. Deventer, The Netherlands.

Tauw Milieu bv. (1998b) *Saneringsonderzoek kleiduivenschietbaan.* Heirweg te Loon, nr. R3657337.N02, Tauw bv. Deventer, The Netherlands.

Tweede Kamer der Staten-Generaal (1997) *Interdepartementaal beleidsonderzoek: bodemsanering 25411*, nr. 1, Tweede Kamer der Staten-Generaal, Den Haag, The Netherlands.

19

Ecotoxicological Risk Assessment of an Explosives-contaminated Site

PIERRE YVES ROBIDOUX[1], JALAL HAWARI[1], SONIA THIBOUTOT[2] AND GEOFFREY I. SUNAHARA [1]

[1]*Applied Ecotoxicology Group, Biotechnology Research Institute, National Research Council of Canada, Montreal, Quebec, Canada*
[2]*Defense Research Establishment Valcartier, Canadian Ministry of National Defense, Val Bélair, Quebec, Canada*

19.1 INTRODUCTION

Nitroaromatic compounds such as 2,4,6-trinitrotoluene (TNT), 1,3,5-trinitro-1,3,5-triazacyclohexane (RDX), octahydro-1,3,5,7-tetranitro-1,3,5,7-tetrazocine (HMX) and their associated byproducts can be released into the environment at factory sites, military areas such as firing ranges and open burning-open detonation (OB/OD) areas and through field use. The extent of sites contaminated by explosives worldwide represents a significant international problem. Different studies have been conducted on the impact of this type of contamination on the environment. Due to the chemical and toxicological properties of these explosives, a number of laboratory and field studies have been carried out (Talmage *et al.* 1999). The well-known explosive TNT and some of its degradation products are toxic and genotoxic at relatively low concentrations to a number of ecological receptors. RDX and HMX are also toxic to some organisms (Talmage *et al.* 1999, Robidoux *et al.* 2000, 2001). Due to their toxicity, these explosives may present a risk to humans and some ecological receptors. To date there is no generic criteria for explosives-contaminated soil, which could be used to manage the future use of decommissioned military training sites. However, screening values have recently been calculated (Talmage *et al.* 1999). In addition, some surface water criteria have been published by different governmental organizations such as the *Ministère de l'Environnement et de la Faune du Québec* (MEFQ 1998).

Ecological risk assessment (ERA; see Chapter 15 for more information) can be used to determine if a site contaminated by explosives exceeds an environmentally acceptable level of protection and may represent a significant risk to the exposed ecological receptor. ERA may also be conducted to protect

Environmental Analysis of Contaminated Sites. Edited by G. I. Sunahara, A. Y. Renoux, C. Thellen, C. L. Gaudet and A. Pilon

biological diversity, to determine the maximum level of contamination that permits an acceptable protection level and to compare treatment technologies or scenarios. This chapter summarizes an approach used to conduct ecological risk assessments of explosives-contaminated sites. The approach presented herein was used to evaluate an explosive production facility having TNT as the monotypic contaminant.

19.2 PROBLEM FORMULATION

The problem formulation provides the conceptual framework for risk assessment (USEPA 1992) and consists of the description of relevant features of the environment, the sources of contamination and the identification of the ecological receptors at the site.

19.2.1 SITE DESCRIPTION

The factory site was heavily contaminated with TNT and its byproducts during the Second World War. This site was decommissioned in the 1970s. Vegetation on the site included graminaceae and trees. Soil invertebrates such as earthworms were observed around the contaminated area. Mammals (e.g., deer and fox) and birds (e.g., passeriformeae) were also present on the site.

19.2.2 NATURE AND SOURCES OF STRESSOR

Preliminary studies have shown that soils and groundwater remain contaminated by TNT. In addition, substances (e.g., dinitrotoluenes) associated with TNT manufacture, as well as TNT degradation products, were also detected. Heavy metals concentrations were shown to be relatively low compared to TNT (maximum of 2 μg Cd L^{-1}, 28 μg Cu L^{-1}, 170 μg Pb L^{-1}, 80 μg Zn L^{-1}) in the groundwater of some areas of the sites. The presence of other contaminants such as hydrocarbons was not suspected to be a significant source of contamination considering the past activities associated with the site. TNT (CAS#118-96-7) is the most widely used military explosive because of its low melting point (80.7 °C; Yinon 1990), its stability, low sensitivity to impact friction and high temperature and its relatively safe method of manufacture.

TNT contamination in soil was heterogeneous at the study site. The contaminated areas were relatively small (ranging from 10 to 100 m^2 dispersed over approximately 5000 m^2) and corresponded to a location where TNT was synthesized (from mononitrotoluene and dinitrotoluene) and later purified (i.e., removal of undesired isomers and residual dinitrated toluenes). Some areas were highly contaminated (up to 259 000 mg TNT kg^{-1} soil). TNT-contaminated soils were generally localized with surrounding areas having low to non-detectable levels of explosive contaminants.

19.2.3 PHYSICAL/CHEMICAL PROPERTIES AND ENVIRONMENTAL BEHAVIOR OF STRESSOR

19.2.3.1 Movement of TNT in the Environment

Volatilization of TNT from surface and groundwater was considered to be negligible since its vapor pressure (1.99×10^{-4} mm at 20 °C; HSDB 1995) and Henry's law constant (4.57×10^{-7} atm m^3 mol^{-1}; HSDB 1995) are low. The moderate water solubility of TNT (130 mg L^{-1} at 20 °C, HSDB 1995) favors the movement of the contaminant. Sorption to a soil matrix is low as indicated by its K_d (range: 4–53) and K_{oc} (525–1585; Spanggord *et al.* 1980, Rosenblatt *et al.* 1991). TNT would not be expected to adsorb strongly to the soil, because of its properties.

19.2.3.2 Transformation of TNT

For environmental abiotic processes (including photolysis), TNT in water has a $t_{1/2}$ in the range 0.5–22 h, based on laboratory studies (Talmage *et al.* 1999). These reactions are typified by the formation of pink-colored products in water. Under aerobic and anaerobic conditions, bacterial and fungal species can transform TNT in water ($t_{1/2}$<6 months; ATSDR 1995) as well as soil ($t_{1/2}$<60 days; Cataldo *et al.* 1989) and in sediment and sludge (Spanggord *et al.* 1980). The detectable degradation products include: 2-hydroxy-aminodinitrotoluene (2-OH-ADNT), 4-hydroxy-aminodinitrotoluene (4-OH-ADNT), 2-aminodinitrotoluene (2-ADNT), 4-aminodinitrotoluene (4-ADNT), 2,4-diaminonitrotoluene (2,4-DANT), 2,6-diaminonitrotoluene (2,6-DANT), 2,4,6-triaminotoluene (TAT) and various tetranitro-azotoluene compounds (Sunahara *et al.* 1998, Talmage *et al.* 1999). Mineralization of TNT is considered low (Hawari *et al.* 1998, 1999).

19.2.4 BIOCONCENTRATION AND BIOACCUMULATION OF TNT

TNT (absorbed by inhalation, ingestion or dermal contact) is metabolized by different mammalian species (ATSDR 1995). Ingested TNT can be excreted within 24 h in urine (50–70%) and feces. Distribution in tissues is less than 1%, indicating a low potential for bioaccumulation. Metabolites recovered in urine include hydroxy-aminodinitrotoluene, diaminonitrotoluene, dinitroaminotoluene as well as glucuronide-conjugated metabolites. Biomonitoring studies (see review by Talmage *et al.* 1999) did not reveal the presence of TNT in tissues (<0.2 mg kg^{-1}) of terrestrial wildlife (deer and small mammals).

TNT is also partially degraded by plants and earthworms (*Eisenia andrei*). Biotransformation products include 2-ADNT and 4-ADNT (Palazzo and Leggett 1986, Cataldo *et al.* 1989, Renoux *et al.* 2000). Calculated and measured log K_{ow} values (1.6–2.7) indicate a low potential for bioconcentration in aquatic organisms (ATSDR 1995). The bioconcentration factor (BCF) calculated for fish using a calculated K_{ow} (2.03) was low (BCF = 20.5, Liu *et al.* 1983). The BCF values measured by Liu *et al.* (1983) in non-steady state conditions ranged from 202 to 453 for aquatic organisms (green alga, water flea, oligochaete and bluegill).

19.2.5 ECOLOGICAL EFFECTS

TNT is toxic ($EC_{50} \geq 0.95$ mg L^{-1}) to aquatic organisms including bacteria (*Vibrio fisheri*; Microtox™), freshwater algae (*Selenastrum capricornutum*), invertebrates (*Ceriodaphnia dubia*) and fathead minnow (*Pimephales promelas*; Smock *et al.* 1976, Dryzyga *et al.* 1995, Sunahara *et al.* 1998, 1999).

TNT-contaminated soils, composted explosives-contaminated soil and soil elutriates have also been found to be toxic ($EC_{50} \geq 1.0$ mg kg^{-1}) for organisms including bacteria (Microtox™; Simini *et al.* 1995, Sunahara *et al.* 1999), subsurface soil microbial communities (Fuller and Manning 1998; Gong *et al.* 1999), plants (*Lactuca sativa*), earthworms (*Eisenia* sp, Phillips *et al.* 1993, Simini *et al.* 1995, Peterson 1996, Jarvis *et al.* 1998, Dodard *et al.* 1999, Robidoux *et al.* 1999, 2000), nematodes and microarthropods (Parmelee *et al.* 1993).

19.2.6 ECOSYSTEMS POTENTIALLY AT RISK

Direct contact organisms such as invertebrates, plants and microorganisms may be affected by the presence of TNT or its (bio)transformation products. Mammals (deer and fox) and birds (passeriformeae) can also be exposed to the toxicants on the site by dermal contact and ingestion. Aquatic organisms may be potentially exposed to TNT or its products via transport by groundwater and surface runoff from contaminated soil to a nearby river.

19.2.7 CONCEPTUAL MODEL

Terrestrial organisms may be exposed to TNT and its transformation products from surface soil areas. These contaminants may cause direct effects on microbial processes, terrestrial plants, soil invertebrates, mammals and birds. TNT may be transformed and/or transferred slowly to the groundwater, and then potentially to the surface water. Thus, aquatic organisms (microorganisms, microphytes, macrophytes, invertebrates and fish) may be directly affected by the presence of these contaminants. A decrease in the population of these organisms may affect other animals such as small mammals and birds. These contaminants may enter into the food chain and affect other sensitive organisms (Suter 1996a).

19.2.8 ENDPOINTS

To evaluate if the selected receptors present on the site may be at risk, this study considered the direct effects of TNT on terrestrial and aquatic organisms. The ability of the soil and the groundwater to sustain life was assessed by considering the potential hazardous effects of the contaminated matrix on soil and aquatic organisms. The assessment endpoints (CCME 1997) for terrestrial and aquatic risks included: effects on microbial processes, effects on survival and reproduction of terrestrial plants, soil invertebrates, mammals (shrew, mouse, fox and deer) as well as birds. For aquatic organisms, the survival and reproduction of microorganisms, microphytes, macrophytes, invertebrates and fish were considered. Measurement endpoints included: analysis of nitroaromatics in the surface soil (0-15 cm), single chemical toxicity (acute and chronic) data for soil microorganisms, terrestrial plants and invertebrates, and aquatic microorganisms, microphytes, macrophytes, invertebrates and fish. In addition, sublethal toxicity (earthworm reproduction) was measured on soil samples, and mesocosms were placed on-site to evaluate lethal and sublethal (using biomarkers) effects under field conditions.

19.3 ASSESSMENT METHOD FOR A TNT-CONTAMINATED SITE

19.3.1 METHODOLOGY FOR EXPOSURE ASSESSMENT

Contaminated areas selected for the assessment (surface areas ranged from approximately 25 to 100 m^2) were defined according to chemical screening characterization data and were split into two zones: one being highly contaminated by TNT (without vegetation) and another which was immediately peripheral (with low vegetation). A reference area (non-contaminated with healthy vegetation) close to the contaminated areas (approximately 30 to 150 m) was also used to define the background toxicity (laboratory and field studies) and contaminant concentrations.

19.3.1.1 Exposure Values

Exposure to TNT by direct contact and ingestion is variable in space and time, and depends on the field conditions. Soil organisms (microorganisms, plants and invertebrates) are relatively immobile and can therefore be exposed to high concentrations of TNT. Considering the distribution and variability of the data, the 95% upper confidence limit (UCL) of the mean TNT concentration was considered as an appropriate conservative estimate of exposure for soil organisms and other receptors used in the screening assessment. In addition, TNT-contaminated groundwater represents the worst-case exposure scenario for aquatic organisms, and the corresponding exposure values were based on measured groundwater concentrations. Thus, the UCL values were selected as the exposure values for contaminated soils and groundwater. For cases in which an adequate number of replicates was lacking and the UCL could not be calculated, the maximal concentration was used. The UCL values were also used for a transport and transformation simulation using the *CalTOX* model (McKone 1993, McKone *et al.* 1997) in order to estimate TNT concentration in surface water. The results of this simulation were also considered in the risk assessment for aquatic organisms.

19.3.1.2 Soil and Water Samples

Surface soil samples were taken from each selected area using a stainless steel tool, transported and homogenized by hand at the laboratory. Groundwater samples were pumped from an observation well. Explosives were extracted from the soil using the acetonitrile-sonication method (USEPA 1997). Concentrations of TNT and some of their degradation products in water or acetonitrile soil extracts were determined by HPLC.

19.3.2 METHODOLOGY FOR EFFECTS ASSESSMENT

19.3.2.1 Single Chemical Toxicity Data

Literature concerning the toxicity of nitroaromatic munition compounds was reviewed by Talmage *et al.* (1999). However, certain data are missing (e.g., DNT and metabolites of TNT) or insufficient to calculate water or soil reference values (criteria or screening benchmarks) for nitroaromatic munition compounds.

19.3.2.2 Reference Values

Since standard reference values for TNT and associated substances were not available, the screening benchmarks proposed by Talmage *et al.* (1999) for aquatic and terrestrial biota were used. In some cases (e.g. earthworm), benchmarks were recalculated and updated in view of new toxicological data. The lowest toxicological values (chronic) for each receptor were then used as reference values.

Contaminated sites are often managed by using generic criteria or, alternatively, by risk-based procedures (CCME 1996, Beaulieu 1998, CEAEQ 1998). For example, the maximal acceptable limit in Quebec (industrial site, level C) for TNT is 1.7 mg kg^{-1} (Beaulieu 1998). However, generic environmental criteria for explosives are currently in development, and some data are missing for many of these compounds. In Quebec, groundwater and surface water contamination are typically managed on a criteria-based approach (Beaulieu 1998, MEFQ 1998). Surface water criteria for TNT, 2,4-dinitrotoluene (2,4-DNT) and 2,6-DNT are 5.5, 910 and 930 $\mu g\ L^{-1}$, respectively.

19.3.2.3 Laboratory Toxicity Assessment

The lethal and sublethal effects of explosive-contaminated soil samples from different areas of the site were assessed using the earthworm (*E. andrei*) reproduction test (ISO 1996). Survival and growth of adult earthworms were determined after 28 days of exposure to soil samples (contaminated and reference areas). Parameters associated with reproduction, including cocoon production and hatching, and juveniles' survival and growth were measured after 56 days of exposure.

19.3.2.4 Field Study: Biological Response and Body Burdens in Earthworms

Earthworms (indigenous and *E. andrei*) were exposed in areas of the site using field soil mesocosms (9.4 L 100-mesh nylon bags containing soil and soil organisms exposed to field conditions) as suggested by Svendsen and Weeks (1997b). Each block of soil was carefully removed from the ground using a square stainless steel tool (30 cm × 30 cm × 15 cm depth) and transferred as a block into the nylon bag. Laboratory earthworms (*E. andrei*; $n = 10$) and indigenous earthworms (single species obtained close to the site and pre-acclimated for 24 h earlier in OECD artificial soil; $n = 10$) were then placed in each nylon bag containing the soil block. Mesocosms ($n = 3$ to 6 units per area) were closed and left in place for 10 days after which the surviving earthworms were counted. Body burdens of TNT and degradation products were estimated in the surviving worms by HPLC using the SW846 Method 8330 (USEPA 1997).

19.3.2.5 Field Study: Biomarkers

At least three surviving earthworms were taken after exposure in mesocosms for the biomarker measurements. The determination of the neutral red retention time (NRRT) was done immediately after the exposure to the toxicant using a histochemical staining technique (Svendsen *et al.* 1996, Svendsen and Weeks 1997a,b). Another biomarker, the total immune activity (TIA), was used as described by others (Goven *et al.* 1993, Bunn *et al.* 1996).

19.3.3 METHODOLOGY FOR RISK ESTIMATION

Concentrations of contaminants in soil and groundwater were screened against toxicity data by using the quotient method (Suter 1996b, CCME 1997). A risk index (RI) was estimated by comparing the exposure value (EV) to the reference value (RV), for each selected area and endpoint. This method permits the identification of the location and contaminants of potential ecological concern, and screening out of those chemicals that do not constitute a potential hazard to the ecological receptors. Thus, estimated risk ($RI = EV \times RV^{-1}$) values lower than 1 would indicate the absence of significant effects on the considered receptors.

19.4 EXPOSURE ASSESSMENT

19.4.1 CONTAMINATED ZONES

The areas selected for the assessment were identified as sites A, B and C (TNT-contaminated areas having no vegetation and corresponding to the production sites). Peripheral contaminated areas having low levels of vegetation were also studied (identified as A′, B′ and C′ in the text below). A non-contaminated reference zone (R) at near proximity of the contaminated areas (approximately 30 m from area A) was also used.

19.4.2 DIRECT CONTACT

19.4.2.1 Soil

Preliminary chemical analysis indicated that the soil contained TNT, trinitrobenzene (TNB), dinitrobenzene (DNB), 2,4-DNT and 2,6-DNT. TNT degradation products such as 2-ADNT and 4-ADNT were also detected. The concentrations of TNT (mg kg^{-1} dry weight) in the soil samples taken from the contaminated area having no observable vegetation reached 17 450 ($n = 10$), 24 883 ($n = 1$) and 258 650 ($n = 3$) in areas A, B and C (highest contaminated area), respectively. For the peripheral contaminated areas, their concentrations (mg kg^{-1} dry weight) only reached 4425 ($n = 10$), 397 ($n = 6$) and 3.8 ($n = 3$) for areas A′, B′ and C′, respectively. The exposure values (95% UCL) for areas A, A′, B, B′, C, C′ were 10 642, 2638, 24 883, 310 and 255 604, 3.8 mg kg^{-1} of TNT, respectively. The concentrations of TNT byproducts (mg kg^{-1}) in soil reached 10.6 (area B; $n = 1$), 6.8 (area B′; $n = 6$), 80 (area A; $n = 10$) and 210 (area A; $n = 10$) for TNB, DNB, 2,4-DNT and 2,6-DNT, respectively. TNT degradation products in soil reached 23.3 (B′; $n = 6$) and 40.5 (B; $n = 1$) for 2-ADNT and 4-ADNT, respectively.

19.4.2.2 Groundwater

Initial studies indicated that the concentrations of TNT in groundwater samples reached 75.5 mg L^{-1} in the highest contaminated area (A), in contrast to the reference area where TNT was not detected. The maximum concentrations (mg L^{-1}) measured in the other areas, i.e., A′, B, B′, C and C′ were 0.10 ($n = 1$), 0.73 ($n = 2$), 24.8 ($n = 1$), 0.31 ($n = 1$) and 0.61 ($n = 1$) of TNT, respectively. The concentrations of explosive byproducts (mg L^{-1}) in groundwater samples taken from area A reached 104.2 for 2,4-DNT ($n = 2$) and 123.6 for 2,6-DNT ($n = 2$). For area B, TNT degradation product concentrations (mg L^{-1}) in groundwater reached 1.0 ($n = 2$) and 0.87 ($n = 1$) for 2-ADNT and 4-ADNT, respectively.

19.4.2.3 Concentrations Estimated Using *CalTOX*

A transport and transformation model, *CalTOX* (McKone 1993, McKone *et al.* 1997) was used to estimate the TNT concentrations in the different environmental compartments (soil, groundwater, plants and surface water, see Appendices I and II). Results of the *CalTOX* simulation show those TNT concentrations in soil and groundwater in the contaminated area would significantly decrease over a 1-year period (Table 19.1).

Table 19.1 TNT concentrations (measured and estimated using the *CalTOX* transport and transformation model) in different environmental compartments of contaminated zones

	TNT[a]/area					
Environmental compartment	A	A′	B	B′	C	C′
Measured concentrations ($t = 0$)[b]						
Ground-surface soil (mg kg^{-1})	10 642	2638	24 883	310.3	255 604	3.8
Groundwater (mg L^{-1})	75.5	0.10	24.8	0.74	0.61	0.31
Estimated concentrations ($t = 365$ days)						
Air (mg m^3)	0.03	0.03	0.03	0.0007	0.028	0.00005
Plants (mg kg^{-1})	13	13	12	2.9	12	0.02
Ground-surface soil (mg kg^{-1})	43	43	43	10	40	0.06
Root-zone soil (mg kg^{-1})	240	240	250	58	240	0.42
Vadose-zone (mg kg^{-1})	0.28	0.28	0.53	0.18	0.28	0.0007
Groundwater (mg L^{-1})	0.0001	0.0001	0.0001	0.00003	0.00005	0.0000001
Surface water (mg L^{-1})	0.0051	0.0051	0.0049	0.001	0.005	0.00008
Sediments (mg kg^{-1})	0.078	0.078	0.075	0.017	0.072	0.0001

[a] Assumptions used for the simulations are given in Appendices I and II.
[b] Upper confidence limit (95%) of measured concentration (or maximum value used if $n \leq 2$).

19.4.3 INGESTION FOR WILDLIFE

19.4.3.1 Mammals

Using a worst-case scenario (i.e., assuming that the animals inhabit the site all the time and consume food only in this area), the exposure to explosives in mammals by ingestion was estimated using the soil concentrations, the available bioconcentration values for plants (maximal concentrations measured in laboratory) and the earthworm tissue residues (based on the biological survey data described in Section 19.5.4). For plants, the BCFs were not available. However, Palazzo and Leggett (1986) reported 714, 614 and 2180 mg kg^{-1} tissue dry weight for TNT, 2-ADNT and 4-ADNT, respectively, in plant roots exposed to 20 mg TNT L^{-1} hydroponic solutions. These values were considered as a maximum concentration that could be accumulated by plants.

It was assumed that vegetation constitutes the total diet of shrew and deer, and was potentially a large part of the diet of mouse ($\leq$67%), whereas invertebrates constitute a large part of the diet of the shrew ($\geq$71%) and the American robin (93%) according to Sample and Suter (1994). The fox also consumes small mammals ($\cong$69%) and birds ($\cong$12%). Because areas A, B and C did not have observable vegetation, area A′ was chosen as the worst-case scenario for exposure. Earthworms were not found in the soil of this area, and probably may not contribute to the total exposure by ingestion by mammalian wildlife. Based on soil concentrations, the calculated maximum exposure concentrations (by ingestion) for TNT for representative mammals and birds ranged from 115

Table 19.2 Maximum exposure concentrations for wildlife (all areas)

Wildlife species	Diet composition (%)[a]	Maximum exposure concentrations by ingestion (mg kg^{-1})						
		TNT	TNB	DNB	2,4-ADNT	2,6-ADNT	2-ADNT	4-ADNT
Short-tailed shrew (*Blarina brevicauda*)	Vegetation (5.4%)	38.6	ND	ND	ND	ND	33.2	117.7
	invertebrates (≥71%)	14.0[b]	ND	ND	ND	ND	123.2[b]	153.1[b]
	soil (13%)	342	0.64	0.86	6.4	9.2	3.0	ND
	total	395	0.64	0.86	6.4	9.2	159	271
White-footed mouse (*Peromyscus leucopus*)	vegetation (≤67%)	478.4	ND	ND	ND	ND	411.4	1461
	arthropods (≥30%)	NA	NA	NA	NA	NA	NA	NA
	soil (<2%)	52.8	0.098	0.13	0.99	1.4	0.46	ND
	total	531	0.098	0.13	0.99	1.4	411.9	1461
Red fox (*Vulpes vulpes*)	vegetation (≤10%)	41.4	NA	NA	NA	NA	61	218.0
	small mammals (69%)	NA	NA	NA	NA	NA	NA	NA
	birds (12%)	NA	NA	NA	NA	NA	NA	NA
	soil (2.8%)	73.9	0.14	0.18	1.4	2.0	0.65	ND
	total	115	0.1	0.18	1.4	2.0	61.7	218
Whitetail deer (*Odocoileus virginiaus*)	vegetation (100%)	714.0	NA	NA	NA	NA	614	2180
	soil (<2%)	52.8	0.098	0.13	0.99	1.4	0.46	ND
	total	767	0.098	0.13	0.99	1.4	614	2180
American robin (*Turdus migratorius*)	invertebrates (93%)	18.3[b]	ND	ND	ND	ND	161.4[b]	200.5[b]
	fruit (7%)	NA	NA	NA	NA	NA	NA	NA
	soil (2.1%)	55.4	0.103	0.134	1.04	1.48	0.49	NA
	total	73.7	0.10	0.14	1.04	1.48	161.8	200.5

Note: Exposure values are based on soil concentrations, maximum BCF or concentrations accumulated in plants (Talmage *et al.* 1999) and invertebrate (this study) tissue residues. NA: not determined (insufficient data). ND: not detected.

[a] Taken from Sample and Suter (1994)

[b] Based on maximum earthworm tissue concentrations measured in the biological survey (19.7, 173.5 and 215.6 mg kg^{-1} dry tissue for TNT, 2-ADNT and 4-ADNT respectively).

to 767 mg kg^{-1} (Table 19.2). Due to the lack of data, the exposure to other nitroaromatics was not determined.

19.4.3.2 Birds

Exposure to TNT (and degradation products) in birds by ingestion was estimated using maximal soil concentrations and the earthworm tissue residues (see biological survey data described in Section 19.5.4). Because of a lack of data, TNT exposure concentrations by eating contaminated fruits (7% of the American robin diet) were not estimated, and thus not considered. Calculated maximum TNT exposure concentrations for birds were 73.7 mg kg^{-1} of soil (Table 19.2).

19.5 EFFECTS ASSESSMENT

19.5.1 EFFECTS OF TNT IN LABORATORY TOXICITY ASSESSMENT

The NOAEL (no observed adverse effect level) in mammals was estimated to be 1.6 mg kg^{-1} day^{-1}, based on the subchronic LOAEL (lowest observed adverse effect level) (160 mg kg^{-1} day^{-1}; testicular atrophy for rat) and an uncertainty factor of 100 (Talmage *et al.* 1999). Study of soils tested with plants (*Phaseolus vulgari*, *Triticum aestivum*) suggests a LOEC (lowest observed effect concentration) of 30 mg kg^{-1} (Cataldo *et al.* 1989). Confidence in this value is considered as moderate and additional studies are needed (Talmage *et al.* 1999). Effects data were not available for representative avian species.

Studies with the earthworms (Robidoux *et al.* 1999, 2000) showed a LOEC value of 110 mg kg^{-1}. Soil microbial processes (including nitrification, nitrogen fixation and dehydrogenase activities) were found to be decreased at 1 mg TNT kg^{-1} of soil (acetonitrile extractable concentration; Gong *et al.* 1999). Figure 19.1 shows the sensitivity distribution of different sublethal effects (LOECs taken from the literature) for soil organisms. Data available were linearized via the cumulative frequency distribution approach and are presented as a function of their rank of sensitivity (%; rank of data/number of data × 100). This method permits one to examine the range of data distribution and the global proportion (percentile) of organisms that may be affected when the toxicity of a chemical is assumed to be a random variable.

Benchmarks are usually based on this approach and are derived by rank-ordering the LOECs and then interpolating a value that approximates the 10th percentile (Efroymson *et al.* 1997a,b). Thus, interpolation of the intercept of the 10th percentile of the chronic sensitivity distribution shows that 2 mg TNT kg^{-1} soil would preserve more than 90% of the endpoints/organisms considered (microorganisms, plants and invertebrates). These data (Figure 19.1) indicate that microorganisms were affected from 1 to 75 mg TNT kg^{-1} in soil, whereas plants were sensitive at concentrations from 30 to 500 mg kg^{-1} and invertebrates from 110 to 880 mg TNT kg^{-1} in soil. Clearly, more toxicological work should be carried out using plants.

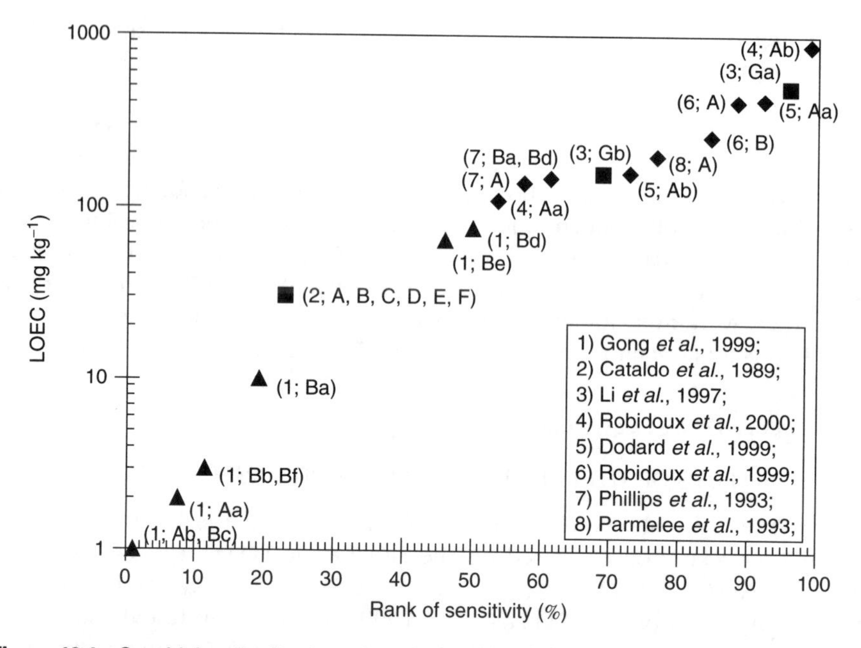

Figure 19.1 Sensitivity distribution of sublethal effects for terrestrial organisms to TNT: microorganisms (▲); plants (■); invertebrates (♦). Data correspond to different experiments or endpoints. Numbers and letters in brackets indicate: reference designated in the inset (number), experiment number (capital letter) and endpoint number (small letter). For the same citation, common capitalized letters (A, B, ...) correspond to same experimental conditions. Common small letters (a, b, ...) correspond to same endpoints. Toxicological data taken from the literature have been linearized using a cumulative frequency distribution approach, for which LOECs are presented as a function of their rank of sensitivity (%; rank of data/number of data $\times$ 100).

Toxicity studies using aquatic organisms demonstrated acute and chronic effects of TNT for microorganisms ($EC_{50} \geq 0.95$ mg L^{-1}; LOEC: 0.03 mg L^{-1}), plants ($EC_{50} \geq 1.1$ mg L^{-1}; LOEC: 1.0 mg L^{-1}), invertebrates ($LC_{50} > 4.4$ mg L^{-1}; LOEC: 1.03 mg L^{-1}) and fish (LC_{50} ranged from 0.8 to 3.7 mg L^{-1}; LOEC: 0.04 mg L^{-1}). Figure 19.2 shows the sensitivity distribution of sublethal effects (LOEC taken from the literature) for aquatic organisms. These data indicate that aquatic microorganisms were affected from 0.03 to 11 mg TNT L^{-1}, whereas microphytes and macrophytes were sensitive from 1 to 25 mg L^{-1}, invertebrates from 1 to 5 mg L^{-1} and fish from 0.04 to 1.4 mg L^{-1} of TNT.

19.5.2 REFERENCE VALUES

The TNT screening benchmarks used for soil organisms for plants, soil invertebrates and soil microbial processes were 30 (Cataldo *et al.* 1989, Talmage *et al.* 1999), 110 (LOEC value from Robidoux *et al.* 2000) and 1 mg kg^{-1} (LOEC value from Gong *et al.* 1999), respectively. The screening benchmarks used for

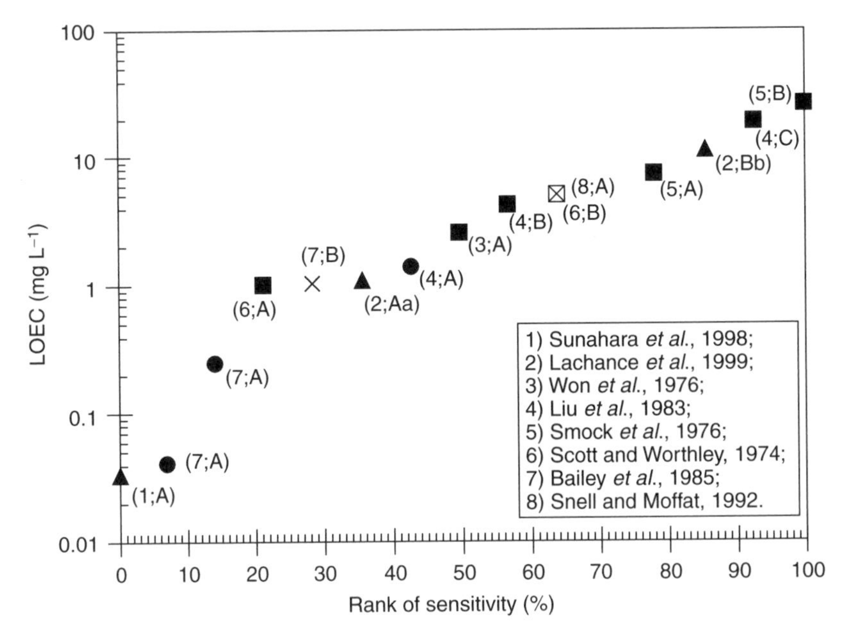

Figure 19.2 Sensitivity distribution of sublethal effects for aquatic organisms to TNT: microorganisms (▲); microphytes (■); macrophytes (□); invertebrates (×); fish (•). Numbers and letters in brackets indicate: reference designated in the inset (number), experiment number (capital letter) and endpoint number (small letter). For the same citation, common capitalized letters (A, B, ...) correspond to same experimental conditions. Common small letters (a, b, ...) correspond to same endpoints. LOEC values taken from literature (reference number in brackets) are presented as a function of their rank of sensitivity (%; rank of data/number of data × 100).

2-ADNT (plants and soil microbial process) were 80 mg kg^{-1} (Pennington 1988, Talmage *et al.* 1999). Because of insufficient data, benchmarks were not determined for other nitroaromatics (TNB, DNB, 2,4-DNT, 2,6-DNT and 4-ADNT). The screening benchmarks used for mammals and birds (TNT, TNB and DNB; Table 19.3) and aquatic organisms (TNT, TNB and 2-ADNT; Table 19.4) were taken from Talmage *et al.* (1999). Because toxicity data were not available for birds, reference values were extrapolated from mammalian toxicity data.

19.5.3 EARTHWORM TOXICITY

The effect of soil samples from different areas ($n = 18$) of the site were assessed. In general, soil samples from areas A and A′ were lethal to earthworms. In these areas, eight out of 10 samples gave 100% mortality. TNT concentrations in these samples ranged from 1146 to 17 063 mg kg^{-1}. A sample containing 116 mg kg^{-1} of TNT was not significantly lethal to earthworms (in agreement with Robidoux *et al.* 1999). However, one sample with low TNT concentration (25 mg kg^{-1})

Table 19.3 Nitroaromatic compounds screening benchmarks for selected wildlife species

Wildlife species		Screening benchmark		
		TNT	TNB	DNB
Short-tailed shrew (*Blarina brevicauda*)	estimated wildlife NOAEL (mg kg^{-1} day^{-1})	3.4	5.8	0.25
	diet (mg kg^{-1} food)	5.6	9.7	0.41
	water (mg L^{-1})[a]	15	26	1.13
White-footed mouse (*Peromyscus leucopus*)	estimated wildlife NOAEL (mg kg^{-1} day^{-1})	3.0	5.3	0.23
	diet (mg kg^{-1} food)	20	34	1.46
	water (mg L^{-1})[a]	10	18	0.75
Red fox (*Vulpes vulpes*)	estimated wildlife NOAEL (mg kg^{-1} day^{-1})	0.8	1.4	0.06
	diet (mg kg^{-1} food)	8.1	14	0.60
	water (mg L^{-1})[a]	9.6	17	0.71
Whitetail deer (*Odocoileus virginianus*)	estimated wildlife NOAEL (mg kg^{-1} day^{-1})	0.4	0.7	0.03
	diet (mg kg^{-1} food)	14	27	1.02
	water (mg L^{-1})[a]	6.6	11	0.49
American robin (*Turdus migratorius*)	estimated wildlife NOAEL (mg kg^{-1} day^{-1})	0.6[b]	ND	ND
	diet (mg kg^{-1} food)	4.9[c]	ND	ND
	water (mg L^{-1})[a]	4.4[d]	ND	ND

Note: Benchmarks are based on NOAEL values (Talmage *et al.* 1999). ND: not determined (insufficient data).
[a]Water concentration that incorporates dietary exposure following both water and food consumption.
[b]Extrapolated from mammal toxicity data: wildlife NOAEL = test NOAEL × (test organism *bw*/wildlife *bw*) × uncertainty factor; *bw* = body weight (rat *bw* = 0.35 kg; mouse *bw* = 0.03 kg).
[c] Concentration = wildlife NOAEL/0.648$(bw)^{0.651}$ (Sample *et al.* 1996).
[d]Concentration = wildlife NOAEL × wildlife *bw*/0.059$(bw)^{0.67}$ (Sample *et al.* 1996).

Table 19.4 Nitroaromatic compounds screening benchmarks for aquatic organisms

Criterion	Screening benchmark		
	TNT	TNB	2-ADNT
Acute water quality criterion (mg L^{-1})	0.57	ND	ND
Chronic water quality criterion (mg L^{-1})	0.09	ND	ND
Secondary acute value (mg L^{-1})	ND	0.06	0.35
Secondary chronic value (mg L^{-1})	ND	0.011	0.02
Lowest chronic value, fish (mg L^{-1})	0.04	0.12	ND
Lowest chronic value, daphnids (mg L^{-1})	1.03	0.75	ND
Lowest chronic value, plants (mg L^{-1})	1.0	0.10	>50
Sediment quality benchmark (mg kg^{-1})[a]	0.09[b]	0.002[c]	ND

Note: Water quality criteria are based on USEPA guidelines (Talmage *et al.* 1999). Secondary chronic values are based on USEPA guidelines (Talmage *et al.* 1999). Benchmarks were not determined for 2,4-DNT, 2,6-DNT and 4-ADNT (insufficient data). ND: not determined (insufficient data).
[a]1% of organic carbon content in the sediment is assumed.
[b]9.2 mg TNT kg^{-1} organic carbon in the sediment.
[c]0.24 mg TNB kg^{-1} organic carbon in the sediment.

in area A′ was significantly lethal (33% mortality). One out of three samples from areas B′ was lethal to earthworms (367 mg TNT kg^{-1}). Others were not lethal ($\leq$130 mg TNT kg^{-1}). Survival rates were 100% for reference samples ($n = 5$).

Soil samples contaminated by explosives from areas A, A′ and B′ caused a number of reproductive effects compared to laboratory controls (OECD soils), as evidenced by the decreased number of total and hatched cocoons and the number and biomass of juveniles (in agreement with Robidoux *et al.* 1999), except for the sample having 25 mg kg^{-1} which showed only a decrease in the total number of cocoons. However, it was noted that the reference soil samples also had effects on reproduction compared to the laboratory control (artificial soil), suggesting the presence of other contaminants, or perhaps an effect of the soil matrix. Nevertheless, the explosive-contaminated samples showed higher reproductive effects compared to reference soils, except for two samples having low concentrations of TNT (25 and 116 mg kg^{-1}, respectively). Growth of surviving earthworms was generally not affected in the TNT-contaminated samples, but it was increased for reference areas compared to controls.

Effects on earthworm reproduction were not correlated at low concentrations ($\leq$130 mg kg^{-1}) of TNT or other related substances. However, a correlation was observed ($R^2 = 0.77$, $n = 4$) from 116 to 367 mg TNT kg^{-1}. Soil samples having TNT concentrations from 370 to 1110 mg kg^{-1} were not available for toxicity assessment. Higher TNT concentrations ($\geq$1146 mg kg^{-1}) were lethal to earthworms. Soil samples from areas B, C and C′ were not tested.

19.5.4 BIOLOGICAL SURVEY DATA AND BIOMARKERS USING THE EARTHWORM

The survival of earthworms (*E. andrei* and indigenous species) exposed to TNT-contaminated soils in mesocosms tended to decrease at increasing concentrations of TNT in soil. No survivors were found at >9033 mg TNT kg^{-1} soil, but most were found at concentrations <1200 mg kg^{-1}. Composting worms *E. andrei* added to the mesocosms tolerated concentrations as high as 4050 mg TNT kg^{-1}, whereas indigenous earthworms were only observed at $\leq$ 1146 mg kg^{-1}. The mortality was not significantly correlated with the soil concentrations of other nitroaromatic compounds. In addition, dried tissue concentrations of TNT and its degradation products in depurgated earthworms increased with soil levels of TNT (<1200 mg kg^{-1} dried soil; data not shown), 2-ADNT (<150 mg kg^{-1} dried soil; data not shown) and 4-ADNT (<220 mg kg^{-1} dried soil; data not shown). However, since the time of depurgation was relatively short (8 h), little interference with contaminated soil may be suspected. A maximum of 19.7 μg TNT g^{-1} dried tissue (in *E. andrei* at 116 mg kg^{-1} TNT in soil) was found in earthworm tissues; whereas a maximum of 173.5 μg 2-ADNT g^{-1} and 215.6 μg 4-ADNT g^{-1} dried tissue (in indigenous earthworms at 1146 mg kg^{-1} TNT in soil) were detected. However, TNT was not detected in earthworms exposed to soils having high concentrations of 2-ADNT and 4-ADNT, suggesting that the parent compound may have been partially degraded *in vivo*. Interestingly, the variability of the survival data was high for the reference mesocosms, ranging from 0 to 90% ($n = 5$) for *E. andrei* and 30 to 70% ($n = 5$) for the indigenous earthworm species, suggesting the presence of other contaminants or an unfavorable effect of the experimental conditions. Effects on reproduction have also been observed in the reference areas (see Section 19.5.3). Mesocosms experiments were not carried out in areas B, C and C′.

The TIA biomarker showed no significant responses in surviving earthworms exposed to the explosives-contaminated soils under field conditions. The NRRT was significantly lower for surviving earthworms (*E. andrei* and indigenous species) exposed to explosives in areas A, A′ and B′ (25 to 367 mg kg^{-1} of TNT) in mesocosms compared to reference soils, except for one mesocosm placed in area B′ containing 116 mg kg^{-1} of TNT.

19.6 RISK CHARACTERIZATION

The potential risk to terrestrial and aquatic receptors was analyzed using a single chemical risk estimation approach (Suter 1996b). Toxicity assessment, biological survey on earthworms and biomarkers were used to support evidence of ecological risk.

19.6.1 RISK ESTIMATION OF SINGLE CHEMICALS

Contaminant concentration (Section 19.4) was screened against benchmarks (Section 19.5). Exposure values used for soil organisms were generally higher than corresponding benchmarks. This would indicate that certain ecological receptors might be at risk. For plants, the RI ranged from 0.13 (area C′) to 8520 (area C). For invertebrates, the RI ranged from 0.03 (area C′) to 2324 (area C). The exposure concentrations were 3.8 (area C′) to 255 604 (area C) times higher than the benchmarks for microbial processes. Considering plants and microbial processes, an RI of <0.5 was estimated for 2-ADNT separately. Because of the lack of data, the risk for soil organisms was not estimated with other explosive-associated compounds. Based on laboratory data (Figure 19.1), all soil organisms (100%) would be affected by the concentrations of TNT in areas A, A′, B and C, whereas $\geq 75\%$ of organisms would be affected in area B′ and $\leq 10\%$ in area C′.

Since exposure values for wildlife (Table 19.2) were also higher than the corresponding benchmarks (Table 19.3), RI values >1 were found for selected species (short-tailed shrew, white-footed mouse, red fox, whitetail deer, American robin). The exposure concentrations of TNT for the overall area were 14.2 (red fox) to 70.5 (short-tailed shrew) times higher than the toxicity benchmarks. The RI associated with DNB ranged from 0.1 (white-footed mouse) to

2.1 (short-tailed shrew); whereas the RI values associated with TNB were ≤ 0.1. Because of the lack of data, risk for wildlife was not estimated for the other explosive-associated compounds.

Groundwater concentrations estimated using *CalTOX* (1-year period) were much lower than the measured or actual groundwater concentrations. However, considering that the concentrations in soil and groundwater were relatively stable over the last years, the concentrations measured during this study were used as a worst-case scenario (Table 19.1). Thus, the concentrations of TNT, especially in areas A and B, may represent a risk for different aquatic organisms. However, TNT may be rapidly diluted and biotransformed to possibly less toxic substances in the nearby river. Figure 19.2 suggests that, based on laboratory data, all aquatic organisms (100%) would be affected by the groundwater TNT concentration from areas A and B, whereas 10 to 40% would be affected by those at sites A′, B′, C and C′. Using the *CalTOX* estimation, less than 10% of the aquatic organisms would be affected.

19.6.2 TOXICITY ASSESSMENT

Toxicity assessment using the earthworm reproduction test data showed that lethal and sublethal toxicities may be associated with exposure to soils from areas A, A′ and B. Soils from sites having $\geq$367 mg TNT kg^{-1} soil may present a risk of lethality, whereas those having $\geq$116 mg kg^{-1} may reduce the reproduction capacity of earthworms. These results agree with earlier studies (Robidoux *et al.* 1999, 2000).

19.6.3 BIOLOGICAL SURVEY

A weak trend between decreased earthworm (*E. andrei* and indigenous species) survival and increasing levels of TNT and associated nitroaromatic degradation products was observed in the mesocosm experiments. No survival of earthworms was found at concentrations >9033 mg TNT kg^{-1} dry soil; whereas lethality was not detected at <1146 mg TNT kg^{-1}. TNT could be partially degraded (to 2-ADNT and 4-ADNT) in the tissue of surviving earthworms, suggesting that these contaminants may be transferred into the food web.

19.6.4 BIOMARKERS

Considering the NRRT responses of surviving earthworms exposed to explosives-contaminated soils under field conditions, sublethal effects may be associated with soils from areas A, A′ and B′ (25 to 367 mg kg^{-1} of TNT) compared to reference soils. These experiments support the data from single chemical and toxicity assessment.

19.6.5 SUMMARY OF RISK CHARACTERIZATION

Contaminants of potential concern for the ecological receptors include the nitroaromatic compounds such as TNT, their degradation (2-ADNT and 4-ADNT) and associated byproducts (TNB, DNB, 2,4-DNT and 2,6-DNT). Based on the risk index, the screening risk assessment indicates that the selected contaminated areas of the heterogeneous site described herein may constitute credible hazards to soil organisms and wildlife when TNT concentrations are ≥ 1 mg kg^{-1}. However, this method does not consider the possibility of synergetic or antagonist chronic effects with the identified contaminants and other compounds. In general, earthworm data from toxicity tests, biological surveys and site-specific biomarker studies showed lethal and sublethal effects and would support the preliminary risk assessment results based on single chemicals. Thus, concentrations of TNT ≥ 100 mg kg^{-1} may constitute a risk for soil organisms and/or wildlife at the study site.

Further studies, such as additional toxicity (or other measures of bioavailability) assessment of TNT degradation products and associated products, and chemical identification of other degradation products, would be necessary before concluding that TNT (over the generic criteria for industrial sites, i.e., 1.7 mg kg^{-1}) and related contaminants at low concentrations (<100 mg kg^{-1} of TNT) constitute a risk for the soil organisms and wildlife inhabiting this site. Field studies are also required before concluding that a risk of effect exists for the aquatic receptors.

19.6.6 UNCERTAINTIES CONCERNING RISK

Ecological risk assessment of nitroaromatic compounds should consider uncertainty associated with different sources, which may limit the conclusion of the evaluation. Sources of uncertainty (Suter 1993) considered for this case study include the following.

19.6.6.1 Ignorance

Multiple Contaminant Exposure

Single chemical toxicity experiments were used to calculate benchmarks. In reality, receptors are exposed to many nitroaromatic contaminants, and may also be exposed at low concentrations to other contaminants such as heavy metals.

Bioavailability of Contaminants in Soil

Solvent extraction methods were used in this study to characterize the total concentrations of (extractable) nitroaromatics, but may not accurately reflect the bioavailable quantities of these contaminants in the soil. In addition, the spiking methods used for toxicity studies may not represent aged soil conditions found in the field.

Transport and Transformation of Contaminants in Water

Soils contaminated by TNT for many years probably constitute the actual source of contamination. Data from this study and others (data not shown) indicate that the groundwater TNT concentrations have been stable for several years. This is not in agreement with the estimation given by *CalTOX* using the theoretical proprieties of TNT (Appendix I), however *CalTOX* was designed for sites $> 1000\ m^2$.

Uptake Factors and Contaminant Concentrations in Unanalyzed Food Types

Uptake factors were calculated from laboratory experiments (literature) or site-specific (mesocosms) data. Because only a few data were available, maximum TNT concentrations in roots were used as the exposure concentrations by plants for mammals. Maximum tissue residues in depurgated earthworms from the mesocosm studies were used as the exposure concentration by invertebrates for mammals. Considering that exposure conditions and food type may be different or change with respect to season, tissue burdens of contaminants are likely to differ between the laboratory and field measurements. The assumption taking the maximum concentration as the exposure value represents the worst-case scenario. Thus, the level of exposure is probably overestimated.

Extrapolation from Published Toxicity Data

Toxicity of nitroaromatic compounds on wildlife species is poorly documented in the open literature. Benchmarks (NOAEL based) were extrapolated from LOAEL for the laboratory rat giving uncertainty on the reference value.

Lack of Benchmarks for Some Nitroaromatic Compounds

Few data are available for nitroaromatic compounds other than TNT. In many cases, benchmarks for TNB, DNB, 2,4-DNT, 2,6-DNT and the TNT degradation products (2-ADNT and 4-ADNT) could not be estimated.

Toxicity Data for Birds

Subchronic or chronic studies were not available for representative avian species. Benchmarks were extrapolated from mammalian toxicity literature.

Confounding Stressors

Toxicity of explosives may be affected by other biotic or abiotic factors. Toxicity of nitroaromatics may be altered by physicochemical interaction in the soil or by field conditions leading to synergetic or antagonistic effects.

Recovery from Exposure

Explosives-contaminated areas are isolated and the receptors may be exposed to the toxicant for only a short-term period without having acute effects. Benchmarks are based on laboratory toxicity data and do not consider the potential recovery if the concentrations of explosives decrease (by natural attenuation) or if the exposure period varies (e.g., seasonal changes in receptor activities).

19.6.6.2 Error

Chemical Analysis

Technical error is relatively low when quality control procedures are applied. The TNT concentrations were generally accurate within ±9.0% for soil and within 15% for groundwater (USEPA 1997).

19.6.6.3 Stochasticity

Sampling Frequency

Because of their limited to low solubilities in water, the concentrations of contaminants in soil may be highly variable. The heterogeneous distribution of TNT in soil ('hot' contaminated soil) and the low number of samples creates uncertainty, especially when exposure concentrations are close to the benchmarks. Toxicant concentrations were, in most cases, clearly different from the benchmarks, reducing the uncertainty on exposure concentrations.

Variable Food and Water Consumption

Food and water consumption by wildlife was assumed to be similar to those reported for the same or related species (Sample and Suter 1994). However, since weather conditions, type of food and behavior may be different in Canada and specific to the study site, food consumption by wildlife at the study site may be slightly different and should be confirmed.

Variable Response to Toxicants

Different species can be exposed to contaminants at different life stages. Data from which benchmarks were derived are usually not based on experiments using the same organisms found in the ecosystem assessed. Because few toxicological data are available for TNT, low to moderate confidence can be placed on the benchmarks.

19.7 CONCLUSION

This chapter describes a preliminary ecotoxicological risk assessment approach used for explosive-contaminated sites. This case study indicates that the selected contaminated areas of the described heterogeneous site constitute credible hazards to soil organisms, mammals and birds. The risk would be mainly associated with TNT exposure. However, cumulative risk associated with byproducts and TNT degradation products cannot be estimated since scant toxicity data are available for these contaminants. Contaminated groundwater may also represent a potential risk for aquatic organisms, considering a worst-case scenario approach. However, the concentrations of TNT may decrease in groundwater using the transport and transformation model *CalTOX*. In addition, TNT would be diluted in the surface water. Thus, the real exposure concentration would be lower than the benchmarks, resulting in a low risk to aquatic organisms. Toxicity tests, biological survey and biomarker data specific to the site using the earthworms (as indicator of exposure and effects) confirmed the potential risk to soil organisms. The uncertainties associated with this study would be reduced at low concentrations of TNT (<100 mg kg^{-1}) following further laboratory (measures of bioavailability such as tissue residues and toxicity tests of the different metabolites of TNT) and field investigations (monitoring and exposure measurement).

19.8 ACKNOWLEDGEMENTS

We thank Manon Sarrazin, Louise Paquet, Serge Delisle and Claude Masson from the Biotechnology Research Institute (BRI) of the National Research Council of Canada, and Sandra Morel from University of Lille 2 (France) for their assistance.

REFERENCES

ATSDR (1995) *Toxicological Profile for 2,4,6-Trinitrotoluene*, Agency for Toxic Substances and Diseases Registry, US Department of Health and Human Services, Public Health Service, Washington, DC.

Bailey HC, Spanggord RJ, Javitz HS and Liu DHW (1985) *Toxicity of TNT Wastewaters to Aquatic Organisms, Vol. III, Chronic Toxicity of LAP Wastewater and 2,4,6-Trinitrotoluene*. Final Report, AD-A164 282, SRI International, Menlo Park, CA.

Beaulieu M (1998) *Politique de Réhabilitation des Sols et de Réhabilitation des Terrains Contaminés*, Les publications du Québec, Sainte-Foy, QC.

Bunn KE, Thompson HM and Tarrant KA (1996) Effects of agrochemicals on the immune systems of earthworms. *Bulletin of Environmental Contamination and Toxicology*, **57**, 632–639.

CCME (1996) *A Framework for Ecological Risk Assessment: General Guidance*, Canadian Council of Ministers of Environment, The National Contaminated Sites Remediation Program, Subcommittee on Environmental Quality Criteria for Contaminated Sites, Winnipeg, Manitoba, 32 pp.

CCME (1997) *A Framework for Ecological Risk Assessment: Technical Appendices*, Canadian Council of Ministers of Environment, The National Contaminated Sites Remediation Program, Subcommittee on Environmental Quality Criteria for Contaminated Sites, Winnipeg, Manitoba, 32 pp.

Cataldo DA, Harvey SD, Fellows RJ, Bean RM and McVeety BD (1989) *An Evaluation of the Environmental Fate and Behavior of Munitions Material (TNT, RDX) in Soil and Plant Systems: Environmental Fate and Behavior of TNT*, US Army Biological Research and Development Laboratory, Pacific Northwest Laboratory, Richland, WA.

CEAEQ (1998) *Procédure d'Évaluation du Risque Écotoxicologique pour la Réhabilitation des Terrains Contaminés*, Centre d'expertise en analyse environnmentale du Québec, Ministère de l'Environnement du Québec, Gouvernement du Québec.

Daniels JI and Knezovich JP (1994) *Human Health Risks from TNT, RDX and HMX in Environmental Media and Consideration of the U.S. Regulatory Environment*, Luxembourg International Congress.

Dodard S, Paquet L, Robidoux P, Powlowski J, Hawari J and Sunahara GI (1999) Ecotoxicological effects of recalcitrant environmental pollutants on two species of soil invertebrates (*Enchytraeus albidus* and *Eisenia andrei*). *Canadian Technical Report of Fisheries and Aquatic Sciences*, **2260**: 89.

Drzyzga O, Gorontzy T, Schmidt A and Blotevogel KH (1995) Toxicity of explosives and related compounds to the luminescent bacterium *Vibrio fischeri* NRRL-B-11177. *Archives of Environmental Contamination and Toxicology*, **28**: 229-235.

Efroymson RA, Will ME, Suter II GW and Wooten AC (1997a) *Toxicological Benchmarks for Screening Contaminants of Potential Concern for Effects on Terrestrial Plants: 1997 Revision*. ES/ER/TM-85/R3. Prepared for the US Department of Energy, Oak Ridge National Laboratory, Oak Ridge, TN.

Efroymson RA, Will ME and Suter II GW (1997b) *Toxicological Benchmarks for Contaminants of Potential Concern for Effects on Soil and Litter Invertebrates Heterotrophic Process: 1997 Revision*. ES/ER/TM-126/R2. Prepared for the US Department of Energy, Oak Ridge National Laboratory, Oak Ridge, TN.

Fuller M and Manning J (1998) Evidence for differential effects of 2,4,6-trinitrotoluene and other munitions compounds on specific subpopulations of soil microbial communities. *Environmental Toxicology and Chemistry*, **17**, 2185-2195.

Gong P, Siciliano SD, Greer CW, Paquet L, Hawari J and Sunahara GI (1999) Effects and bioavailability of 2,4,6-trinitrotoluene in spiked and field contaminated soils to indigenous microorganisms. *Environmental Toxicology and Chemistry*, **18**, 2681-2688.

Goven AJ, Eyambe GS, Fitzpatrick LC, Venables BJ and Cooper EL (1993) Cellular biomarkers for measuring toxicity of xenobiotics: effects of polychlorinated biphenyls on earthworm *Lumbricus terrestris* coelomocytes. *Environmental Toxicology and Chemistry*, **12**: 863-870.

Hawari J, Spencer B, Halasz A, Thiboutot S and Ampleman G (1998) Biotransformation of TNT with anaerobic sludge: the role of triaminotoluene. *Applied Environmental Microbiology*, **64**, 2200-2206.

Hawari J, Halasz A, Beaudet A, Ampleman G and Thiboutot S (1999) Biotransformation routes of 2,4,6-trinitrotoluene (TNT) by *Phanerochaete chrysosporium* in agitated cultures at pH 4.5: a time course study. *Applied Environmental Microbiology*, **65**, 2977-2986.

Howard PH, Boethling RS and Davis WF (1991), *Handbook of Environmental Degradation Rates*, Lewis, Chelsea, MI.

HSDB (1995) *2,4,6-Trinitrotoluene*, Hazardous Substances Data Bank, MEDLARS Online Information Retrieval System, National Library of Medicine, Bethesda, MD.

ISO (1996) *Soil quality — Effects of pollutants on earthworms (Eisenia fetida fetida, E. fetida andrei). Part 2: Determination of effects on reproduction*. ISO DIS 11268-2.2, Draft, International Standards Organization, Geneva.

Jarvis SA, McFarland VA and Honeycutt ME (1998) Assessment of the effectiveness of composting for the reduction of toxicity and mutagenicity of explosives soil. *Ecotoxicology and Environmental Safety*, **39**, 131-135.

Lachance B, Robidoux PY, Hawari J, Ampleman G, Thiboutot S and Sunahara GI (1999) Cytotoxic and genotoxic effects of energetic compounds on bacterial and mammalian cells *in vitro*. *Mutation Research*, **444**, 25-39.

Li ZM, Peterson MM, Comfort SD, Horst GL, Shea PJ and Oh BT (1997) Remediating TNT-contaminated soil by soil washing and Fenton oxidation. *The Science of the Total Environment*, **204**, 107 - 115.

Liu D, Spanggord R, Bailey HC, Javizt H and Jones D (1983), *Acute Toxicity of LAP Wastewater and 2,4,6-Trinitrotoluene*, Vol. I, SRI International, Menlo Park, CA.

Loranger S and Courchesne Y (1997) Health risk assessment of an industrial site contaminated with polycyclic aromatic hydrocarbons using CalTOX, an environmental fate/exposure model. *SAR and SAR in Environmental Research*, **6**, 81 - 104.

McKone TE (1993). *CalTOX, a multimedia total-exposure model for hazardous-waste sites*, UCRL-CR-111456PtI, Lawrence Livermore National Laboratory, Livermore, CA.

McKone TE, Hall D and Kastenberg WE (1997), *CalTOX Version 2.3 — Description of Modification and Revisions*, University of California, Berkeley, CA.

MEFQ (1998) *Critères de Qualité de l'Eau de Surface au Québec*. Direction des écosytemes aquatiques, Ministère de l'Environnement et de la Faune du Québec.

Palazzo AJ and Leggett DC (1986) Effects and disposition of TNT in a terrestrial plant. *Journal of Environmental Quality*, **15**, 49 - 52.

Parmelee RW, Wentsel RS, Phillips CT, Simini M and Checkai RT (1993) Soil microcosm for testing the effects of chemical pollutants on soil fauna communities and trophic structure. *Environmental Toxicology and Chemistry*, **12**, 1477 - 1486.

Pennington JC (1988) *Soil Sorption and Plant Uptake of 2,4,6-Trinitrotoluene*, US Army Biomedical Research and Development Laboratory, Fort Detrick, Frederick, MD.

Peterson MM (1996) TNT and 4-amino-2,6-dinitrotoluene influence on germination and early seedling development of tall fescue. *Environmental Pollution*, **93**, 57 - 62.

Phillips CT, Checkai RT and Wentsel RS (1993) *Toxicity of Selected Munitions and Munition-Contaminated Soil on the Earthworm (Eisenia foetida)*, US Army Chemical and Biological Defense Agency, Edgewood Research, Development & Engineering Center, Research and Technology Directorate, Aberdeen Proving Ground, MD.

Renoux AY, Sarrazin M, Hawari J and Sunahara GI (2000) Transformation of 2,4,6-trinitrotoluene (TNT) in soil in the presence of the earthworm *Eisenia andrei*. *Environmental Toxicology and Chemistry*, **19**, 1473 - 1480.

Robidoux PY, Hawari J, Thiboutot S, Ampleman G and Sunahara GI (1999) Acute toxicity of 2,4,6-trinitrotoluene (TNT) in the earthworms (*Eisenia andrei*) *Ecotoxicology and Environmental Safety*, **44**, 311 - 321.

Robidoux PY, Svensen C, Caumartin J, Hawari J, Thiboutot S, Ampleman G, Weeks JM and Sunahara GI (2000) Chronic toxicity of energetic compounds in soil using the earthworm (*Eisenia andrei*) reproduction test. *Environmental Toxicology and Chemistry*, **19**, 1764 - 1773.

Robidoux PY, Hawari J, Thiboutot S, and Sunahara GI (2001) Chronic toxicity of octahydro-1,3,5,7-tetranitro-1,3,5,7-tetrazocine (HMX) in soil using the earthworm (*Eisenia andrei*) reproduction test. *Environmental Pollution*, **111**, 283 - 292.

Rosenblatt DH, Burrows EP, Mitchell WR and Parmer DL (1991) *The Handbook of Environmental Chemistry*, vol. **3**, Hutzinger O (ed), Springer-Verlag, Berlin, pp. 196 - 234.

Sample BE and Suter II GW (1994) *Estimating Exposure of Terrestrial Wildlife to Contaminants*. ES/ER/TM-125. Prepared for the US Department of Energy, Oak Ridge National Laboratory, Oak Ridge, TN.

Sample BE, Opresco DM and Suter II GW (1996) *Toxicological Benchmarks for Wildlife: 1996 Revision*. ES/ER/TM-86/R3. Prepared for the US Department of Energy, Oak Ridge National Laboratory, Oak Ridge, TN.

Scott CD and Wortlhley EG (1974). *The Toxicity of TNT and Related Wastes to an Aquatic Flowering Plant, Lemna perpusilla Torr*. AD-778 158, Aberdeen Proving Ground, MD.

Simini M, Wentsel RS, Checkai R, Phillips C, Chester NA, Major MA and Amos JC (1995) Evaluation of soil toxicity at Joliet Army Ammunition Plant. *Environmental Toxicology and Chemistry*, **14**, 623 - 630.

Smock LA, Stoneburner DL and Clark JR (1976) The toxic effects of trinitrotoluene (TNT) and its primary degradation products on two species of algae and the fathead minnow. *Water Research*, **10**, 537-543.

Snell TW and Moffat BD (1992) A 2-d life cycle test with the rotifer *Brachionus calyciflorus*. *Environmental Toxicology and Chemistry*, **11**, 1249-1257.

Spanggord RJ, Mill T, Chuo T, Mabey W, Smith J and Lee S (1980) *Environmental Fate Studies on Certain Munition Wastewater Constituents, Final Report, Phase I—Literature Review*, SRI International, Menlo Park, CA.

Sunahara GI, Dodard S, Sarrazin M, Paquet L, Ampleman G, Thiboutot S, Hawari J and Renoux AY (1998) Development of a soil extraction procedure for ecotoxicity characterization of energetic compounds. *Ecotoxicology and Environmental Safety*, **39**, 185-194.

Sunahara GI, Dodard S, Sarrazin M, Paquet L, Hawari J, Greer C, Ampleman G, Thiboutot S and Renoux AY (1999) Ecotoxicological characterization of energetic substances using a soil extraction procedure. *Ecotoxicology and Environmental Safety*, **43**, 138-148.

Suter II GW (1993) *Ecological Risk Assessment*, Lewis, Chelsea, MI.

Suter II GW (1996a) *Guide for Developing Conceptual Models for Ecological Risk Assessments. ES/ER/TM-186. Prepared for the US Department of Energy*, Oak Ridge National Laboratory, Oak Ridge, TN.

Suter II GW (1996b) *Risk Characterization for Ecological Risk Assessments of Contaminated Sites* ES/ER/TM-200. Prepared for the US Department of Energy, Oak Ridge National Laboratory, Oak Ridge, TN.

Svendsen C, Meharg AA, Freestone P and Weeks JM (1996) Use of an earthworm lysosomal biomarker for the ecological assessment of pollution from an industrial plastics fire. *Applied Soil Ecology*, **3**, 99-107.

Svendsen C and Weeks JM (1997a) Relevance and applicability of a simple earthworm biomarker of copper exposure. I. Links to ecological effects in a laboratory study with *Eisenia andrei*. *Ecotoxicology and Environmental Safety*, **36**, 72-79.

Svendsen C and Weeks JM (1997b) Relevance and applicability of a simple earthworm biomarker of copper exposure. II. Validation and applicability under field conditions in a mesocosm experiment with *Lumbricus rubellus*. *Ecotoxicology and Environmental Safety*, **36**, 80-88.

Talmage SS, Opresko DM, Maxwell CJ, Welsh CJE, Cretella FM, Reno PH and Daniel FB (1999) Nitroaromatic munition compounds: environmental effects and screening values. *Reviews of Environmental Contamination and Toxicology*, **161**, 1-156.

USEPA (1992) *Framework for Ecological Risk Assessment*, United States Environmental Protection Agency, Risk Assessment Forum, Washington, DC.

USEPA (1997) Method 8330: Nitroaromatics and nitroamines by high performance liquid chromatography (HPLC). In *Test Methods for Evaluating Solid Waste*. SW-846, Update III, Part 4:1 (B), Office of Solid Waste, United States Environmental Protection Agency, Washington, DC.

Won WD, DiSalvo LH and Ng J (1976) Toxicity and mutagenicity of 2,4,6-trinitrotoluene and its microbial metabolites. *Applied and Environmental Microbiology*, **31**, 576-580.

Yinon YJ (1990) *Toxicity and Metabolism of Explosives*, CRC Press, Boca Raton, FL.

APPENDIX I

Table A19.1 Chemical properties of TNT used for *CalTOX* simulations

Properties	Value	CV (%)	Reference
Molecular weight (g mol^{-1})	227.2	0.1	Rosenblatt *et al.* (1991)
Octanol–water coefficient (K_{ow})	100	150	Rosenblatt *et al.* (1991)
Melting point (K)	353.75	0.68	Rosenblatt *et al.* (1991)
Vapor pressure (Pa)	0.026	1.5	ATSDR (1995)
Solubility (mol m^3)	0.572	55	ATSDR (1995)
Henry's law constant (K_h; Pa m^3 mol^{-1}, 25 °C)	0.011	25	Rosenblatt *et al.* (1991)
Diffusion coefficient; pure air (m^2 day^{-1})	5.098	5	Daniels and Knezovich (1994)
Diffusion coefficient; pure water (m^2 day^{-1})	5.01×10^{-4}	25	Daniels and Knezovich (1994)
Organic carbon partition coefficient (K_{oc})	525	60	Rosenblatt *et al.* (1991)
Partition coefficient; ground-root soil layer	2–11	150	Talmage *et al.* (1999)
Partition coefficient; vadose-zone soil layer	5.5–22.2	150	Talmage *et al.* (1999)
Half-life in air (days)	0.47	100	Howard *et al.* (1991)
Half-life in surface soil (days)	180	100	Howard *et al.* (1991)
Half-life in root-zone soil (days)	180	100	Howard *et al.* (1991)
Half-life in vadose-zone soil (days)	180	100	Howard *et al.* (1991)
Half-life in groundwater (days)	360	100	Howard *et al.* (1991)
Half-life in surface water (days)	0.0533	100	Howard *et al.* (1991)
Half-life in sediments (days)	180	100	Talmage *et al.* (1999)

APPENDIX II

Table A19.2 Landscape characteristics of the TNT-contaminated area used for *CalTOX* simulations

Landscape characteristics	Value used[a]/area			CV(%)
	A and A′	B and B′	C and C′	
Contaminated area (m^2)[b]	100	50	25	10
Annual average concentration (m day^{-1})	3.31×10^{-3}	3.31×10^{-3}	3.31×10^{-3}	10
Atmospheric dust load (kg m^3)	5.4×10^{-8}	5.4×10^{-8}	5.4×10^{-8}	20
Deposition velocity of air particles (m day^{-1})	86.4	86.4	86.4	30
Thickness of ground soil (m)	0.015	0.015	0.015	100
Soil particle density (kg m^3)	2000	3000	2000	5
Water content in surface soil (vol. fraction)	0.164	0.131	0.164	20
Air content in surface soil (vol. fraction)	0.25	0.10	0.25	20
Erosion in surface soil (kg m^2 day^{-1})	2.74×10^{-5}	2.74×10^{-5}	2.74×10^{-5}	20
Thickness of root-zone (m)	5	5	5	20
Water content of root-zone soil (vol. fraction)	0.155	0.155	0.155	30
Air content of root-zone (vol. fraction)	0.25	0.17	0.25	30
Thickness of vadose-zone soil (m)	11	5	11	10
Water content of root-vadose (vol. fraction)	0.28	0.3	0.28	20
Air content of root-vadose (vol. fraction)	0.2	0.17	0.2	20
Thickness of aquifer (m)	0.5	0.5	0.5	30
Solid material density in aquifer (kg m^3)	1420	2650	1420	5
Porosity of aquifer zone (vol. fraction)	0.33	0.33	0.33	20
Ambient environmental temperature (K)	277	277	277	2
Organic carbon fraction in upper-zone soil	0.035	0.015	0.035	100
Organic carbon fraction in vadose-zone	0.035	0.01	0.035	100
Organic carbon fraction in aquifer zone	0.0001	0.0001	0.0001	100
Boundary layer thickness in air above soil (m)	0.005	0.005	0.005	20
Yearly average wind speed (m day^{-1})	3.6×10^5	3.6×10^5	3.6×10^5	24

[a]Values taken from McKone (1993) or Loranger and Courchesne (1997).
[b]Approximate value.

Influence of Soil Remediation Techniques on the Bioavailability of Heavy Metals

C.A.M. VAN GESTEL[1], L. HENZEN[2], E.M. DIRVEN-VAN BREEMEN[3] AND J.W. KAMERMAN[4]

[1] *Institute of Ecological Science, Vrije Universiteit, Amsterdam, The Netherlands*
[2] *TNO Nutrition and Food Research Institute, Department of Environmental Toxicology, Delft, The Netherlands*
[3] *Laboratory for Ecotoxicology, National Institute of Public Health and the Environment, Bilthoven, The Netherlands*
[4] *Province of Gelderland, Department of Environment and Water, Sub Department of Soil Remediation, Arnhem, The Netherlands*

20.1 INTRODUCTION

Industrial and mining activities contribute to metal pollution (Salomons *et al.* 1995). This pollution can pose an environmental and human health risk. Therefore, clean-up or remediation of metal-polluted soils is needed. Currently, a number of remediation techniques are available (Assink 1988). However, they are often not efficient enough to reduce concentrations of metals present in polluted soils below non-polluted background levels (Versluis *et al.* 1988). The question that needs to be addressed becomes: what is the potential ecological risk of the remaining metal residues in remediated soils?

Bioassays have received attention as tools to assess the potential ecological risk of polluted soils (Keddy *et al.* 1995, Van Gestel 1997) and may also be used for the evaluation of decontaminated soils and their potential suitability for (re)use in land management practice (DECHEMA 1995). This chapter describes two bioassay-based studies on the bioavailability of heavy metals in polluted and remediated soils. Both studies focus on bioaccumulation of metals in test organisms (earthworms, plants) exposed to soils for a certain period of time. Effects on survival, body weight changes (earthworms) and growth (plants) are also examined.

This chapter is dedicated to the late Mrs D.M.M. Adema, who was responsible for the bioassays performed with *Cyperus esculentus* at the TNO Institute.

Environmental Analysis of Contaminated Sites. Edited by G. I. Sunahara, A. Y. Renoux, C. Thellen, C. L. Gaudet and A. Pilon

The first study was performed at the end of the 1980s, and compares bioavailability of metals in soils decontaminated by applying different techniques on a laboratory scale. Comparison of bioavailability data for remediated soils was hampered by the lack of a database for background metal concentrations in test organisms exposed to non-polluted soils. Therefore a comparison was only possible for test organisms exposed to soils before and after remediation, using metal concentrations in the test organisms and bioaccumulation factors.

The second study determined 'normal' background concentrations for metals in the representative test organisms. The established background concentrations were then used to evaluate results of similar bioassay methods applied to other polluted soils remediated by commercially available remediation processes. Bioassay methods used in the first study were further improved in the second study to allow for routine application of the test methods on polluted and remediated soils.

20.2 MATERIALS AND METHODS

20.2.1 BIOASSAYS ON LABORATORY-SCALE REMEDIATED SOILS (STUDY #1)

20.2.1.1 Site Selection and Soil Sampling

Six sites were selected based upon:

1. Metal pollution considered problematic because of high toxicity or frequency of occurrence;
2. Ability to obtain a non-polluted control soil having more or less the same soil characteristics as the polluted site;
3. Sites representing different soil types;
4. Low levels of other pollutants present.

We describe results obtained on four of these selected soils. Table 20.1 provides the main characteristics of the polluted and corresponding control soils and the levels of the most important polluting metals. Soils I and VI are loamy sands, taken from a galvanic industry, soil III is a loamy (loess type) soil from a pigment factory (ceramic industries) and soil V is a humic sand from an old pesticide industry.

Soil samples were taken from the top layer to a depth of <30 cm, from randomly chosen locations on polluted and control sites. Samples were air dried for one week and homogenized by sieving (4 mm) and mixing the soil at least three times.

Soil remediation was performed in the laboratory. For each remediation experiment, a batch of 15–20 L of soil was processed by the following commonly used methods:

- base fractionation with diluted NaOH (pH 11) in an upstream column;
- three-step extraction with mild acid (HCl; pH 3);
- three-step extraction with strong acid (HCl; pH 1);

Table 20.1 Characteristics and metal concentrations of polluted and corresponding control soils used in the first study. Metals used for the evaluation of bioassay results are shown in bold (see Figures 20.1–20.3). Also included are background levels of metals calculated for the polluted and control soil applying the soil-type correction model developed in the Dutch soil protection policy (Vegter 1995)

	pH-KCl		% OC		% Clay		CEC (mmol kg^{-1})			Concentration (mg kg^{-1} dry soil)		Background level mg kg^{-1} dry soil	
Soil no.	Control	Polluted	Control	Polluted	Control	Polluted	Control	Polluted	Metal	Control	Polluted	Control	Polluted
I	3.4	6.9	1.0	1.8	3.6	3.8	23	64	Sn	26	880	[a]	[a]
									Cd	<4.0	17	0.49	0.46
									Cu	69	128	19.1	18.2
									Pb	455	560	56.9	55.3
III	7.3	6.1	2.1	2.3	12.9	16.7	98	147	**As**	<10	52	23.2	21.6
									Cd	<1.0	23	0.60	0.57
									Pb	112	1520	70.6	66.5
									Zn	260	1770	106	94.1
V	4.8	6.5	5.8	1.6	3.5	9.8	102	70	**Hg**	<0.008	159	0.22	0.24
VI	5.3	4.0	1.0	1.9	5.3	6.2	95	41	Cd	<1.0	33	0.51	0.47
									Cu	64	198	20.7	20.3
									Cr	21	4800	60.6	62.4
									Ni	22	770	15.3	16.2
									Pb	51	88	59.4	58.9

[a] No equation available to calculate background levels for Sn.

- extraction with nitrilo-triacetic acid (NTA) as a complexing agent (15 g H_3NTA kg^{-1} soil at pH 6);
- froth-flotation (treatment with foam in an upstream column).

Following remediation, soils were washed with water to remove any residual extracting agent. Original soil pH was restored after which soils were washed again to remove buffering salts.

20.2.1.2 Plant Bioassays

Cyperus esculentus (yellow nut sedge) was chosen for plant bioassays. This species reproduces asexually, and grows under dry and wet (flooded or wetland) soil conditions (Marquenie *et al.* 1998, Jenner *et al.* 1992). A plant nutrient solution was prepared by adding 10 ml of solutions A (68 g $Ca(NO_3)_2 \cdot 4H_2O$ and 62 g KNO_3 in 1 L groundwater) and B (46 g $MgSO_4 \cdot 7H_2O$ and 13.6 g KH_2PO_4 in 1 L groundwater) and 1 ml of solutions C (2.69 g H_3BO_3, 2.0 g $MnSO_4 \cdot H_2O$, 0.51 g $ZnSO_4 \cdot 7H_2O$, 0.13 g $Na_2MoO_4 \cdot 2H_2O$ and 0.08 g $CuSO_4 \cdot 5H_2O$ in 1 L groundwater) and D (500 ml D1 (16.66 g $C_{16}H_{14}O_8N_2Na_2 \cdot 2H_2O$ and 2.91 g KOH in 500 ml distilled water) and 200 ml D2 (12.44 g $FeSO_4 \cdot 7H_2O$ and 2 ml 0.5 M H_2SO_4 in 200 ml distilled water) in 1 L distilled water) to 1 L ground water.

Bioassays were performed according to Marquenie *et al.* (1988). Tests under upland conditions were carried out in Plexiglas cylinders (diameter 15 cm; height 30 cm) with a gauze bottom (mesh width 0.5 mm). Cylinders were filled with a 20-cm soil layer and placed in a dish containing 2–3 cm of the nutrient solution, allowing a moisture gradient to develop in the test soil. Before the start of the bioassays, soils were saturated with nutrient solution; this was considered necessary to support plant growth and to reduce the possible impact on the bioassay resulting from differences in the nutritional status of the test soils.

Since metal speciation and bioavailability may be dependent upon the oxygen status of soils, tests were also performed under wetland conditions. For these tests, Plexiglas cylinders were placed into buckets filled with nutrient solution to a level 1.5 cm above the soil surface. Soils were saturated with nutrient solution before they were flooded.

Water loss was compensated for by adding distilled water (three to five times per week). Only one replicate cylinder was used per test soil and moisture condition. Following pre-incubation, five shoots of *C. esculentus* were planted into each test soil.

Plants were housed in a climate room at 25±2 °C, 70% relative humidity and a light intensity of 6400 lux during 16 h per day. After 45 days, shoots were harvested and fresh weight determined. Shoots were frozen for further analysis.

20.2.1.3 Earthworm Bioassays

Eisenia fetida were obtained from our cultures (on a substrate of sphagnum peat, potting soil and horse manure); all animals were adult with a well-developed clitellum and were at the age of 6.5–15.5 weeks.

The remediated soils had a very sandy, coarse texture and were considered unsuitable for earthworms, therefore soils were amended with 2.5% sphagnum peat (2 mm sieved; the pH of the peat was adjusted to the pH of the test soil by the addition of $CaCO_3$). The bioassay method follows Marquenie *et al.* (1987). Plexiglas cylinders (diameter 15 cm; height 30 cm) were filled with a 20 - 24-cm layer of the test soils and placed on a Petri dish filled with demineralized water to allow for the development of a moisture gradient. Cylinders were covered with gauze (0.1 mm) and incubated in a climate-controlled room at 14 - 17 °C under continuous illumination (420 - 540 lux). After 7 days pre-incubation, 20±0.5 g biomass of earthworms (fresh weight), corresponding to 60 - 70 individuals, was added to each cylinder. After four weeks of exposure, cylinders were emptied; earthworms were recovered by hand sorting and weighed. Worms were placed on wet filter paper to allow emptying of their guts for 48 h, weighed again and frozen until further analysis. Half of the earthworms were used for determination of dry weight (after 16 h drying at 105 °C) and ash content (after ashing for 4 h at 600 °C).

One replicate test container was used per test soil. The fine fraction resulting from soil fractionation was not used for the earthworm bioassay, because the resulting muddy substrate was considered unsuitable for earthworms.

20.2.1.4 Chemical Analysis

Metal analysis in soil samples was performed according to Dutch standard procedures NEN 6465, 6464 or 6447. For most metals, soil was digested with HNO_3 and H_2SO_4. For As and Hg, $K_2S_2O_8$ was added to the digestion solution. For Zn analysis, digestion with HNO_3 and HCl was applied. Digests were analysed by flame atomic absorption spectrometry (AAS), in the case of Hg applying a hydride technique, or by graphite furnace AAS (As).

For plant metal analysis, dry plant material was digested with H_2SO_4 and H_2O_2. For Hg analysis, plant material was digested with HNO_3 in closed Teflon containers. Most metals were analysed by graphite furnace AAS; for As analysis, hydride AAS was applied, while Hg was measured using inductively coupled plasma - atomic emission spectrometry (ICP - AES). Zn was measured with flame AAS. Detection limits ranged between 0.005 mg kg^{-1} dry weight for Cd and 2.0 mg kg^{-1} dry weight for Zn. Quality control was performed by analysing NBS-SRM 1571 orchard leaves; results were in agreement with certified values.

Earthworms were lyophilized and digested in Teflon digestion tubes with HNO_3. For determination of Hg, digestion with HNO_3 was performed on wet earthworm samples. Zn was analysed by flame AAS, Hg by cold vapour AAS; all other metals were analysed by graphite furnace AAS. Standards and matrix modifiers were applied to cope with matrix effects. Detection limits ranged from 0.005 mg kg^{-1} dry weight for Hg to 5 mg kg^{-1} dry weight for Zn.

20.2.2 DEVELOPMENT OF A REFERENCE SYSTEM FOR THE EVALUATION OF BIOACCUMULATION BIOASSAYS (STUDY #2)

Based on earlier studies by Edelman and De Bruin (1986) and with support of the Institute of Forestry and Nature Research, 19 Dutch nature reserve areas were selected. Soil samples were obtained using a split corer. The litter layer was removed and the top 10-cm mineral soil layer was sampled. On each site, 20 samples were randomly taken from two to three plots. Samples were air dried, sieved (2.8 mm) and homogenized.

20.2.2.1 Plant Bioassays

Raphanus sativus cv Novired (radish) and *Lactuca sativa* cv Ravel RZ (lettuce) seeds were obtained from commercial seed producers.

A plant nutrient solution was prepared as a 1:1 mixture of dilutions (275 times) of stock solutions I and II. Stock solution I contains 86.4 g L^{-1} $Ca(NO_3)_2 \cdot 4H_2O$, 7.9 g L^{-1} KNO_3, 2 g L^{-1} NH_4NO_3 and 4.36 g L^{-1} 7% Fe-DTPA and stock solution II contains 60.44 g L^{-1} KNO_3, 17 g L^{-1} KH_2PO_4, 33.9 g L^{-1} $MgSO_4 \cdot 7H_2O$, 0.17 g L^{-1} $MnSO_4 \cdot H_2O$, 0.145 g L^{-1} $ZnSO_4 \cdot 7H_2O$, 0.24 g L^{-1} $Na_2B_4O_7 \cdot 10H_2O$, 0.019 g L^{-1} $CuSO_4 \cdot 5H_2O$ and 0.012 g L^{-1} $Na_2MoO_4 \cdot 2H_2O$. To allow for a comparison of plant and earthworm bioassays, this nutrient solution was also used in the bioassays with earthworms (see below).

To increase soil porosity and enhance seed germination, test soils were mixed with perlite, a stony, non-adsorbing substrate (80 ml perlite L^{-1} air dried soil). Test soils were brought to field capacity with nutrient solution and placed into Plexiglas cylinders (diameter 10 cm; height 10.5 cm); two cylinders were filled per test soil. After a 5-day pre-incubation period, 10 plant seeds were sown (at a depth of 0.5 cm below the soil surface) in each container. Moisture loss was corrected for by reweighing the containers daily and replenishing the weight loss with nutrient solution.

Cylinders with *L. sativa* were incubated in a climate-controlled room at $20 \pm 3\,^{\circ}C$, light/dark regime of 16/8 h and a light intensity of 9000 lux. For *R. sativus*, a higher light intensity was used (20 000 lux), a day temperature of $15 \pm 2\,^{\circ}C$ and a night temperature of $10 \pm 2\,^{\circ}C$.

After two weeks, the number of plants was reduced to five per cylinder by randomly removing additional plants. This procedure should minimize potential shading effects of developing plant leaves.

Bioassays were terminated after four weeks. Radish plants were pulled out of the soil, washed with demineralized water and shoots were separated from the roots using titanium scissors. Lettuce shoots were harvested by cutting them just above the soil surface, using the same pair of titanium scissors. Shoots and roots were weighed, frozen in 10% HNO_3 washed plastic bottles and lyophilized. Plant material was finely ground in a Teflon mill under liquid nitrogen. Bioassays with plants were run twice, each time with two replicate containers per soil type and plant species.

20.2.2.2 Earthworm Bioassays

To obtain earthworms (*E. fetida*) with reduced metal levels, animals were cultured on a metal-poor substrate of perlite, sphagnum peat, lettuce and potatoes.

For earthworm bioassays, soils were mixed with perlite (0.24 L in 3 L air dried soil) to increase soil porosity. The resulting substrate was moistened with the nutrient solution used for the plant bioassays (see above) and placed into Plexiglas cylinders with a gauze bottom (diameter 15 cm; height 30 cm) to obtain a 20-cm soil layer. Dried cow dung, 70 g, mixed with 70 ml demineralized water was placed on top of the soil as a food source for the earthworms; to avoid excessive fungal growth, the cow dung was covered with a 1 - 2-cm layer of soil.

Cylinders were placed into a Petri dish containing 2 - 3 cm of nutrient solution and incubated in a climate-controlled room at $20 \pm 2\,^{\circ}C$ with continuous illumination (400 lux). After one week pre-incubation, approximately 70 earthworms were introduced into each cylinder; 25% of these earthworms were adults, the remaining were juveniles. Cylinders were covered with cheesecloth to prevent earthworms from escaping. During the test, a constant level of nutrient solution was maintained in the Petri dishes.

The bioassay was terminated after four weeks by emptying the cylinders and recovering the surviving earthworms. Earthworms were weighed, kept for 48 h on wet filter paper, reweighed, frozen, lyophilized and separated into two portions. One portion was used for chemical analysis, the other to determine dry weight. For metal analysis, lyophilized portions were finely ground (under liquid nitrogen) in a Teflon grinding mill. As a control, earthworms taken from the culture were analysed as well.

20.2.2.3 Chemical Analysis

Soil, plant and earthworm tissues were digested in closed Teflon containers in concentrated HNO_3/HCl using a microwave. Concentrations of Al, Ba, Ca, Co, Fe, Mg, Mo, Na, Sn and Zn were determined by ICP, K by flame AAS, As, Cd, Cr, Cu, Ni, Pb, Sb, Se and V by graphite furnace AAS and Hg by cold vapour AAS. Detection limits for analysis of the biological tissues were dependent on the amount of material available for the analysis. BCR 143 'Sewage sludge amended soil 'was used as a reference material. Results were generally in agreement with certified values (deviation less than 10% for Cd, Cu, Pb, Zn and Hg).

20.2.3 APPLICATION OF THE REFERENCE SYSTEM TO INDUSTRIAL-SCALE REMEDIATED SOILS

Metal-polluted soil (Budel) (Table 20.5), from a nearby zinc smelter, was subjected to commercially available remediation techniques and used for the bioassays in this study. Remediation techniques used were the flotation method

(Budel-1 and Budel-2 soils) and an extractive clean-up method (Budel-3 and Budel-4 soils). The polluted soil Budel-1 was taken from the input of the flotation factory; since the factory was not functioning at the time of sampling, remediated soil had to be taken from the available stock of remediated soil. The polluted and remediated soils Budel-2, 3 and 4 were sampled on the inlet and outlet, respectively of the cleaning factory. All soil samples were air dried and sieved (2.8 mm). Part of the remediated soils was amended with compost to compensate for the loss of organic matter during the clean-up process (4.30, 4.68, 4.21 and 6.09% of dry weight, respectively for the soils Budel-1 to Budel-4). The compost contained 15 mg As kg^{-1}, 1 mg Cd kg^{-1}, 40 mg Cu kg^{-1}, 75 mg Pb kg^{-1} and 200 mg Zn kg^{-1} dry soil. Considering metal levels in the compost, these amounts did not significantly change the metal levels in the remediated soils. Amended soils were then incubated in closed containers outdoors for 6 months to allow for equilibration and ageing before starting the bioassays.

Bioassays on polluted, remediated and aged remediated soils were performed following the same procedures as described above for the non-polluted reference soils. Different containers were used in plant bioassays: 250-ml plastic boxes instead of the larger Plexiglas cylinders. Metal analysis of soils and biological tissues followed the same procedures as described above, but was restricted to As, Cd, Cu, Pb and Zn.

20.3 RESULTS

20.3.1 BIOASSAYS ON LABORATORY-SCALE REMEDIATED SOILS (STUDY #1)

Control soils were not always as clean as expected and the physico-chemical characteristics (pH, organic matter, clay content, soil texture) of these soils differed from those of the corresponding polluted soils (Table 20.1). During remediation, fine soil particles (clay, organic matter) were removed. As a consequence, the physical characteristics of the remediated soils differed from the control and polluted soils, even after amending with sphagnum peat for the purpose of the earthworm bioassays.

Plant growth on the acid (pH 1) extracted soil III was strongly inhibited. On the polluted and NTA or pH 3 decontaminated soils VI, most earthworms died and plant growth was strongly inhibited. On the pH 3 remediated soil VI, all surviving earthworms died after incubation on wet filter paper. Only fractionation appeared to improve the conditions for earthworm survival in soil VI.

Figures 20.1 - 20.3 show the main results of the bioassays with earthworms and plants on the polluted and decontaminated soils. In The Netherlands, a soil-type correction equation is used to calculate background concentrations for individual metals in soils with differing physico-chemical properties. These equations have been developed on the basis of background concentrations found by Edelman and De Bruin (1986) in non-polluted soils from nature reserve areas. In these equations, metal concentrations are related to organic matter and/or clay content of the soil (Vegter 1995). Applying these equations to the current soils tested, background levels were calculated for the polluted soils and are given in Table 20.1 and indicated by horizontal lines in Figures 20.1-20.3. Background levels for metals in earthworms are taken from the second study described in this chapter (Table 20.4). Also these values are indicated in Figures 20.1-20.3. In some cases, the background levels are much lower than

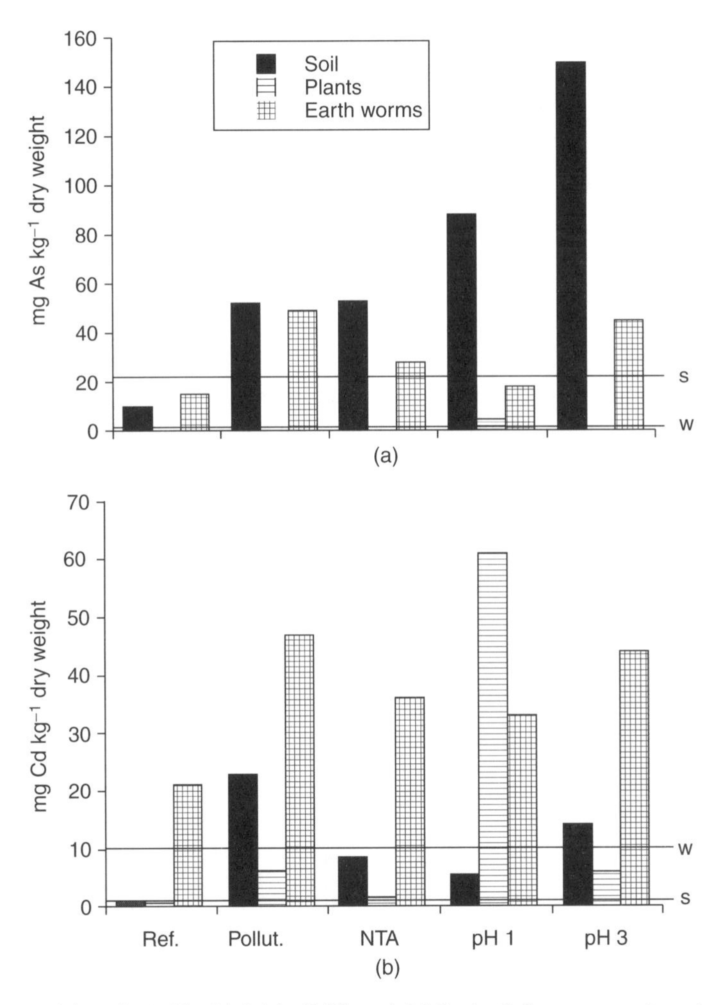

Figure 20.1 (a) As (soil III), (b) Cd (soil III) and (c) Cu (soil I) concentrations in control (reference), polluted and remediated soils and in earthworms (*E. fetida*) and plants (*C. esculentus*) kept on these soils for four and six weeks, respectively. See text for further explanation of remediation techniques depicted on the X-axis. Horizontal lines indicate background metal levels in soils (S) and earthworms (W), taken from Tables 20.1 and 20.3, respectively.

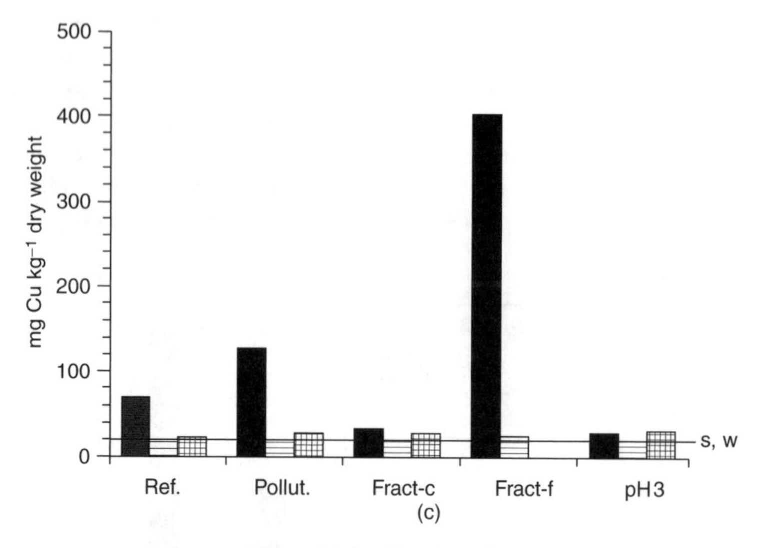

Figure 20.1 (*Continued*).

the metal levels found; in that case the background levels are given in the legend rather than by inserting a horizontal line in Figures 20.1–20.3.

From Table 20.1 and Figures 20.1–20.3, it is apparent that metal levels in the control soils often exceed the background level. In general, remediation

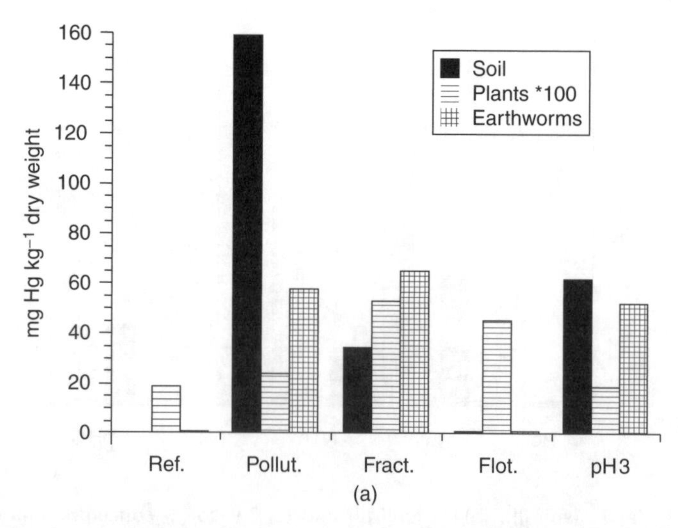

Figure 20.2 (a) Hg (soil V), (b) Pb (soil III) and (c) Zn (soil III) concentrations in control (reference), polluted and remediated soils and in earthworms (*E. fetida*) and plants (*C. esculentus*) kept on these soils for four and six weeks, respectively. See Figure 20.1 for further explanation. For Hg background levels are much lower than the values presented in the figures: 0.20–0.24 and 0.1 mg kg^{-1} dry weight for soil and earthworms, respectively.

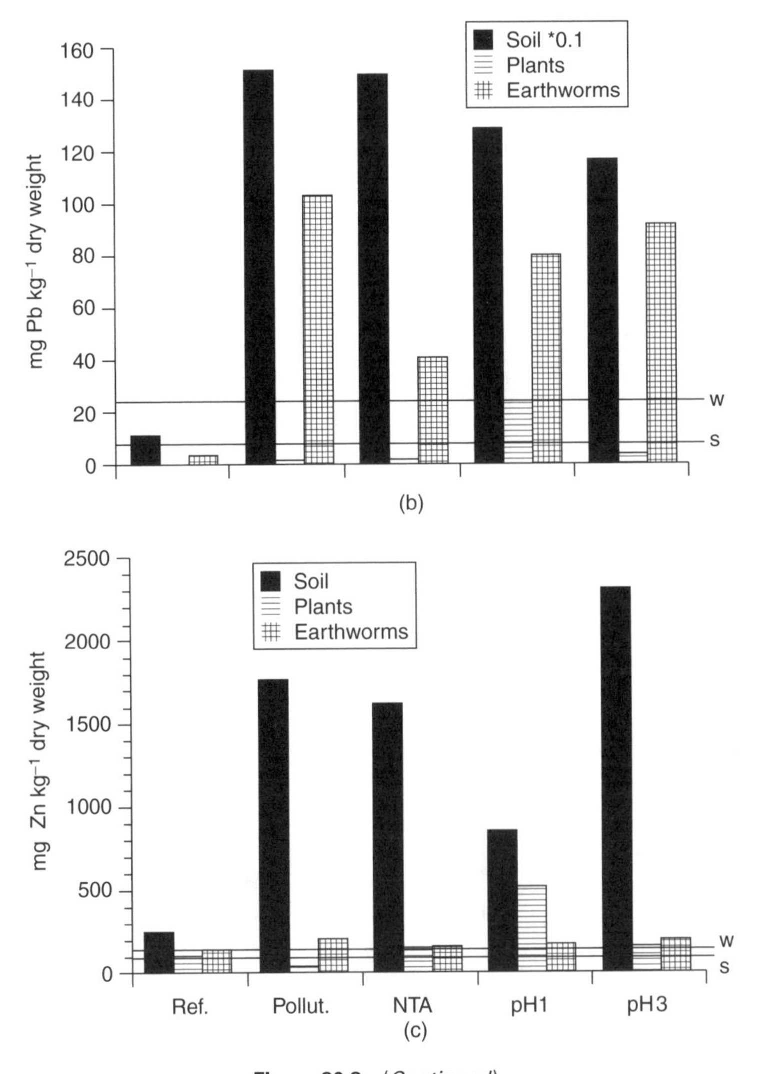

Figure 20.2 (*Continued*).

techniques applied resulted in a reduction of soil metal levels by 50–90%. Froth-flotation appeared to be very efficient in the removal of Hg from soil V: 99.4%. Nevertheless, remediation techniques were not sufficient to reduce metal concentrations to background levels. However, for Cu concentrations in the soils remediated by fractionation (I and VI) and extraction with NTA (VI) were below background levels. In some cases, removal efficiency was very low: Zn and Pb in soil III (0–52%). Concentrations of As in soil III appeared to increase upon decontamination. Apparently, As is mainly bound by or incorporated in the coarse soil particles. As a consequence, removal of fine soil particles upon

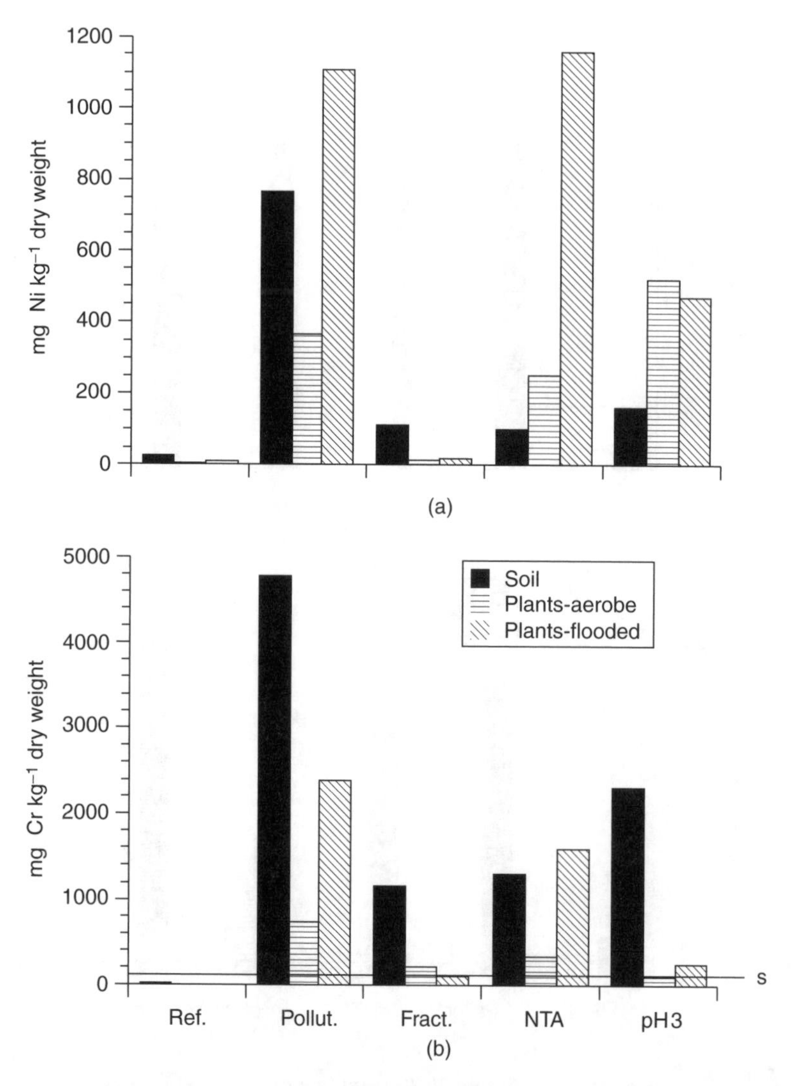

Figure 20.3 (a) Ni and (b) Cr concentrations in control (reference), polluted and remediated soils VI and in plants (*C. esculentus*) kept on these soils for six weeks under upland (aerobe) or flooded conditions, respectively. See Figure 20.1 for further explanation. The background level for Ni in soil is 15–16 mg kg^{-1} dry weight.

soil remediation leads to an increased concentration in the remaining (coarse) soil. Fractionation leads to an accumulation of metals in the fine fraction (fract-f) while the remaining remediated coarse fraction (fract-c) contained much lower metal concentrations, as shown for Cu in Figure 20.1.

Concentrations of most metals in earthworm and plant tissues were higher when exposed to polluted soils compared to the control soils

(Figures 20.1–20.3). In many cases, metal concentrations in both or one of the test organisms exposed to remediated soils remained the same as exposed to the polluted soils. In a few cases, metal levels decreased in test organisms exposed to remediated soils. Metal concentrations in plants exposed to upland and wetland conditions were generally in agreement with each other. Plants from wetland conditions contained higher levels of Cr and Ni in polluted and NTA extracted soil VI (Figure 20.3). Poor plant growth on these soils may have influenced these results.

When expressed as biota-to-soil accumulation factors (BSAF) (Table 20.2), metal uptake in both plants and earthworms was often higher on the remediated soils, suggesting an increased bioavailability. Cu and Zn levels in plants and earthworms showed less variation (Figures 20.1 and 20.2) and did not seem to be affected by soil metal levels. These metals are essential elements and may be physiologically regulated by organisms. *E. fetida* and *C. esculentus* exposed to polluted and remediated soils were able to regulate the levels of these essential metals in their tissues.

20.3.2 DEVELOPMENT OF A REFERENCE SYSTEM FOR THE EVALUATION OF BIOACCUMULATION BIOASSAYS (STUDY #2)

Table 20.3 shows the characteristics of the 19 soils taken from non-polluted nature reserve areas. The soils selected cover a large range of pH, organic carbon and CEC values. Similarly, large variations are observed in elemental concentrations. Mo, Sb and Se were not detected in these reference soils and are therefore not included in Table 20.3. Detection limits for these metals were 4.0, 0.6 and 0.6 mg kg^{-1} dry soil, respectively.

Earthworms survived quite well on all these natural, non-polluted soils, except one soil, where a 10% mortality was recorded, maybe due to the high clay content of this soil. Earthworm biomass increased in all test soils (44.9 to 84.8%) and was not correlated with soil characteristics. Dry weight was typically 14–17% of fresh weight in all soils.

Elemental concentrations measured in the earthworms are summarized (mean$\pm$standard deviation) in Table 20.4. Fe and Al were not measured in the test organisms, as concentrations in earthworms were around or below the detection limit of 0.52–0.74 mg kg^{-1} dry weight. For vanadium, a concentration of 2.0 mg kg^{-1} dry weight was measured in earthworms from only one soil; on all other soils, levels in the earthworms were below the detection limit of 1.8–2.5 mg kg^{-1} dry weight. Sb and Mo were not detected in the earthworms (detection limits 0.52–0.74 and 3.7–5.0 mg kg^{-1} dry weight, respectively). Cr, Ni, Se and Sn were only detected in earthworms recovered from 9, 12, 8 and 7 out of the 19 test soils, respectively and in very low concentrations (Table 20.4). Elemental concentrations in the earthworms taken from cultures were lower than in earthworms exposed to the 19 reference soils. Except for K and Mg, whose levels decreased rather than increased upon exposure to the reference soils.

Table 20.2 BSAF values for the bioaccumulation of metals by earthworms (*E. fetida*) and plants (*C. esculentus*) after four weeks exposure to polluted and remediated soils in study #1. Fract. = fractionation with NaOH, only data for tests on the coarse fraction are given. Flot. = flotation technique

		BSAF in *E. fetida*						BSAF in *C. esculentus*					
Soil no.	Metal	Polluted	Fract	pH 1	pH 3	NTA	Flot	Polluted	Fract	pH 1	pH 3	NTA	Flot
I	Cu	0.22	0.71	n.d.	0.76	n.d.	n.d.	0.093	0.41	n.d.	0.41	n.d.	n.d.
	Cd	0.54	14	n.d.	3.0	n.d.	n.d.	3.3	6.1	n.d.	2.6	n.d.	n.d.
	Sn	0.020	0.021	n.d.	0.025	n.d.	n.d.	<	<	n.d.	<	n.d.	n.d.
III	Pb	0.068	n.d.	0.062	0.079	0.027	n.d.	0.0010	n.d.	0.018	0.0032	0.0012	n.d.
	Zn	0.12	n.d.	0.20	0.087	0.10	n.d.	0.077	n.d.	0.61	0.070	0.097	n.d.
	Cd	2.0	n.d.	6.0	3.1	4.2	n.d.	0.27	n.d.	11.1	0.44	0.19	n.d.
	As	0.94	n.d.	0.20	0.30	0.53	n.d.	0.018	n.d.	0.056	0.0049	0.0055	n.d.
V	Hg	0.36	1.86	n.d.	0.84	n.d.	0.74	0.0015	0.015	n.d.	0.0031	n.d.	0.046
VI	Cu	—	0.61	n.d.	—	—	n.d.	0.036	0.16	n.d.	0.11	0.50	n.d.
	Cd	—	3.9	n.d.	—	—	n.d.	0.58	1.2	n.d.	3.1	2.7	n.d.
	Cr	—	0.076	n.d.	—	—	n.d.	0.15	0.18	n.d.	0.056	0.25	n.d.
	Ni	—	0.070	n.d.	—	—	n.d.	0.48	0.12	n.d.	3.2	2.5	n.d.

n.d. not determined
— no data because of large mortality
< below detection limit

Table 20.3 Ranges of soil characteristics and metals concentrations in 19 non-polluted reference soils sampled in Dutch nature reserve areas. Mean and corresponding standard deviation and minimum and maximum values are given

Parameter	Mean	Std.dev.	Minimum	Maximum
pH-KCl	3.4	0.8	2.5	6.3
pH-H_2O	4.3	0.9	3.6	6.8
% OC	8.4	8.8	2.0	33.1
% Clay	10.4	14.4	0.7	47.0
CEC (mmol kg^{-1})	223	200	34	780
Al (mg kg^{-1} dry soil)	7724	9629	369	30 498
As	11.3	12.0	0.9	40.9
Ba	52.4	76.8	3.6	243
Ca	2086	2344	90	7437
Cd	0.35	0.36	<0.02	1.30
Co	3.7	4.5	<0.2	13.3
Cr	13.5	14.9	1.0	47.0
Cu	9.4	9.3	0.4	31.0
Fe	11 268	12 020	474	33 832
Hg	0.20	0.09	0.06	0.38
K	772	798	148	2411
Mg	1447	2314	60	7093
Na	350	321	59	950
Ni	10.7	14.6	0.3	43.0
Pb	42.0	27.6	<1.0	94.7
Sn	6.4	9.1	0.6	28.1
V	18.3	17.4	2.3	53.8
Zn	5.2	38.8	0.6	127

Based on the concentrations found in the earthworms after four weeks exposure to 19 selected non-polluted soils, representative background values for elemental concentrations in *E. fetida* are proposed in Table 20.4. These values are the maximum levels encountered in the bioassays on the non-polluted reference soils.

Germination and growth of lettuce appeared to be strongly affected by low soil pH, where germination and growth were poor or absent. At pH<3.5 growth was strongly reduced. The plant tissue obtained showed large fluctuations in dry weight, ranging from 6 to 22% of fresh weight. When plotted on a log–log scale, dry-to-fresh weight ratios showed a negative correlation with plant growth (Figure 20.4; $r^2 = 0.830$; $n = 33$). Compared to literature data, elemental concentrations in lettuce were high (see Van Gestel *et al.* 1992 for a review). Since elemental concentrations in plants usually show a strong correlation with dry weight, this may be explained from the poor growth. Given this fact, and since many elements could not be analysed due to the lack of sufficient biomass, it was decided not to present metal levels found in lettuce in this paper.

Growth of radish was also affected by soil pH: on soils with pH-KCl<3.0 growth was reduced. On soils with pH 3.5, more than 0.5 g biomass per plant (roots+shoots) was obtained. Similarly to lettuce, the dry-to-fresh weight ratio

Table 20.4 Mean (mg kg^{-1} dry matter±standard deviation) metal levels in earthworms (*E. fetida*) and plants (*R. sativus*; roots and shoots) exposed to 19 non-polluted soils taken from Dutch nature reserve areas and proposed 'normal' background values (reference value) in these organisms to be used for the evaluation of bioassays on polluted and remediated soils

Metal	Earthworms (mean±std. dev)	Proposed ref. value	radish roots (mean±std. dev)	Proposed ref. value	radish shoots (mean±std. dev.)	Proposed ref. value
As	0.65[a]	1.0	–±–	0.5[b]	–±–	0.55
Ba	3.4±3.3	12	30.6±20.2	70	37.4±19.6	80
Ca	4792±372	6000	3312±879	5100	15 307±12 153	41 000
Cd	3.2±1.9	10	0.86±0.62	2.0	1.7±1.0	4.0
Co	3.3±2.0	8.0	2.0±0.9	4.0	2.0±1.5	6.0
Cr	0.9±1.0	3.5	1.3±0.4	1.8	– ± –	0.6
Cu	10.1±2.8	20	10.6±12.3	50	5.4±2.7	10
Hg	0.09±0.06	0.1	0.3±0.4	1.3	0.07±0.09	0.4
K	12 591±587	14 000	40 219±10 556	65 000	37 006±10 731	57 000
Mg	934±85	1500	1319±226	1750	3241±1028	5600
Mo	– ± –	4.0[b]	– ± –	3.0[b]	3.6±1.8	4.2[b]
Na	5202±392	6000	4492±7478	35 000	4784±2621	11 000
Ni	1.1±1.0	3.0	2.6±1.6	5.0	1.8±1.1	4.0
Pb	6.2±6.5	25	10.7±11.7	37	1.6±1.2	5.0
Sb	– ± –	0.5[b]	– ± –	0.5[b]	– ± –	0.5[b]
Se	0.8±0.3	1.5	–±–	0.06[b]	– ± –	0.5[b]
Sn	1.9±1.0	2.5	– ± –	0.5[b]	– ± –	0.5[b]
V	– ± –	2.0[b]	– ± –	2.0[b]	– ± –	2.0[b]
Zn	98±4.5	150	100±65	210	216±119	450

[a]Only one value available.
[b]Detection limit.

for radish roots increased with decreasing growth rate (Figure 20.4; $r^2 = 0.724$; $n = 19$). Dry weight fluctuated between 9 and 15% of the fresh weight for radish shoots and between 7 and 16% for radish roots.

Elemental concentrations in radish roots and shoots are summarized in Table 20.4. Values in this table are the mean of individual results ($N = 2$) per soil type. As, Mo, Sb, Se, Sn and V were not detected in radish roots; detection limits for these elements were 0.47–6.8, 3.2–46, 0.47–6.8, 0.5–6.8, 0.47–6.8 and 1.6–23 mg kg^{-1} dry weight, respectively. The large variation in detection limits is caused by the large difference in the amounts of plant biomass available for the analysis. Cr and Co were only detected in radish root samples from 3 and 7 out of the 19 test soils (detection limits 0.16–0.94 and 0.95–11 mg kg^{-1} dry weight, respectively). In radish shoots, As, Cr, Sb, Se, Sn and V were not detectable at detection limits of 0.54–2.9, 0.18–0.97, 0.54–2.9, 0.50–2.9, 0.54–2.9 and 1.8–9.7 mg kg^{-1} dry weight, respectively. Mo and Co were only detected in radish shoot samples grown on 4 and 9 out of the 19 test soils (detection limits 3.6–19 and 0.90–4.8 mg kg^{-1} dry weight, respectively).

Based on the concentrations found in the plant tissues after growth on the 19 selected reference soils, representative background values for elemental concentrations in *R. sativus* roots and shoots used for the bioassays with

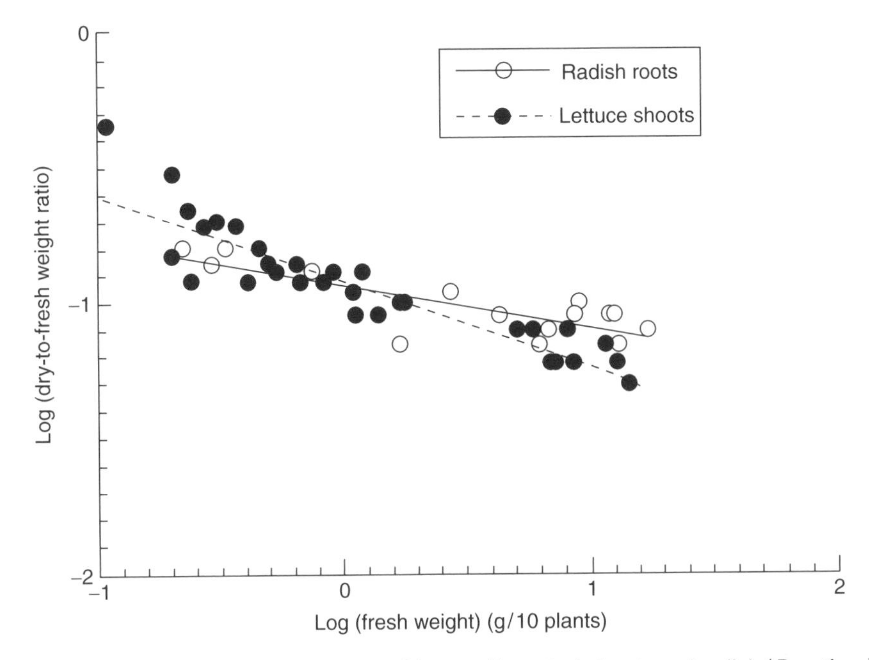

Figure 20.4 Dry-to-fresh weight ratios of lettuce (*L. sativa*) shoots and radish (*R. sativus*) roots as a function of fresh weight after four weeks growth on non-polluted soils from Dutch nature reserve areas (log values).

polluted or remediated soils are proposed in Table 20.4. These values are the maximum concentrations that should be encountered in the bioassays with the non-polluted reference soils.

20.3.3 APPLICATION OF THE REFERENCE SYSTEM TO INDUSTRIAL-SCALE DECONTAMINATED SOILS

Table 20.5 shows the main characteristics and metal concentrations of the four selected polluted and corresponding remediated soils. From this table, it is obvious that the soils contained increased levels of Cd and Zn. Remediation led to a strong decrease of clay and organic matter contents. Except for Budel-3, pH values of the remediated soils were higher than those of the polluted soils. Compared to the remediated soils, pH values of the aged soils were very high.

Remediation was most efficient for Cd (69–80% removal), followed by Zn (66–74%), Pb (57–66%), Cu (54–72%) and As (56–62%). Cu removal from soil Budel-1 was, however, negligible. Metal levels in the remediated soils exceeding the background level are shown in bold. The corresponding background levels, also given in Table 20.5, are calculated on the basis of soil characteristics applying the soil-type correction models developed in the Dutch soil protection policy (Vegter 1995) (see above). From Table 20.5 it is obvious that Cd and Zn

Table 20.5 Main characteristics and metal concentrations in four polluted soils taken from the Budel area before and after remediation by a commercial soil remediation company. Metal concentrations in bold exceed the background level calculated on the basis of soil characteristics applying the soil-type correction model developed in the Dutch soil protection policy (Vegter 1995)

				Concentration(mg kg^{-1} dry soil)				
Soil	pH-KCl	% OC	% Clay	As	Cd	Cu	Pb	Zn
Polluted soils								
Budel-1	5.7	2.6	1.24	8.7	4.5	25	115	680
Budel-2	6.1	2.2	1.57	12.1	5.0	37	149	787
Budel-3	5.7	4.5	1.67	8.6	4.8	24	127	690
Budel-4	6.1	4.5	1.93	11.7	6.4	50	170	1005
Remediated soils								
Budel-1	6.9	0.45	0.31	3.3	**1.4**	**24**	50	**189**
Budel-2	6.5	0.82	0.19	5.3	**1.5**	**17**	**64**	**266**
Budel-3	5.2	0.61	0.19	3.6	**1.1**	8	47	**177**
Budel-4	6.6	0.82	0.36	5.1	**1.3**	14	**57**	**300**
Background levels				17–19	0.5–0.6	18–21	55–60	60–70

concentrations exceed these background levels in all soils, whereas for Cu and Pb exceedance occurs in two out of the four soils.

Unexpectedly, earthworms showed high mortality on the aged remediated soil Budel-3. In all other soils, earthworm mortality was less than 10%. It should be noted that compared to all other soils, conductivity measured in pore water extracted from remediated and aged Budel-3 soils was much higher: 11 800 - 14 000 $\mu S\ cm^{-1}$ in these two soils compared with 2100 - 6700 $\mu S\ cm^{-1}$ on all other soils. This might have affected earthworm survival. Earthworm wet weights increased by 4.9 - 17% on polluted soils, 7.4 - 15% on remediated soils and 22 - 53% on aged remediated soils. In fact, the highest individual weight increase was observed on the aged remediated soil Budel-3, suggesting that the few surviving earthworms had a high weight gain. Dry weight was 15 - 17% of fresh weight on the polluted and remediated soils, but was lower on the aged remediated soils (12 - 13%).

Radish did not grow on remediated Budel-3 soil, which may be attributed to the reduced soil pH. On the corresponding aged remediated Budel-3 soil, growth was somewhat better, although roots remained fairly small. In general, plant growth was best on the aged remediated soils, with on average 1.0 (Budel-3) to 3.17 g fresh biomass per plant (roots+shoots) after four weeks. On the polluted and remediated soils, biomass production was between 0.37 and 1.29 g fresh weight per plant. Dry weight to fresh weight ratios of shoots were higher on remediated soils (12 - 13% of fresh weight) compared to polluted (8 - 11%) and aged remediated soils (8 - 9%). Root dry weights did not show a difference between soils and amounted to 8 - 12% of fresh weight. As indicated above (Figure 20.4), high dry-to-fresh weight ratios may be indicative of poor plant growth conditions.

Lettuce did not grow on both the remediated and aged remediated Budel-3 soils. On all other remediated soils, growth was much better than on the polluted soils. Dry weights of the lettuce shoots were fairly high on the polluted soils (5–20% of fresh weight), which is indicative of poor growth conditions. On the remediated soils, dry weights were much lower: 7–9% of fresh weight.

Concentrations of Cd and Zn are given in Figures 20.5 and 20.6 for earthworms, radish roots and shoots exposed to polluted, remediated and aged

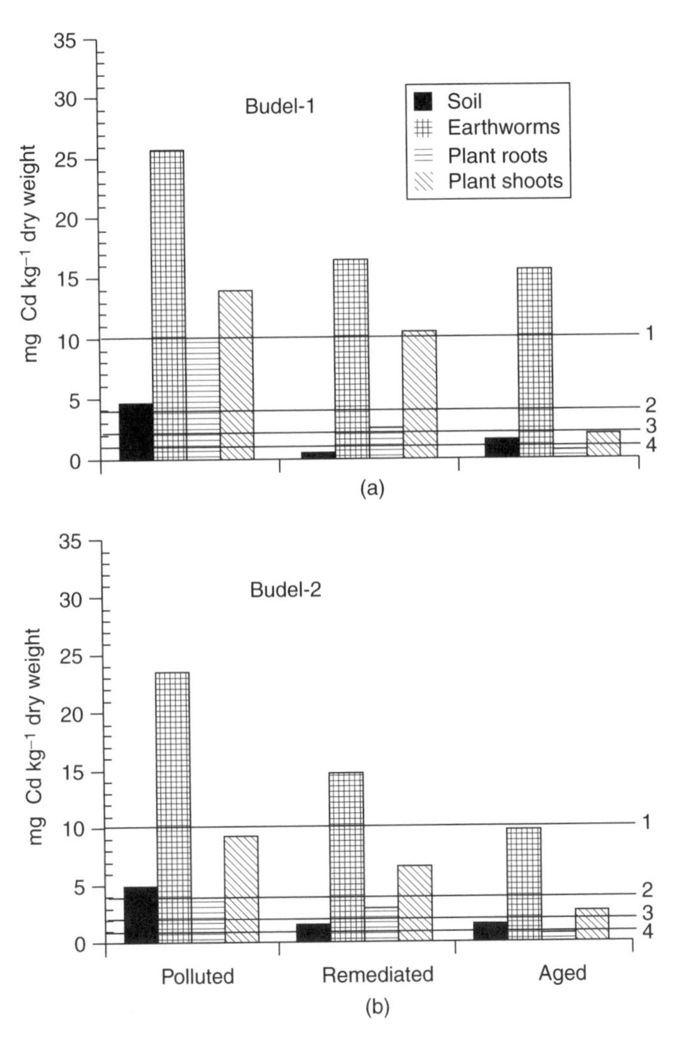

Figure 20.5 Cd concentrations in polluted, remediated and aged remediated Budel soils and in plants (*R. sativus*; roots and shoots) and earthworms (*E. fetida*) kept on these soils for four weeks. Horizontal lines indicate background levels of metals in earthworms (1), plant shoots (2), plants roots (3) and soil (4), taken from Tables 20.4 and 20.5.

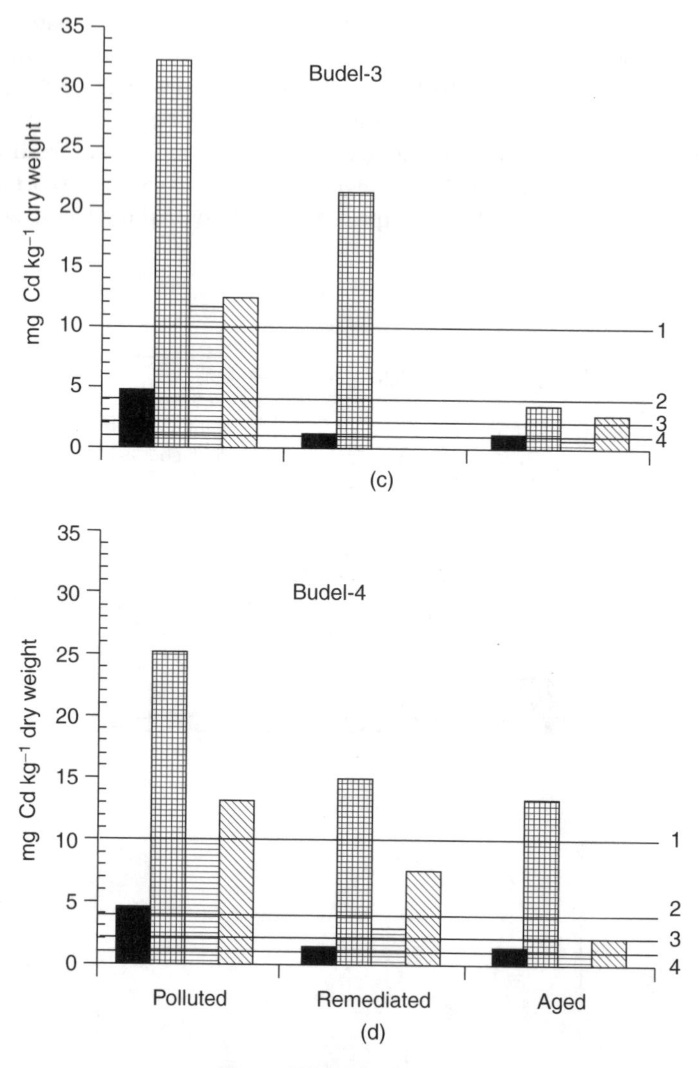

Figure 20.5 (*Continued*).

remediated soils along with metal concentrations in soil. Data for lettuce is not included in these figures, since lettuce did not grow on the aged remediated soils. The metal levels in lettuce shoots on the other soils were similar to those in the radish shoots.

From Figure 20.5, it can be concluded that Cd levels in earthworm and plant tissues were lower when exposed to the remediated soils than to the polluted soils. Ageing of the remediated soils with compost further reduced bioavailability. Cd levels in earthworms exposed to all soils exceeded the established background value of 10 mg kg^{-1} dry weight estimated from the results of bioassays with non-polluted soils (Table 20.4). For plant roots and shoots, the 'normal' background values derived from bioassays on non-polluted reference soils (Table 20.4) were exceeded on the polluted and remediated soils, but not for the aged remediated soils. For Zn (Figure 20.6) similar conclusions can be drawn. In the earthworms, Zn seems to be regulated on the polluted soils. Zn levels in the earthworms were still somewhat increased (156–228 mg kg^{-1} dry weight), but for worms exposed to the remediated soils the Zn levels were just above the background value of 150 mg kg^{-1} derived from the results of bioassays with non-polluted reference soils. They were well below this value in

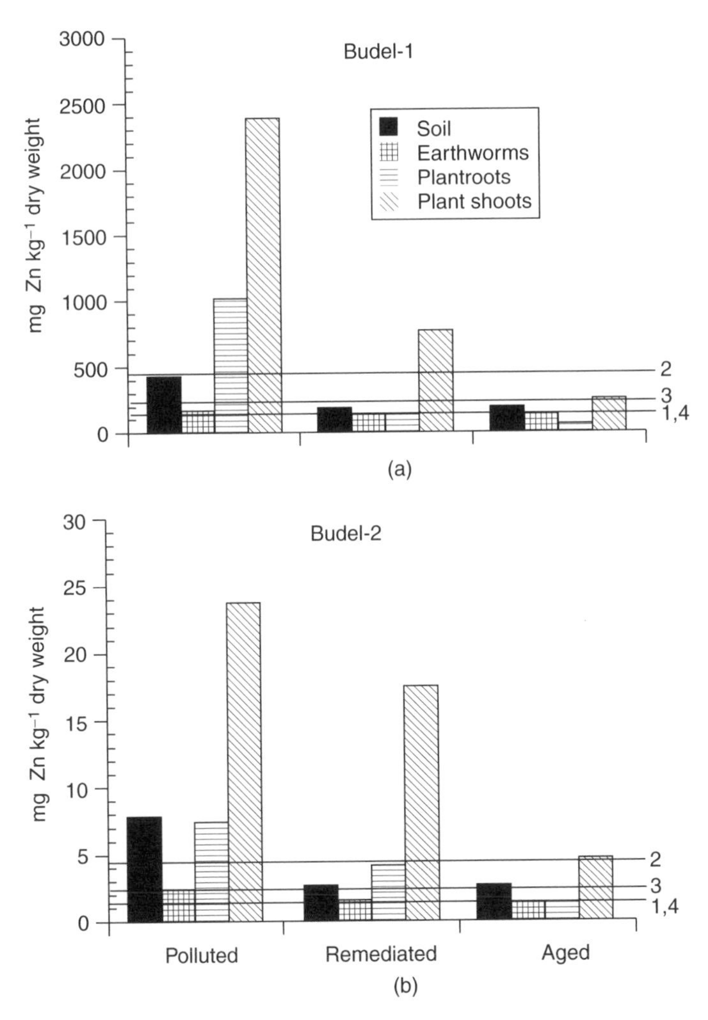

Figure 20.6 Zn concentrations in polluted, remediated and aged remediated Budel soils and in plants (*R. sativus*; roots and shoots) and earthworms (*E. fetida*) kept on these soils for four weeks. See Figure 20.5 for further explanation.

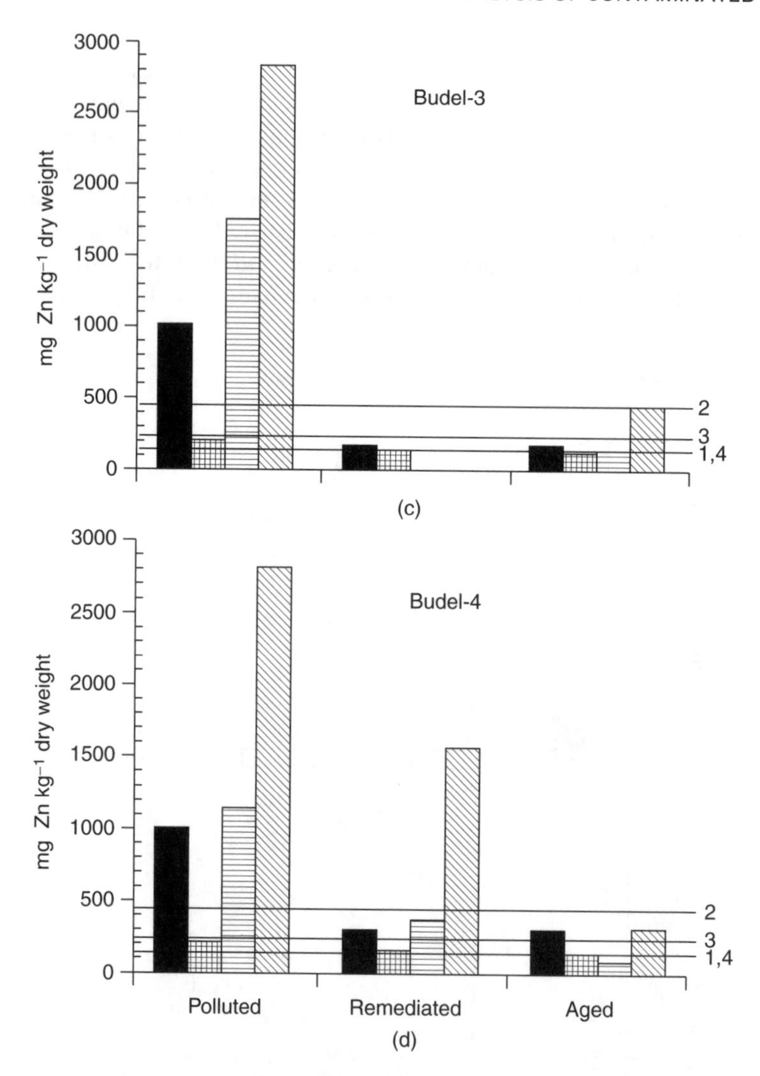

Figure 20.6 (*Continued*).

earthworms exposed to the aged remediated soils. For Pb and Cu similar results were obtained (not shown).

When expressing the uptake of Cd, Cu and Zn in the test organisms as BSAF values (Table 20.6), for all organisms (earthworms, radish roots and shoots, lettuce shoots) these BSAFs are higher on the remediated soils than on the polluted soils, suggesting an increased bioavailability on the remediated soils. BSAFs on the aged remediated soils are lower when compared to remediated soils, but higher than for the corresponding polluted soils. Only for Pb do BSAF values remain similar or decrease on remediated and aged soils.

Table 20.6 BSAF values for the bioaccumulation of metals by earthworms (*E. fetida*) and plants (*R. sativus*) after four weeks exposure to polluted, remediated and aged remediated soils in study #2

		BSAF values											
		E. fetida				*R. sativus* roots				*R. sativus* shoots			
Soil no.	Treatment	Cd	Cu	Pb	Zn	Cd	Cu	Pb	Zn	Cd	Cu	Pb	Zn
Budel-1	polluted	5.7	0.90	0.057	0.23	3.1	0.59	0.038	3.5	2.2	0.60	0.17	1.5
	remediated	12	0.93	0.030	0.72	7.5	0.53	0.026	4.1	1.8	0.34	0.056	0.77
	aged	11	0.77	0.036	0.70	1.4	0.017	0.032	1.3	0.36	0.18	0.028	0.34
Budel-2	polluted	4.7	0.46	0.065	0.29	1.8	0.37	0.026	3.0	0.76	0.52	0.11	0.93
	remediated	9.8	1.2	0.064	0.58	4.4	1.1	0.030	6.6	1.7	0.11	0.097	1.6
	aged	6.5	0.72	0.023	0.52	1.8	0.082	0.017	1.8	0.33	0.18	0.023	0.34
Budel-3	polluted	6.7	0.88	0.094	0.30	2.6	0.25	0.035	4.1	2.4	3.0	0.38	2.6
	remediated	19	2.0	0.085	0.86	–	–	–	–	–	–	–	–
	aged	3.2	0.50	0.053	0.72	2.5	0.78	0.017	2.6	0.55	0.69	0.094	0.63
Budel-4	polluted	3.9	0.42	0.074	0.20	2.1	0.35	0.046	2.8	1.4	0.59	0.15	1.1
	remediated	11	1.6	0.18	0.55	5.8	1.0	0.037	5.2	2.2	0.41	0.091	1.3
	aged	10	1.0	0.040	0.48	1.5	0.014	0.021	1.1	0.31	0.18	0.054	0.26

– No data available because plants did not grow.

20.4 DISCUSSION

20.4.1 INFLUENCE OF SOIL REMEDIATION ON METAL BIOAVAILABILITY

Results of both experiments described in this chapter clearly demonstrate that remediation of metal-polluted soils does not necessarily lead to reduced metal bioavailability. BSAF values for metal levels in earthworms and plants were usually higher on the remediated soils than on the corresponding polluted soils (Tables 20.2 and 20.6). This can be explained by the fact that extractive remediation of the soils mainly led to a removal of the fine soil particles. In addition, extracting agents (NTA, acid, soap) may have increased solubility of the metals. Consequently, metal was solubilized and the solubilized metals were not completely removed upon extraction or washing. Also, free metal species were not bound to sorption sites (since clay and organic matter have been washed away). This hypothesis was confirmed by chemical analysis and speciation studies on the soils investigated in the first study (data not shown). It also appeared that part of the metal residues was still available in the remediated soils. The lack of ageing may have contributed to the increased bioavailability of metals in remediated soils. In the first study, soils were subjected to the bioassays shortly after remediation and as a consequence equilibration may not have been achieved when the bioassays started.

To reduce the bioavailability of metals in remediated soils, the loss of organic matter due to extractive remediation was compensated for in the second study by the addition of compost and a 6-month equilibration period. This clearly reduced metal bioavailability (Figures 20.5 and 20.6), and in many cases metal levels in the test organisms were reduced below the 'normal' background values defined in this study (Table 20.4).

20.4.2 SUITABILITY OF TEST SPECIES FOR BIOASSAYS

Earthworms and plants are important components of terrestrial ecosystems and are therefore considered relevant test organisms for bioassays on polluted and remediated soils. Both organisms are easy to handle and standardized test methods have been described for determining toxicity of chemicals (OECD 1984a,b). From this study, it appears that the earthworm *E. fetida* and the plant species *C. esculentus* and *R. sativus* are suitable test organisms for bioassays on polluted and remediated soils. *L. sativa* appeared to be less suitable, mainly because of its poor growth on acid soils; for less acid soils, *L. sativa* may also be a suitable test species. *C. esculentus* seems to be fairly robust and not affected by soil characteristics or high metal levels, making this species suitable for bioaccumulation studies. *C. esculentus* can also be tested under both upland and wetland conditions (Marquenie *et al.* 1988, Jenner *et al.* 1992). Tests with radish and lettuce can be performed in smaller test containers, increasing the number of replicates. Well standardized testing methods exist for phytoxicity testing (OECD 1984b). Since metal levels in lettuce were generally in agreement with those in radish, the latter species may be preferred. An additional advantage of *R. sativus* is that both roots and shoots can be harvested. This may provide additional information on the potential shoot transfer of soil pollutants.

Metal levels in the earthworms and plants on the non-polluted reference soils are difficult to compare with literature data. For earthworms, a comparison with literature data is hampered by the fact that only few data are available for the same species (*E. fetida*) used in this study and indications that bioaccumulation of metals may be species-specific (see, e.g., Terhivuo *et al.* 1994). Nevertheless, elemental levels found in *E. fetida* appeared to be in agreement with those found by other authors in earthworms sampled from or exposed to non-polluted soils (see Van Gestel *et al.* 1992 for a review).

For lettuce, elemental levels found in the literature were generally lower than those found in our study and more in agreement with the levels found in radish. This can be attributed entirely to the fact that growth of lettuce was severely hampered by low soil pH. Only for the essential elements Ca, Cu, K, Mg, Na and Zn were concentrations measured in this study in agreement with values reported in the literature (Van Gestel *et al.* 1992). Elemental levels found for radish showed good agreement with values reported in the literature (Van Gestel *et al.* 1992).

The test period of four weeks used for the bioassays in this study seems to be long enough to establish an equilibrium (Marquenie and Simmers 1988), although Corp and Morgan (1991) concluded that for earthworms this period may not be long enough to simulate field exposure conditions. For the bioassays with plants, a period of four weeks seems to be long enough to obtain sufficient biomass for metal analysis.

Besides bioaccumulation, other endpoints may be included in the bioassays. This study has shown effects on earthworm and plant growth. The dry-to-fresh weight ratio of plants also appeared to be a good indicator of reduced growth conditions (Figure 20.4). A proper reference or control is needed to allow for inclusion of other sublethal endpoints in bioassays (see below).

20.4.3 BIOACCUMULATION AS AN ENDPOINT IN BIOASSAYS ON POLLUTED AND REMEDIATED SOILS

Several bioassay methods may be applied to polluted and remediated soils, as described in DECHEMA (1995) and Keddy *et al.* (1995). Many different parameters may be used, ranging from survival or germination to sublethal endpoints such as growth and reproduction. Bioassays may be applied to intact soil samples and soil extracts. In the latter case, aquatic test organisms can also be used for the evaluation of polluted and remediated soils, thus expanding the battery of potential test organisms (DECHEMA 1995, Keddy *et al.* 1995).

A proper risk assessment of soil pollution should preferably focus on sublethal endpoints in bioassays applied to intact soil samples. The presence of soil particles may strongly affect the response of test organisms (Forge *et al.* 1993), and therefore it is considered inappropriate to rely solely on tests on extracts (Van Gestel 1997). However, there are only few toxicity tests available with soil organisms that may be used for the evaluation of soils (Keddy *et al.* 1995). Recently developed test methods (Løkke and Van Gestel 1998) are not yet standardized or have not yet been fully investigated for their applicability to naturally polluted soils.

When focusing on sublethal endpoints, it should be realized that the performance of test organisms may be affected by soil characteristics. This is shown in the second study described here: in the non-polluted soils from nature reserve areas plant growth was strongly influenced by soil pH, and in the (aged) remediated soils unexplained mortality of earthworms and/or inhibition of plant growth was observed. This may seriously hamper a proper evaluation of responses observed on polluted or remediated soils, especially when proper control soils are not available. The first study clearly demonstrated the difficulties of finding proper control soils. When considering the polluted soils, control soils either had different physico-chemical characteristics or contained increased levels of other or the same pollutants (Table 20.1). Texture may be affected by remediation techniques to such an extent that finding a corresponding control soil may be impossible.

Effects of toxicants may be related to internal concentrations in the test organisms (McCarty and Mackay 1993). Thus bioaccumulation of contaminants can serve as both a measure of bioavailability and an indicator of potential risk of pollutants in soil. In the case of metal-polluted soils, bioaccumulation may therefore be an alternative option for evaluating exposure and effect (Van Wensem *et al.* 1994, Van Straalen 1996), and the reason why we have chosen to focus on bioaccumulation.

Bioaccumulation of metals can be considered in two ways. In the first study, we compared bioaccumulation levels of metals in earthworms and plants exposed to polluted and decontaminated soils. This demonstrated that although metal levels in these test organisms were often lower on the remediated soils, bioavailability was higher: BSAF values were always higher on the remediated compared to corresponding polluted soils (Tables 20.2 and 20.6). The use of BSAF values does, however, have some drawbacks (Chapman *et al.* 1996). For metals, BSAF values might increase with decreasing soil concentration. For essential metals, such as Zn and Cu, body concentrations may be regulated over a large range of exposure concentrations as demonstrated in earthworms by Morgan and Morgan (1988), Van Gestel *et al.* (1993) and in this study (Figures 20.1, 20.2 and 20.6). Plants appear to be less capable of regulating internal Zn levels (Figures 20.2 and 20.6). Further, BSAF values cannot give an indication of potential risks; for that purpose it is essential to compare concentrations in the test organisms with internal effect concentrations (Van Wensem *et al.* 1994, Van Straalen 1996).

For remediated soils to be safe, it is essential that bioavailability be reduced to background levels. Considering this, and the fact that internal concentrations may give a better indication of risk than external concentrations or BSAF values, the second study was undertaken. The 'normal' background values of metals in earthworms and plants defined on the basis of this study (Table 20.4) might offer a useful tool to evaluate the results of bioassays involving metal-polluted and remediated soils.

20.5 ACKNOWLEDGEMENTS

W.A. Van Dis is acknowledged for performing the earthworm bioassays in the first study.

Thanks are also due to our colleagues at the TNO Institute and at the National Institute of Public Health and the Environment, who assisted in the metal analysis and characterization of the test soils.

REFERENCES

Assink JW (1988) Physico-chemical treatment methods for soil remediation. In *Contaminated Soil '88*, Wolf K, Van den Brink WJ and Colon FJ (eds), Kluwer Academic Publishers, Dordrecht, pp. 861-870.

Chapman PM, Allen HE, Godtfredsen K and Zgraggen MN (1996) Evaluation of bioaccumulation factors in regulating metals. *Environmental Science and Technology*, **30**, A448-A452.

Corp N and Morgan AJ (1991) Accumulation of heavy metals from polluted soils by the earthworm *Lumbricus rubellus*: can laboratory exposure of control worms reduce biomonitoring problems? *Environmental Pollution*, **74**, 39-52.

DECHEMA (1995) *Biologische Testmethoden für Böden*. Bericht des Interdisciplinären Arbeitskreises 'Umweltbiotechnologie — Boden' 4: Adhoc-Arbeitsgruppe 'Methoden zur Toxikologischen/Ökotoxikologischen Bewertung von Böden', Deutsche Gesellschaft für Chemisches Apparatewesen, Chemische Technik und Biotechnologie eV, Frankfurt am Main.

Edelman T and De Bruin M (1986) Background values of 32 elements in Dutch topsoils, determined with non-destructive neutron activation analysis. In *Contaminated Soil*, Assink JW and Van den Brink WJ (eds), Martinus Nijhoff Publishers, Dordrecht, pp. 89-99.

Forge TA, Berrow ML, Darbyshire JF and Warren A (1993) Protozoan bioassays of soil amended with sewage sludge and heavy metals, using the common soil ciliate *Colpoda steinii*. *Biology and Fertility of Soils*, **16**, 282-286.

Jenner HA, Janssen-Mommen JPM and Koeman JH (1992) Effects of coal gasification slag as a substrate for the plant *Cyperus esculentus* and the worm *Eisenia fetida*. *Ecotoxicology and Environmental Safety*, **24**, 46-57.

Keddy CJ, Greene JC and Bonnell MA (1995) Review of whole-organism bioassays: soil, freshwater sediment and freshwater assessment in Canada. *Ecotoxicology and Environmental Safety*, **30**, 221-251.

Løkke H and Van Gestel CAM (eds) (1998) *Handbook of Soil Invertebrate Toxicity Tests*, John Wiley & Sons, Chichester.

Marquenie JM and Simmers JW (1988) A method to assess potential bioavailability of contaminants. In *Earthworms in Waste and Environmental Management*, Edwards CA and Neuhauser EF (eds), SPB Academic Publishers, The Hague, pp. 367-375.

Marquenie JM, Simmers JW and Kay SH (1987) *Preliminary Assessment of Bioaccumulation of Metals and Organic Contaminants at the Times Beach Confined Disposal Site, Buffalo, NY.* Miscellaneous Paper EL-87-6, US Army Corps of Engineers Waterways Experimental Station.

Marquenie JM, Crawley DK, De Jong P and Jenner HA (1988) Growth and metal uptake of the plant *Cyperus esculentus* and the worm *Eisenia fetida* in a worst-case experiment on coal fly-ash and Rhine sediment. *Kema Scientific and Technical Reports*, **6**(4), 113-121.

McCarty LS and Mackay D (1993) Enhancing ecotoxicological modeling and assessment. Body residues and modes of toxic action. *Environmental Science and Technology*, **27**, 1719-1728.

Morgan JE and Morgan AJ (1988) Earthworms as biological monitors of cadmium, copper, lead and zinc in metalliferous soils. *Environmental Pollution*, **54**, 123-138.

OECD (1984a). *OECD Guideline for Testing of Chemicals 207. Earthworm, Acute Toxicity Tests*, Organization for Economic Co-operation and Development, Paris.

OECD (1984b). *OECD Guideline for Testing of Chemicals 208. Terrestrial Plants, Growth Test*, Organization for Economic Co-operation and Development, Paris.

Salomons W, Förstner U and Mader P (eds) (1995) *Heavy Metals — Problems and Solutions*, Springer-Verlag, Berlin.

Terhivuo J, Pankakoski E, Hyvärinen H and Koivisto I (1994) Pb uptake by ecologically dissimilar earthworm (Lumbricidae) species near a lead smelter in south Finland. *Environmental Pollution*, **85**, 87-96.

Van Gestel CAM (1997) Scientific basis for extrapolating results from soil ecotoxicity tests to field conditions and the use of bioassays. In *Ecological Risk Assessment of Contaminants in Soil*, Van Straalen NM and Løkke H (eds), Chapman & Hall, London, pp. 25-50.

Van Gestel CAM, Dirven-Van Breemen EM and Kamerman JW (1992) *Beoordeling van gereinigde grond. IV. Toepassing van bioassays met planten en regenwormen op referentiegronden*. Report no. 216402004, National Institute of Public Health and the Environment, Bilthoven, The Netherlands.

Van Gestel CAM, Dirven-Van Breemen EM and Baerselman R (1993) Accumulation and elimination of cadmium, chromium and zinc and effects on growth and reproduction in *Eisenia andrei* (Oligochaeta, Annelida). *The Science of the Total Environment*, **Supplement**, 585-597.

Van Straalen NM (1996) Critical body concentrations: their use in bioindication. *Bioindicator Systems for Soil Pollution*, Van Straalen NM and Krivolutsky DA (eds), Kluwer Academic Publishers, Dordrecht, pp. 5-16.

Van Wensem J, Vegter JJ and Van Straalen NM (1994) Soil quality criteria derived from critical body concentrations of metals in soil invertebrates. *Applied Soil Ecology*, **1**, 185-191.

Vegter JJ (1995) Soil protection in the Netherlands. In *Heavy Metals — Problems and Solutions*, Salomons W, Förstner U and Mader P (eds), Springer-Verlag, Berlin, pp. 79-100.

Versluis CW, Aalbers TG, Adema DMM, Assink JW, Van Gestel CAM and Anthonissen IH (1988) Comparison of leaching behaviour and bioavailability of heavy metals in contaminated soils and soils cleaned up with several extractive and thermal methods. In *Contaminated Soil '88*, Wolf K, Van den Brink JW and Colon FJ (eds), Kluwer Academic Publishers, Dordrecht, pp. 11-21.

21

Contaminated Sites Redevelopment Using the Risk Assessment/Risk Management Approach: The Montreal Experience

CHANTAL GUAY AND SERGE BARBEAU

Ville de Montréal — Service des travaux publics et de l'environnement — Laboratoire, Montréal, Québec, Canada

21.1 INTRODUCTION

In Canada, thousands of old industrial sites (brownfields) are left unused due to their suspected contamination. These sites, which may pose a risk to human health and the environment, have important economical repercussions for industrial cities where they are mainly found. In fact, it is estimated that the cost of the brownfield phenomenon to all government levels in Canada represents many billions of dollars in lost taxes (NRTEE 1998). Following the publication of the *Politique de réhabilitation des terrains contaminés* (Contaminated Sites Rehabilitation Guidelines) by the Ministry of Environment of Quebec (Québec Environment) in 1988, the City of Montreal first underlined this problem. It is also at its request that the Interministerial Committee on the Montreal Island Contaminated Sites Problem was created (Ville de Montréal 1993). Recommendations of this committee dedicated to finding solutions to the contaminated sites problem led to the creation of a programme called *Projet-pilote d'aide aux municipalités de la région de Montréal propriétaires de terrains contaminés* (Pilot project to aid municipalities owning contaminated sites in the Montreal area). The Pilot Project was first aimed at helping to clean sites using new technologies that were to be evaluated and tested and, consequently, at facilitating the emergence of technologies well adapted to the nature of the contamination found on urban sites.

Another main objective of the Pilot Project was to elaborate and validate new approaches of contaminated sites remediation that would provide an

Environmental Analysis of Contaminated Sites. Edited by G. I. Sunahara, A. Y. Renoux, C. Thellen, C. L. Gaudet and A. Pilon

adequate level of protection to human health and the environment without being as costly as the generic criteria approach proposed by Quebec Environment (Ville de Montréal 1995). In achieving that goal, the Pilot Project was to test the new Québec Environment risk assessment approach, which was developed to evaluate the impacts of major construction projects on human health and the environment. For the Pilot Project this new procedure was to be applied to contaminated sites and compared against the method proposed by the USEPA for toxicological risk assessment, and the ecotoxicological risk assessment methodology developed especially for the Pilot Project by the City of Montreal and the consulting firm D'Aragon, Desbiens, Halde Associates (DDH). Hence in Quebec, the Pilot Project posed most of the difficulties related to using the new contaminated sites remediation approach proposed by Québec Environment. This chapter presents the difficulties and questions raised by the use of the risk assessment/risk management (RA/RM) approach in the City of Montreal's perspective, without review or rebuttal by Quebec Environment. Results of the Pilot Project encouraged the Centre for Soil Cleanup (*Centre d'assainissement des sols*) to conduct a study aimed at defining the implications of using risk assessment for contaminated sites redevelopment. The Centre regroups landowners such as the City of Montreal, Bell Canada, Canada Lands Corporation and other parties, who are concerned by the contaminated sites issue. It constitutes a forum to exchange and discuss this problem and its solutions. The main objectives of the Centre for Soil Cleanup study were to describe the legal context surrounding the use of the RA/RM approach to site rehabilitation and to learn more about the reactions and behaviour of third parties implicated in contaminated sites redevelopment projects, which include the financial services sector, the government and the public.

Conclusions of the Centre study conducted by KPMG and Lapointe Rosenstein (1997) are to the effect that the main advantage of the RA/RM approach is that it allows an important reduction of the cleanup costs when compared to an excavation/landfilling approach based on generic criteria. Consequently, if this economic benefit makes the redevelopment of a larger number of contaminated sites possible, an environmental gain will certainly be achieved. For example, if old industrial sites located in the centre of the city are reused instead of 'green' sites in the suburbs or the countryside, then urban sprawl and its severe consequences on the environment will be reduced.

Unfortunately, the negative perception of the risk assessment approach by the parties involved places it in a weak position compared to the generic criteria approach. In fact in Quebec, the generic approach is far more used and known by the many parties involved and generally considered as 'user friendly'. Thus, it is not surprising that it appears to be more reassuring compared to the new risk assessment procedure that is just starting to be used. Moreover, the risk assessment approach is often perceived as esoteric, since many experts using highly technical language are required to complete such studies.

This perception is cause for concern to many parties, including the government authorities, when a developer or an owner suggests using the risk assessment approach to clean up a site. This unrest is usually manifested by doubt and conservatism toward this approach, which appears to many people as much more risky than the generic criteria approach.

But contrary to this perception, the risk assessment approach allows a much better definition and more rigorous estimation of the potential risk to humans and to the environment than the generic criteria approach. Nevertheless, the Montreal experience shows that the acceptance of this new approach to contaminated sites redevelopment lies in the improvement of the preconceived perception that many parties have of the risk assessment procedure.

In fact, most of the discussions that were held between the City of Montreal and Québec Environment during the course of the Pilot Project were directly related to the differences in perception of the risk assessment approach. The perception held by environmental authorities had important consequences on the remediation plans that were finally authorized, and therefore on the benefits that may follow from using the new approach to contaminated sites redevelopment. In particular, the divergence between the City and Quebec Environment as per the importance of the risk when using such an approach, and the mechanisms necessary to manage these risks, is reflected in the number and type of controls which are imposed to ensure the integrity and durability of the mitigation measures (Guay 1999).

Based on the experience acquired during the Pilot Project and the results of the Centre for Soil Cleanup study, and because the implementation of institutional controls within the risk assessment approach to contaminated sites should be viewed as part of the larger brownfield problem, the work of three major organizations greatly interested in this question has been examined. The RA/RM approach to contaminated sites cleanup will have a greater chance of succeeding if the institutional controls that must be used with such an approach favour contaminated sites redevelopment.

Since the beginning of the 1990s, the Canadian Council of Ministers of the Environment (CCME), the National Round Table on Environment and Economy (NRTEE) and the Canada Mortgage and Housing Corporation (CMHC) have been looking into the brownfield problem. The work carried out by the CCME led to 13 recommendations to ensure a coherent approach to the redevelopment of contaminated sites in Canada. The main objectives of these recommendations were to clarify who is responsible for remediation of contaminated sites and to assist government authorities to elaborate laws, regulations and guidelines regarding these issues. Principles such as 'the polluter pays', fairness and sustainable development comprise these recommendations, as well as specific prescriptions regarding the identification and determination of responsible (and non-responsible) parties, and the definition of 'what is a contaminated site?' (CCME 1993).

The NRTEE has been looking into the obstacles to brownfields redevelopment. A series of multilateral consultations made throughout Canada confirmed that the implementation of the 13 principles identified by the CCME is essential to a coherent approach to contaminated sites redevelopment. The NRTEE workgroup has also emphasized the most acute concerns of the financial services sector toward contaminated sites (NRTEE 1998). These concerns include: the legal uncertainty regarding the liability of contaminated sites, the lack of reliable rules and standards to evaluate cleanup costs, the paucity of insurance vehicles for site contamination, the scarcity of databases and registries of contaminated sites and the lack of consensus regarding the information that should be included in such documents, the negative perception of the general population toward contaminants and their consequences, and the dearth of financial or economic incentives within the actual regulations and guidelines for contaminated sites rehabilitation.

Finally, the CMHC, which has been forced into dealing with the contaminated sites issues due to the nature of its activities, has recently conducted a review of Canadian brownfield redevelopment case studies. The conclusions of this study have allowed the CMHC to issue 22 recommendations aimed at improving brownfield redevelopment in Canada (Delcan *et al.* 1996, 1997). These suggestions, which are of legislative, administrative, methodological and political order, comprise numerous mechanisms that government authorities should implement to facilitate contaminated sites redevelopment. Such recommendations include: the 'user pays' principle in the approval and validation process of a remediation plan, the elaboration of cleanup criteria based on RA/RM approaches and taking into account site function and depth of contaminants, the option of reusing less contaminated soils on more contaminated sites, the issuing of a certificate of approval after completion of the cleanup activities according to the authorities' requirements and the creation of databanks and registries of contaminated sites.

21.2 THE CONTAMINATED SITES PHENOMENON IN MONTREAL

Montreal, which was the cradle of industrialization in Canada, must now face the consequences of such a history: a large number of lots are left vacant and contaminated after the departure of commercial and industrial facilities. Thus, one may find in this city numerous small to medium size sites ($<20\ 000\ m^2$) which are contaminated above the levels considered acceptable according to the provincial policy. In general, the contaminants come from a large number of sources and comprise organic and inorganic components. Moreover, because of the diversified use of the sites and the very frequent existence of uncontrolled backfilling operations, the type of contaminants, their spatial distribution and concentrations vary greatly from one site to another. It is also important to note that the activities that were carried out on these sites have mainly polluted the first few metres of soil.

The high costs of cleanup of these contaminated sites according to the generic criteria of the policy constitute a serious obstacle for the old industrial cities in this province who want to reuse those vacant lots, as well as for the many promoters who wish to redevelop these sites for industrial, commercial or residential purposes (Ville de Montréal 1995).

Québec Environment proposes three different levels of generic criteria for many substances. The level 'A' is considered to represent the background level of inorganic compounds and the limit of detection for organic compounds. The 'B' level is the maximum acceptable limit for residential, recreative and institutional use. Finally, the 'C' level is the maximum acceptable limit for commercial or industrial sites. In that context, the use of the generic criteria approach to site cleanup means that all contaminated soils above the applicable criteria will be excavated and disposed of off-site, otherwise the site would have to be treated until the concentrations of contaminants in the soils left in place are below the criteria. In the Montreal context, this latter option is rarely used due to the absence of efficient treatment technologies that can be adapted to the nature of the contaminants (mixture of organic and inorganic compounds) and their presence in all soil fractions.

An estimate of the cleanup costs (off-site disposal) that would result from applying the generic criteria approach to all sites proposed for development between 1993 and 1998 in Montreal (approximately 390 ha) amounted to $320 000 000 (Ville de Montréal 1993). Considering the current price of contaminated soil disposal, these costs would now be estimated at $70 000 000. In the context of limited financial resources and very expensive cleanup costs, the Government of Quebec and the City of Montreal started, in March 1995, the Pilot Project. An important feature of this project was to apply and evaluate the new RA/RM approach proposed by Quebec Environment during the course of the revision of its 1988 Contaminated Sites Guidelines.

The Québec Environment RA/RM approach was to be compared to the risk assessment methodology developed especially for the Pilot Project by the City of Montreal and DDH, whose services were retained to carry out the risk assessments in the Pilot Project (Desbiens 1995). The ultimate objective of these evaluations was to create a new management tool dedicated to urban contaminated sites redevelopment.

21.3 APPLICATION OF THE RA/RM APPROACH TO MONTREAL'S CONTAMINATED SITES

21.3.1 CONTEXT

It is important to note that in Quebec, the owner of a contaminated site (a site that contains contaminated soils which in Quebec are excluded from the hazardous wastes list) has no legal obligation to perform any cleanup activities with the exception of leakage of underground storage tanks and accidental spills of hazardous wastes (KPMG and Lapointe Rosenstein 1997). Nevertheless, each year the City of Montreal applies the 'due diligence' principle and carries out a large number of rehabilitation projects according to the Québec Environment guidelines.

During the Pilot Project, the City used the new RA/RM approach to site remediation under the guidance and approval of Québec Environment. Once it was decided that a contaminated site would be submitted to the RA/RM approach, the City would proceed through the following steps.

1. Meet with Québec Environment in order to decide on the nature and scope of the studies which have to include three parts: a toxicological risk assessment, an ecotoxicological risk assessment and an impact assessment of the contaminants on groundwater.
2. Prepare a detailed report and remediation plan (including the mitigation measures to be implemented, the monitoring programme for these measures and any land restrictions) to be submitted to Québec Environment for approval.
3. After completion of the site remediation, provide a report to Québec Environment attesting the conformity of the approved cleanup plan in order that a certificate of compliance can be issued. The certificate would attest to the presence of residual contamination above the accepted levels for the proposed land use, the mitigation measures, the monitoring programme and the land use restrictions.
4. Record the presence of residual contamination above the accepted levels for the proposed land use, the mitigation measures, the monitoring programme and the land use restrictions in a notice at the Land Records Office.

21.3.2 CHARACTERISTICS OF THE SITES SUBMITTED TO THE RA/RM APPROACH

The RA/RM approach was used on a total of six sites during the course of the Pilot Project. Most of these were small to medium sized sites (1200 to 14 000 m^2) and were located in old industrial areas of Montreal. Two of these lots were to be redeveloped for residential purposes and the four others were to receive commercial or industrial facilities.

Generally, the site assessment of the contaminated lots indicated the presence of heavy metals (Cu, Pb, Zn) and polycyclic aromatic hydrocarbons (PAHs), in concentrations above the proposed levels for residential or industrial land use. The estimated cleanup costs (excavation and landfilling) of the six sites based on the site assessments and generic criteria (at that time) would be a total of \$9 600 000 (between \$170 000 and \$4 340 000 per site; Guay *et al.* 1997).

21.3.3 DEVELOPMENT SCENARIOS USED FOR THE RISK ASSESSMENTS

During the course of the risk assessment procedure, each site was evaluated for a minimum of three different development scenarios. First, the sites, in their actual state, were studied in order to appraise the necessity of short-term intervention. Second, the sites were submitted to the assessment according to the future development plan, i.e., for residential, commercial or industrial purposes. In the evaluation, only the elements that were strictly related to the type of development were to be implemented on the sites. Finally, the third scenario included the replacement of the first metre of soil by clean material (Desbiens 1997).

21.4 RISK ASSESSMENT PROCEDURES

During the course of the Pilot Project and for purposes of comparison, two different risk assessment procedures were used. The first procedure, developed by DDH and the City of Montreal, comprised a toxicological risk assessment based on the USEPA approach (USEPA 1989) and the ecotoxicological risk evaluation elaborated by DDH and the City (DDH - City approach). The second procedure proposed by Québec Environment had three steps which included a toxicological risk assessment carried out according to a method developed by Québec Environment in June 1996 (Québec Environment 1996a,b), an ecotoxicological risk assessment (DDH - City approach) and an impact study of the contaminants on groundwater, also carried out according to the method developed by Québec Environment (Guay *et al.* 1997). The important steps of each assessment procedure and their results were reported by Desbiens (1996) and are presented in the following sections.

21.4.1 HUMAN TOXICOLOGICAL RISK ASSESSMENT

The basic procedure comprises five general steps: the scope of work, the data collection and evaluation, the hazard identification and toxicity assessment, the exposure assessment and the risk characterization. Based on the data collected from the site assessments, it was established that metals such as cadmium, copper, lead, tin and zinc, and petroleum hydrocarbons such as monocyclic aromatic hydrocarbons (MAHs) and PAHs, were considered to be the stress agents. In fact, each of these contaminants was present on the sites at concentrations exceeding 'A' criteria (background levels).

Once the stress agents were identified, their toxicity had to be evaluated, and risk factors such as cancer potency factors for carcinogenic substances and reference doses for non-carcinogenic substances had to be determined. Documents such as USEPA IRIS and HEAST and the Québec Environment databanks were consulted to carry out this task.

Once the toxicity evaluation of the substances was complete, estimation of the site-specific intake by the population of carcinogenic and non-carcinogenic substances was carried out. All possible exposure pathways were considered and characteristics of the population were assessed during this stage. Concentrations of the contaminants in all applicable environmental media were measured or calculated using computer simulations. Integration of the toxicity and intake exposure was finally carried out to verify the presence of a potential risk, and on that basis elaborate a corrective action.

The non-cancer risk was estimated by dividing the intake by the reference dose (considered a safe dose) for each chemical and individual exposure route. The obtained hazard quotients for each substance were then added and the resulting hazard index was compared to the acceptable value of 1. All hazard indexes greater than 1 were considered problematic.

The cancer risk represents the probability of an individual developing cancer as a result of his lifetime exposure to a carcinogenic substance. The risk level was calculated by multiplying the lifetime average daily dose by the cancer potency factor (slope factor) and compared to the acceptable risk level of one new case of cancer per 1 000 000 individuals.

21.4.2 ECOTOXICOLOGICAL RISK EVALUATION

The second part of the risk assessment procedure used during the Pilot Project was the ecotoxicological risk evaluation (DDH - City procedure). This evaluation comprised the following steps: the scope of work, the identification of ecological receptors and stress agents, the confirmation of the exposure, the intake estimation, the estimation of the reference concentration and the risk characterization. This procedure was developed especially for the Pilot Project with a specific application to urban sites by taking many factors into account:

- Urban sites are deteriorated compared to country sites or wild areas. For example, the animal and vegetal species are less diversified in the city than in the countryside;
- Generally, the studied sites are destined to be used by humans and will not become conservation areas.

Nevertheless, the observed degradation of the environment in urban areas must be reduced; it appears important to proceed with an ecotoxicological risk evaluation and, if necessary, corrective actions should be taken. The DDH - City procedure was used for that purpose.

To carry out the second part of the procedure (i.e., the receptor identification), the City proposed a list of sensitive components to be protected or enhanced: recognized nesting, mating or resting areas for birds or wildfowl; recognized spawning areas, rivers or marshes with protected, threatened or endangered species; protected, threatened or endangered vegetal species; species used for hunting; species used for commercial or sports fishing; exceptional trees or woodland; landscaping elements (actual or future); vegetable gardens (now or in the future); as well as other sensitive habitats.

The data collection on each site was achieved, in order to verify if any of the proposed sensitive elements were present on-site or in its surroundings.

Generally, the sensitive elements registered were large trees and the landscaping elements and vegetable gardens planned in the redevelopment scenarios.

As in the human health risk assessment, the stress agents were found to be metals (such as arsenic, cadmium, lead, nickel and tin) and petroleum hydrocarbons (such as MAHs and PAHs). All substances found in excess of the 'A' criteria (background levels) were selected to carry out the risk evaluation. The validation of the exposure to the stress agents was then accomplished. Since the sensitive receptors are in direct contact with the contaminated soils (i.e., by their roots), the exposure was confirmed.

The intake was subsequently estimated by considering the concentrations measured in the first metre of soil. This layer was considered to be the most critical to ensure healthy vegetation and trees at the surface, according to discussions with experts of McGill University, Jardin Botanique de Montréal, Alberta Environment and Université du Quebéc à Montréal, and a review of literature (Power *et al.* 1981, Oddie and Bailey 1988, Geiger *et al.* 1993, Tassé *et al.* 1993). The risk index was finally appraised by comparing the intake concentrations to reference concentrations developed and used in The Netherlands. A risk index of less than 1 was considered acceptable, as proposed by many organizations such as CCME (see Chapter 13) and Quebec Environment.

21.4.3 IMPACT STUDY ON GROUNDWATER

The standard procedure proposed by Québec Environment (1996 c,d) included four stages.

1. Establish if the groundwater is contaminated based on the background level or detection limits established by Québec Environment.
2. If the presence of contaminants is confirmed (concentrations above the background level or detection limit) and the active source can be identified, remove the source.
3. In order to decide if a corrective action is required, evaluate if the contamination has real or possible impacts (by comparing the concentrations of contaminants to Québec Environment generic criteria established for different discharge points such as rivers, sewers or wells).
4. Finally, if present, recover any floating phases.

Application of this procedure revealed that, in general, the surface groundwater present in the fill materials and the natural soil was slightly contaminated (below the 'B' generic criteria) with heavy metals, mineral oil and grease. All contaminating activities were already ceased on the study sites and no active source of contamination could be identified. The study also showed that all water run-offs would be collected by the existing sewer system and that the groundwater was not discharging into any wells or water bodies. In that context, it was assumed, for evaluation purposes, that the groundwater would discharge into the sewer system if breaches were present.

21.5 RESULTS OF THE RISK ASSESSMENTS AND PROPOSED CORRECTIVE ACTION PLAN

The results of the risk assessments based on the USEPA procedure (1989) showed that the sites selected for the Pilot Project constitute a potentially significant risk for human health when the population is in direct contact with the contaminated soil. All sites with contaminated soils at the surface and all sites that would be developed for residential purposes, including vegetable gardens planted directly in the contaminated soils, represent a significant risk (Guay *et al.* 1997).

It is important to note that only the results of the USEPA procedure were used to elaborate the remedial scenarios. This is because, when using the human health risk assessment procedure proposed by Québec Environment which includes the estimation of the complete exposure dose (i.e., both the background dose and the dose specific to the site), the background levels of many non-carcinogenic substances such as toluene, xylene, ethylbenzene, heavy metals and PAHs exceeded the reference dose (established safe dose). The values of background and reference doses used in this procedure were provided in the guidelines prepared by the Risk Assessment Group of Québec Environment (1996b). Thus, the risk index for these substances, no matter which site was being assessed, was greater than 1 and considered significant. During the Pilot Project, Québec Environment, which was to approve the remedial scenarios based on the results of the risk assessment, did not allow the contaminants on the sites to contribute to any increase of the exposure dose. Since non-carcinogenic PAHs were present on all studied sites in concentrations above the 'A' generic criteria and no mitigation measures could completely eliminate the exposure to the PAHs, it would have been necessary to remove all soils contaminated with these substances. This action appeared unnecessary and excessive, considering that a remedial plan based on the generic criteria would only require the removal of soils with contaminants in concentrations above the 'B' or 'C' generic criteria, which are generally 5 to 100 times the background level.

Therefore, the results of the USEPA procedure were used since it allows calculation of the risk posed by the contaminated site itself, that is, the supplemental health risk to the population caused by the presence of that site. It is important to note that the Quebec Health and Social Services Ministry (MSSS) is now in charge of approving the human health risk assessments. This Ministry does not impose, *a priori*, restrictions when the background dose of a contaminant exceeds its reference dose. In that case, the background doses used are the ones established by Québec Environment, but the reference doses can be obtained from IRIS, HEAST, ATSDR, Health Canada or WHO. However, the MSSS requires that the risk indexes be calculated using the complete exposure doses and that the risk indexes caused by the background doses and by the site-specific doses be calculated, compared and discussed (MSSS 1999b).

The ecotoxicological risk evaluations carried out showed that the identified receptors (mature trees, vegetable gardens, landscaping elements) are potentially at risk when in direct contact with the contaminated soils.

Finally, the studies on the impact on groundwater showed that there is no concern since it is considered to infiltrate into the sewers and the contaminant concentrations are well below the acceptable sewer discharge levels.

Based on the results presented above, remedial scenarios proposed include:

- Management of excavated material for construction purposes according to the Québec Environment guidelines;
- Covering of contaminated soils left in place by clean soils, asphalt (parking areas) or concrete (buildings slabs) in the final stage of the redevelopment plan;
- Planting the landscape elements (trees and shrubs) in clean soil pits (minimal depth of 1 m);
- For one of the studied sites, installation of a biogas recovery system underneath the proposed building;
- Finally, in order to ensure the durability and integrity of the proposed mitigation measures and that any future owners will be aware of them, it was recommended that all the redevelopment conditions and land use restrictions be put down in writing in the sales contract.

If these measures were to be applied, it was estimated, at that time, that the cleanup costs of the six sites would range from \$60 000 to \$1 000 000 for a total of \$2 700 000, an economy of more than 70% compared to the generic criteria-based cleanup estimated costs.

21.6 STEPS TOWARD APPROVAL OF RA/RM-BASED CLEANUP PLANS

As discussed previously, once the risk assessments are complete, corrective action plans are elaborated and all resulting documents are presented to the authorities for approval and issuing of applicable cleanup authorizations.

Consequently, all required documents were presented for approval in June 1996. To date, the City has received approval and authorization to proceed with two of the six studied sites intended for industrial use. One of the authorizations was obtained after many discussions with a technical evaluation committee (GTE), and after the proposed cleanup plan was drastically changed. The essential elements of the many discussions between the City and the GTE are presented in the next section.

21.7 APPLICATION OF RA/RM PROCEDURES TO AN INDUSTRIAL SITE

On 18 April, 1997, the City of Montreal recommended that one of the sites studied during the Pilot Project, which was to be sold to a transportation company, be developed according to the RA/RM results. Apart from the management of the excavated material for construction purposes according to the Québec Environment guidelines, the remedial scenario included capping of the contaminated soils left in place by the following elements:

- the concrete slab-on-grade of the building;
- the granular foundation and asphalt (for a total of a minimum of 30 cm) in the access road and parking area;
- a minimum layer of 15 cm of clean soil in the grassed areas;
- a minimum of 1 m of clean soil for the other landscape elements (trees, shrubs).

In order to ensure the durability of these measures and their knowledge by all future owners, the transfer act was to include the following declarations and obligations of the buyer:

- That the Buyer has received all the studies, documents and correspondence relative to the site and declares that he has read it and is satisfied with it.
- That the authorization to proceed with the site cleanup according to the RA/RM procedure necessitates the following land use restrictions:
  - (i) At the location of the building and parking area, the soils always be capped with concrete and asphalt;
  - (ii) The strip fronting the street always has 15 cm of clean soil and is covered by grass;
  - (iii) In order to plant trees, a layer of clean soil of a minimum of 1 m in thickness be present on top of soils contaminated above the 'B' generic criteria.
- That the authorization to proceed with the site cleanup according to the RA/RM procedure necessitates the following commitments:
  - (i) That the Buyer commits himself to all soils excavated for future development purposes of part or all the property being managed according to the applicable Québec Environment's guidelines;
  - (ii) That if, once the property has been developed according to the requirements called for by the RA/RM procedure, the Buyer was to decide to modify part or all of the implemented elements, he would have to obtain the Québec Environment's authorization before carrying out the modifications;
  - (iii) That if part or all of the property was to be sold, the Buyer commits to record in the transfer documents all the applicable declarations regarding the land use restrictions and commitments mentioned above.

However, the proposed measures were not considered sufficiently protective. It was requested that the cleanup plan also include the following:

- That soils contaminated by arsenic above the 'C' generic criteria be eliminated off-site and replaced by clean soils;
- That in the road access and parking area, the contaminated soils left in place be covered by a granular foundation followed by a special mix of asphalt (cold mix) for a total of 455 mm. This type of structure was chosen for its superior flexibility and resistance to cracking;
- In the grassed area, that a minimum layer of 1 m of clean soil be provided;
- That a monitoring programme of the asphalt surface be implemented and all required repairs carried out once a year by the Buyer;
- That groundwater be monitored for a minimum of 1 year;
- That a copy of the land transfer act containing all the land use restrictions and buyer commitments be sent to Québec Environment.

It was also required that the Buyer and the Vendor sign an agreement with Québec Environment. In addition to a detailed description of the cleanup plan, the agreement includes a Buyer's commitment to the effect that, when this mechanism becomes available, he will have to record a notice stating the contamination level of the site at the Land Records Office. This new mechanism is part of the measures proposed in the Soil Protection and Contaminated Sites Rehabilitation Guidelines (*Politique de protection des sols et de réhabilitation des terrains contaminés*) when the RA/RM procedure is applied.

These new guidelines have been in effect since June 1998, but many of the mechanisms included in this document will require further laws and regulation modifications to take full effect.

21.8 DIFFICULTIES OF IMPLEMENTING THE RA/RM PROCEDURE TO URBAN SITES CLEANUP IN MONTREAL

Considering the significant financial resources required to clean up the very numerous contaminated sites in Montreal, one of the main objectives of the Pilot Project was to develop new approaches to contaminated sites cleanup that would provide an adequate level of protection without being as expensive as the generic criteria approach. Based on the results of the six studies carried out during the Pilot Project, it appears that the RA/RM approach meets this requirement. However, many difficulties were encountered due to the framework proposed for this contaminated site cleanup approach (Guay *et al.* 1997). Some of the obstacles encountered during the Pilot Project have been smoothed away, but many difficulties have yet to find solutions.

In the 1998 Québec Environment Contaminated Site Cleanup Guidelines, it was proposed that only the methods developed by Quebec Environment for the ecotoxicological risk assessment and the impact study on groundwater, and by the MSSS for the toxicological risk assessment, can be used.

There are many advantages to the use of well-defined methods; but risk assessment is, by definition, a site-specific study that has to permit some latitude to the authors of such assessments. Actually, with parameters varying from one site to another (target population, contaminants and exposure paths), the assessment method has to be adaptable, allowing modifications based on the professional judgment of the experts carrying out the study.

The proposed method for ecotoxicological risk assessment, which was developed after the Pilot Project, is complex and very extensive. In fact, one has to question the necessity of such a study in an urban context. Some developers have responded to this question negatively and consequently have excluded this part of the risk assessment procedure for urban sites (Belles-Isles and Sarrailh 1997). However, considering the limited experience with the Québec Environment ecotoxicological risk assessment method, it appears premature to exclude it. Only after using and testing it a number of times will it be possible to decide if this method is applicable to urban sites and, if not, a dedicated procedure must be developed for urban sites, as was done during the Pilot Project.

As for the impact study on groundwater, many problems are present in an urban context. First, how can the background levels be established in an area where numerous sources of contaminants exist? Likewise, how can it be certain that the contaminants present in the groundwater are related to the contaminated soils of the studied site, since groundwater travels from one site to another without any consideration of their respective limits? The eventual publication of a groundwater protection and conservation guideline by Québec Environment may bring some answers to the many questions raised.

In the fall of 1996, the human health risk assessment group of Québec Environment was abolished. From that time on, the MSSS became the only body responsible for the human health risk assessments carried out for contaminated sites. Although the MSSS has been designated as responsible for this part of the risk assessment studies, the approval procedure for cleanup plans based on the RA/RM method has not changed. It is the Québec Environment regional office that issues the applicable authorizations, according to the recommendations of a technical review group (GTE) from Québec Environment and the MSSS.

This ambiguity in the roles of Québec Environment and the MSSS in the approval procedure for cleanup plans based on the RA/RM method raises many questions. Will the approval procedure be revised altogether since the MSSS could be willing to play a more important role in this field? Which human health risk assessment method will the MSSS propose? Will they rely on existing methods such as the USEPA or will they elaborate their own? MSSS, which has formed a work group on the guiding principles to be used in the context of toxicological risk management, issued in 1999 principles for human health risk assessment (MSSS 1999a) as well as a guide for carrying out such an evaluation in the context of site cleanup and environmental impact studies (MSSS 1999b). These documents brought some answers to many of the questions previously described. It is interesting to note that the MSSS has initiated, by the creation of this work group, an important reflection not only on the use of risk assessment in a general context, but also in the perspective of contaminated sites management. In the end, this reflection process, which includes hearings of all interested parties, should clarify the context of application of the RA/RM procedure and, in a larger sense, begin the debate on the assessment of the impacts of environmental problems on public health in order to establish the priorities of our interventions (Nantel 1997).

The Québec Environment Guidelines propose a restricted use of the RA/RM approach on residential sites. It considers that it will be difficult to follow up on the residents' activities and will only permit this procedure to be used for soils below a depth of 2 m, when in fact the majority of the contamination is found in the upper 2 m on urban sites. According to Québec Environment, the majority of the resident activities will be held within the upper 2 m layer. In addition, the RA/RM approach will only be permitted on large sites. This is because Québec Environment (1998) considers that the cleanup of small to medium sized sites will be faster and less costly in the long run using the criteria approach. These two restrictions greatly reduce the number of sites where the RA/RM approach can be used in Montreal, and greatly minimize its advantages.

Even though the province of British Columbia's experience has shown that the RA/RM approach is more profitable for larger projects and sites (G. Fox, personal communication), it is unfortunate that Québec Environment is compelled to interfere with the developer's decision-making process regarding the applicability of the RA/RM approach for small contaminated sites.

In fact, in this era of budget and personnel cuts in the public sector, some of the restrictions imposed by Québec Environment on the application of this new approach could be perceived as attempts to reduce the number of RA/RM corrective action plans that they would have to approve.

As for the 2 m restriction on residential sites (where residents can have access to individual yards), no other Canadian jurisdictions have implemented such a measure, relying instead on institutional controls. This restriction imposed by Québec Environment overrules the conclusions of the risk assessment approach, which is, by nature, site-specific and, if completed with the risk management considerations, should dictate the appropriate mitigation measures. Québec Environment (1998) considers that without that protective layer, it will be difficult to ensure the security of the residents.

Consequently, developers who wish to use the RA/RM approach to clean up contaminated sites for residential purposes will have to face two constraints: they will have to respect the 'B' generic criterion for the first 2 m of soil, and institutional controls will be in effect for the rest of the site in order to ensure the durability of the implemented risk management measures.

The government position to adopt safe and cautious lines of conduct is justified and indisputable. However, the restrictions imposed when using the

RA/RM approach to contaminated sites cleanup brings one to reflect on the necessity of such constraints for an activity aimed at reducing the potential risk posed by a contaminated site (in a redevelopment context), especially when compared to measures implemented for other activities that are evidently increasing the risk to the population and the environment (KPMG and Lapointe Rosenstein 1997). For example, in order to obtain the permits to implement a new gas station (i.e., a new source of contamination), the promoter will be under no obligation to carry out any risk assessment when it is well known that such installations release large quantities of carcinogenic and toxic fumes (BTEX), and are responsible for the contamination of soils and groundwater from leaking underground tanks or from surface spills.

The Pilot Project has illustrated that many successful applications of the RA/RM approach by developers will be required in order to alleviate any doubts about the process in order to convince the authorities to reduce the number of restrictions. Some of these restrictions appear to be a reflection of fears and apprehension toward the new and complex risk assessment approach, as well as a lack of resources to monitor the performance of this approach.

In fact, the generic criteria approach to this problem is now so familiar and reassuring for all interested parties (developers, financial services sectors, authorities) that any new method is considered with suspicion even if it allows a more rigorous and thorough assessment of the risk posed by contaminants present on a site.

The confidence in the generic criteria approach is so absolute that there is little, if any, concern for the potential residual risk posed by the contaminants left in place. In most cases, this approach will target contaminated soils above 'B' level for residential use or 'C' level for commercial or industrial use, while leaving in place contaminated soils above background concentrations but below 'B' or 'C' level.

On the other hand, the confidence is not the same for the residual risk that follows from the RA/RM approach, which is subject to numerous control and monitoring measures. The reticence of Québec Environment toward this approach has led them to impose a large number of restrictions when used on sites intended for residential, commercial or industrial purposes.

Thus, the main problem is one of perception of the risk posed by this new approach which has not been used often, is not well known, requires additional data that are not available and appears risky in comparison to the familiar, reassuring, easy to manage and very widely applied generic criteria approach (KPMG and Lapointe Rosenstein 1997).

This perception problem is also present in the general context of site contamination and has been noted by many organizations interested in this issue, such as the NRTEE and the CHMC. Its resolution will only be achieved through information and education of all interested parties, including the public (Delcan *et al.* 1996, 1997), and must be addressed promptly by all parties and integrated into the authorities' lines of action.

In other respects, the developer who resorts to the RA/RM approach to clean up a site will have to record a notice at the Land Records Office. The notice will include information to the effect that contaminants in concentrations above the land use generic criteria are still present, the type of mitigation measures put in place, the monitoring programme to be implemented and the land use restrictions. This document will impact the property since all of the above-mentioned information will be available to anyone searching through the Land Records. This measure will place a stigma on the property, the importance of which will be directly related to the message carried by such a notice: if the message is negative, chances are the stigma will be greater than if the message is positive.

How will a property value be affected where contaminants are still present above the land use generic criteria? If the application of the RA/RM approach brings an important reduction of the cleanup costs but, in return, is subjected to land use restrictions (e.g., no vegetable gardens), how much will the value of the property be affected compared to one that bears no conditions? Will potential buyers be interested in a well-located property in an urban area but with land use restrictions, when they could very easily buy a 'green' site in the suburbs?

One of the main objectives of any rehabilitation action on brownfields is to enable selling of the cleaned site so it can be redeveloped. How will the RA/RM procedure be perceived by the financial services sector, lawyers and other parties who have just got used to the generic criteria approach and often apply it to the letter? They will certainly have to be educated again to take into account the new approach.

The financial institutions will be more likely to consider a new approach to contaminated sites rehabilitation if it is supported and approved by the authorities. Nevertheless, they will always be interested in the consequences of sharing the responsibility in the case of legal action, as well as in the reaction of local groups who could have a considerable financial impact on the project, or the corporate image of the developer. Thus, even if the CCME principles regarding the sharing of responsibility toward contaminated sites are integrated into the authorities' policies and applied, the problem of the perception of the risk and confidence between the parties involved or touched by the redevelopment project will remain (KPMG and Lapointe Rosenstein 1997).

21.9 CONCLUSIONS

In a context of significant financial resources required to clean up a large number of brownfields, the Pilot Project has allowed the City of Montreal to apply an RA/RM approach to contaminated sites rehabilitation, which proved to be adaptable to the City's problems and capable of greatly reducing the cleanup costs as well as protecting human health and the environment (Guay *et al.* 1997).

Nevertheless, the applicability of this new rehabilitation approach to urban contaminated sites within the framework proposed by Québec Environment is greatly reduced. Many obstacles threatening the utilization of this approach were identified during the course of the Pilot Project.

On residential sites, the RA/RM approach will only be authorized from a depth of 2 m, the first 2 m of soil having to comply with the residential generic criteria (Québec Environment 1998). Also, the authorities will not permit the use of the RA/RM approach on small sites since they consider that it will not be profitable (Québec Environment 1998). These two conditions will have the effect of greatly diminishing the applicability of the RA/RM approach for urban contaminated sites.

Although the results of the six risk assessments carried out during the Pilot Project have shown that the elimination of the contact between the contaminants and the exposed population (human and the environment) using the development components allows the reduction of the risks to an adequate level (Desbiens 1997), additional measures were required in order to accept the remedial plans, considering that the ones proposed were not sufficient. The concerns were mostly related to the integrity and durability of the proposed measures. Thus, it was necessary to implement more conservative plans than those which were dictated by the risk assessment results, diminishing at the same time the economic advantages of such plans, especially for small to medium size sites. This new approach to the contaminated sites problem, based on complex tools and methodologies and requiring additional resources, worries the authorities, who in return translate their concerns into conservatism in the risk management process.

It is important to state that the apprehension towards this new approach is shared by many of the interested parties, mainly the future owners, the financial services sector and local groups (KPMG and Lapointe Rosenstein 1997). Obviously, the attitude of the authorities toward this method will greatly influence the perception of the other parties.

In that context, it appears essential that all concerned parties contribute to finding ways to reduce the reluctance expressed toward this new approach to the contaminated sites issue. The CCME, NRTEE and the CMHC have been working on that question and are proposing means to diminish the adverse reaction of many parties. In particular, they emphasize that it is essential that all parties improve their knowledge of the brownfield problem, its consequences and possible solutions. The greater their understanding of the issue, the less chance the reactions toward it will be dictated by fear of the unknown. Likewise, the more the RA/RM approach for contaminated sites cleanup plans is used, the more familiar the procedure will become to concerned parties, providing them with opportunities to better comprehend its advantages and limitations, as they have experienced with the generic criteria approach.

Fortunately, concepts such as information availability regarding contaminated sites and public consultations within the approval procedures of cleanup plans based on the RA/RM approach have been included into the new Soil Protection and Contaminated Site Cleanup Guidelines. Also, mechanisms such as the Site Registry, certificate of compliance, site environmental profile and deed on land title will be used in the near future.

Nevertheless, to ensure their efficiency, these concepts and mechanisms must be combined with an extensive education and information campaign on the RA/RM approach to the contaminated sites problem. Only the open-minded attitude of all parties will ensure the future of this new solution to the contaminated sites problem.

It should also be stated that many of the tools related to the RA/RM approach proposed by Québec Environment in its 1998 Guidelines are just starting to be used or will be implemented in the near future (assessment methods, approval procedure, certificate of compliance, financial incentives and deed restrictions). In the context of inexperience, it is difficult to predict the consequences of such instruments. However, it is obvious that the particular nature of the chosen tools and the way they will be implemented will have definite impacts on the success of this new approach.

Finally, it should be underlined that since the completion of the Pilot Project, other solutions to the contaminated sites problem, which is inherently economic, have been examined by the City of Montreal. One of the answers has been the creation by the Quebec Government of a brownfields redevelopment grant programme aimed at helping developers and owners who need to clean up contaminated sites as part of their construction projects. This new $40 million grant programme will support the efforts of redevelopment of the City of Montreal and the City of Quebec until the year 2003. This programme was recently extended to all other municipalities in the Province of Quebec, with $50 million additional funding by the government.

Consequently, and considering that in the last two years prices to dispose of contaminated soils have been reduced drastically in Quebec due to the introduction of new authorized landfills, the City of Montreal has seen the solutions to its contaminated sites problem change from sole and unaffordable, to numerous and less costly, and thus could be compelled to use the RA/RM approach for those sites where its economical advantages outweigh the costs of current practice. Only the future will tell.

REFERENCES

Belles-Isles J-C and Sarrailh J (1997) Le recouvrement des terrains résidentiels contaminés: une solution sécuritaire? Analyse et gestion de risques: réalités et tendances. In *Texte des conférences du IXᵉ Colloque régional sur l'environnement*, 24 octobre 1997, Ordre des ingénieurs du Québec et Association des biologistes du Québec, Québec, pp. 26–35.

CCME (1993) *Contaminated Sites Liability Report — Recommended Principles for a Consistent Approach across Canada*, Canadian Council of Ministers of the Environment, Winnipeg, 18 pp.

Delcan, Golder Associates Ltd. and McCarthy-Tétreault (1996) *Removing Barriers to the Redevelopment of Contaminated Sites for Housing*, Report to the Canadian Mortgage and Housing Corporation, Ottawa, 127 pp.

Delcan, Golder Associates Ltd. and McCarthy-Tétreault (1997) *Urban Brownfields: Case Studies for Sustainable Economic Development — The Canadian example*, Report to the Canadian Mortgage and Housing Corporation, Ottawa, 107 pp.

Desbiens R (1995) *Détermination des approches méthodologiques — Rapport d'étape — Projet-pilote d'aide aux municipalités de la région de Montréal propriétaires de terrains contaminés*, Rapport à la Ville de Montréal, D'Aragon, Desbiens, Halde Associés ltée, Montréal, 45 pp.

Desbiens R (1996) *Projet-pilote d'aide aux municipalités de la région de Montréal propriétaire de terrains contaminés — Propriété X — Rapports: analyse de risques à la santé, évaluation écotoxicologique, évaluation de l'impact sur les eaux souterraines, scénario de restauration proposé*, Rapport à la Ville de Montréal, D'Aragon, Desbiens, Halde Associés ltée, Montréal, 149 pp.

Desbiens R (1997) *Projet-pilote d'aide aux municipalités de la région de Montréal propriétaire de terrains contaminés — Rapport intérimaire*, Rapport à la Ville de Montréal, D'Aragon, Desbiens, Halde Associés ltée, Montréal, 86 pp.

Geiger G, Federer P and Sticher H (1993) Reclamation of heavy metal contaminated soil: field studies and germination experiments. *Journal of Environmental Quality*, **22** (1): 201-207.

Guay C (1999) *Difficultés et contraintes liées à l'utilisation de l'analyse de risques comme outil de gestion des terrains contaminés en milieu urbain*. MEnv Thesis, Sherbrooke University, 82 pp.

Guay C, Barbeau S, Morin D and Desbiens R (1997) Application de l'analyse de risques comme outil de gestion des terrains contaminés à Montréal. In *Compte rendu du 6e symposium sur les eaux souterraines et les sols contaminés — Americana 1997*, 18 au 21 mars 1997, Montréal, pp. 359-367.

KPMG and Lapointe Rosenstein (1997) *Évaluation des implications de l'utilisation de l'analyse de risques pour la réhabilitation des terrains contaminés au Québec*, Rapport au Centre d'assainissement des sols, Montréal, 145 pp.

MSSS (1999a) *Évaluation et gestion du risque toxicologique au Québec*, Principes directeurs d'évaluation du risque toxicologique pour la santé Québec, Ministère de la santé et des services sociaux, 57 pp.

MSSS (1999b) *Évaluation du risque toxicologique au Québec — Lignes directrices pour la réalisation des évaluations du risque toxicologique pour la santé humaine dans le cadre de la procédure d'évaluation et d'examen des impacts sur l'environnement et l'examen de réhabilitation de terrains contaminés*. Document de consultation Québec, Ministère de la santé et des services sociaux, 90 pp.

Nantel A (1997) L'analyse de risque en santé publique: un dossier en pleine mutation. Analyse et gestion de risques: réalités et tendances. In *Texte des conférences du IXᵉ Colloque régional sur l'environnement*, 24 octobre 1997, Ordre des ingénieurs du Québec et Association des biologistes du Québec, Québec, pp. 10-14.

NRTEE (1998) *Greening Canada's Brownfield Sites*, National Round Table on the Environment and the Economy, Ottawa, 72 pp.

Oddie TA and Bailey AW (1988) Subsoil thickness effects on yield and soil water when reclaiming sodic minespoil. *Journal of Environmental Quality*, **17** (4), pp. 623-627.

Power JF, Sandoval FM, Ries RE and Merill SD (1981) Effects of topsoil and subsoil thickness on soil water content and crop production on a disturbed soil. *Soil Science Society American Journal*, **45**, 124-129.

Québec Environment (1996a) *Lignes directrices pour la réalisation des analyses des risques toxicologiques Version préliminaire pour consultation*, Ministère de l'Environnement et de la Faune du Québec — Groupe d'analyse de risque — Direction des laboratoires, Québec, 48 pp.

Québec Environment (1996b) *Guide technique pour la réalisation des analyses préliminaires des risques toxicologiques — Version préliminaire pour consultation*, Ministère de l'Environnement et de la Faune du Québec — Groupe d'analyse de risque — Direction des laboratoires, Québec, 763 pp.

Québec Environment (1996c) *Projet de politique de protection et de conservation des eaux*, Ministère de l'Environnement et de la Faune du Québec, Québec, 36 pp.

Québec Environment (1996d) *Plan d'action pour la mise en œuvre de la Politique de protection et de conservation des eaux souterraines*, Projet Ministère de l'Environnement et de la Faune du Québec, Québec, 89 pp.

Québec Environment (1998) *Politique de protection des sols et de réhabilitation des terrains contaminés*, Ministère de l'Environnement et de la Faune du Québec Service des lieux contaminés — Direction des Politiques du secteur industriel, Québec, 95 pp.

Tassé N, Cyr J, Beauchemin S, Gasser MO, Benoît P and Marcoux A (1993) *Problématique de la revégétation des haldes minières à caractère neutre en présence de résidus forestiers*, INRS Géoressource.

USEPA (1989) *Risk Assessment Guidance for Superfund. Volume 1: Human Health Evaluation Manual.* Oswer directive 9285.7-01A, United States Environmental Protection Agency, Office of Emergency and Remedial Response, Washington, DC.

Ville de Montréal (1993) *Document de travail relatif aux coûts d'application de la Politique de réhabilitation des terrains contaminés à Montréal*, Service du génie — Division du Laboratoire, Service des affaires institutionnelles — Division des analyses d'impacts et de l'environnement, Montréal, 22 pp. (unpublished).

Ville de Montréal (1995) *Projet-pilote d'aide aux municipalités de la région de Montréal propriétaires de terrains contaminés Stratégie d'intervention*, Service du génie — Division du Laboratoire, Montréal, 11 pp. (unpublished).

22

Environmental Fate and Human Exposure Modeling of the Residual TPH Contamination in a Bioremediated Petroleum Storage Site

SYLVAIN LORANGER[1], YVAN POULIOT[2], SÉBASTIEN SAUVÉ[1], LUC DUSSAULT[2] AND YVON COURCHESNE[1]

[1] *QSAR Risk Assessment Service Inc., Montreal, Quebec, Canada*
[2] *Biogénie Inc., Sainte-Foy, Québec, Canada*

22.1 INTRODUCTION

Environmental contamination and human exposure to petroleum products are a widespread and frequent problem (see also Chapter 23). However, the assessment of the potential human health risks of such complex mixtures containing hundreds of individual organic chemicals has proven difficult, elusive and imprecise. Among the difficulties associated with the risk assessment, one needs to consider the compositional change of spilled products arising from weathering and aging processes (e.g., volatilization, hydrolysis, photolysis and biodegradation) and transport throughout environmental compartments (e.g., vadose-zone soil, groundwater and air).

The lack of toxicological data for each individual compound, notably for the weathered products, further complicates the analysis of the links between exposure dose and risk. Many approaches have been proposed to address this question. For example, the Massachusetts Department of Environmental Protection (MADEP) and the Total Petroleum Hydrocarbon Criteria Working Group (TPHCWG), rather than quantifying the entire range of petroleum hydrocarbons, suggest dividing petroleum hydrocarbons into subgroups of compounds based on their number of carbon atoms. This results in a more manageable number of fractions and allows the selection of a representative toxicological value (e.g., reference dose or RfD) for each subgroup using a

Environmental Analysis of Contaminated Sites. Edited by G. I. Sunahara, A. Y. Renoux, C. Thellen, C. L. Gaudet and A. Pilon

well-characterized reference compound or surrogate (MADEP 1994, TPHCWG 1997a). This approach offers an interesting alternative to cleanup standard guidelines, traditionally based on a substance-by-substance evaluation, which do not consider the risk that the mixture may pose as a whole.

In Canada, the Canadian Council of Ministers of the Environment (CCME) has recommended the use of fractions and surrogates for managing ecological and human health risk related to total petroleum hydrocarbon (TPH) contamination and exposure (CCME 1997, 1999, 2000, 2001).

In the province of Quebec, the Ministry of Environment came out with new guidelines for the remediation of contaminated sites (MEF, 1998). These guidelines use ecological and human risk assessment as tools for managing the remediation of industrial, commercial or residential sites as well as the traditional chemical-specific soil and groundwater cleanup levels. However, sites regulated under the Petroleum Products Regulation (Gouvernement du Québec 1999), such as service stations and storage sites, are still evaluated on a criteria-based approach unless excavation is technically impractical or in the case of *in situ* treatment under infrastructures. In this latter case only, a risk analysis, which includes elements such as risk assessment, risk management and risk communication, could be conducted for site management.

Based on the original paradigm described for the first time by the National Research Council (NRC 1983), the risk assessment process is divided into four steps: hazard identification, exposure assessment, toxicity assessment and risk characterization. Each step necessitates specific data, which are interrelated in the risk assessment process. Hazard identification aims to identify the source of contamination as well as the chemicals of concern, the adverse health effects related to these substances and the potential routes and media of human exposure. This first step consists essentially in gathering and evaluating all the available data (e.g., characterization studies) related to the source of contamination. The exposure assessment is the estimation of the magnitude, frequency, duration and route of exposure of a target population to the selected contaminants. The toxicity assessment is the process of quantitatively evaluating the toxicological information about selected contaminants on carcinogenic and non-carcinogenic effects (i.e., reference doses and carcinogenic slope factors) and characterizing the relationship between the exposure or absorbed dose and the incidence of deleterious effects in the exposed population. Finally, risk characterization summarizes and combines the results of the exposure and toxicity assessments, in order to give an estimation of the potential risk for the human population and evaluate the uncertainty related to these estimations. In general, this whole process will give risk estimates, in terms of probability (cancer risk) or ratio (non-cancer risk), for a long-term exposure period (i.e., over a human lifetime).

Using the extractable petroleum hydrocarbon (EPH) fraction approach, our objective is to illustrate an application of the risk assessment procedure for the evaluation of the human health risk related to the residual contamination of a former petroleum storage site in a residential area near Quebec City. We also aim to measure and quantify the relationship between TPH and EPH fractions, and use the predicted EPH values to evaluate different remedial scenarios using an environmental fate and multimedia exposure model, CalTOX (McKone 1993).

22.2 MATERIALS AND METHODS

22.2.1 SITE CHARACTERISTICS

The contaminated site is located on the shore of the St. Lawrence River in Sillery (Quebec, Canada). It is owned by the Canadian National Railways and covers an area of approximately 46 100 m^2. Over a period of almost 60 years (1935 - 1993), the industrial activities related to petroleum storage and handling facilities left a significant soil contamination. More than 20 aboveground and underground storage tanks containing gasoline, bunker C, No. 2 fuel, diesel, waste oil, kerosene and fuel additives were scattered around the site. Pipelines and railway and truck loading racks were also part of the equipment. Between 1993 and 1996 all the facilities, including administrative buildings, were dismantled. In July 1995, a vast on-site bioremediation program was undertaken. Several remedial technologies including air sparging, bioventing, biopile treatment station and off-site disposal were used for the decontamination of more than 40 000 m^3 of soil (Biogénie Inc. 1998). This program led to a significant reduction of BTEX and TPH ($C_{10}-C_{50}$) concentrations in the soil and in the groundwater. In December 1997, on average, all the substances measured in the soil (0-4.5 m depth), including oil and grease, TPH, polycyclic aromatic hydrocarbons (PAHs), metals and polychlorinated biphenyls (PCBs) were below the provincial generic criterion 'B' for residential land use. However, in some areas, some discrete soil samples still showed TPH concentrations exceeding the criterion.

22.2.2 EPH DETERMINATIONS

To refine the characterization of the TPH soil contamination, different EPH fractions (aliphatic: $C_{12}-C_{18}$, $C_{19}-C_{36}$; aromatic: C_{11}-C_{22}) were measured in about 15% of the TPH soil samples (30 analyses/207 soil samples). The EPH concentrations were measured using the analytical procedure described by the MADEP (1994) approach. However, due to the loss of the C_9-C_{11}aliphatic fraction during the analysis, only the $C_{12}-C_{18}$ fraction was quantified instead of C_9-C_{18}.

22.2.3 CalTOX MODEL

CalTOX is a compartmental model, which combines two major components: a multimedia transport and transformation model and a multimedia, multiple pathway exposure model (McKone 1993, McKone *et al.* 1997). The environmental fate component is based on mass balance and fugacity principles, while the exposure component includes deterministic equations (USEPA 1989) which estimate the intake through ingestion, inhalation and dermal contact of the chemical of concern in the potentially contaminated air, water, soil and food.

The model comprises seven environmental compartments: air, ground-surface soil (air/soil interface), root-zone soil, vadose-zone soil (unsaturated zone), plants, surface water and sediments. The current version of CalTOX does not consider the groundwater as a specific compartment. It considers instead the concentration of the contaminant in the water leaching from the vadose-zone soil as an input to the groundwater zone.

Source input can be assigned to five of the CalTOX compartments: in the root zone and the vadose zone, the source must be specified as being an initial concentration (mg kg^{-1}) at time 0 (t_0), while in air, ground-surface soil and surface water, the source can be specified as a continuous input (mol day^{-1}).

CalTOX calculates the average or the maximum daily exposure dose (mg kg^{-1} body weight day^{-1}) per media and pathway for the selected target group (i.e., adults and children). For this study, only the maximum daily doses were used to calculate the non-carcinogenic risk for the EPH fractions, since no carcinogenic substance (e.g., benzene) were retained for the analysis.

22.2.4 MODEL PARAMETERS

CalTOX uses three sets of input data: the exposure factors of each target group (Table 22.1), the landscape characteristics of the site (Table 22.2), and the physical and chemical properties of the substance (Table 22.3). In addition to these parameters, CalTOX requires an estimation of the coefficient of variation (CV, defined as the arithmetic standard deviation divided by the arithmetic mean) for each parameter to perform a sensitivity or uncertainty analysis. Although it is difficult to calculate the CV from the data found in the literature, we defined an arbitrary CV range which corresponds to parameters with relatively low (CV = 20%) to high (CV = 1000%) variability or uncertainty depending on the reliability of the estimation methods and the source of the information (McKone 1994a,b).

22.2.4.1 Exposure Factors and Scenarios

According to the development plan of the site, about 78% (36 100 m^2) of the area will be used for a housing development and 22% (10 000 m^2) will remain for industrial and commercial activities. For the purpose of this study, risk estimations were based on an environmental exposure to the aliphatic and aromatic EPH fractions of two target groups: adults and children (0–4 years old), without sex discrimination and only in the residential area (Table 22.1).

Table 22.1 Human exposure factors

Exposure factors		Adult (≥20 years)	Toddler (0–4 years)	CV (%)	Reference
Body weight (kg)	BW	70	16	50	MSSS (1999)
Surface area (m^2)	Sab	1.8	0.6	50	MSSS (1999)
Active breathing rate (m^3 day^{-1})[a]	Bra	27	17	50	MEF (1996)
Resting breathing rate (m^3 day^{-1})	BRr	12	2	50	MEF (1996)
Fruit and vegetable intake (g day^{-1})	Ifv	436	314	100	Health Canada (1994)
Soil ingestion (mg day^{-1})	Isl	20	100	100	MSSS (1999)
Fraction of fruits and vegetables that are 'exposed produce'	fabvgrdv	0.08	0.08	50	MEF (1996)
Fraction of fruits and vegetables local[b]	flocal_v	0.1	0.1	50	MSSS (1999)
Plant–air partition factor, particles (m^3 kg^{-1} [FM])	Kpa_part	3300	3300	500	Mc Kone (1993)
Rainsplash (mg kg^{-1} [plant FM] × mg kg^{-1} [soil])	rainsplash	0.0034	0.0034	100	Mc Kone (1993)
Exposure time, active indoors (h day^{-1})	Etai	4.5	8	50	USEPA (1997)
Exposure time, outdoors at home (h day^{-1})	Etao	0.3	0.6	50	USEPA (1997)
Exposure time, indoors resting (h day^{-1})	Etri	11.5	15	50	USEPA (1997)
Indoor dust load (kg m^{-3})	dust	3×10^{-8}	3×10^{-8}	50	Mc Kone (1993)
Exposure frequency to soil on skin (day $year^{-1}$)	Efsl	214	214	50	c
Soil adherence to skin (mg cm^{-1})	Slsk	0.1	0.5	100	MEF (1996)
Ratio of indoor gas conc. to soil gas conc.	Airsoil	0.0001	0.0001	100	Mc Kone (1993)

[a] Light activity, rapid walk.
[b] Urban area.
[c] Exposure from April to October inclusive.

Table 22.2 Landscape characteristics of the contaminated site (Anse au Foulon, Sillery, QC)

Landscape characteristics		With mitigation	Without mitigation	CV (%)	Reference
Contaminated area (m^2)	Area	36 100	36 100	20	Biogénie Inc. (1998)
Annual average precipitation (m day^{-1})	rain	3.31×10^{-3}	3.31×10^{-3}	20	Environment Canada (1993)
Atmospheric dust load (kg m^{-3})	Rhob_a	5.40×10^{-8}	5.40×10^{-8}	20	MEF (1992)
Ambient environmental temperature (K)[a]	Temp	288	288	20	Environment Canada (1993)

(*continued overleaf*)

Table 22.2 (*continued*)

Landscape characteristics		With mitigation	Without mitigation	CV (%)	Reference
Yearly average wind speed (m day^{-1})	v_w	360 000	360 000	20	Environment Canada (1993)
Deposition velocity of air particles (m day^{-1})	v_d	86.4	86.4	20	Mc Mahon and Denison (1979)
Surface water into landscape (m day^{-1})	Inflow	10^{-6}	10^{-6}	100	b
Land surface runoff (m day^{-1})	runoff	8.2×10^{-4}	8.2×10^{-4}	100	MEF (1993)
Groundwater recharge (m day^{-1})	recharge	10^{-4}	10^{-4}	500	b
Plant dry mass inventory (kg dry matter m^{-2})	bio_inv	1.5	1.5	50	MEF (1996)
Plant dry-mass fraction	bio_dm	0.2	0.2	50	Baes *et al.* (1984)
Plant fresh-mass density (kg m^{-3})	rho_p	1000	1000	50	Mc Kone (1993)
Thickness of the ground soil layer (m)	d_g	0.01	0.01	50	Mc Kone (1993)
Soil particle density (kg m^{-3})	rhos_s	2000	2000	50	Droppo *et al.* (1994)
Water content in surface soil (vol. fraction)	beta_g	0.15	0.2	50	Droppo *et al.* (1994)
Air content in surface soil (vol. fraction)	alpha_g	0.25	0.2	50	Droppo *et al.* (1994)
Erosion of surface soil (kg m^{-2} day^{-1})	er_g	2.74×10^{-5}	2.74×10^{-5}	50	Agriculture Canada (1990)
Thickness of root-zone soil (m)	d_s	1	0.3	20	b
Water content of root-zone soil (vol. fraction)	beta_s	0.15	0.2	50	Droppo *et al.* (1994)
Air content of root-zone soil (vol. fraction)	alpha_s	0.25	0.2	50	Droppo *et al.* (1994)
Thickness of vadose-zone soil (m)	d_v	4	5	20	b
Water content of vadose-zone soil (vol. fraction)	beta_v	0.15	0.2	50	Droppo *et al.* (1994)
Air content of vadose-zone soil (vol. fraction)	alpha_v	0.25	0.2	50	Droppo *et al.* (1994)
Organic carbon fraction in root and vadose-zone soil	foc	0.035	0.035	100	Droppo *et al.* (1994)
Boundary layer thickness in air above soil (m)	del_ag	0.005	0.005	20	Mc Kone (1993)
Thickness of aquifer layer (m)	d_q	1	1	20	b
Solid material density in aquifer (kg m^{-3})	rhos_q	1420	1420	20	Droppo *et al.* (1994)
Porosity of aquifer zone	beta_q	0.33	0.33	20	Droppo *et al.* (1994)

[a] Average temperature in Quebec City from April to October inclusive between 1961 and 1990.
[b] Estimated (see text for explanation).

Table 22.3 Physical and chemical properties of the EPH fractions

Parameters		Aliphatic		Aromatic		
		C_{12}–C_{18}	C_{19}–C_{36}	C_{11}–C_{22}	CV (%)	Reference
Equivalent carbon number	*EC*	15	28	16	50	[a]
Molecular weight (g mol^{-1})	MW	210	285	160	100	TPHCWG (1997b)
Octanol–water partition coefficient	K_{ow}	2×10^7	5×10^8	10^5	1000	TPHCWG (1997b)
Melting point (K)	T_m	273	273	273	500	[a]
Boiling point (°C)	Bp	270	320	280	1000	TPHCWG (1997b)
Vapor pressure (atm)	VP	2.09×10^{-5}	4.37×10^{-10}	9.12×10^{-6}	500	TPHCWG (1997b)[b]
Solubility (mg L^{-1})	*S*	1.78×10^{-4}	1.26×10^{-11}	2.19	500	TPHCWG (1997b)[b]
Henry's law constant (cm^3 cm^3)	*H*	79.4	1.45×10^2	1.05×10^{-2}	1000	TPHCWG (1997b)[b]
Organic carbon partition coefficient K_{oc} (ml g^{-1})	K_{oc}	1.51×10^7	2.00×10^{13}	7.94×10^3	500	TPHCWG (1997b)[b]
Diffusion coefficient in pure air (m^2 day^{-1})	D_{air}	8.64×10^{-1}	8.64×10^{-1}	8.64×10^{-1}	100	TPHCWG (1997b)
Diffusion coefficient in pure water (m^2 day^{-1})	D_{water}	8.64×10^{-5}	8.64×10^{-5}	8.64×10^{-5}	100	TPHCWG (1997b)[a]
Partition coefficient in ground, root and vadose-zone soil layers	K_d	5.34×10^5	7.04×10^{11}	2.80×10^2	100	Karickhoff (1981)
Partition coefficient in plant aboveground/soil (kg [s] kg $^{-1}$ [pFM])	K_{ps}	5.48×10^{-4}	7.22×10^{-5}	9.92×10^{-3}	100	Mc Kone (1993)[d]
Biotransfer factor, plant/air (m^3 [a] kg^{-1} [pFM])	K_{pa}	4.35×10^3	7.94×10^4	2.19×10^5	100	Mc Kone (1993)[d]
Skin permeability coefficient (cm h^{-1})	K_{pw}	1.00	42.17[a]	1.00	100	Mc Kone (1993) USEPA (1992)[d]
Fraction dermal uptake from soil	D_s	5.31×10^{-7}	1.70×10^{-11}	2.20×10^{-3}	100	Mc Kone (1993)[d]
Reaction half-life in air (day)	Half_a	7	7	7	100	[a]
Reaction half-life in surface soil (day)	Half_g	730	730	730	100	[a]

(continued overleaf)

Table 22.3 (*continued*)

Parameters		Aliphatic		Aromatic		
		C_{12}–C_{18}	C_{19}–C_{36}	C_{11}–C_{22}	CV (%)	Reference
Reaction half-life in root-zone soil (day)	Half_s	730	730	730	100	a
Reaction half-life in vadose-zone soil (day)	Half_v	730	730	730	100	a
Reaction half-life in groundwater (day)	Half_q	730	730	730	100	a

[a] Estimated (see text for explanation).
[b] $\log_{10} VP = -0.36 \times EC + 0.72$ (aliphatic and aromatic); $\log_{10} S = -0.55 \times EC + 4.5$ (aliphatic); $\log_{10} S = -0.21 \times EC + 3.7$ (aromatic); $\log_{10} H = 0,02 \times EC + 1.6$ (aliphatic); $\log_{10} H = -0.23 \times EC + 1.7$ (aromatic); $\log_{10} K_{oc} = 0.45 \times EC + 0.45$ (aliphatic); $\log_{10} K_{oc} = 0.1 \times EC + 2.3$ (aromatic).
[c] $K_d = K_{oc} \times f_{oc}$.
[d] $K_{ps} = 7.7 \times K_{ow}^{-0.578}$; $K_{pa} = (0.5 + (0.4 + 0.01 \times K_{ow}) \times R \times Temp/H) \times 0.001$, R = universal gas constant, 831 Pa m^3 mol^{-1} K^{-1}; $K_{pw} = MW^{-0.6} \times ((0.0000024 + 0.00003 \times K_{ow}^{0.8})/\delta_{skin})$ for $MW \leq 280$; $K_{pw} = 0.0019 \times K_{ow}^{0.71} \times 10^{(-0.0061 \times MW)}$ for $MW > 280$; $D_s = (1 - \exp(-(0.5 \times (ETai + ETao)) \times (K_{pw}/(K_d \times 1.7))/(Slsk/0.015)))$.

Two scenarios were retained: one with the bioremediated soil in place (scenario without mitigation) and another with 30 cm of clean topsoil over the bioremediated soil (scenario with mitigation). We assumed that each individual is exposed to EPH fractions through inhalation of gas and particles (indoor and outdoor), ingestion of homegrown food products, ingestion of soil and dermal contact of soil. Finally, the exposure would begin 1 year after the end of the site restoration. Most of the exposure factors used for simulations were based on several governmental documents (Health Canada 1994, MEF 1996, Richardson 1997, USEPA 1997, MSSS 1999) and the CalTOX database (McKone 1993).

22.2.4.2 Landscape Characteristics

For modeling purposes, the contaminated area (36 100 m^2) corresponds to the residential zone of the site (Table 22.2). The depths of contamination in the root zone and the vadose zone are, respectively, 1 and 4 m before mitigation, and 0.3 and 5 m after mitigation. Meteorological and hydrogeological data were taken from governmental publications, while plant and soil characteristics were estimated from published data. Groundwater recharge was estimated as 20% of the land surface runoff.

In the present case, only the root zone and the vadose zone were selected as input source. For the unmitigated scenario, the root-zone (depth = 1 m) and the vadose-zone (depth = 4 m) concentrations for each EPH fraction were estimated from the linear regression based on the 95% upper bound limit of the average TPH concentration measured above and below the first meter of contaminated soil, respectively. For the scenario with mitigation, the EPH concentrations in each zone were also based on the linear regression (all depths), except that we used half of the TPH detection limit (50 mg kg^{-1}) for the root zone covered with 30 cm of clean topsoil and the 95% upper bound limit of the average TPH concentration measured over all depths for the vadose zone (depth = 5 m).

22.2.4.3 Physical and Chemical Properties of the EPH Fractions

Complex mixtures, such as petroleum hydrocarbons, do not behave in the environment as single pure liquids. Moreover, chemical fate or transport data for these compounds are almost non-existent. Based on the assumption that chemicals with similar nature (aliphatic or aromatic) and boiling point behave similarly, the TPHCWG proposed an approach to define specific TPH fractions based on their physical and chemical properties (TPHCWG 1997b). In fact, each fraction is divided on the basis of an equivalent carbon number index (*EC*) which is related to the boiling point of individual constituents. Fraction-specific properties (e.g., vapor pressure) are then estimated using empirical relations (Table 22.3). It must be mentioned that the TPHCWG proposed more TPH fractions or *EC* ranges for aliphatic and aromatic chemicals than the MADEP.

Since we chose the MADEP methodology to measure the EPH fractions in soil, we estimate the *EC* as the mid-point of each aliphatic and aromatic fraction. For example, the C_{12}–C_{18} aliphatic fraction is represented by an *EC* of 15. These *EC*s were then used as the independent variable to estimate the vapor pressure, solubility, Henry's law constant and organic carbon partition coefficient (K_{oc}). The molecular weight and the octanol–water partition coefficient (K_{ow}) of the EPH fractions were based on the average value of the fraction-specific properties defined by the TPHCWG for our aliphatic and aromatic carbon range. For melting point temperature, values for EPH may vary from −200 to 100 °C (TPHCWG 1997b). Consequently, we assumed that the average value would be around 0 °C (273 K) with a CV of 500%. Other coefficients, such as soil–solution partition coefficients (K_d), plant-soil partition coefficients (K_{ps}) and skin permeability (K_{pw}) were estimated from equations found in the CalTOX documentation (McKone 1993). Since there are no half-life values in the literature for the selected EPH, we also assume conservatively that the half-life of each EPH fraction will be 7 days in air, and 2 years in soil and groundwater.

22.2.5 SENSITIVITY AND UNCERTAINTY ANALYSIS

The sensitivity analysis is used to calculate changes which occur in the model response (i.e., Y = hazard ratio, HR) when there are variations (usually small) in the individual model parameters (i.e., X = half-life in root-zone layer). This analysis enables us to evaluate the influence of each parameter on the model output and to rank all these parameters on the basis of their relative contribution to the variance in the model responses.

This analysis has been done with the add-in forecasting and risk analysis software called Crystal Ball (Decisioneering Inc. 1996). The program uses the CalTOX output spreadsheet which contains all the input parameters (assumption cell) and output variables (forecast cell). The purpose of this analysis was to evaluate the sensitivity of the HR (Y) related to the exposure of toddlers to each EPH fraction in the unmitigated scenario. The Crystal Ball software calculates the relative contribution of each parameter to the variance in the model response. Here, the term 'variance' does not correspond to its general statistical definition, but is defined as an approximation calculated by squaring the rank correlation coefficients and normalizing to 100%.

On the other hand, uncertainty analysis is used to evaluate the variation or the imprecision of the output or predicted variable (Y) based on the collective variation of the model parameters (instead of parameter-by-parameter as in the sensitivity analysis). This analysis was used to predict, in the mitigated and unmitigated scenarios, the HR related to the exposure of the target groups (toddlers and adults) to EPH fractions. It also permits one to compare the result of these simulations to the estimated value (point estimate) generated by CalTOX using the reasonable maximum exposure (RME) estimation determined traditionally by the USEPA (1989). Using the Crystal Ball software, a Monte Carlo simulation was performed to generate a probability density function (PDF) for each HR calculated from the unmitigated scenario. Statistics of the predicted variable (mean, standard deviation, percentiles, etc.) were calculated after 1000 trials.

22.2.6 TOXICOLOGICAL REFERENCE VALUES

TPHCWG (1997b) and MADEP (1994) have proposed fraction-specific reference doses with respect to concentrations for oral and inhalation routes of exposure (Table 22.4). These values were used to estimate HR for each EPH fraction (i.e., exposure dose divided by the reference value) and to evaluate the potential

Table 22.4 Fraction-specific reference dose per exposure route (mg kg^{-1} day^{-1})

Agency	Fraction	Ingestion	Inhalation[b]
MADEP	aliphatic C_9–C_{18}	0.6[a]	0.6
	aliphatic C_{19}–C_{36}	6	non-volatile
	aromatic C_{11}–C_{22}	0.03	0.02
TPHCWG	aliphatic C_9–C_{16}	0.1	0.3
	aliphatic C_{17}–C_{35}	2	non-volatile
	aromatic C_9–C_{16}	0.04	0.06
	aromatic C_{17}–C_{35}	0.03	non-volatile

[a] Underlined values were used for risk calculation.
[b] RfD calculated from the RfC based on an individual of 70 kg inhaling 20 m^3 of air per day.

of the TPH mixture. It must be emphasized that each reference value is used to evaluate only the non-carcinogenic effects of exposure to a given fraction, but not the carcinogenic endpoints. Although the fraction-specific values given by the TPHCWG are not exactly in the same *EC* range as the MADEP, we have calculated the HR using both approaches for each target group.

22.3 RESULTS

22.3.1 SOIL AND GROUNDWATER CONTAMINATION

On average, PAH, MAH, TPH and metal concentrations in soil over the site (Table 22.5) were below the provincial 'B' criterion for a residential site (Figure 22.1), at least 6 months before the end of the bioremediation program (June 1998). However, hot spots were still present in some areas of the site, mainly below the 1-m depth. These areas were decontaminated during the last phase of the remediation program (January to June 1998) through soil venting, sparging, digging and groundwater pumping. In groundwater, BTEX and TPH contamination was below the criterion for wastewater and surface water uses (Table 22.6).

The average C_{12}-C_{18} and C_{19}-C_{36} aliphatic EPH fractions in soil were respectively 195 and 299 mg kg^{-1} (Table 22.5). No statistical difference was noted between the root zone (<1 m) and the vadose zone (>1 m) for each EPH fraction ($p>0.05$), due mainly to the sample heterogeneity as shown by the high CV (57 to 123%, Table 22.7).

22.3.2 EPH REGRESSIONS

Simple linear regressions between TPH (independent variable) and the other EPH fractions (Figure 22.2) indicate that the relationships for the complete data set are statistically significant (with r^2 varying between 53 and 72%;

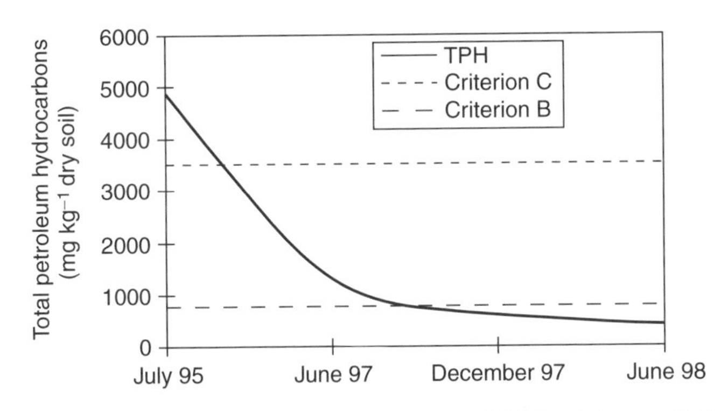

Figure 22.1 Average TPH concentrations in soil during bioremediation.

Table 22.5 Concentration of organic and inorganic substances in soil

Substance	*n*	Arithmetic mean[a]	Standard deviation	95% UL	99th Percentile	Maximum	Criteria[b] A	B	C
Polycyclic aromatic hydrocarbon									
Acenaphthene	29	0.42	0.79	0.72	3.40	4.10	<0.1	10	100
Acenaphthylene	29	0.07	0.05	0.09	0.24	0.30	<0.1	10	100
Anthracene	29	0.16	0.18	0.23	0.83	1.00	<0.1	10	100
Benzo(*a*)anthracene	29	0.16	0.17	0.23	0.76	0.90	<0.1	1	10
Benzo(*a*)pyrene	29	0.13	0.10	0.17	0.44	0.50	<0.1	1	10
Benzo(*b*)fluoranthene	29	0.14	0.10	0.18	0.44	0.50	<0.1	1	10
Benzo(*k*)fluoranthene	29	0.15	0.19	0.23	0.86	1.00	<0.1	1	10
Benzo(*g,h,i*)perylene	29	0.13	0.11	0.17	0.49	0.60	<0.1	1	10
Chrysene	29	0.19	0.13	0.24	0.47	0.50	<0.1	1	10
Dibenzo(*a,h*)anthracene	29	0.07	0.03	0.08	0.17	0.20	<0.1	1	10
Fluoranthene	29	0.30	0.33	0.42	1.48	1.70	<0.1	10	100
Fluorene	29	0.80	1.61	1.41	6.93	8.50	<0.1	10	100
Indeno(1,2,3-cd)pyrene	28	0.11	0.09	0.15	0.42	0.50	<0.1	1	10
Naphthalene	29	1.35	2.02	2.11	7.49	8.30	<0.1	5	50
Phenanthrene	29	1.29	2.09	2.08	9.12	11.00	<0.1	5	50
Pyrene	29	0.35	0.35	0.49	1.64	2.00	<0.1	10	100
Monocyclic aromatic hydrocarbon									
Benzene	151	0.06	0.10	0.08	0.40	1.20	<0.3	0.5	5
Ethylbenzene	151	0.11	0.19	0.14	1.10	1.30	<0.4	3	30
Toluene	151	0.08	0.20	0.11	0.75	2.30	<0.5	5	50
Xylenes	151	0.34	1.48	0.58	3.90	17.00	<0.7	5	50
Oil and grease (O&G)	53	209	310	294	1338	1900	None	None	None
Total petroleum hydrocarbon	207	604	504	674	2000	2500	<100	700	3500

Extractable petroleum hydrocarbon									
Aliphatic fraction (C_{12}–C_{18})	30	195	176	258	718	750	None	None	None
Aliphatic fraction (C_{19}–C_{36})	30	299	358	427	1561	1900	None	None	None
Aromatic fraction (C_{11}–C_{22})	30	104	86	135	380	400	None	None	None
Metals									
Cadmium	9	0.89	1.76	2.25	4.00	4.00	1.5	5	20
Chromium	9	22.67	7.48	28.42	31.68	32.00	85	250	800
Copper	9	32.89	16.83	45.82	72.80	76.00	40	100	500
Nickel	9	28.44	6.15	33.17	36.00	36.00	50	100	500
Lead	9	56.67	30.13	79.83	108.56	110.00	50	500	1000
Zinc	9	86.67	25.30	106.11	118.40	120.00	100	500	1500

Limit of detection (LOD): PAH, MAH = 0.1 mg kg^{-1}; O&G, TPH = 100 mg kg^{-1}; Cr, Ni = 0.5 mg kg^{-1}; Cu = 0.25 mg kg^{-1}; Pb = 1.5 mg kg^{-1}; Zn = 0.15 mg kg^{-1}; Cd = 0.05 mg kg^{-1}.

[a] Values under the LOD were replaced by 0.5 × LOD.

[b] Based on MEF (1998): criterion A = background (metals), LOD (organics); criterion B = residential, commercial uses; criterion C = industrial, commercial uses.

Table 22.6 Concentration of BTEX, oil and grease, and TPH in groundwater

Substance	n	Arithmetic mean	Standard deviation	95% UL	99th Percentile	Maximum	Provincial Criteria[a] Drinking water	Provincial Criteria[a] Surface waste water
Monocyclic aromatic hydrocarbon								
Benzene	40	25	63	44	256	260	5	590
Ethylbenzene	40	42	6	44	27	33	2.4	420
Toluene	41	3	7	5	28	30	24	200
Xylenes	39	24	99	55	448	570	300	820
Oil and grease (O&G)	16	273	402	470	1507	1700	None	None
TPH (C_{10}–C_{50})	20	474	1655	1199	6130	7500	None	3500

Limits of detection: MAH = 0.1 mg kg^{-1}; O&G, TPH = 100 mg kg^{-1}.
[a] Based on MEF (1998).

Table 22.7 EPH concentration in soil

EPH fraction	Depth (m)	n	Arithmetic mean	Standard deviation	95% UL	99th Percentile	Maximum	CV (%)
Aliphatic C_{12}–C_{18}	<1	13	236	237	245	737	750	100
Aliphatic C_{19}–C_{36}	<1	13	400	491	667	1760	1900	123
Aromatic C_{11}–C_{22}	<1	13	134	115	196	392	400	123
Aliphatic C_{12}–C_{18}	>1	17	164	109	216	347	350	67
Aliphatic C_{19}–C_{36}	>1	17	221	191	312	652	660	86
Aromatic C_{11}–C_{22}	>1	17	82	47	104	206	220	57

Table 22.8 Hazard ratio calculated by Monte Carlo simulations for mitigated and unmitigated scenarios

Substance target group	Aliphatic C_{12}–C_{18} Adult	Aliphatic C_{12}–C_{18} Toddler	Aliphatic C_{19}–C_{36} Adult	Aliphatic C_{19}–C_{36} Toddler	Aromatic C_{11}–C_{22} Adult	Aromatic C_{11}–C_{22} Toddler
No. of iterations	1000	1000	1000	1000	1000	1000
Mean	2.00×10^{-2}	6.30×10^{-2}	3.30×10^{-5}	1.70×10^{-4}	4.60	14.0
Median	4.00×10^{-3}	1.30×10^{-2}	8.30×10^{-6}	5.70×10^{-5}	1.10	3.00
95th Percentile	8.70×10^{-2}	2.50×10^{-1}	1.30×10^{-4}	7.00×10^{-4}	19.0	62.0
CV (%)	331	349	248	228	287	306
Minimum	5.30×10^{-6}	1.90×10^{-5}	1.90×10^{-8}	7.00×10^{-8}	6.10×10^{-7}	1.40×10^{-6}
Maximum	1.60	5.40	1.00×10^{-3}	5.10×10^{-3}	3.00×10^{2}	9.90×10^{2}
Point estimate—CalTOX	4.40×10^{-3}	1.50×10^{-2}	1.10×10^{-4}	5.20×10^{-4}	9.80×10^{-1}	3.10

$p<0.001$). These relationships are even stronger if we consider only the 1-m surface-soil samples ($r^2 = 0.72$–0.79; Table 22.8). These regressions permitted the parameterization of the model for specific EPH fractions from the TPH data set.

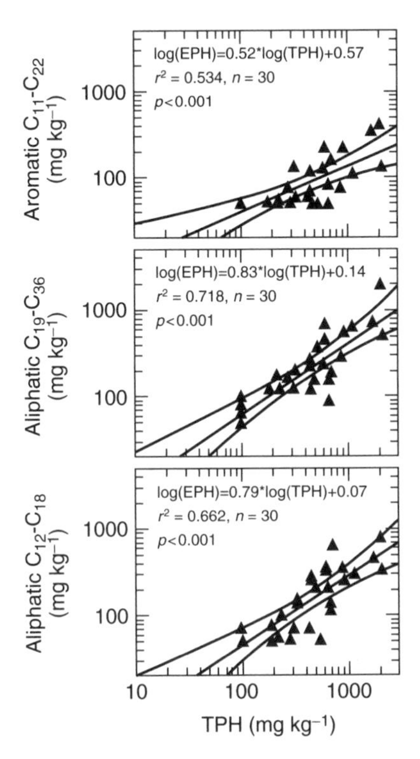

Figure 22.2 Relationships between aliphatic (C_{12}–C_{18}, C_{19}–C_{36}) and aromatic (C_{11}–C_{22}) EPH fractions *vis-à-vis* the TPH fraction. The linear regressions are illustrated with a 95% confidence interval.

22.3.3 MULTIMEDIA CONCENTRATIONS

The EPH fractions estimated by CalTOX for each environmental compartment are illustrated in Figure 22.3. Based on the initial root-zone and vadose-zone EPH concentrations (t_0), the predicted values in these compartments remain almost constant, even after an exposure duration of 70 years. However, the addition of clean topsoil in the root zone (scenario with mitigation) reduced only slightly the EPH concentration estimated in the vadose zone. On the other hand, it limits considerably the volatilization of aromatic fractions from the soil underneath to the air and its intake by plants.

For the surface water and groundwater compartments, the EPH contamination is related essentially to the aromatic fractions due to its high solubility in water compared to aliphatic fractions. It must be noted that the higher concentration in the groundwater after mitigation is due to the inclusion of all soil samples for the initial vadose-zone concentration. On the other hand, in the scenario with mitigation, the new vadose zone incorporates all the data of the site, including the highly contaminated surface soil samples.

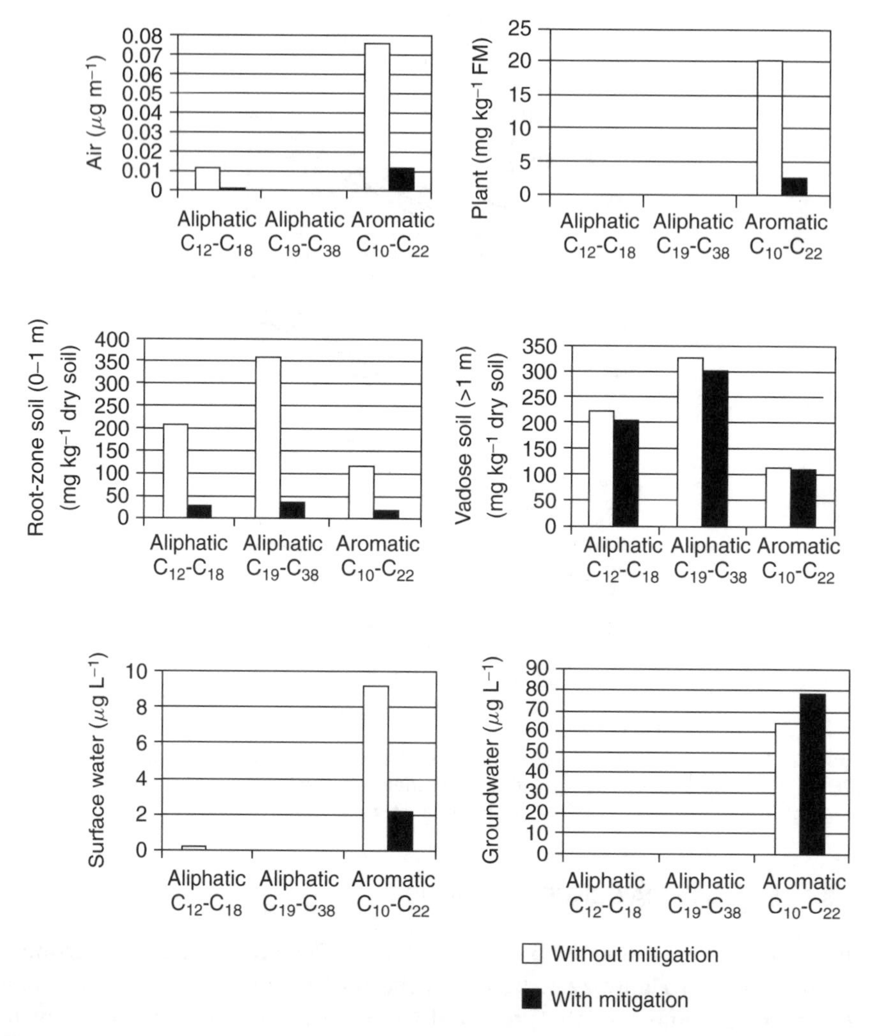

Figure 22.3 EPH concentrations predicted by CalTOX for selected environmental compartments.

22.3.4 EXPOSURE DOSE

Environmental media concentrations are used by CalTOX to estimate exposure media concentrations which in turn serve to calculate exposure doses for each specific pathway. In our case, the ingestion of homegrown food and surface soil is the main route of exposure to aromatic EPHs. Other fractions, as well as other exposure pathways (i.e., inhalation, dermal contact) could be considered negligible. For toddlers, the ingestion of soil is particularly important, contributing to about three times the dose estimated for an adult.

The addition of clean topsoil (scenario with mitigation) reduced significantly the exposure dose, notably for the aromatic fraction.

22.3.5 RISK CHARACTERIZATION

The HR for each target group has been calculated using the TPHCWG and MADEP reference doses (Figure 22.4). For the scenario without mitigation, the HD for the aromatic fraction exceeds the unity (3.1) for toddlers, indicating a potential risk (Table 22.8). For adults, this ratio is close to the limit value (HR = 0.97). As mentioned earlier, the addition of clean topsoil reduced significantly the exposure and thus the risk. The HR for toddlers and adults then becomes 0.50 and 0.26, respectively. The use of either the MADEP or TPHCWG reference value had no impact on the risk estimations, both values yielded similar HR values.

Finally, it must be noted that, for the unmitigated scenario, the ingestion of locally-grown fruits and vegetables containing aromatic EPHs significantly contributes to the exposure dose predicted by the model, and to the risk of both target groups. If this exposure media is not considered (as it might not be applicable), the HR would thus be below one (Figure 22.5).

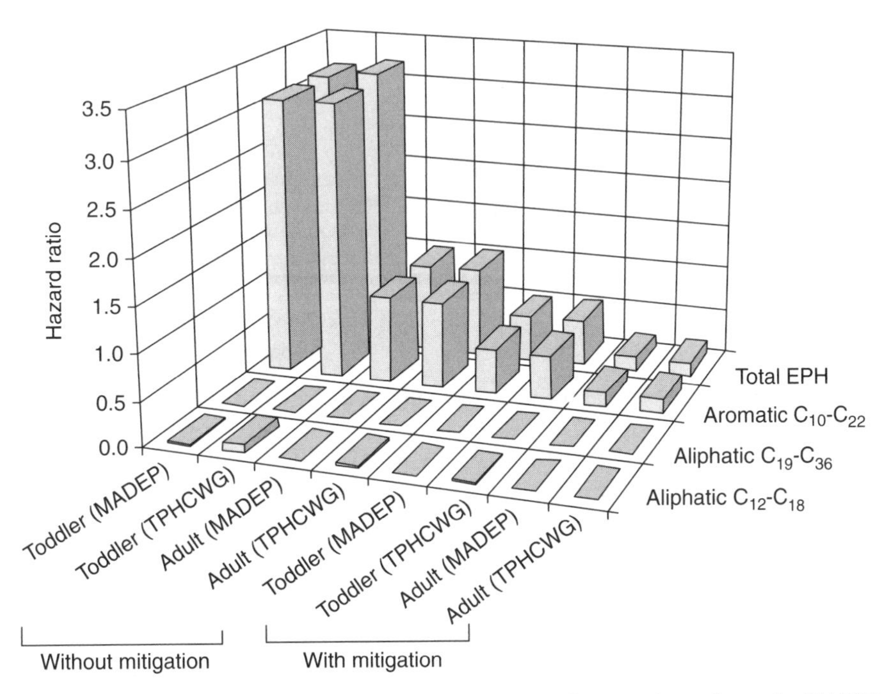

Figure 22.4 Hazard ratio calculated by CalTOX using reference doses from the MADEP and TPHCWG for each target group.

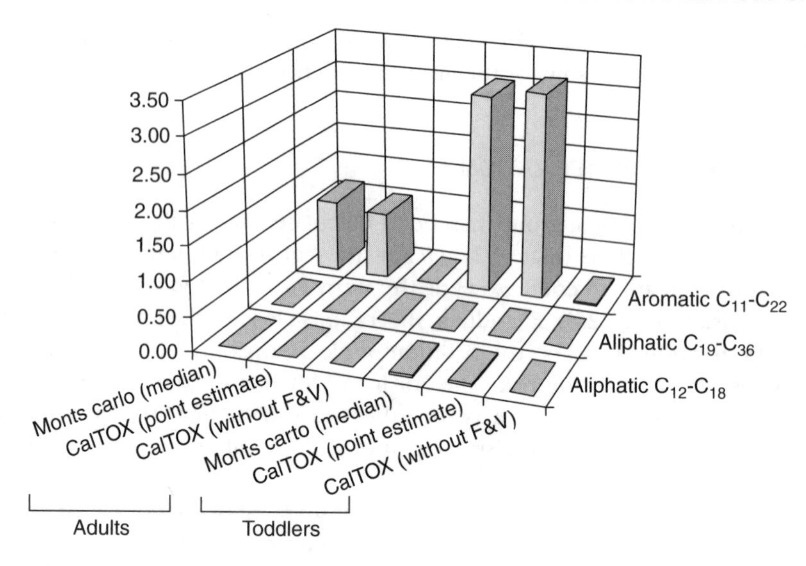

Figure 22.5 Hazard ratio calculated by Monte Carlo simulation and CalTOX for the unmitigated scenario.

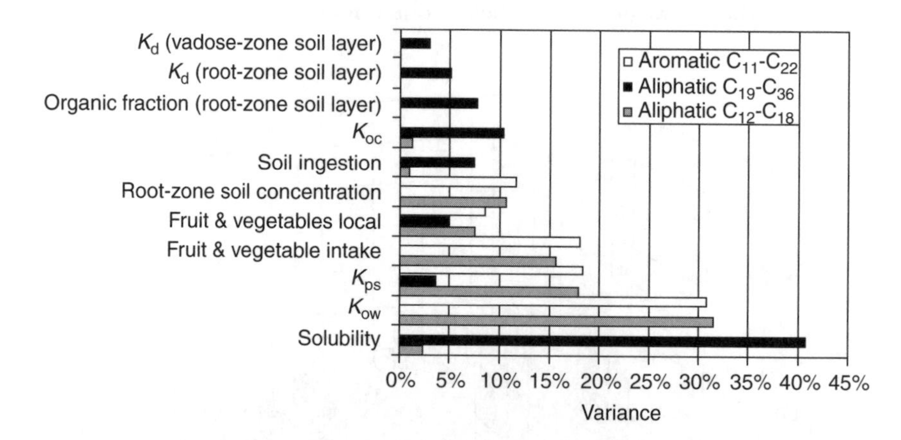

Figure 22.6 Sensitivity analysis for the unmitigated scenario.

22.3.6 SENSITIVITY AND UNCERTAINTY ANALYSIS

As shown in Figure 22.6, the sensitivity of parameters with almost the same *EC* range, i.e., aliphatic $C_{12}-C_{18}$ ($EC = 15$) and aromatic $C_{11}-C_{22}$ ($EC = 16$), is quite similar. In this case, the HR is affected by the K_{ow} (~30% of the variance), by the K_{sp} (~17%), which is itself related to K_{ow} (Table 22.3), and by the fruit and vegetable intake (~15%). On the other hand, the HR for the aliphatic $C_{19}-C_{36}$ fraction ($EC = 28$) is influenced mainly by the solubility (~40%) and, to a lesser extent, by K_d and K_{oc} values (<10%). As for the other partition coefficients, these values have been estimated with regression equations (Table 22.3).

HR (median values) calculated with Monte Carlo simulations using PDF for all the parameters are similar to the deterministic point estimates calculated by CalTOX for each target group (Table 22.8). The CVs for each EPH fraction and target group varied between 228 and 349% after 1000 trials.

22.4 DISCUSSION AND CONCLUSION

Despite a reduction of more than 80% of the TPHs ($C_{10}-C_{50}$) in the soil after an extensive bioremediation program, some areas of the former storage site still exceeded the provincial generic criterion 'B' for residential use. A risk assessment was conducted to evaluate the potential impact of this residual contamination on the health of the future residents of the site. Based upon the toxicological evaluation of EPH fractions, aliphatic ($C_{12}-C_{18}$, $C_{19}-C_{36}$) and aromatic ($C_{11}-C_{22}$) fractions were measured for about 15% of the TPH soil samples. This allowed the use of linear regressions to derive a simple empirical model to predict various EPH fractions from the TPH data, with r^2 varying from 54 to 72% ($p<0.001$). The CalTOX model was then used to estimate the environmental and exposure media concentrations, exposure doses and potential risk related to different remedial scenarios. Monte Carlo simulations showed that much of the variability in the exposure estimation is related to partition coefficients such as K_{ow}, K_{oc} and K_d.

The results of the exposure model suggest a potential EPH health risk to the future residents arising mainly from the ingestion of locally grown fruits and vegetables contaminated by petroleum hydrocarbons in the root-zone soil. On the other hand, volatilization and diffusion of EPH fractions from the root-zone soil to air can be considered insignificant, based on the low volatility of these fractions and the relatively low air concentration predicted by the model. Consequently, the contribution of the inhalation pathway to the total exposure dose and to the risk is negligible.

The scenario with mitigation was based on an addition of 30 cm of clean topsoil, which may be sufficient to prevent health risk by limiting the root absorption of contaminants by plants and direct soil ingestion by humans. In most soil the roots of plants are confined within the first meter of depth, while in agricultural lands the depth of plowing is 15 to 25 cm. Moreover, the diffusion depth, which is the depth below which a contaminant is unlikely to escape by diffusion, is in the order of a meter or less for the most volatile contaminants (Jury *et al.* 1990). For semi-volatile or non-volatile substances such as EPHs, the diffusion depth of each fraction is probably less than a meter. Only a direct field sampling and laboratory study using clean soil spiked with a known quantity of a mixture, or using soil from a contaminated site, would validate the multiphase partitioning of these fractions in soil. In addition, the direct measurement of EPH fractions in plants would give a better idea of the potential contamination of the TPHs in the root-zone soil. It would also help to quantify more accurately the human exposure to homegrown fruits and vegetables.

Despite the fact that the site owner had to respect the actual governmental guidelines and reduce the TPH concentrations according to generic criteria, the use of a risk-based approach shows an interesting alternative to traditional criteria-based remediation and may help to define site-specific cleanup levels.

22.5 ACKNOWLEDGEMENTS

We are grateful to Biogénie inc. for financial help and laboratory analysis, and to Canadian National Railways for allowing us access to their data. We also thank Paul-André Biron, Jacques Saulnier and Dany Dumont for maps and data analysis.

REFERENCES

Agriculture Canada (1990) *Soil Landscapes of Canada (Quebec — Southwest)*, Center for Land and Biological Resources Research, Ottawa, Ontario (map).

Baes CF, Sharp RD, Sjorren AL and Shor RW (1984) *A Review and Analysis of Parameters for Assessing Transport of Environmentally Released Radionuclides through Agriculture*. ORNL-5786, NTIS #DE85-000287, Oak Ridge National Laboratory, Oak Ridge, TN.

Biogénie Inc. (1998) *Réhabilitation environnementale du terrain de l'ancien dépot pétrolier Shell à Sillery*, Ste-Foy, Québec, 30 pp.

CCME (1997) *Report from the Canadian Council of Ministers of the Environment Workshop on Petroleum Hydrocarbons in Soil, Burlington, Ontario, (15-17 October 1997)*, 30 pp.

CCME (1999) *Report from the Canadian Council of Ministers of the Environment Workshop on Development of Canada-wide Standards for Petroleum Hydrocarbons in Soil, Edmonton, Alberta, (28-30 April 1997)*, 30 pp.

CCME (2000) *Canada-wide Standards for Petroleum Hydrocarbons (PHCs) in Soil: Scientific Rationale, Supporting Technical Document*, Winnipeg, Manitoba, 312 pp.

CCME (2001) *Canada-wide Standards for Petroleum Hydrocarbons (PHCs) in Soil: User Guidance*, Winnipeg, Manitoba, 42 pp.

Decisioneering Inc. (1996) *Crystal Ball User Manual Version 4.0*, Decisioneering, Inc., Denver, CO, 286 pp.

Droppo JG, Strenge DL, Buck JW, Hoopes BL, Brockhaus RD, Walter MB and Whelan G (1994) *Multimedia Environmental Pollutant Assessment System (MEPAS), Application Guidance*, Vols. 1 and 2, Pacific Northwest Laboratories, Richland, WA.

Environment Canada (1993) *Canadian Climate Normals (1961-90)*, Atmospheric Environment Service, Montreal, Quebec.

Gouvernement du Québec (1999) *Règlement sur les produits pétroliers (c.U-1.1, r.1)*, Éditeur Officiel du Québec, Québec, 61 pp.

Health Canada (1994) *Human Health Risk Assessment for Priority Substances*, Canada Communication Group, Ottawa, Ontario, 36 pp.

Jury WA, Russo D, Streile G and El Abd H (1990). Evaluation of volatilization of organic chemicals residing below the soil surface. *Water Resources Research*, **26**, 13-20.

Karickhoff SW (1981) Semi-empirical estimation of sorption of hydrophobic pollutants on natural sediments and soils. *Chemosphere*, **10**, 833-846.

MADEP (1994) *Interim Final Petroleum Report: Development of Health-based Alternative to the Total Petroleum Hydrocarbon (TPH) Parameter*, Massachusetts Department of Environmental Protection, Bureau of Waste Site Cleanup, Boston, MA.

McKone TE (1993) *CalTOX, A Multimedia Total Exposure Model for Hazardous Waste Sites (Parts I, II, & III)*. UCRL-CR-111456PtI/PtII/PtIII, Department of Toxic Substances Control, Lawrence Livermore National Laboratory, Livermore, CA, 196 pp.

McKone TE (1994a) *A Multimedia Total Exposure Model for Hazardous Waste Sites — Spreadsheet User's Guide*, Department of Toxic Substances Control, Lawrence Livermore National Laboratory, Livermore, CA, 45 pp.

McKone TE (1994b) Uncertainty and variability in human exposures to soil contaminants through homegrown food: a Monte Carlo assessment. *Risk Analysis*, **14**, 449-463.

McKone TE, Hall D and Kastengerg WE (1997) *CalTOX Version 2.3, Description of Modifications and Revisions*. NTIS PB97185060, California Environmental Protection Agency, Sacramento, CA, 31 pp.

McMahon TA and Denison PJ (1979) Empirical atmospheric deposition parameters — a survey. *Atmospheric Environment*, **13**, 571-585.

MEF (1992) *Programme de surveillance de la qualité de l'atmosphère — sommaires annuels*, Ministère de l'Environnement du Québec, Québec.

MEF (1993) *Annuaire hydrologique 1991-1992*, Direction du réseau hydrique, Ministère de l'Environnement du Québec, Québec.

MEF (1996) *Guide technique pour la réalisation des analyses préliminaires des risques toxicologiques*, Groupe d'analyse de risque, Direction des Laboratoires, Ministère de l'Environnement du Québec, Québec, 771 pp.

MEF (1998) *Politique de protection des sols et de réhabilitation des terrains contaminés*, Service des lieux contaminés, Ministère de l'Environnement du Québec, Québec, 124 pp.

MSSS (1999) *Lignes directrices pour la réalisation des évaluations du risque toxicologique pour la santé humaine dans le cadre de la procédure d'évaluation et d'examen des impacts sur l'environnement et l'examen de réhabilitation de terrains contaminés — Document de consultation*, Groupe de travail technique sur les méthodologies d'évaluation du risque, Ministère de la Santé et des Services Sociaux du Québec, Montréal, 90 pp.

NRC (1983) *Risk Assessment in the Federal Government: Managing the Process*, National Academy Press, Washington, DC.

Richardson GM (1997) *Compendium of Canadian Human Exposure Factors for Risk Assessment*, O'Connor Associates Environmental Inc., Ottawa, Ontario, 71 pp.

TPHCWG (1997a) *Development of Fraction-Specific Reference Dose (RfD) and Reference Concentration (RfC) for Total Petroleum*, Vol. 4, Total Petroleum Hydrocarbon Criteria Working Group, Amherst Scientific Publishers, Amherst, MA, 137 pp.

TPHCWG (1997b) *Selection of Representative TPH Fractions Based on Fate and Transport Consideration*, Vol. 3, Total Petroleum Hydrocarbon Criteria Working Group, Amherst Scientific Publishers, Amherst, MA, 102 pp.

USEPA (1989) *Risk Assessment Guidance for Superfund*, Vol. I, *Human Health Evaluation Manual (Part A)*. EPA/540/1-89/002, US Environmental Protection Agency, Office of Emergency and Remedial Response, Washington, DC.

USEPA (1992) *Dermal Exposure Assessment: Principles and Applications*, EPA/600/8-91/011B, US Environmental Protection Agency, Office of Research and Development, Washington, DC.

USEPA (1997) *Exposure Factors Handbook*, Vol. I, II & II. EPA/600/P-95/002Fa, US Environmental Protection Agency, Office of Research and Development, Washington, DC.

An Application of Risk-based Corrective Action (Atlantic RBCA) for the Management of Two Sites Impacted with Residual Hydrocarbons in Groundwater

CLAUDE CHAMBERLAND[1], ADRIEN PILON[2], SIMON PARADIS[1], AND STEPHEN ESPOSITO[3]

[1]*Shell Canada Products, Calgary, Alberta, Canada*
[2]*National Research Council Canada, Biotechnology Research Institute, Montreal, Quebec, Canada*
[3]*Jacques Whitford Environment Ltd, Dartmouth, Nova Scotia, Canada*

23.1 INTRODUCTION

23.1.1 BACKGROUND

The four Canadian Atlantic provinces (New Brunswick, Prince Edward Island, Nova Scotia and Newfoundland) are often depicted as beautiful, small communities located along the eastern Canadian seashore. In fact, a large proportion of the population lives in rural or semi-urban communities in these provinces and relies greatly on groundwater for potable water use.

Accordingly, environmental regulators in the Atlantic provinces have been proactive in implementing regulations requiring underground storage tank upgrades in the 1980s. This forced owners to replace unprotected steel storage tanks in favor of protected steel or fiberglass tanks, often with mandatory secondary liner systems. The regulators also adopted stringent provincial cleanup criteria for soil and groundwater impacted with hydrocarbons, mainly for the protection of human health.

Although the environmental issues were almost similar amongst the four provinces, the remedial criteria were different from one province to another. This led the regulators to consider harmonization of remedial criteria for petroleum-impacted sites in the early 1990s. Initial efforts were directed at

Environmental Analysis of Contaminated Sites. Edited by G. I. Sunahara, A. Y. Renoux, C. Thellen, C. L. Gaudet and A. Pilon

obtaining consensus on a new set of generic criteria that could be applied in all four provinces. Moreover, during this harmonization process, the concept of risk assessment and risk management gained favor. This was due, in part, to significant advances in risk assessment methodology and limitations in remediation technology that prevented the achievement of the existing generic criteria at many 'pump and treat' sites.

In 1995, the inter-provincial collaboration on petroleum-impacted site management was reformulated into a multi-stakeholder committee of provincial regulators, members of the Canadian Petroleum Products Institute (CPPI) and environmental consultants. The committee, referred to as Atlantic Partnership in RBCA Implementation (Atlantic PIRI), elected to use the ASTM Risk-based Corrective Action (RBCA) standard E1739-95 (ASTM 1995) as the foundation for incorporating risk assessment into a harmonized technical approach for Atlantic Canada.

23.1.2 HOW CLEAN IS CLEAN?

The use of site-specific remedial criteria has been accepted at numerous contaminated sites in Canada over the past 10 years (MOEE 1996). The Canadian Council of Ministers of the Environment (CCME) has developed documentation describing a process for deriving risk-based remedial criteria for individual sites (CCME 1996, 2000). Historically, sites which have benefited from site-specific risk-based criteria have tended to be those with complex issues and multiple contaminants such as pesticides, heavy metals, polycyclic aromatic hydrocarbons (PAHs) and chlorinated hydrocarbons.

Chronic leakage from storage systems has resulted in numerous petroleum hydrocarbon-contaminated sites which are a significant expenditure for remediation of sites in Canada. The Atlantic PIRI committee wanted to ensure that any process for development of remedial criteria for petroleum-impacted sites would be transparent, scientifically defensible and user friendly. In addition, the following requirements were established:

1. A method that dealt with either specific components of petroleum hydrocarbons or with mixtures of hydrocarbons such as gasoline, diesel and furnace oil;
2. A method that dealt with weathered product;
3. A method that would have continuing improvement through continuous research and development;
4. A method that was efficient, applicable to small and medium sized problems;
5. A method that could easily be utilized by site professionals.

23.1.3 RBCA APPLICATION

In reviewing the numerous options, the Atlantic PIRI committee chose the ASTM RBCA standard (ASTM 1995) as the preferred method for the management of petroleum-impacted sites in Atlantic Canada. Previous regulatory applications of the RBCA standard in American jurisdictions such as Texas and Michigan were limited to applying remedial criteria to individual chemicals such as benzene, toluene, ethylbenzene and xylenes (BTEX), naphthalene and PAHs. Historically, the Atlantic Canadian regulators have applied the total petroleum hydrocarbon (TPH) criterion, and wished to continue its use.

At about the same time, the Total Petroleum Hydrocarbon Criteria Working Group (TPHCWG) was finalizing their documents that included a methodology for the use of the TPH criterion in the RBCA methodology (TPHCWG 1997-1999). The method divides a petroleum mixture into discrete fractions based on equivalent carbon number ranges. Physical and chemical properties for each of the carbon fractions were determined and toxicological properties were defined for each fraction based on available data such as Volume 4 of TPHCWG (1997-1999). The TPHCWG had high peer acceptance due to the participation of regulators, academics, the USEPA, the American Petroleum Institute and numerous consultants.

The Atlantic PIRI committee combined the ASTM RBCA standard with the TPHCWG methodology to produce an Atlantic RBCA tool capable of meeting the requirements noted above (Atlantic PIRI 1999). A customized user-friendly software package, the 'Atlantic RBCA tool kit', was developed to support the framework. The committee updated the tool kit with Canadian human exposure parameters and with additional toxicological data available from national sources.

Since RBCA only considers human receptors, an ecological risk evaluation matrix was also developed to be used in conjunction with the Atlantic PIRI tool kit during site assessments (Atlantic PIRI 1999). The result is a flexible tool capable of considering conservative scenarios for both on-site and off-site exposure pathways Tier I and Tier II risk assessment levels.

23.1.4 THE LEGISLATIVE CONTEXT

The four provinces signed a memorandum of understanding (MOU) in 1997. The purpose of this MOU was to endorse a regional multi-stakeholder consultation project on the adoption of a risk-based management process for use on petroleum-contaminated sites.

The introduction of RBCA and risk assessment/management at petroleum sites represent two major improvements in the management of impacted sites. The first is the acceptance by provincial regulators in the Atlantic provinces of the principles of risk assessment in their respective site management processes. The second major improvement is the provision of a tool, namely Atlantic RBCA, which the regulatory authorities accept and endorse (Atlantic PIRI 1999). These improvements have resulted in a standard and cost-effective method of evaluating human health risk at petroleum release sites.

Atlantic PIRI has developed a Tier I look-up table that provides harmonized remedial criteria for various land uses, soil types and petroleum compounds. The Tier I look-up tables are intended to replace the previous generic remedial criteria for soil and groundwater in all the provinces, and New Brunswick, Nova Scotia and Prince Edward Island have already adopted the new criteria (Atlantic PIRI 1999).

If the site exceeds Tier I criteria, the options are to either remediate to the Tier I criteria or to undertake a site-specific assessment. For Tier II, data is collected for calculation of site-specific target levels (SSTLs) based on different points of exposure (see Section 23.3). These will serve as remedial criteria for that site only. If the site still exceeds the SSTLs, a Tier III risk assessment can be done which requires a high level of site-specific information and the application of more sophisticated fate and transport modeling expertise.

Each province is in the process of updating its individual contaminated site management guidelines to include site-specific risk-based criteria as part of an acceptable remediation strategy. The case studies discussed in this chapter were carried out in the Canadian province of Prince Edward Island (PEI). At the time that these studies were evaluated, the Prince Edward Island Department of Technology and Environment (PEIDTE) petroleum-contaminated site remediation guidelines had established three soil and groundwater protection classes:

- Class 'A' refers to highly sensitive sites, where a third party well was located within 150 m of source in the recharge area or within 50 m of source in the discharge area. Generic guideline PEIDTE Level A Groundwater Guidelines: 0.3 mg L^{-1} for TPH, 0.005 mg L^{-1} for benzene;
- Class 'B' means no wells within 150 m of source in the recharge area or within 50 m of source in the discharge area. Generic guideline PEIDTE Level B Groundwater Guidelines: 3.0 mg L^{-1} for TPH, 0.05 mg L^{-1} for benzene;
- Class 'C' represents a location with no wells within a 250-m radius.

PEIDTE recommended the use of these guidelines for site classification and cleanup; however, site-specific remediation guidelines with evaluation of human and/or ecological risks were accepted if developed within the RBCA Atlantic framework. Since then, new Tier I look-up tables have replaced the PEIDTE guidelines as mentioned earlier (Atlantic PIRI 1999).

23.2 CASE STUDIES

23.2.1 SITE BACKGROUNDS

23.2.1.1 SITE A

The site is located in an urban area in Prince Edward Island. A service station operates on the site. The general topography of the site is very gently sloping (1%

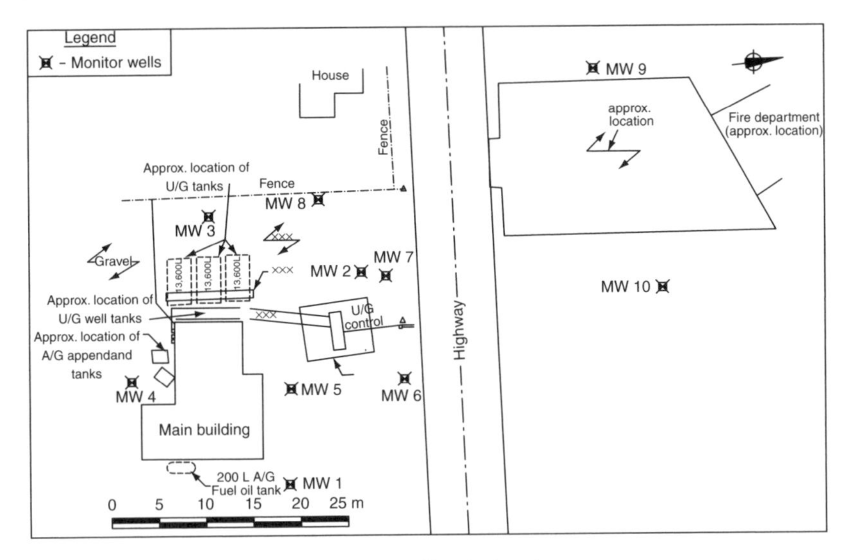

Figure 23.1 Site A site plan.

grade) towards a bay near the ocean, located approximately 190 m southwest of the site.

Adjacent property uses include a fire station and post office north of the site, across the main street; commercial buildings east of the site; and residential houses west and south of the site. The closest residential house is located approximately 8 m west of the property line of the site. Figure 23.1 shows the site and surrounding area. Due to the fact that no potable wells are located within 150 m of the site, generic 'Class B' groundwater remediation guidelines would generally apply to the site.

Site assessment work carried out at the site identified hydrocarbon impacts in the soil and groundwater adjacent to the pump island. The assessment indicated that groundwater from two of the wells exceeded the Generic Class B Groundwater Guidelines, and soil samples collected from one on-site borehole exceeded the applicable soil guidelines. Since remediation to the generic criteria would be very costly, a site-specific approach was adopted for this site.

23.2.1.2 SITE B

The site is located in a semi-rural area of Prince Edward Island. No buildings are currently present on the site itself and three previously existing underground fuel storage tanks were removed in 1994. The general local topography of the site is very gently sloping towards the northeast. Figure 23.2 shows a detailed site plan. To the west of the site, across a road, is a building formerly used as a general store. To the south of the site, across the highway, are several residential houses and a commercial guest house. Immediately to the north and northeast of the site are open grassy fields.

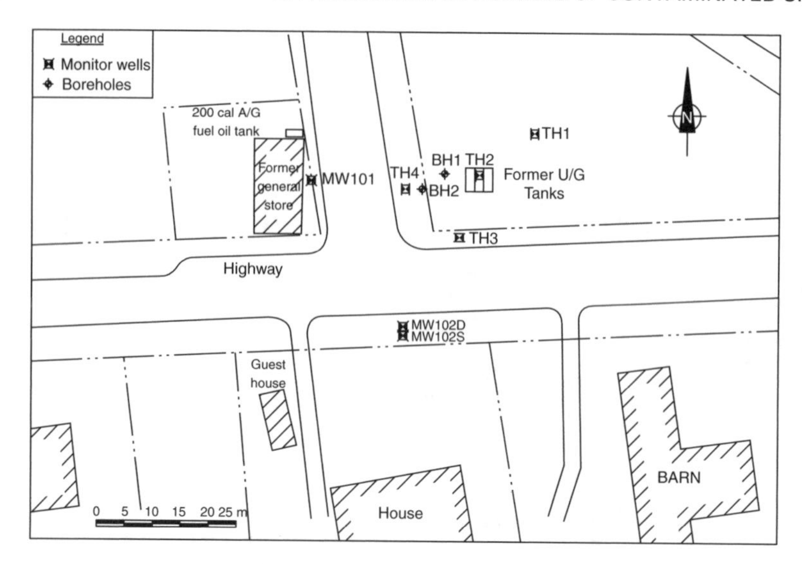

Figure 23.2 Site B site plan.

The impacts are related to the subsurface release of hydrocarbons from three steel underground fuel storage tanks which had been removed from the site in 1994. At the time of the storage tank removal, hydrocarbon-impacted soil from the tank excavation had also been removed. The subsurface hydrocarbon impacts mainly affect, and are persistent in, the groundwater. Indications are that residual hydrocarbons are present in fractured bedrock at the site and are therefore difficult to remediate to regulatory groundwater objectives.

Due to the proximity of potable water wells, this site would be classified as 'Class A' under the existing Prince Edward Island remediation guidelines. Since the remedial excavation, no soil samples have exceeded the generic soil quality criteria; however, three of the four monitoring wells continue to exceed water quality criteria. For this reason it was desired to develop site-specific groundwater quality criteria.

23.2.2 GEOLOGY

At both sites, bedrock in this area of Prince Edward Island is Permo-Pennsylvanian 'Red Beds', consisting of varying amounts of sandstone, siltstone, claystone and conglomerate, with bedding dips generally less than 5%. The hydrostratigraphic unit is at least 150 m thick and may be partially confined by overlying clayey till. Typically, an upward-fining sequence is observed, grading from relatively coarse conglomerate at the base, through coarse to fine sandstone alternating with siltstone, to very fine sandstone and siltstone at the top (Van de Poll 1983). It is likely that there are semi-confining claystone and siltstone beds within the aquifer. Site-specific geological information for both sites is shown in Table 23.1.

Table 23.1 Site geology and hydrogeology summary

	Site A	Site B
Site geology	Bedrock typically consists of fresh to highly weathered, red brown, medium grained sandstone with occasional mudstone seams. Overburden was typically found to consist of 1.0 m to 2.0 m of silty sand with a trace gravel and/or sand fill	Bedrock typically consists of fresh to highly weathered, red brown, medium grained sandstone with occasional mudstone seams. Approximately 3 m of silty sand till overlies the bedrock
Water levels	Water levels at the site are typically about 10 m below grade. Temporal groundwater changes show typically higher levels in the spring and lower water levels in the fall of each year	Water levels at the site are typically about 3.5 m below grade
Flow direction and gradient	The groundwater direction flows towards the municipal pumping well in a northerly direction with a low horizontal hydraulic gradient (i_h) of 0.02 (2%)	Groundwater flows in a westerly direction with a low horizontal hydraulic gradient (i_h) of 0.005 (0.5%). A low vertical hydraulic gradient (i_v) was calculated from the well pair MW-102 S/D to be $i_v = 0.010$ (1%) downward (i.e., recharge conditions)
Hydraulic conductivity	K_h values of between 2.6×10^{-4} cm s^{-1} and 6.5×10^{-4} cm s^{-1} for wells MW-4, MW-6 and MW-8 screened in the shallow bedrock. For the purpose of this study, a hydraulic conductivity value of 4×10^{-4} cm s^{-1} is adopted	K_h value of 10^{-3} cm s^{-1} for groundwater flow in the vicinity of the shallow bedrock/overburden interface; K_h of 9×10^{-5} cm s^{-1} for groundwater flow at depth in the bedrock aquifer (11 m)

23.2.3 HYDROGEOLOGY

Work carried out at a variety of sites by different investigators on PEI indicate hydraulic conductivities ranging from 10^{-1} cm s^{-1} near the bedrock surface, decreasing to 10^{-5} cm s^{-1} at depths greater than 35 m. Matrix hydraulic conductivities ranged from 10^{-6} cm s^{-1} to 5×10^{-5} cm s^{-1}. From this, it is concluded that the fracture systems, rather than the matrix, likely control groundwater flux and velocity. Site-specific hydrogeological information for both sites is also shown in Table 23.1.

23.3 METHODOLOGY

Risk assessment examines three basic elements which constitute risk: the chemical hazard, the receptor and the exposure pathway by which the receptor is exposed to the hazard (Figure 23.3). Risk management strategies can be used to reduce risks by reducing the severity of the hazard, the presence of the receptor or the degree of exposure.

Risk assessment for contaminated sites includes the following components.

- Hazard Identification: identification of the environmental hazards that may pose a health risk (e.g., chemicals of concern).

Figure 23.3 Risk assessment.

- Receptor Characterization: identification of the human receptors that may be exposed to the above hazards, and estimation of their characteristics.
- Exposure Assessment: qualitative and quantitative evaluation of the likelihood or degree to which the receptors would be exposed to the hazard.
- Risk Characterization: quantitative assessment of the actual health risk of each hazard to each receptor, based on the degree of exposure.
- Uncertainty Assessment: review of the uncertainty associated with the above risk elements in the risk estimation process.
- Target Level Determination: the determination of concentrations at the site below which no adverse effects would be expected.

The methodology for this assessment was as follows.

1. The chemicals of concern and their respective concentrations at the site were identified based on comparing the field program results to the generic soil and groundwater quality criteria published by PEIDTE (1992).
2. A receptor survey was carried out to identify the most sensitive receptors for the hazards identified at the site.
3. For the exposure assessment, all potential exposure pathways were identified for each hazard-receptor combination. The exposure scenarios that have been qualitatively considered for human receptors at the sites include:
   - Ingestion/dermal contact with soil;
   - Inhalation/ingestion/dermal contact with dust;
   - Ingestion of vegetation or garden produce grown in contaminated soil or irrigated with contaminated groundwater;
   - Ingestion/dermal contact with surface water;
   - Ingestion/dermal contact with groundwater;
   - Inhalation of hydrocarbon vapors.
4. From this list, a qualitative assessment of the likelihood of exposure (and thus with the highest likelihood to contribute a health risk) was carried forward for further quantitative analysis. The likelihood of exposure is considered and evaluated in terms of the following series of definitions:

- *Very unlikely likelihood of exposure* — level of exposure that could result in adverse effects is not expected;
- *Unlikely likelihood of exposure* — level of exposure that could result in adverse effects would probably not occur;
- *Possible likelihood of exposure* — level of exposure that could result in adverse effects might be expected;
- *Likely likelihood of exposure* — level of exposure that could result in adverse effects is expected. Exceedance of this exposure level might be expected;

5. Using the RBCA tool kit, the concentrations of the contaminants of concern at the point of exposure (POE) were determined. For direct exposure pathways such as soil ingestion and dermal contact, the concentration at the POE was the concentration of the contaminant of concern at the source. For indirect pathways such as soil to outdoor air, simple and predictive models were used to calculate an attenuation coefficient between the source zone and the POE.
6. Chemical intake to receptors was estimated.
7. The health risks resulting from this chemical intake were determined.
8. SSTLs were determined by back-calculating the concentrations below which no adverse health effects would be expected. Based on the approach and toxicity values determined by the TPHCWG and standard Canadian default, exposure limits were derived for a residential child, a residential adult and a commercial adult. For non-carcinogenic substances, a target hazard index (HI) of 1.0 was used to derive all exposure limits (for reference, the total allowable hazard index, or sum of the hazard quotients considered protective, is 1.0). For carcinogenic substances, an incremental excess lifetime cancer risk (IELCR) of 10^{-5} was used in the analyses.

Petroleum hydrocarbon products such as gasoline and diesel are a complex mixture of hundreds of different types of hydrocarbon compounds. Risk assessment for petroleum hydrocarbons is a complex undertaking since each hydrocarbon compound has a particular set of physical behavior characteristics (such as solubility, sorption, etc.) and each may have a particular toxicological effect. The TPHCWG (1997-1999) has developed a methodology which considers petroleum hydrocarbons as 13 separate fractions, each having similar physical and toxicological properties. The health risk for TPH is calculated using representative concentrations and physical and toxicological properties for each of the 13 fractions. From this, a TPH SSTL can be calculated that is protective of human health. A specific laboratory analysis procedure is required to determine the concentration of the individual TPH fractions. Note that, by definition, the TPHCWG methodology incorporates toxicity values for toluene, ethylbenzene and xylenes. As such, determining a TPH exposure limit in addition to the BTEX limits is an added level of conservatism.

23.4 SITE A – RBCA ASSESSMENT

The following sections provide a general overview of the hazard identification, receptors considered, exposure assessment and determination of site-specific target levels for Site A.

23.4.1 HAZARD IDENTIFICATION

Table 23.2 shows the chemical parameters carried forward in the assessment. The results suggest that the chemical hazards are related mainly to BTEX and TPH in groundwater. Benzene, for instance, exceeds the Canadian drinking water guidelines in monitoring wells MW-2 and MW-7 on the site, but is not detectable in monitoring wells outside the property, or in the service station water well. Other organic contaminants, PAHs, polychlorinated biphenyls (PCBs), chlorinated solvents and pesticides, were not suspected based on historical use of the site and therefore not tested. Inorganic parameters (mg L^{-1}) were

Table 23.2 Hazard identification for Site A

Potential hazards	Potential source	Tests performed?	Maximum measured concentration[a] (groundwater, mg L^{-1})	Carried forward in risk assessment?
Explosion/ flammability	hydrocarbon liquids (gasoline)	yes	none	no
Petroleum Hydrocarbons				
Benzene	low concentration in fuel	yes	0.096	yes
Toluene	low concentration in fuel	yes	0.882[b]	yes
Ethylbenzene	fuel component	yes	0.071[b]	yes
Xylenes	fuel component	yes	0.971[b]	yes
TPH (undifferentiated)	yes	yes	10.5	yes
Polycyclic aromatic hydrocarbons	not documented/not suspected	no	not tested	no
Polychlorinated biphenyls	not documented/not suspected	no	not tested	no
Chlorinated organic solvents	not documented/not suspected	no	not tested	no
Pesticides	not documented/not suspected	no	not tested	no
Metals	not documented/not suspected	no	lead 0.001	no
Inorganic parameters	hydrocarbon biodegradation parameters/or fuel related	yes (selected parameters)	nitrate + nitrite 4.8, sulfate 126, iron 8.6, manganese 0.76	no

[a]Maximum parameter concentrations in groundwater based on 31 January 1998 Alberta MUST analysis.
[b]If results from the TPH fractionation analysis are higher than Alberta MUST results, these are used as the maximum values (conservative approach – highest value from any analysis).

also found in groundwater (nitrates + nitrites, 4.8; sulfates, iron, 8.6; maganese, 0.76; lead, 0.001).

23.4.2 RECEPTOR IDENTIFICATION

The potential human receptors which may be affected by the potential hazards include: (1) humans within residences near the site (off-site residential); (2) humans working on the site (on-site commercial). For the purpose of this qualitative evaluation, the human receptor is conservatively characterized as an average healthy child or adult.

Existing and intended land use is an important factor in evaluating the potential exposures and estimating risk. The exposure assessment has been performed considering that the layout of the buildings and groundwater use near the site will not change in the future.

The site is located within an urban community and is surrounded by residential and commercial properties. There are no natural ecological habitat areas within the immediate vicinity of the site. The nearest natural ecological habitats are located in a bay approximately 190 m to the southwest. Given that the fuel release occurred in the subsurface, it is unlikely that local ecological receptors will be affected by the contamination. The greatest potential for the contaminant impacts to affect an ecological receptor is related to the potential discharge of surface water to the bay southwest of the site, via groundwater discharge. However, since this scenario is considered unlikely based on the distance of required subsurface flow along which natural dilution and biodegradation factors will act, ecological receptors were not considered further in the risk assessment.

23.4.3 EXPOSURE ASSESSMENT

Potential sources for the subsurface hydrocarbon impacts include residual hydrocarbon impacts from spills in the vicinity of the underground fuel storage tanks and distribution lines at the site. The site and surrounding area is serviced by the town municipal water supply. The municipal water supply is fed by groundwater wells located north and northwest of the site.

Each potential exposure scenario listed in the methology above has been evaluated qualitatively. From this evaluation, it was estimated unlikely that humans would be exposed to contaminants through ingestion of groundwater from the municipal well 280 m north of the site. However, a quantitative analysis was recommended. A possible exposure due to inhalation of vapors (indoors, residences) was also considered and quantitative analysis was also recommended. However, since groundwater impacts are located at 10 m or more, the potential for vapor emissions from hydrocarbons-impacted groundwater to nearby residences was considered low.

As such, the conceptual model developed for evaluating the quantitative exposure of the hypothetical potential receptors and the pathways carried forward for quantitative analysis are as follows.

Table 23.3 Site-specific target levels for Site A

Parameter	Maximum site concentration (mg L^{-1})	Minimum calculated SSTL (mg L^{-1})	Site concentration exceeded?
Aromatic fractions			
C_6–C_7	0.09	>sol	N
>C_7–C_8	0.77	450	N
>C_8–C_{10}	3.11	>sol	N
>C_{10}–C_{12}	1.19	>sol	N
>C_{12}–C_{16}	0.62	>sol	N
>C_{16}–C_{21}	0.27	>sol	N
>C_{21}–C_{32}	<0.1	>sol	N
Aliphatic fractions			
C_6–C_8	0.1	>sol	N
>C_8–C_{10}	0.05	>sol	N
>C_{10}–C_{12}	<0.01	>sol	N
>C_{12}–C_{16}	<0.05	>sol	N
>C_{16}–C_{21}	<0.05	>sol	N
>C_{21}–C_{32}	<0.1	>sol	N
TPH	6.62	N/C	N/A
Benzene	0.560	5.5	N

Notes: >Sol indicates that acceptable concentrations exceed the fraction's solubility. N/C = not calculated. N/A = Not available.

1. Hydrocarbon vapors from the impacted groundwater diffuse upwards through the soils into off-site residential buildings. Residents may inhale the vapors.
2. The dissolved phase hydrocarbon parameters are transported via groundwater flow in the shallow bedrock to the municipal well located 280 m north of the site. Residents may ingest the groundwater. Transport of dissolved phase hydrocarbons downward through the leaky aquifer into the deeper and more permeable aquifer that supplies the majority of the groundwater to the municipal wells was also considered, but it is a less direct pathway and therefore presents a less conservative model.

23.4.4 SITE-SPECIFIC CRITERIA

Maximum site concentration and SSTLs in groundwater for specific aliphatic and aromatic hydrocarbon fractions, TPH and benzene developed for Site A using the RBCA tool kit are shown in Table 23.3.

23.5 SITE B – RBCA ASSESSMENT

The following sections provide a general overview of the hazard identification, receptors considered, exposure assessment and determination of SSTLs for Site B.

23.5.1 HAZARD IDENTIFICATION

Table 23.4 shows the chemical parameters carried forward in the assessment. The results suggest that the chemical hazards are related mainly to BTEX and TPH in groundwater. The hydrocarbons were found mainly near the source and were not detected in the wells on adjacent properties. No other contaminants were tested, since historical data did not suggest their potential presence at the site. High values of metal parameters are related to high turbidity in water samples.

23.5.2 RECEPTOR IDENTIFICATION

Petroleum installations at Site B have been decommissioned, therefore the land could potentially be used in the future for residential or commercial use. Currently, the potential human receptors which may be affected by the potential hazards include: (1) humans within residences down-gradient of the site; (2) humans working on buried utilities in the vicinity of the site; and (3) potential future human residents on-site (if the land is redeveloped).

Table 23.4 Hazard identification for Site B

Potential hazards	Potential source	Tests performed?	Maximum measured concentration[a] (groundwater, mg L^{-1})
Explosion/flammability	hydrocarbon liquids (fuel oil)	absence confirmed	None
Petroleum hydrocarbons			
Benzene	low concentration in fuel	yes	0.025
Toluene	low concentration in fuel	yes	0.015[b]
Ethylbenzene	fuel component	yes	0.06[b]
Xylenes	fuel component	yes	0.225[b]
TPH—fuel oil	yes	yes	2.78
Polycyclic aromatic hydrocarbons	not documented/not suspected	no	not tested
Polychlorinated biphenyls	not documented/not suspected	no	not tested
Chlorinated organic solvents	not documented/not suspected	no	not tested
Pesticides	not documented/not suspected	no	not tested
Metals	not documented/not suspected	no	lead 33[c]
Inorganic parameters	hydrocarbon biodegradation parameters/or fuel related	yes (selected parameters)	chloride 112, sulfate 40, iron 9.02[c], manganese 53.4[c]

[a]Maximum parameter concentrations in groundwater based on 11 September, 1997 Alberta MUST analysis.
[b]If results from the TPH fractionation analysis are higher than Alberta MUST results, these are used as the maximum values (conservative approach – highest value from any analysis).
[c]High values for metal parameters reflect unfiltered, turbid samples (adsorption to soil particles).

For the purpose of this qualitative evaluation the 'human receptor' is conservatively characterized as a healthy child or adult. The site is located within a semi-rural community and is surrounded by open grassy fields and/or residential yards. There are no natural ecological habitat areas within the immediate vicinity of the site. The nearest natural ecological habitats are located near a small river, approximately 250 m to the north.

Given that the fuel release occurred in the subsurface, it is unlikely that local ecological receptors will be affected by the contamination. The greatest potential for the contaminant impacts to affect an ecological receptor is related to the potential discharge of surface water into a marsh or to the sea, both to the south of the site, via groundwater discharge. However, these scenarios are considered unlikely, based on the distance of required subsurface flow along which natural dilution and biodegradation factors will act. Site groundwater chemistry data also indicate that groundwater recharge of hydrocarbon impacts to down-gradient surface waters is unlikely. As such, ecological receptors have not been considered further.

23.5.3 EXPOSURE ASSESSMENT

No residual hydrocarbons have been noted in the subsurface soils. Each potential exposure scenario listed in the methology above has been evaluated qualitatively. The qualitative risk evaluation has identified that the greatest potential for risk to human health is through ingestion of hydrocarbons in groundwater. Potential additional risk is through inhalation of hydrocarbon vapors diffusing from impacted groundwater. Ecological receptors are not considered to be at risk and are therefore not evaluated further. As such, the conceptual model developed for evaluating the quantitative exposure of the hypothetical potential receptors and the pathways carried forward for quantitative analysis are as follows.

- Residual hydrocarbons from the vicinity of the former site underground storage tanks contribute dissolved components to the groundwater. The dissolved hydrocarbon parameters are transported via groundwater flow to nearby residential wells. Residents may ingest the groundwater.
- Potential future use of the site could involve on-site residents using groundwater.
- Hydrocarbon vapors from the impacted groundwater diffuse upwards through the soils into potential future on-site residential buildings. Potential residents may inhale the vapors.

23.5.4 SITE-SPECIFIC CRITERIA

The SSTLs developed for Site B using the RBCA tool kit are shown in Table 23.5. The SSTLs in this table only consider off-site groundwater use. SSTLs developed

Table 23.5 Site-specific target levels for Site B

Parameter	Maximum site concentration (mg L^{-1})	Minimum calculated SSTL (mg L^{-1})	Site concentration exceeded?
Aromatic Fractions			
C_6–C_7	<0.05	1.8	N
>C_7–C_8	<0.05	1.8	N
>C_8–C_{10}	1.2	42.0	N
>C_{10}–C_{12}	0.17	6.1	N
>C_{12}–C_{16}	0.14	5.0	N
>C_{16}–C_{21}	0.1	>sol	N
>C_{21}–C_{32}	<0.1	>sol	N
Aliphatic fractions			
C_6–C_8	0.17	>sol	N
>C_8–C_{10}	0.36	>sol	N
>C_{10}–C_{12}	<0.01	>sol	N
>C_{12}–C_{16}	<0.05	>sol	N
>C_{16}–C_{21}	<0.05	>sol	N
>C_{21}–C_{32}	<0.1	>sol	N
TPH	2.78	N/C	N/A
Benzene	0.025	1.6	N

Notes: >sol indicates that acceptable concentrations exceed the fraction's solubility. N/C = not calculated. N/A = not available.

for the site considering on-site groundwater use are lower than the concentrations observed at the site, suggesting adverse health effects may be possible if future use of the site involves using groundwater.

23.6 RESULTS OF RISK ANALYSIS

The objective of these studies was to evaluate whether known concentrations of petroleum hydrocarbons in the subsurface at the site presented a risk to human health or to the environment, and to determine an appropriate risk management solution, if required. The approach taken was to initially evaluate all potential risk scenarios using a qualitative methodology. Those pathways considered to pose the greatest potential risk were further evaluated quantitatively to determine whether or not actual risks were likely.

23.6.1 SITE A

A risk assessment has been performed of subsurface hydrocarbon impacts at Site A.

- No health risks from the subsurface hydrocarbon contamination are expected for humans on-site or off-site (HI<1 and IELCR$<10^{-5}$).
- Based on a qualitative analysis, the subsurface hydrocarbon contamination is not likely to present a risk to ecological receptors.

- Acceptable on-site concentrations of hydrocarbon in groundwater are at or near the solubility level with the exception of benzene, which is 5.5 mg L^{-1}. The highest observed benzene concentration at the site had been 0.560 mg L^{-1}, as such, no further remedial action was required.

23.6.2 SITE B

A risk assessment has been performed of subsurface hydrocarbon impacts at Site B.

- No health risks from the subsurface hydrocarbon contamination are expected for humans off-site.
- The subsurface hydrocarbon contamination is not likely to present a risk to ecological receptors.
- No health risks from vapor inhalation are expected from future use of the site, assuming that a standard slab-on-grade structure was constructed near the impacts.
- Adverse health effects may be possible if groundwater at the site (and in a defined 'groundwater exclusion zone') is used for human consumption for well drawing from the shallow groundwater aquifer (i.e., a dug well) without treatment. The groundwater exclusion zone was defined using the RBCA tool kit.

23.7 CONCLUSIONS

The RBCA methodology developed for the Atlantic provinces was used successfully to determine appropriate corrective action for two sites in Prince Edward Island; since then the regulatory authority has accepted the conclusions of these two analyses. The use of risk-based methods such as RBCA for the management of contaminated sites is gaining favor in many North American jurisdictions. The use of site-specific remedial objectives at petroleum-contaminated sites has allowed the continued operation or re-use of sites that may have been required to close or remain dormant under previous site management guidelines.

REFERENCES

ASTM (1995) *Standard Guide for Risk-based Corrective Action Applied at Petroleum Release Sites*. E1739-95, American Standard for Testing Materials.

Atlantic PIRI (1999) *Atlantic RBCA Reference Documentation*, Atlantic Partnership in RBCA Implementation. www.atlanticrbca.com.

CCME (1996) *A Protocol for the Derivation of Environmental and Human Health Soil Quality Guidelines*, Canadian Council of Ministers of the Environment, Manitoba Statutory Publications, Winnipeg, Manitoba.

CCME (2000) *Canada-wide Standards for Petroleum Hydrocarbons (PHCs) in Soil: Scientific Rationale, Supporting Technical Document*, Canadian Council of Ministers of the Environment.

MOEE (1996) *Guideline for Use at Contaminated Sites in Ontario*, Ministry of Environment and Energy of Ontario, Queen's Printer for Ontario.

PEIDTE (1992) *Prince Edward Island Petroleum Contaminated Site Remediation Guidelines*, Prince Edward Island Department of Technology and Environment.

TPHCWG (1997 - 1999) *Total Petroleum Hydrocarbon Criteria Working Group Series*, Vols. 1 - 5, Amherst Scientific Publishers, Amherst, MA.

Van de Poll HW (1983) *Geology of Prince Edward Island*. Report 83-1, Prince Edward Island Department of Energy and Forestry, Energy Minerals Branch.

PART FOUR

Closing Remarks

24

Closing Remarks

GEOFFREY I. SUNAHARA[1], CLAUDE THELLEN[2], CONNIE L. GAUDET[3], ADRIEN PILON[1] AND AGNÈS Y. RENOUX[1]

[1]Biotechnology Research Institute, National Research Council of Canada, Montreal, Quebec, Canada
[2]Ministère de l'Environnement du Quebec, Centre d'expertise en analyse environnementale, Quebec, Canada
[3]Environment Canada, Environmental Quality Branch, Ottawa, Ontario, Canada

24.1 INTRODUCTION

We are at the beginning of the third millennium, and we find the contaminated site assessment sector moving from 'technology and management driven' decisions of the 1960s and early 1970s through the use of 'numerical criteria based on best professional judgement', to a comprehensive suite of 'effects-based tools' such as bioassays and risk assessment approaches, for conducting scientific, risk-based assessments to support the management of contaminated sites. From a field that was in its scientific infancy some 10 years ago, when the 'how clean is clean' controversy erupted, the scientific tools and approaches for contaminated site assessment have now expanded to encompass an impressive arsenal of effects-based techniques for assessing the significance of contamination as well as comprehensive risk assessment procedures.

This book is intended as a reference for scientists and managers alike, to understand the interrelated scientific, technical and socio-economic issues and concepts for addressing contaminated land issues from the points of view of human and ecological health. These concerns are being shared at the international level (North American and European perspectives), as evidenced by the contributions found in Part Three of this book (Case Studies).

The overall acceptance of an integrated approach will ultimately be determined by the 'comfort level of understanding' by those implied in this process, i.e., ecotoxicologists, risk assessors and stakeholders (risk managers, contaminated site owners and regulators). Although more and more ecotoxicity methods and approaches are being developed for contaminated site assessment, some difficulties still exist in their understanding and acceptance by the stakeholders. The following will discuss, by use of different examples, some of the essential

Environmental Analysis of Contaminated Sites. Edited by G. I. Sunahara, A. Y. Renoux, C. Thellen, C. L. Gaudet and A. Pilon

factors that can increase this 'comfort level of understanding' and general acceptance of ecotoxicological and human health (where applicable) risk assessment for contaminated site management. One important element towards this acceptance will be the building of better communication pathways and understanding between the players and stakeholders involved in this issue.

24.2 AVOIDING MISUNDERSTANDINGS BETWEEN ECOTOXICOLOGISTS, ENVIRONMENTAL HEALTH SCIENTISTS AND MANAGERS

Ecotoxicity tests are used at different times and for different purposes, e.g., the detection of a contaminant through the measurement of toxicity and the quality of soil, as well as the risk assessment of indigenous species. These differences are not always clear to everyone involved in the contaminated site risk assessment and management process.

A common question asked by some managers is 'why do ecotoxicologists measure effects using species which are not present on the site?'. This is a question of the appropriateness of the test species used for the study site assessment. However, the ecological representativity may not be the issue, but rather whether a contaminated site is considered 'toxic' (using a surrogate species). The use of bioassays can be different depending upon the need, e.g., the use and generation of benchmarks of toxicity (using pure compounds) or the estimation of the risk to a specific ecological receptor compared to the detection of overall toxicity of mixtures of contaminants in unknown samples. For benchmark use and determination of risk, the test species have to be representative, i.e., ecologically relevant, for the site assessed, although this is less important if one is only considering overall toxicity determinations. These differences may be obvious to the ecotoxicologist, but this subtle distinction may not be clear to the stakeholders.

This type of misunderstanding would lead to confusion and greater lack of confidence (decreased comfort level) between the players and stakeholders (e.g., scientist/service provider and manager/client). The challenge for the scientific and risk assessment service providers will be to explain to the risk manager (client) in a clear and simple fashion the technical limitations of the ecotoxicity bioassays, as well as the numerous assumptions underlying the ecotoxicological risk assessment procedures.

The management and regulatory needs of these ecotoxicity methods and approaches can often change depending on the circumstance. And, as such, can be viewed by the scientist as a 'moving target', in that these needs are often forgotten by the risk managers, long before the scientists have successfully developed the appropriate tools to resolve the original problem. Thus, earlier and greater communication, confidence and planning are required between the different players and stakeholders.

It may take some time before implementation of environmental health (ecological and human health) risk assessment can be fully established, and there is a greater need for a common vocabulary (agreement and clarification, definition of all terms, concepts, management needs and technical limitations) between the scientists, the assessors and the managers. For example, the lack of a definition of an 'urban' ecosystem reflects this need. What are the receptors that require protection? Defining the quality of an urban ecosystem in generic terms may not be an easy question for the contaminated site risk assessor. For the restoration of an urban industrial site, in addition to the citizens (human health risk assessment), what other components of the ecosystem (ERA) do we wish to protect (*vis-à-vis* the trees, birds, etc.)? Since the function and quality of a soil in an industrial environment are not the same as those for agricultural or residential use, what biological index should be measured for regulatory use? A consensus of concepts and terminology between players and stakeholders is thus needed.

24.3 GROWING ACCEPTANCE OF EFFECT-BASED MEASUREMENTS FOR CONTAMINATED SITE RISK ASSESSMENT

The appropriate interpretation of biological data remains a challenging task for the laboratory and field (eco)toxicologist. In addition, there are a number of uncertainties (based on assumptions and lack of scientific data) within the risk assessment process that would limit accurate predictions of risk. However, these technical and theoretical limitations should not discourage the manager from making risk-based decisions.

Uncertainty exists at all levels of the contaminated site characterization and risk assessment processes. Although it is often considered at the laboratory level (Part One of this book), it is also found at different steps of the risk assessment processes (Part Two), as presented in case studies (Part Three). The ultimate goal of science is to reduce uncertainty in the estimate of risk at contaminated sites, and thus make more effective decisions for reducing risk. The scientist who will be assessing contamination at a site, the manager who will be making final decisions on the remediation of a site, and the public and stakeholders with a vested interest in the level to which the site is cleaned up will, in their own way, balance the benefits and costs of gathering more data to reduce uncertainty against that of taking action in a timely manner. However, in complex ecological systems (including urban scenarios), uncertainty can never be eliminated and therefore becomes a major driver in decision making.

In order to improve communications (as discussed above), another important element in this complex problem will be the improvement of the base level of knowledge (ecological and toxicological) of the stakeholders. Scientists should improve their ability to promote the use of scientific tools and continuously inform the users (risk assessors, managers and regulators) about this application. It is most probable that the current and growing movement towards ecological and human health-based approaches in risk assessment will greatly improve the understanding and acceptance of toxicity effects measurements (as illustrated in Part One of this book) and their use in contaminated site assessment (Parts Two and Three).

24.4 TRANSITION AND THE NEED FOR HARMONIZATION

There are a number of reasons for the evolving trend that criteria are becoming more 'effect-based' and are more likely to be used as part of the supporting tools for ecological and human health risk assessment rather than in isolation, i.e., more of a merging than a diverging type of tool. This is illustrated in Chapter 16 which shows, for example, that most American states rely on a tiered approach that uses criteria in the first tier. This is also true of The Netherlands, as well as many Canadian jurisdictions. Yet, at the same time, their approaches incorporate much more information on effects, risk, exposure, etc. They also include increased regulatory awareness and pressure towards the incorporation of scientifically defensible estimates of risk into their management decisions. This has led to reduced cost of effects assessment, better timelines to allow successful site restoration and the advent of new scientific tools.

Up to now, regulation has been the major driver for this transition. The scientists have facilitated this process by offering the regulators different types of tools to validate and give credibility to 'science-based' policy decisions. Furthermore, the integration of (eco)toxicity data at the (eco)toxicological risk assessment level is also being encouraged and is now becoming a reality in a number of jurisdictions. As evidenced by the contributions in this book, these technical and theoretical advancements in contaminated site assessment have evolved considerably and are becoming more acceptable for regulatory use.

With such a rapid evolution in approaches and thinking around contaminated site remediation, there has been a correspondingly rapid growth in the literature, e.g., textbooks, guidance manuals, policies, scientific papers, as well as proceedings, which deal with different aspects of the issue. Thus, a concerted effort should be undertaken to unite the research, risk assessment and risk management fields. This will ultimately forge stronger links and partnerships across these often disparate disciplines as a basis for dealing with complex contamination issues.

Index

Note: Page references in *italics* refer to Figures; those in **bold** refer to Tables